Applied Mathematical Sciences
Volume 99

Applied Mathematical Sciences

1. *John:* Partial Differential Equations, 4th ed.
2. *Sirovich:* Techniques of Asymptotic Analysis.
3. *Hale:* Theory of Functional Differential Equations, 2nd ed.
4. *Percus:* Combinatorial Methods.
5. *von Mises/Friedrichs:* Fluid Dynamics.
6. *Freiberger/Grenander:* A Short Course in Computational Probability and Statistics.
7. *Pipkin:* Lectures on Viscoelasticity Theory.
8. *Giacoglia:* Perturbation Methods in Non-linear Systems.
9. *Friedrichs:* Spectral Theory of Operators in Hilbert Space.
10. *Stroud:* Numerical Quadrature and Solution of Ordinary Differential Equations.
11. *Wolovich:* Linear Multivariable Systems.
12. *Berkovitz:* Optimal Control Theory.
13. *Bluman/Cole:* Similarity Methods for Differential Equations.
14. *Yoshizawa:* Stability Theory and the Existence of Periodic Solution and Almost Periodic Solutions.
15. *Braun:* Differential Equations and Their Applications, 3rd ed.
16. *Lefschetz:* Applications of Algebraic Topology.
17. *Collatz/Wetterling:* Optimization Problems.
18. *Grenander:* Pattern Synthesis: Lectures in Pattern Theory, Vol. I.
19. *Marsden/McCracken:* Hopf Bifurcation and Its Applications.
20. *Driver:* Ordinary and Delay Differential Equations.
21. *Courant/Friedrichs:* Supersonic Flow and Shock Waves.
22. *Rouche/Habets/Laloy:* Stability Theory by Liapunov's Direct Method.
23. *Lamperti:* Stochastic Processes: A Survey of the Mathematical Theory.
24. *Grenander:* Pattern Analysis: Lectures in Pattern Theory, Vol. II.
25. *Davies:* Integral Transforms and Their Applications, 2nd ed.
26. *Kushner/Clark:* Stochastic Approximation Methods for Constrained and Unconstrained Systems.
27. *de Boor:* A Practical Guide to Splines.
28. *Keilson:* Markov Chain Models—Rarity and Exponentiality.
29. *de Veubeke:* A Course in Elasticity.
30. *Shiatycki:* Geometric Quantization and Quantum Mechanics.
31. *Reid:* Sturmian Theory for Ordinary Differential Equations.
32. *Meis/Markowitz:* Numerical Solution of Partial Differential Equations.
33. *Grenander:* Regular Structures: Lectures in Pattern Theory, Vol. III.
34. *Kevorkian/Cole:* Perturbation Methods in Applied Mathematics.
35. *Carr:* Applications of Centre Manifold Theory.
36. *Bengtsson/Ghil/Källén:* Dynamic Meteorology: Data Assimilation Methods.
37. *Saperstone:* Semidynamical Systems in Infinite Dimensional Spaces.
38. *Lichtenberg/Lieberman:* Regular and Chaotic Dynamics, 2nd ed.
39. *Piccini/Stampacchia/Vidossich:* Ordinary Differential Equations in $\mathbf{R}^n$.
40. *Naylor/Sell:* Linear Operator Theory in Engineering and Science.
41. *Sparrow:* The Lorenz Equations: Bifurcations, Chaos, and Strange Attractors.
42. *Guckenheimer/Holmes:* Nonlinear Oscillations, Dynamical Systems and Bifurcations of Vector Fields.
43. *Ockendon/Taylor:* Inviscid Fluid Flows.
44. *Pazy:* Semigroups of Linear Operators and Applications to Partial Differential Equations.
45. *Glashoff/Gustafson:* Linear Operations and Approximation: An Introduction to the Theoretical Analysis and Numerical Treatment of Semi-Infinite Programs.
46. *Wilcox:* Scattering Theory for Diffraction Gratings.
47. *Hale et al:* An Introduction to Infinite Dimensional Dynamical Systems—Geometric Theory.
48. *Murray:* Asymptotic Analysis.
49. *Ladyzhenskaya:* The Boundary—Value Problems of Mathematical Physics.
50. *Wilcox:* Sound Propagation in Stratified Fluids.
51. *Golubitsky/Schaeffer:* Bifurcation and Groups in Bifurcation Theory, Vol. I.

(continued following index)

Jack K. Hale Sjoerd M. Verduyn Lunel

Introduction to Functional Differential Equations

With 10 Illustrations

Springer-Verlag
New York Berlin Heidelberg London Paris
Tokyo Hong Kong Barcelona Budapest

Jack Hale
School of Mathematics
Georgia Institute of Technology
Atlanta, GA 30332
USA

Sjoerd M. Verduyn Lunel
Vrije Universiteit Amsterdam
De Boelelaan 1081a
1081 HV Amsterdam
The Netherlands

Mathematics Subject Classification (1991): 34K20, 34A30, 39A10

Library of Congress Cataloging-in-Publication Data
Hale, Jack K.
Introduction to functional differential equations/Jack K. Hale, Sjoerd M. Verduyn Lunel.
p. cm. — (Applied mathematical sciences ; v.)
Includes bibliographical references.
ISBN 0-387-94076-6
1. Functional differential equations. I. Verduyn Lunel, S. M. (Sjoerd M.) II. Title. III. Series: Applied mathematical sciences (Springer-Verlag New York Inc.) ; v.
QA1.A647
[QA372]
510 s—dc20
[515'.35] 93-27729

Printed on acid-free paper.

Production managed by Natalie Johnson; manufacturing supervised by Jacqui Ashri.
Photocomposed copy produced using the authors' TeX files.
Printed and bound by R. R. Donnelley & Sons, Harrisonburg, VA.
Printed in the United States of America.

9 8 7 6 5 4 3 2 1

ISBN 0-387-94076-6 Springer-Verlag New York Berlin Heidelberg
ISBN 3-540-94076-6 Springer-Verlag Berlin Heidelberg New York

Preface

The present book builds upon an earlier work of J. Hale, "Theory of Functional Differential Equations" published in 1977. We have tried to maintain the spirit of that book and have retained approximately one-third of the material intact. One major change was a complete new presentation of linear systems (Chapters 6–9) for retarded and neutral functional differential equations. The theory of dissipative systems (Chapter 4) and global attractors was completely revamped as well as the invariant manifold theory (Chapter 10) near equilibrium points and periodic orbits. A more complete theory of neutral equations is presented (see Chapters 1, 2, 3, 9, and 10). Chapter 12 is completely new and contains a guide to active topics of research. In the sections on supplementary remarks, we have included many references to recent literature, but, of course, not nearly all, because the subject is so extensive.

Jack K. Hale
Sjoerd M. Verduyn Lunel

Contents

Introduction

In many applications, one assumes the system under consideration is governed by a principle of causality; that is, the future state of the system is independent of the past states and is determined solely by the present. If it is also assumed that the system is governed by an equation involving the state and rate of change of the state, then, generally, one is considering either ordinary or partial differential equations. However, under closer scrutiny, it becomes apparent that the principle of causality is often only a first approximation to the true situation and that a more realistic model would include some of the past states of the system. Also, in some problems it is meaningless not to have dependence on the past. This has been known for some time, but the theory for such systems has been extensively developed only recently. In fact, until the time of Volterra [1] most of the results obtained during the previous 200 years were concerned with special properties for very special equations. There were some very interesting developments concerning the closure of the set of exponential solutions of linear equations and the expansion of solutions in terms of these special solutions. On the other hand, there seemed to be little concern about a qualitative theory in the same spirit as for ordinary differential equations.

In his research on predator-prey models and viscoelasticity, Volterra [1, 2] formulated some rather general differential equations incorporating the past states of the system. Also, because of the close connection between the equations and specific physical systems, Volterra attempted to introduce a concept of energy function for these models. He then exploited the behavior of this energy function to study the asymptotic behavior of the system in the distant future. These beautiful papers were almost completely ignored by other workers in the field and therefore did not have much immediate impact.

In the late thirties and early forties, Minorksy [1], in his study of ship stabilization and automatic steering, pointed out very clearly the importance of the consideration of the delay in the feedback mechanism. The great interest in control theory during these and later years has certainly contributed significantly to the rapid development of the theory of differential equations with dependence on the past state.

In the late forties and early fifties, a few books appeared which presented the current status of the subject and certainly greatly influenced later developments. In his book, Mishkis [1] introduced a general class of equations with delayed arguments and laid the foundation for a general theory of linear systems. In their monograph at the Rand Corporation, Bellman and Danskin [1] pointed out the diverse applications of equations containing past information to other areas such as biology and economics. They also presented a well organized theory of linear equations with constant coefficients and the beginnings of stability theory. A more extensive development of these ideas is in the book of Bellman and Cooke [1]. In his book on stability theory, Krasovskii [1] presented the theory of Liapunov functionals emphasizing the important fact that some problems in such systems are more meaningful and amenable to solution if one considers the motion in a function space even though the state variable is a finite-dimensional vector.

With such clear indications of the importance of these systems in the applications and with the number of interesting mathematical problems involved, it is not surprising that the subject has undergone a rapid development in the last forty years. New applications also continue to arise and require modifications of even the definition of the basic equations. We list below a few types of equations that have been encountered merely to give an idea of the diversity and give appropriate references for the specific applications.

The simplest type of past dependence in a differential equation is that in which the past dependence is through the state variable and not the derivative of the state variable, the so-called retarded functional differential equations or retarded differential difference equations. For a discussion of the physical applications of the differential difference equation

$$\dot{x}(t) = F(t, x(t), x(t-r)), \qquad \dot{x} = \frac{dx}{dt}$$

to control problems, see Minorsky [2, Ch. 21]. Lord Cherwell (see Wright [1, 2]) has encountered the differential difference equation

$$\dot{x}(t) = -\alpha x(t-1)[1 + x(t)]$$

in his study of the distribution of primes. Variants of this equation have also been used as models in the theory of growth of a single species (see Cunningham [1]). Dunkel [1] suggested the more general equation

$$\dot{x}(t) = -\alpha \left[\int_{-1}^{0} x(t+\theta)\, d\eta(\theta) \right] [1 + x(t)]$$

for the growth of a single species.

In his study of predator-prey models, Volterra [1] had earlier investigated the equations

$$\dot{x}(t) = \left[\epsilon_1 - \gamma_1 y(t) - \int_{-r}^{0} F_1(\theta) y(t+\theta) d\theta\right] x(t)$$

$$\dot{y}(t) = \left[-\epsilon_1 + \gamma_2 x(t) + \int_{-r}^{0} F_2(\theta) x(t+\theta) d\theta\right] y(t)$$

where x and y are the number of prey and predators, respectively, and all constants and functions are non-negative. For similar models, Wangersky and Cunningham [1, 2], have also used the equations

$$\dot{x}(t) = \alpha x(t) \left[\frac{m - x(t)}{m}\right] - bx(t)y(t)$$

$$\dot{y}(t) = -\beta y(t) + cx(t-r)y(t-r)$$

for predator-prey models.

In an attempt to explain the circummutation of plants (and especially the sunflower), Israelson and Johnsson [1, 2] have used the equation

$$\dot{\alpha}(t) = -k \int_{1}^{\infty} f(\theta) \sin \alpha(t - \theta - t_0)\, d\theta$$

as a model, where α is the angle the top of the plant makes with the vertical (see also Klein [1]). For other applications, see Johnson and Karlsson [1].

Under suitable assumptions, the equation

$$\dot{x}(t) = \sum_{i=0}^{N} A_i x(t - T_i)$$

is a suitable model for describing the mixing of a dye from a central tank as dyed water circulates through a number of pipes. An application to the distribution in man of labeled albumin as it circulates from the blood stream through the interstitial fluids and back to the blood stream is discussed by Bailey and Reeve [1] (see also Bailey and Williams [1]). Boffi and Scozzafava [1, 2] have also encountered this equation in transport problems.

In an attempt to describe the spread of measles in a metropolitan area, London and Yorke [1] have encountered the equation

$$\dot{S}(t) = -\beta(t)S(t)[2\gamma + S(t-14) - S(t-12)] + \gamma$$

where $S(t)$ is the number of susceptible individuals at time t, γ is the rate at which individuals enter the population, $\beta(t)$ is a function characteristic of the population, and an individual exposed at time t is infectious in the time interval $[t-14, t-12]$.

In an analysis of gonorrhea, Cooke and Yorke [1] have studied the equation

$$\dot{I}(t) = g(I(t - L_1)) - g(I(t - L_2))$$

where I represents the number of infectious individuals and g is a non-negative function vanishing outside a compact interval.

A more general equation describing the spread of disease taking into account age dependence was given by Cooke [1] and Hoppenstadt and Waltman [1]. For other equations that occur in the theory of epidemics, see Waltman [1]. For other models in the biomedical sciences, see Banks [1]. Grossberg [1, 2] has encountered interesting differential equations in the theory of learning.

The equation

$$\dot{x}(t) = -\int_{t-r}^{t} a(t-u)g(x(u))\,du$$

was encountered by Ergen [1] in the theory of a circulating fuel nuclear reactor and has been extensively studied by Levin and Nohel [1]. In this model, x is the neutron density. It is also a good model in one-dimensional viscoelasticity in which x is the strain and a is the relaxation function.

Taking into account the transmission time in the triode oscillator, Rubanik [1, p. 130] has encountered the van der Pol equation

$$\ddot{x}(t) + \alpha\dot{x}(t) - f(x(t-r))\dot{x}(t-r) + x(t) = 0$$

with the delayed argument r. Taking into account the retarded connections between oscillating systems, Starik [1] has encountered the system

$$\begin{aligned}
&\ddot{y}(t) + [\omega^2 + \epsilon\lambda\sin\phi(t-r_1)]y(t) = -\epsilon[h\dot{y}(t) + \gamma y^3(t-r^2)]\\
&I\ddot{\phi}(t) = \epsilon\big[L(\dot{\phi}(t)) - H(\phi(t) - \sigma_1^2 y^2(t-r_3)\cos\phi(t))\\
&\qquad - \sigma_2\sin\phi(t) - \sigma_3\cos\phi(t)\big].
\end{aligned}$$

In the theory of optimal control, Krasovskii [2] has studied extensively the system

$$\begin{aligned}
\dot{x}(t) &= P(t)x(t) + B(t)u(t)\\
y(t) &= Q(t)x(t)\\
\dot{u}(t) &= \int_{-r}^{0} d[\eta(t,\theta)]y(t+\theta) + \int_{-r}^{0} d[\mu(t,\theta)]u(t+\theta).
\end{aligned}$$

There are also a number of applications in which the delayed argument occurs in the derivative of the state variable as well as in the independent variable, the so-called neutral differential difference equations. Such problems are more difficult to motivate but often arise in the study of two or more simple oscillatory systems with some interconnections between them. For simplicity, it is usually assumed that the interaction of the components of the coupled systems takes place immediately. In many cases, the time for the interaction to take place is important even in determining the qualitative behavior of the system. It often occurs that the connection between the coupled systems can be adequately described by a system of linear hyperbolic partial differential equations with the motion of each individual system being described by a boundary condition. In some cases, the connection through the partial differential equations (considered as a connection by a traveling wave) can be replaced by connections with delays. Generally, the resulting ordinary differential equations involve delays in the highest-order derivatives. A general discussion of when this process is valid may be found in Rubanik [1] and Cooke and Krumme [1].

For example, Brayton [1] considered the lossless transmission line connected as shown in Fig. 0.1, where $g(v)$ is a nonlinear function of v and gives the current in the indicated box in the direction shown. This problem may be described by the following system of partial differential equations

$$L\frac{\partial i}{\partial t} = -\frac{\partial v}{\partial x}, \qquad C\frac{\partial v}{\partial t} = -\frac{\partial i}{\partial x}, \qquad 0 < x < 1, \ \ t > 0,$$

with the boundary conditions

$$E - v(0,t) - Ri(0,t) = 0, \qquad C_1\frac{dv(1,t)}{dt} = i(1,t) - g(v(1,t)).$$

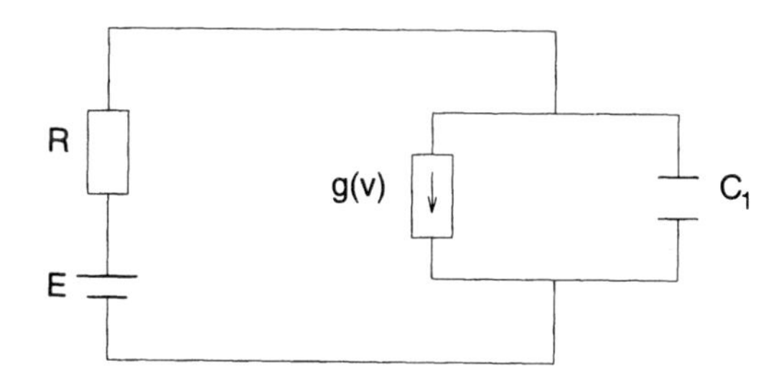

Fig. 0.1.

We now indicate how one can transform this problem into a differential equation with delays. If $s = (LC)^{-1/2}$ and $z = (L/C)^{1/2}$, then the general solution of the partial differential equation is given by

$$v(x,t) = \phi(x - st) + \psi(x + st)$$

$$i(x,t) = \frac{1}{z}[\phi(x - st) - \psi(x + st)]$$

or

$$2\phi(x - st) = v(x,t) + zi(x,t)$$

$$2\psi(x + st) = v(x,t) - zi(x,t).$$

This implies

$$2\phi(-st) = v\big(1, t + \frac{1}{z}\big) + zi\big(1, t + \frac{1}{s}\big)$$

$$2\psi(st) = v\big(1, t - \frac{1}{z}\big) - zi\big(1, t - \frac{1}{s}\big).$$

Using these expressions in the general solution and using the first boundary condition at $t - (1/s)$, one obtains

$$i(1,t) - Ki\big(1, t - \frac{2}{s}\big) = \alpha - \frac{1}{z}\, v(1,t) - \frac{K}{z}\, v\big(1, t - \frac{2}{s}\big)$$

where $K = (z-R)/(z+R)$, $\alpha = 2E/(z+R)$. Inserting the second boundary condition and letting $u(t) = v(1,t)$, we obtain the equation

$$\dot{u}(t) - K\dot{u}\big(t - \frac{2}{s}\big) = f\big(u(t), u\big(t - \frac{2}{s}\big)\big)$$

where $s = \sqrt{LC}$,

$$C_1 f(u(t), u(t-r)) = \alpha - \frac{1}{z}\, u(t) - \frac{K}{z}\, u(t-r) - g(u(t)) + Kg(u(t-r)),$$

all constants are positive and depend on the parameters in the original equations. Also, if $R > 0$, then $K < 1$.

If generalized solutions of the original partial differential equation were considered, the delay equation would require differentiating the difference $u(t) - Ku(t - (2/s))$ rather than each term separately; that is, one would consider the equation

$$\frac{d}{dt}\big[u(t) - Ku\big(t - \frac{2}{s}\big)\big] = f\big(u(t), u\big(t - \frac{2}{s}\big)\big).$$

The prescription for passing from a linear partial differential equation with nonlinear boundary conditions to a delay equation is certainly not unique and other transformations may be desirable in certain situations. This fact is illustrated following the ideas of Lopes [1]. Let $|K| < 1$ (i.e., $R > 0$) and let p be any solution of the difference equation

$$p(t) - Kp\big(t - \frac{2}{s}\big) = -b(t), \quad b(t) = \frac{z}{z+R} E\big(t - \frac{1}{s}\big).$$

If E is periodic, one can choose p periodic with the same period. Using the first boundary condition at $t - (1/s)$ and the general solution, one obtains

$$\phi(1 - st) = b(t) - K\psi(st - 1).$$

If $w(t) = \psi(1 + st) - p(t)$, then evaluation in the general solution gives

$$v(1,t) = w(t) - Kw(t - r)$$

$$i(1,t) = -\frac{1}{z}\, w(t) - \frac{K}{z}\, w(t-r) + q$$

where $r = 2/s$, $zq(t) = -p(t) - Kp(t-r) + b(t)$. Using the second boundary condition one obtains the equation

$$\begin{aligned} C_1 \frac{d}{dt}[w(t) - Kw(t-r)] = q - \frac{1}{z}w(t) - \frac{K}{z}w(t-r) \\ - g(w(t) - Kw(t-r)). \end{aligned}$$

In his consideration of shunted transmission lines, Lopes [2] encountered equations of the preceding type with two delays.

Sometimes boundary control of a linear hyperbolic equation can be more effectively studied by investigating the corresponding control problem for the transformed equations (see Banks and Kent [1]).

Another similar equation encountered by Rubanik [1] in his study of vibrating masses attached to an elastic bar is

$$\ddot{x}(t) + \omega_1^2 x(t) = \epsilon f_1(x(t), \dot{x}(t), y(t), \dot{y}(t)) + \gamma_1 \ddot{y}(t-r),$$

$$\ddot{y}(t) + \omega_2^2 x(t) = \epsilon f_2(x(t), \dot{x}(t), y(t), \dot{y}(t)) + \gamma_2 \ddot{x}(t-r).$$

In studying the collision problem in electrodynamics, Driver [1] considered systems of the type

$$\dot{x}(t) = f_1(t, x(t), x(g(t))) + f_2(t, x(t), x(g(t)))\dot{x}(g(t))$$

where $g(t) \leq t$. In the same problem, one encounters delays g that also depend on x.

El'sgol'tz [1, 2], Sabbagh [1], and Hughes [1] have considered the variational problem of minimizing

$$V(x) = \int_0^1 F(t, x(t), x(t-r), \dot{x}(t), \dot{x}(t-r))\, dt$$

over some class of functions x. Generally, the Euler equations are of the form

$$\ddot{x}(t) = f(t, x(t), x(t-r), \dot{x}(t), \dot{x}(t-r), \ddot{x}(t-r))$$

with some appropriate boundary conditions.

In the slowing down of neutrons in a nuclear reactor, the asymptotic behavior as $t \to \infty$ of the equation

$$x(t) = \int_t^{t+1} k(s)x(s)\, ds$$

or

$$\dot{x}(t) = k(t+1)x(t+1) - k(t)x(t)$$

seems to play an important role (see Slater and Wilf [1]). The state at time t depends on the future state of the system. This can be considered as a special case of the retarded equation if we replace t by $-\tau$ and investigate solutions in the direction of decreasing τ.

In his study of the dynamics of certain types of elastic materials, Volterra [3] suggested that an appropriate model for the system would be partial differential equations of the following type:

$$u_t(x,t) - \Delta u(x,t) + \int_0^t \sum_{i,j=1}^n u_{x_i x_j}(x,\tau)\phi_{i,j}(t,\tau)\,d\tau = f(x,t)$$

$$u_t(x,t) - \Delta u(x,t) + R(x,t)u(x,\omega(t)) = f(x,t)$$

where $R(x,t)$ is a linear differential operator in x of first order and $\omega(t) \in [0,t]$ for each t. Systems of this type and even more general ones have been studied by Artola [1], Baiocchi [1, 2], Dafermos [1, 2], MacCamy [1, 2, 3], and Slemrod [1].

There are many other systems for which the future behavior depends on the past and yet there are no derivatives involved at all. One of these is the difference equation

$$x(t) = g\big(t, x(t-1), x(t-2), \ldots, x(t-N)\big),$$

so important in problems in economics and models of heredity.

Finally, the Volterra integral equations

$$x(t) = f(t) + \int_0^t a(t,s,x(s))\,ds$$

occur often in applied mathematics. For specific references and many more examples and applications, see the Miscellaneous Exercises and Research Problems in Bellman and Cooke [1] and the books of Miller [1], Corduneanu [1], Halanay [1], and Grippenberg et al. [1].

The preceding examples have amply illustrated the importance and frequency of occurrence of equations that depend on past history. The diversity of the different types of equations makes it seem at first glance to be almost impossible to find a class of equations that contains all of these and is still mathematically tractable and interesting. Of course, one could write an equivalent integral equation for all differential equations and then consider general operator equations to obtain existence, uniqueness, etc. Some such general papers have appeared (see, in particular, Tychonov [1] and Neustadt [1]) and include a few of these types. The difficulty in this approach is to incorporate into the resulting functional equation all of the distinct properties associated with the original differential equation. One obtains a general existence theorem for a functional equation and it becomes a major task to verify that one of the special equations satisfies all of the hypotheses. But more importantly, some of the dynamics and geometry of the original problem are lost.

In this book, we emphasize the dynamics and the resulting flow induced by the equations. Our objective is to obtain a theory for classes of equations that begins to be as comprehensive as the available theory for ordinary differential equations. We continually attempt to emphasize the underlying ideas involved, hoping that further research in similar directions will lead to extensions of more complicated problems that occur in the applications. To accomplish our objective, we first discuss at great length equations with no delays in the derivatives (the so-called retarded functional differential equations). We then introduce a class of equations with delays in the derivatives (neutral equations), which includes many but not all of these special types. We hope in this way to isolate a class of equations that is small enough to have a rich mathematical structure and yet is large enough to include many interesting applications. As remarked earlier, the experience and information gained by this approach are useful in the discussion of other types of equations.

A brief description of the organization of the book follows. The first chapter introduces the subject through linear differential difference equations of retarded and neutral type with constant coefficients. In this way, the reader becomes familiar at an elementary level with the characteristic equation, the fundamental solution, and the role of the fundamental solution in determining precise exponential bounds on the solutions of the homogeneous equation as well as the behavior of the solutions of nonhomogeneous equations.

For a rather general class of retarded and neutral functional differential equations, Chapter 2 contains the basic theory of existence, uniqueness, continuation, and continuous dependence on parameters and initial data.

Chapter 3 is fundamental for an understanding of some of the differences between ordinary differential equations and functional differential equations. This chapter contains numerous examples depicting various types of behavior of the solutions. A careful study of these examples will develop the type of intuition that should allow the reader to avoid pitfalls as well as make sensible conjectures of what to expect in specific problems. In addition, in Chapter 3, we discuss exponentially small solutions for linear autonomous equations and give a very useful characterization of the solution operator of nonlinear systems in terms of a contracting semigroup and a completely continuous operator.

Chapter 4 contains an abstract theory of dissipative processes, concentrating particularly on maximal compact invariant sets and compact global attractors. This theory leads to a procedure for comparing the flows of infinite-dimensional systems. It also has applications to stability theory as well as the existence of periodic solutions of periodic functional differential equations.

Chapter 5 is an extensive study of the theory of stability. We emphasize and contrast the method of Liapunov functionals and the method of Razumikhin. Many examples are given illustrating the results.

Chapters 6–8 deal with linear systems of retarded FDE. Chapter 6 contains the general theory of time-varying systems, including the variation-of-constants formula, formal adjoint equations, and boundary-value problems. Some relationships are also given between the various types of stability for linear systems. Chapter 7 contains the fundamental theory of linear autonomous equations. It shows how the theory for functional differential equations is related to the theory of linear ordinary differential equations with constant coefficients, including a decomposition analogous to the Jordan block decomposition for matrices. These results are fundamental in the study of perturbed linear systems as well as the generic theory. Chapter 8 deals with the same questions as Chapter 7 except for periodic systems.

Chapter 9 is devoted to topics in neutral functional differential equations, especially the theory for linear systems analogous to Chapters 6–8 for retarded equations and the stability theory analogous to Chapter 5 for retarded equations.

Chapter 10 is devoted to the discussion of stable, unstable, and center manifolds near an equilibrium point for retarded and neutral functional differential equations. For retarded equations, similar results are given for periodic orbits of autonomous systems. Analogous but weaker results are given for periodic orbits of neutral equations.

Chapter 11 deals with the existence of nonconstant periodic solutions of autonomous equations. The Hopf bifurcation is given as well as a general method for determining periodic solutions. The latter method is probably the most powerful one available at the present time. Various illustrations are given.

In Chapter 12, we present selected directions in which the field of functional differential equations has been going in recent years. The topics in Sections 12.1–12.9 are self-explanatory and are presented in some detail with few proofs.

Each chapter has a section called "Supplementary Remarks" that contains many other directions with references in which the field of functional differential equations is going.

The Appendix contains the classical procedure for determining when the roots of a characteristic equation are in the left half-plane. The examples needed in the text are discussed in detail.

There are now several books devoted to FDE and their applications that complement the present text (for example, see Diekmann, van Gils, Verduyn Lunel, and Walther [1], Driver [3], El'sgol'tz and Norkin [1], Gosplamy [1], Kolmanovskii and Nosov [1], and Kolmanovskii and Myshkis [1]).

1
Linear differential difference equations

In this chapter, we discuss the simplest possible differential difference equations; namely, linear equations with constant coefficients. For these equations, a rather complete theory can be developed using very elementary tools. The chapter serves as an introduction to the more general types of equations that will be encountered in later chapters. It also is intended to bring out the roles of the characteristic equation and the Laplace transform and to emphasize some of the differences between retarded and neutral equations. Since ordinary differential equations and difference equations are special cases of the theory, we begin the discussion with the latter.

1.1 Differential and difference equations

Let $\mathbb{R} = (-\infty, \infty)$, $\mathbb{R}^n$ be any real n-dimensional normed vector space. For the scalar differential equation

$$\dot{x} = Ax, \tag{1.1}$$

where A is a constant, all solutions are given by $(\exp At)c$, where c is an arbitrary constant. In case A is an $n \times n$ matrix and x is an n-vector, the same result is true, of course, with c an n-vector. Each column of $\exp At$ has the form $\sum p_j(t) \exp \lambda_j t$, where each $p_j(t)$ is an n-vector polynomial in t and each λ_j is an eigenvalue of the matrix A; that is, each λ_j satisfies the characteristic equation

$$\det(\lambda I - A) = 0. \tag{1.2}$$

The coefficients of the polynomials p_j are determined from the generalized eigenvectors of the eigenvalue λ_j. As a consequence of this representation of the solutions of Equation (1.1), complete information of the solution is obtained from the eigenvalues and eigenvectors of the matrix A. Note that the characteristic equation (1.2) can be obtained by trying to find nontrivial solutions of Equation (1.1) of the form $(\exp \lambda t)c$.

Now consider the nonhomogeneous equation

$$\dot{x} = Ax + f(t) \tag{1.3}$$

where f is a given continuous function from $\mathbb{R}$ to $\mathbb{R}^n$. It is well known that the solution of Equation (1.3) with $x(0) = c$ is given by the variation-of-constants formula

$$x(t) = e^{At}c + \int_0^t e^{A(t-s)} f(s)\, ds. \tag{1.4}$$

The derivation is easily obtained by making the transformation of variables $x(t) = (\exp At)y(t)$ and observing that

$$\dot{y} = e^{-At} f(t).$$

Representation (1.4) will be needed later in this chapter to prove an existence theorem for a differential difference equation.

The theory for simple difference equations follows closely in spirit the theory for Equation (1.1). Consider the difference equation

$$x(t) = Ax(t-1) + Bx(t-2) \tag{1.5}$$

where A and B are constants. By introducing an additional variable $y(t) = x(t-1)$, this scalar equation (1.5) is equivalent to the two-dimensional equation

$$z(t) = Cz(t-1), \qquad z = \begin{bmatrix} x \\ y \end{bmatrix}, \qquad C = \begin{bmatrix} A & B \\ 1 & 0 \end{bmatrix}. \tag{1.6}$$

To obtain a solution of Equation (1.6) defined for all $t \geq 0$, one must specify a 2-vector function ϕ on $[-1, 0]$. For any $\theta \in [-1, 0]$, the solution z of Equation (1.6) is given by

$$z(t) = C^{t-\theta}\phi(\theta), \qquad t = \theta, \theta + 1, \ldots, \theta + k, \ldots. \tag{1.7}$$

Thus we see that the behavior of the solutions is determined by the eigenvalues and eigenvectors of the matrix C. The eigenvalues of C are the roots of the characteristic equation

$$\rho^2 - A\rho - B = 0. \tag{1.8}$$

Even though an initial function on $[-r, 0]$ is needed to have Equation (1.6) define a function on $[0, \infty)$, the problem actually is finite dimensional since the detailed structure of the solution is determined by the mapping C on the plane. As for the differential equation (1.1), notice that the characteristic equation (1.8) can be obtained by seeking nontrivial solutions of Equation (1.5) of the form $x(t) = \rho^t c$, where c is a nonzero constant.

In this form, Equation (1.5) seems to be no more complicated than Equation (1.1) since it is very similar to a linear map of the plane into itself. However, the analysis of this equation is very sensitive to the numbers 1 and 2 on the right-hand side. If we consider the equation

$$x(t) = Ax(t-r) + Bx(t-s) \tag{1.9}$$

where r/s is irrational, $s > r > 0$, the problem is completely different. In contrast to the case $r = 1$, $s = 2$, one cannot obtain any solution of Equation (1.9) by specifying initial values only at $x(-r), x(-s)$. The problem is basically infinite-dimensional and one sees that a reasonable initial-value problem for Equation (1.9) is to specify an initial function on $[-s, 0]$ and use Equation (1.9) to determine a solution for $t \geq 0$.

In analogy to the preceding cases, one would expect the characteristic equation for Equation (1.9) to be obtained by seeking nontrivial solutions of Equation (1.9) of the form $x(t) = \rho^t c$, where $c \neq 0$ is constant. The resulting equation is

$$\rho^s - A\rho^{s-r} - B = 0. \tag{1.10}$$

This equation for r/s irrational has infinitely many solutions. Therefore, it is not obvious that the solutions of Equation (1.9) can be obtained as linear combinations of the characteristic functions. Even without discussing the question of representation of solutions in series, it is not even obvious that the asymptotic behavior (stability, etc.) of the solutions of Equation (1.9) is determined by the solutions of the characteristic equation (1.10). Both of these problems have a positive solution and, as we shall see, one method of attack is through the Laplace transform.

A generalization of Equation (1.9) would be the equation

$$x(t) = \int_{-\infty}^{0} d[\mu(\theta)]x(t+\theta)$$

where μ is a function of bounded variation.

1.2 Retarded differential difference equations

The simplest linear retarded differential difference equation has the form

$$\dot{x}(t) = Ax(t) + Bx(t-r) + f(t) \tag{2.1}$$

where A, B, and r are constants with $r > 0$, f is a given continuous function on $\mathbb{R}$, and x is scalar.

The first question is the following: what is the initial-value problem for Equation (2.1)? More specifically, what is the minimum amount of initial data that must be specified in order for Equation (2.1) to define a function for $t \geq 0$? A moment of reflection indicates that a function must be specified on the entire interval $[-r, 0]$. In fact, let us prove

Theorem 2.1. *If ϕ is a given continuous function on $[-r, 0]$, then there is a unique function $x(\phi, f)$ defined on $[-r, \infty)$ that coincides with ϕ on $[-r, 0]$ and satisfies Equation (2.1) for $t \geq 0$. Of course, at $t = 0$, the derivative in Equation (2.1) represents the right-hand derivative.*

Proof. If x is a solution of Equation (2.1) which coincides with ϕ on $[-r, 0]$, then the variation-of-constants formula (1.4) implies that x must satisfy

$$\begin{aligned} &x(t) = \phi(t), \qquad t \in [-r, 0], \\ &x(t) = e^{At}\phi(0) + \int_0^t e^{A(t-s)}[Bx(s-r) + f(s)]\,ds, \qquad t \geq 0. \end{aligned} \tag{2.2}$$

Also, if x satisfies Equations (2.2), then x must satisfy Equation (2.1). It is only necessary to show that Equations (2.2) have a unique solution. But this is trivial to demonstrate since we may explicitly calculate the solution by the method of steps. In fact, on the interval $0 \leq t \leq r$, the function x is given uniquely by

$$x(t) = e^{At}\phi(0) + \int_0^t e^{A(t-s)}[B\phi(s-r) + f(s)]\,ds.$$

Once x is known on $[0, r]$ and since it is continuous on this interval, the second of Equations (2.2) can be used to obtain x on $[r, 2r]$. The process may be continued to prove the theorem. □

If f is not continuous but only locally integrable on $\mathbb{R}$, then the same proof yields the existence of a unique solution $x(\phi, f)$. Of course, by a solution, we mean a function that satisfies Equation (2.1) almost everywhere.

Theorem 2.2. *If $x(\phi, f)$ is the solution of Equation (2.1) defined by Theorem 2.1, then the following assertions are valid.*

(i) *$x(\phi, f)(t)$ has a continuous first derivative for all $t > 0$ and has a continuous derivative at $t = 0$ if and only if $\phi(\theta)$ has a derivative at $\theta = 0$ with*

$$\dot{\phi}(0) = A\phi(0) + B\phi(-r) + f(0). \tag{2.3}$$

If f has derivatives of all orders, then $x(\phi, f)$ becomes smoother with increasing values of t.

(ii) *If $B \neq 0$, then $x(\phi, f)$ can be extended as a solution of Equation (2.1) on $[-r - \epsilon, \infty)$, $0 < \epsilon \leq r$, if and only if ϕ has a continuous first derivative on $[-\epsilon, 0]$ and Equation (2.3) is satisfied. Extension further to the left requires more smoothness of ϕ and f and additional boundary conditions similar to Condition (2.3).*

Proof. Part (i) is obvious from Equation (2.1). The necessity of Part (ii) is also obvious. To prove the sufficiency, simply observe that the extension to the left of $-r$ can be accomplished by using the formula

$$x(t-r) = \frac{1}{B}[\dot{x}(t) - Ax(t) - f(t)] \tag{2.4}$$

by the method of steps; that is, if ϕ satisfies the stated conditions, then the right-hand side is known explicitly for $t \in [-\epsilon, 0]$ and, therefore, $x(s)$ is known for $s \in [-r-\epsilon, -r]$. If ϕ has a first derivative on $[-r, 0]$, then Relation (2.4) defines the solution on $[-2r, -r]$ and, therefore the solution is extended to $[-2r, \infty)$.

To extend the solution to the interval $[-2r-\epsilon, \infty)$, $0 < \epsilon \le r$, requires that the function $x(s)$, $s \in [-r-\epsilon, -r]$, defined by Formula (2.4) be continuously differentiable and satisfies

$$\dot{x}(-r) = Ax(-r) + Bx(-2r) + f(-r). \tag{2.5}$$

This requires the right-hand side of Formula (2.4) to be continuously differentiable on $[-\epsilon, 0]$, which imposes conditions on f and ϕ as well as boundary conditions at 0. It is sufficient for f to be differentiable on $[-\epsilon, 0]$, ϕ to have two continuous derivatives on $[-\epsilon, 0]$ and to satisfy some additional boundary conditions obtainable from Formulas (2.4) and (2.5) and the relation

$$\dot{\phi}(-r) = \frac{1}{B}\left[\ddot{\phi}(0) - A\dot{\phi}(0) - \dot{f}(0)\right]. \tag{2.6}$$

This completes the proof of the theorem. □

Due to the smoothing property (i) of Theorem 2.2, many results from ordinary differential equations are valid for retarded equations. The similarities will become more apparent as further results are obtained.

1.3 Exponential estimates of $x(\phi, f)$

In this section, we derive an estimate on how the solution $x(\phi, f)$ of Equation (2.1) depends on ϕ and f. These estimates are basic to the application of the Laplace transform and to obtaining an analogue of the variation-of-constants formula.

To obtain these estimates, we need the following fundamental lemma.

Lemma 3.1. *If u and α are real-valued continuous functions on $[a, b]$, and $\beta \ge 0$ is integrable on $[a, b]$ with*

$$u(t) \le \alpha(t) + \int_a^t \beta(s)u(s)\,ds, \qquad a \le t \le b, \tag{3.1}$$

then

$$u(t) \le \alpha(t) + \int_a^t \beta(s)\alpha(s)\left[\exp \int_s^t \beta(\tau)\,d\tau\right] ds, \qquad a \le t \le b. \tag{3.2}$$

If, in addition, α is nondecreasing, then

$$u(t) \leq \alpha(t) \exp\Big(\int_a^t \beta(s)\, ds\Big), \qquad a \leq t \leq b. \tag{3.3}$$

Proof. Let $R(t) = \int_a^t \beta(s)u(s)\, ds$. Then

$$\frac{dR}{dt} = \beta u \leq \beta\alpha + \beta R$$

and

$$\frac{d}{ds}\big[R(s) \exp\big(-\int_a^s \beta\big)\big] \leq \beta(s)\alpha(s) \exp\big(-\int_a^s \beta\big).$$

Integrating from a to t, one obtains

$$R(t) \leq \int_a^t \beta(s)\alpha(s) \exp\Big(\int_s^t \beta\Big)\, ds$$

and Inequality (3.1) yields Inequality (3.2). If α is nondecreasing, then Inequality (3.2) yields

$$u(t) \leq \alpha(t)\Big[1 + \int_a^t \beta(s)\big(\exp \int_s^t \beta\big)\, ds\Big]$$

for $t \in [a, b]$. A direct integration gives Inequality (3.3) and the lemma is proved. □

Theorem 3.1. *Suppose $x(\phi, f)$ is the solution of Equation* (2.1) *defined by Theorem* 2.1. *Then there are positive constants a and b such that*

$$|x(\phi, f)(t)| \leq a e^{bt}\Big(|\phi| + \int_0^t |f(s)|\, ds\Big), \qquad t \geq 0 \tag{3.4}$$

where $|\phi| = \sup_{-r\leq\theta\leq 0} |\phi(\theta)|$.

Proof. Since $x = x(\phi, f)$ satisfies Equation (2.1) for $t \geq 0$,

$$x(t) = \phi(0) + \int_0^t [Ax(s) + Bx(s - r) + f(s)]\, ds$$

for $t \geq 0$ and $x(t) = \phi(t)$ for $t \in [-r, 0]$. Therefore, for $t \geq 0$,

$$\begin{aligned} |x(t)| &\leq |\phi| + \int_0^t |f(s)|\, ds + \int_0^t |A||x(s)|\, ds + \int_{-r}^t |B||x(s)|\, ds \\ &\leq (1 + |B|r)|\phi| + \int_0^t |f(s)|\, ds + \int_0^t (|A| + |B|)|x(s)|\, ds. \end{aligned}$$

Applying Lemma 3.1, Formula (3.3), to this inequality, we obtain

$$|x(t)| \le \left[(1+|B|r)|\phi| + \int_0^t |f(s)|\,ds\right]\exp(|A|+|B|)t.$$

Since $|B| \ge 0$, it follows that Inequality (3.4) is satisfied with $a = 1+|B|r$ and $b = |A|+|B|$. □

Some immediate implications of Theorem 3.1 are the following. Since Equation (2.1) is linear and solutions are uniquely defined by ϕ, it is obvious that the solution $x(\phi, 0)$ of the homogeneous equation

$$\dot{x}(t) = Ax(t) + Bx(t-r) \tag{3.5}$$

that coincides with ϕ on $[-r, 0]$ is linear in ϕ; that is, $x(\phi+\psi, 0) = x(\phi, 0) + x(\psi, 0)$ and $x(a\phi, 0) = ax(\phi, 0)$ for any continuous functions ϕ and ψ on $[-r, 0]$ and any scalar a. For $f = 0$, Inequality (3.4) implies that $x(\phi, 0)(t)$ is continuous in ϕ for all t; that is, $x(\cdot, 0)(t)$ is a continuous linear functional on the space of continuous functions on $[-r, 0]$. The Riesz representation theorem then implies that the solution can be represented as a Stieltjes integral.

By the remark after Theorem 2.1 and the proof of Theorem 3.1, Estimate (3.4) is valid for locally integrable functions f. Using the same reasoning as earlier, the function $x(0, \phi)$ is a solution of the nonhomogeneous equation (2.1) with zero initial data. Estimate (3.4) shows that $x(0, \cdot)(t)$ is a continuous linear functional on the locally integrable functions. Therefore, an integral representation of $x(0, f)(t)$ is known to exist. Such a representation corresponds to the variation-of-constants formula.

This functional analytic approach will be exploited in detail in later chapters for more general problems. However, for this simple equation, the use of the Laplace transform is just as convenient and brings in some other concepts that are important in the general theory.

1.4 The characteristic equation

The *characteristic equation* for a homogeneous linear differential difference equation with constant coefficients is obtained from the equation by looking for nontrivial solutions of the form $e^{\lambda t}c$ where c is constant. For example, the scalar equation

$$\dot{x}(t) = Ax(t) + Bx(t-r) \tag{4.1}$$

has a nontrivial solution $e^{\lambda t}c$ if and only if

$$h(\lambda) \stackrel{\text{def}}{=} \lambda - A - Be^{-\lambda r} = 0. \tag{4.2}$$

We need some properties of the characteristic equation (4.2). In the following, Re λ designates the real part of λ.

Lemma 4.1. *If there is a sequence* $\{\lambda_j\}$ *of solutions of Equation* (4.2) *such that* $|\lambda_j| \to \infty$ *as* $j \to \infty$, *then* $\text{Re } \lambda_j \to -\infty$ *as* $j \to \infty$. *Thus, there is a real number* α *such that all solutions of Equation* (4.2) *satisfy* $\text{Re } \lambda < \alpha$ *and there are only a finite number of solutions in any vertical strip in the complex plane.*

Proof. For any solution λ of Equation (4.2),

$$|\lambda - A| = |B|e^{-r\text{Re}\lambda}.$$

Consequently, if $|\lambda| \to \infty$, then $\exp(-r\text{Re }\lambda) \to \infty$, which implies the first statement of the lemma. This also implies the existence of α as in the lemma. Since $h(\lambda)$ is an entire function, there can be only a finite number of zeros of $h(\lambda)$ in any compact set. These facts imply there are only a finite number in any vertical strip in the complex plane and the lemma is proved. □

Theorem 4.1. *Suppose* λ *is a root of multiplicity* m *of the characteristic equation* (4.2). *Then each of the functions* $t^k \exp \lambda t$, $k = 0, 1, 2, \ldots, m-1$, *is a solution of Equation* (4.1). *Since Equation* (4.1) *is linear, any finite sum of such solution is also a solution. Infinite sums are also solutions under suitable conditions to ensure convergence.*

Proof. If $x(t) = t^k e^{\lambda t}$, then

$$\begin{aligned} e^{-\lambda t}[\dot{x}(t) - Ax(t) - Bx(t-r)] &= t^k\lambda + kt^{k-1} - At^k - B(t-r)^k e^{-\lambda r} \\ &= \sum_{j=0}^{k} \binom{k}{j} t^{k-j} h^{(j)}(\lambda) \end{aligned}$$

where the last expression is obtained by expanding $(t-r)^k$ by the binomial theorem and making the observation that the coefficients are related to the derivatives $h^{(j)}(\lambda)$, $h^{(0)}(\lambda) = h(\lambda)$, of the function $h(\lambda)$ as indicated.

If λ is a zero of $h(\lambda)$ of multiplicity m, then $h(\lambda) = h^{(1)}(\lambda) = \cdots = h^{(m-1)}(\lambda) = 0$. Consequently, $x(t) = t^k e^{\lambda t}$ is a solution of Equation (4.1) for $k = 0, 1, \ldots, m-1$. This proves the first part of the theorem. The last assertions are obvious and the theorem is proved. □

1.5 The fundamental solution

In the next section, we are going to apply the Laplace transform to obtain the solution of the initial-value problem for the nonhomogeneous equation (2.1). The characteristic equation (4.2) arises naturally as does the need for a function whose Laplace transform is $h^{-1}(\lambda)$. It is actually possible to find a solution of the homogeneous equation (4.1), called the *fundamental solution* of Equation (4.1), whose Laplace transform is $h^{-1}(\lambda)$. This function will now be defined without further motivation.

Let $X(t)$ be the solution that satisfies Equation (4.1) for $t \geq 0$ and satisfies the initial condition

$$X(t) = \begin{cases} 0, & t < 0, \\ 1, & t = 0. \end{cases} \tag{5.1}$$

Our basic existence theorem does not apply to the initial data (5.1), but the same proof as in Theorem 2.1 may be used to prove existence. Furthermore, the exponential estimate in Theorem 3.1 is likewise valid. Also, it is clear that $X(t)$ is of bounded variation on any compact set.

Our next objective is to apply the Laplace transform to $X(t)$. The following two lemmas needed from the theory of the Laplace transform are stated without proof.

Lemma 5.1. (Existence and convolution of Laplace transform). *If $f : [0,\infty) \to \mathbb{R}$ is measurable and satisfies*

$$|f(t)| \leq ae^{bt}, \qquad t \in [0,\infty), \tag{5.2}$$

for some constants a and b, then the Laplace transform $\mathcal{L}(f)$ defined by

$$\mathcal{L}(f)(\lambda) = \int_0^\infty e^{-\lambda t} f(t)\, dt \tag{5.3}$$

exists and is an analytic function of λ for Re $\lambda > b$. *If the function $f * g$ is defined by $f * g(t) = \int_0^t f(t-s)g(s)\, ds$, then $\mathcal{L}(f * g) = \mathcal{L}(f)\mathcal{L}(g)$.*

The following notation is used

$$\int_{(c)} = \lim_{T \to \infty} \frac{1}{2\pi i} \int_{c-iT}^{c+iT}$$

where c is a real number.

Lemma 5.2. (Inversion theorem). *Suppose $f : [0,\infty) \to \mathbb{R}$ is a given function, $b > 0$ is a given constant such that f is of bounded variation on any compact set, and $t \mapsto f(t)\exp(-bt)$ is Lebesgue integrable on $[0,\infty)$. Then, for any $c > b$,*

$$\int_{(c)} \mathcal{L}(f)(\lambda) e^{\lambda t} d\lambda = \begin{cases} \frac{1}{2}[f(t+) + f(t-)], & t > 0, \\ \frac{1}{2} f(0+), & t = 0. \end{cases} \tag{5.4}$$

Theorem 5.1. *The solution $X(t)$ of Equation* (4.1) *with initial data* (5.1) *is the fundamental solution; that is,*

$$\mathcal{L}(X)(\lambda) = h^{-1}(\lambda). \tag{5.5}$$

Also, for any $c > b$,

$$X(t) = \int_{(c)} e^{\lambda t} h^{-1}(\lambda)\, d\lambda, \qquad t > 0 \tag{5.6}$$

where b is the exponent associated with the bound on $X(t)$ in Theorem 3.1, $|X(t)| \leq a\exp(bt)$, $t \geq 0$.

Proof. Since $X(t)$ satisfies the exponential bounds in Theorem 3.1, $\mathcal{L}(X)$ exists and is analytic for Re $\lambda > b$. Multiplying Equation (4.1) by $e^{-\lambda t}$, integrating from 0 to ∞, and integrating the first term by parts, one obtains Equation (5.5). Since $X(t)$ is of bounded variation on compact sets and continuous for $t \geq 0$, Relation (5.6) follows from the inversion formula (5.4). The theorem is proved. □

Our next objective is to obtain very precise exponential bounds on $X(t)$ in terms of the maximum of the real parts of the solutions of the characteristic equation.

Theorem 5.2. *If $\alpha_0 = \max\{\text{Re}\ \lambda : h(\lambda) = 0\}$, then, for any $\alpha > \alpha_0$, there is a constant $k = k(\alpha)$ such that the fundamental solution $X(t)$ satisfies the inequality*

$$|X(t)| \leq ke^{\alpha t}, \qquad t \geq 0. \tag{5.7}$$

Proof. From Theorem 5.1, we know that

$$X(t) = \int_{(c)} e^{\lambda t} h^{-1}(\lambda)\, d\lambda$$

where c is some sufficiently large real number. We may take $c > \alpha$. We want to prove first that

$$X(t) = \int_{(\alpha)} e^{\lambda t} h^{-1}(\lambda)\, d\lambda. \tag{5.8}$$

To show this, consider the integration of the function $e^{\lambda t}h^{-1}(\lambda)$ around the boundary of the box Γ in the complex plane with boundary $L_1 M_1 L_2 M_2$ in the direction indicated, where the segment L_1 is the set $\{c + i\tau : -T \leq \tau \leq T\}$, the segment L_2 is the set $\{\alpha + i\tau : -T \leq \tau \leq T\}$, the segment M_1 is the set $\{\sigma + iT : \alpha \leq \sigma \leq c\}$, and the segment M_2 is the set $\{\sigma - iT, \alpha \leq \sigma \leq c\}$ (see Fig. 1.1). Since $h(\lambda)$ has no zeros in this box, it follows that the integral over the boundary is zero. Therefore, Relation (5.8) will be verified if we show that

$$\int_{M_1} e^{\lambda t} h^{-1}(\lambda)\, d\lambda, \quad \int_{M_2} e^{\lambda t} h^{-1}(\lambda)\, d\lambda \to 0$$

as $T \to \infty$.

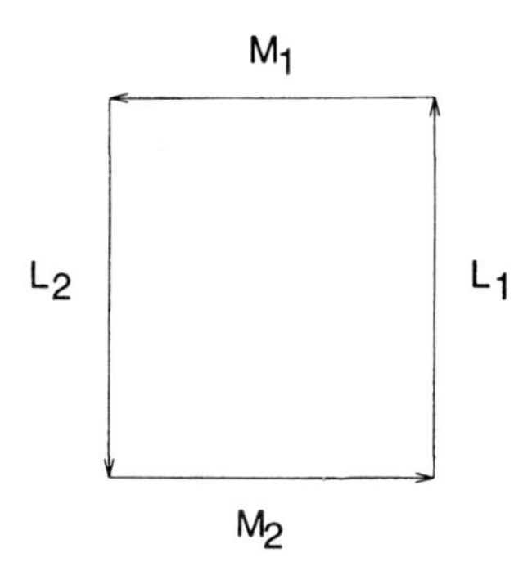

Fig. 1.1.

Choose T_0 so that

$$\left(1 + \frac{\alpha^2}{T^2}\right)^{1/2} - \frac{1}{T}(|A| + |B|e^{-\alpha r}) \geq \frac{1}{2}$$

for all $T \geq T_0$. If $T \geq T_0$ and $\lambda \in M_1$; that is, $\lambda = \sigma + iT$, $\alpha \leq \sigma \leq c$, and $T \geq T_0$, then

$$|h^{-1}(\lambda)| \leq \frac{1}{(\sigma^2 + T^2)^{1/2} - |A| - |B|\exp(\sigma r)} \leq \frac{2}{T}.$$

Therefore,

$$\left|\int_{M_1} e^{\lambda t} h^{-1}(\lambda)\, d\lambda\right| \leq \frac{2}{T} e^{ct}(c - \alpha) \to 0 \quad \text{as } T \to \infty.$$

In the same way, the integral over $M_2 \to 0$ as $T \to \infty$. This proves Equation (5.8).

Suppose T_0 as above. If $g(\lambda) = h^{-1}(\lambda) - (\lambda - \alpha_0)^{-1}$, then

$$|g(\lambda)| = \left|\frac{A + Be^{-\lambda r} - \alpha_0}{\lambda - \alpha_0} h^{-1}(\lambda)\right| \leq \frac{2}{T^2}(|A| + |B|e^{-\alpha r} + |\alpha_0|)$$

for $\lambda = \alpha + iT$, $|T| \geq T_0$. Therefore,

$$\int_{(\alpha)} |g(\lambda)|\, d\lambda < +\infty \quad \text{and} \quad \left|\int_{(\alpha)} e^{\lambda t} g(\lambda)\, d\lambda\right| \leq K_1 e^{\alpha t}, \quad t > 0,$$

where K_1 is a constant. Furthermore, the integral $\int_{(\alpha)} e^{\lambda t}(\lambda - \alpha_0)^{-1}\, d\lambda$ is known to exist and admit an estimate of the form

$$\left|\int_{(\alpha)} e^{\lambda t}(\lambda - \alpha_0)^{-1}\, d\lambda\right| \leq K_2 e^{\alpha t}, \qquad t > 0.$$

Consequently, using Equation (5.8), the proof of the theorem is complete with the constant k given by $K_1 + K_2$. □

The proof of Theorem 5.2 does not only provide a precise exponential bound for the fundamental solution $X(t)$, but also a method to expand $X(t)$ as a series of characteristic solutions $P_j(t)e^{\lambda_j t}$. Here P_j is a polynomial and λ_j a root of the characteristic equation.

We proceed as follows. In the region $\{\lambda \in \mathbb{C} : |\lambda - A| > |B|e^{-\operatorname{Re}\lambda r}\}$ the characteristic equation has no roots. So, if we choose the box Γ in the preceding proof such that

$$(\alpha - A)^2 + T^2 > B^2 e^{-2\alpha r}, \tag{5.9}$$

then, along the segments M_1, M_2, Estimate (5.9) holds. So

$$\Big|\int_{M_1} h^{-1}(\lambda)e^{\lambda t}\,d\lambda\Big| \le \frac{2}{T}e^{ct}(c-\alpha) \to 0 \qquad \text{as} \quad T \to \infty.$$

In the same way, the integral over $M_2 \to 0$ as $T \to \infty$. Since the zeros of $h(\lambda)$ have an asymptotic distribution, we can choose a sequence α_m, $m = 1, 2, \ldots$, so that the line $\operatorname{Re}\lambda = \alpha_m$ does not contain a root of the characteristic equation. Therefore, the Cauchy theorem of residues implies

$$\begin{aligned} X(t) &= \int_{(c)} e^{\lambda t}h^{-1}(\lambda)\,d\lambda \\ &= \int_{(\alpha_m)} e^{\lambda t}h^{-1}(\lambda)\,d\lambda + \sum_{j=1}^{k_m} \operatorname*{Res}_{\lambda=\lambda_j} e^{\lambda t}h^{-1}(\lambda), \end{aligned} \tag{5.10}$$

where $\lambda_1, \ldots, \lambda_{k_m}$ are the roots of the characteristic equation such that $\operatorname{Re}\lambda_j > \alpha_m$. It is easy to see that

$$\operatorname*{Res}_{\lambda=\lambda_j} e^{\lambda t}h^{-1}(\lambda) = e^{\lambda_j t}P_j(t)$$

and that $P_j(t)e^{\lambda_j t}$ is a solution of Equation (4.1) with initial function $\phi_j(\theta) = P_j(\theta)e^{\lambda_j \theta}$, $-r \le \theta \le 0$, $j = 1, 2, \ldots$. Since Equation (4.1) is linear,

$$Y_\alpha(t) = \int_{(\alpha)} e^{\lambda t}h^{-1}(\lambda)\,d\lambda$$

is a solution of Equation (4.1) as well. Further, in the same way as in the proof of Theorem 5.2,

$$|Y_{\alpha_m}(t)| \le Ke^{\alpha_m t}.$$

To analyze the series as $\alpha \to -\infty$, it suffices to analyze the solution

$$Y_0(t) = \lim_{m\to\infty} Y_{\alpha_m}(t), \qquad t \ge 0.$$

In this particular case, one can prove that $Y_0(t) = 0$ for $t > 0$. In general, $Y_0(t) \neq 0$, which is surprising since Y_0 is a solution that decays faster than any exponential. Such solutions are called small solutions and will be studied in Chapter 3. Series expansions will be studied in Chapter 7.

1.6 The variation-of-constants formula

In this section, we obtain a representation of the solution $x(\phi, f)$ of the nonhomogeneous equation (2.1) in terms of the fundamental solution $X(t)$ of the previous section. In particular, it will be shown that the solution is given by

$$(6.1) \qquad x(\phi, f)(t) = x(\phi, 0)(t) + \int_0^t X(t-s)f(s)\,ds, \qquad t \geq 0.$$

The representation in this form will be referred to as the *variation-of-constants formula.* This terminology is not a precise description but it is easy to remember because of the similarity to ordinary differential equations.

It is also possible to represent the solution $x(\phi, 0)$ of the homogeneous equation in terms of the fundamental solution $X(t)$. The precise result is contained in the following theorem.

Theorem 6.1. *The solution $x(\phi, f)$ of the nonhomogeneous equation* (2.1) *can be represented in the form* (6.1). *Furthermore, for $t \geq 0$,*

$$(6.2) \qquad x(\phi, 0)(t) = X(t)\phi(0) + B\int_{-r}^0 X(t-\theta-r)\phi(\theta)\,d\theta.$$

Proof. As remarked earlier, there are many ways to obtain this result. The present proof is based on the Laplace transform and other proofs in more general situations will be given later. To apply the Laplace transform, the function f should be bounded by an exponential function. For any compact interval $[0, T]$, one can redefine f as a continuous function so that it is zero outside the interval $[0, T+\epsilon]$, $\epsilon > 0$. Then f will be bounded by an exponential function. If we prove the theorem for this case, then the theorem will be valid for $0 \leq t \leq T$. Since T is arbitrary, the result will be proved. Therefore, we may assume f is bounded by an exponential function.

From Theorem 3.1, it is valid to apply the Laplace transform to each term in Equation (2.1). Let $x = x(\phi, f)$ be a solution of this equation. Multiplying this equation by $e^{-\lambda t}$, Re $\lambda > c$, sufficiently large, integrating from 0 to ∞, and integrating the first term by parts, we obtain

$$h(\lambda)\mathcal{L}(x)(\lambda) = \phi(0) + Be^{-\lambda r}\int_{-r}^0 e^{-\lambda\theta}\phi(\theta)\,d\theta + \int_0^\infty e^{-\lambda t}f(t)\,dt.$$

Using the inversion formula in Lemma 5.2, the expression for $x(t)$ is

$$x(t) = \int_{(c)} e^{\lambda t} h^{-1}(\lambda) \big[\phi(0) + Be^{-\lambda r} \int_{-r}^{0} e^{-\lambda \theta} \phi(\theta) d\theta + \int_0^\infty e^{-\lambda t} f(t)\, dt\big]\, d\lambda. \tag{6.3}$$

Evaluation of the last term in this expression is easy since it represents merely the inverse Laplace transform of X from Theorem 5.1. Therefore, the convolution theorem says this term is $\int_0^t X(t-s)f(s)\, ds$. Since the first two terms depend only on ϕ, this proves that the variation-of-constants formula (6.1) is valid.

To prove that the first two terms in Equation (6.3) coincide with Equation (6.2), we proceed as follows. The first term is $X(t)\phi(0)$ by the inversion formula and Theorem 5.1.

If $\omega : [-r, \infty) \to [0,1]$ is defined by $\omega(\theta) = 0$, $\theta \geq 0$, $\omega(\theta) = 1$ for $\theta < 0$ and the definition of ϕ is extended to $(-r, \infty)$ by defining $\phi(\theta) = \phi(0)$ for $\theta \geq 0$, then

$$\begin{aligned} e^{-\lambda r} \int_{-r}^{0} e^{-\lambda \theta} \phi(\theta)\, d\theta &= \int_0^\infty e^{-\lambda s} \phi(-r+s)\omega(-r+s)\, ds \\ &= \mathcal{L}(\phi(-r+\cdot)\omega(-r+\cdot)). \end{aligned}$$

Consequently, the second term in Equation (6.3) is B multiplied by the inverse Laplace transform of the product of the Laplace transforms of X and $\phi(-r+\cdot)\omega(-r+\cdot)$. The convolution theorem, therefore, implies this second term for $t > 0$ is equal to

$$B \int_0^t X(t-s)\phi(-r+s)\omega(-r+s)\, ds = B \int_0^r X(t-s)\phi(-r+s)\, ds.$$

The latter relation follows from the definition of X and ω. Letting $s = r+\theta$, one obtains the complete formula (6.2) and the theorem is proved. □

Theorem 6.1 has many interesting implications. For example, it implies the nontrivial result that the exponential behavior of the solutions of the homogeneous equation (4.1) is determined by the characteristic equation (4.2) as stated in the following result.

Theorem 6.2. *Suppose* $\alpha_0 = \max\{\text{Re}\, \lambda : h(\lambda) = 0\}$ *and* $x(\phi)$ *is the solution of the homogeneous equation* (4.1), *which coincides with* ϕ *on* $[-r, 0]$. *Then, for any* $\alpha > \alpha_0$, *there is a constant* $K = K(\alpha)$ *such that*

$$|x(\phi)(t)| \leq Ke^{\alpha t}|\phi|, \qquad t \geq 0, \quad |\phi| = \sup_{-r \leq \theta \leq 0} |\phi(\theta)|. \tag{6.4}$$

In particular, if $\alpha_0 < 0$, *then one can choose* $\alpha_0 < \alpha < 0$ *to obtain the fact that all solutions approach zero exponentially as* $t \to \infty$.

Proof. The proof is an obvious application of the exponential bounds for X in Theorem 5.2 and Formula (6.2). □

Knowing that the asymptotic behavior of the solutions of the homogeneous differential difference equation is governed by the solutions of the characteristic equation, it is possible to use the variation-of-constants formula to determine the asymptotic behavior of perturbed nonlinear systems of the form

$$\dot{x}(t) = Ax(t) + Bx(t-r) + f(x(t), x(t-r)). \tag{6.5}$$

In fact, if we assume existence for the initial-value problem, then the solution $x = x(\phi)$ of Equation (6.5) with initial data ϕ on $[-r, 0]$ is given by

$$x(t) = y(t) + \int_0^t X(t-s) f(x(s), x(s-r))\, ds \tag{6.6}$$

where $y = y(\phi)$ is the solution of the homogeneous equation (4.1) and X is the fundamental solution of that equation.

If $f(x, y)$ satisfies $f(0,0) = 0$, $\partial f(0,0)/\partial(x, y) = 0$ and $\alpha_0 < 0$ in Theorem 6.2, then one can prove the analogue of the Poincaré-Liapunov theorem on stability with respect to the first approximation. The method of proof is similar to the proof of the corresponding result for ordinary differential equations. The proof is omitted since a more general result will be given later and our main purpose in this chapter is only to acquire intuition.

Finally, we remark that other perturbation results can be obtained in a similar manner.

1.7 Neutral differential difference equations

In this section, we introduce another class of equations depending on past and present values but that involve derivatives with delays as well as the function itself. Such equations historically have been referred to as *neutral differential difference equations.* The presentation will not be as detailed as the one for the retarded equations of the previous section. We concentrate only on those proofs that are significantly different from the ones for retarded equations. We also point out some of the differences between neutral equations and retarded equations.

The model nonhomogeneous equation is

$$\dot{x}(t) - C\dot{x}(t-r) = Ax(t) + Bx(t-r) + f(t). \tag{7.1}$$

where A, B, C, and r are constants with $r > 0$, $C \neq 0$ and f is a continuous function on $\mathbb{R}$. The corresponding homogeneous equation is

$$\dot{x}(t) - C\dot{x}(t-r) = Ax(t) + Bx(t-r). \tag{7.2}$$

For this equation, it is a little more difficult to define the concept of a solution and the appropriate space of initial data. In the retarded case, the initial space did not play an important role in the initial-value problem since the solution becomes continuously differentiable for $t \geq 0$ and, in particular, it is continuous. For Equation (7.1) with $C \neq 0$, we will see that such smoothing of the solution does not occur with increasing t.

Let us define the initial-value problem as follows. Suppose ϕ is a given continuously differentiable function on $[-r, 0]$. A solution $x = x(\phi, f)$ of Equation (7.1) through ϕ is a continuous function x defined on $[-r, \infty)$ with $x(t) = \phi(t)$ for $t \in [-r, 0]$, x continuously differentiable except at the points kr, $k = 0, 1, 2, \ldots$ and x satisfying Equation (7.1) except at these points.

Using the same proof as in the proof of Theorem 2.1, one can show there always exists a solution of Equation (7.1) through ϕ.

For $C \neq 0$, Equation (7.1) implies that the solution can never have more derivatives than the initial function ϕ. Also, the solution $x = x(\phi, f)$ will in general have a discontinuous derivative at kr, $k = 0, 1, 2, \ldots$. In fact, suppose

$$\dot{\phi}(0) \neq C\dot{\phi}(-r) + A\phi(0) + B\phi(-r) + f(0).$$

Then the solution $x(t)$ has a discontinuous derivative at $t = 0$. From Equation (7.1), this implies $\dot{x}(r)$ is discontinuous since $C \neq 0$. The same reasoning applies for all values kr, $k = 0, 1, 2, \ldots$.

On the other hand, if

$$\dot{\phi}(0) = C\dot{\phi}(-r) + A\phi(0) + B\phi(-r) + f(0) \tag{7.3}$$

then the solution $x(t)$ has a continuous derivative at $t = 0$. Equation (7.1) can again be used to show that $x(t)$ has a continuous derivative for all $t \geq -r$. Consequently, Relation (7.3) is necessary and sufficient for the solution x to have a continuous derivative for all $t \geq -r$.

Notice also that $C \neq 0$ implies that one can obtain a unique solution of Equation (7.1) on $(-\infty, 0]$, which coincides with ϕ on $[-r, 0]$. In fact, if $y(t) = x(t-r)$, then y satisfies the differential equation

$$\dot{y}(t) = -\frac{B}{C}\, y(t) - \frac{A}{C}\, x(t) + \frac{1}{C}\, \dot{x}(t) - \frac{1}{C}\, f(t).$$

One may prove existence with the argument in the proof of Theorem 2.1 using the variation-of-constants formula in the backward direction. These two results are summarized in the following theorem.

Theorem 7.1. *If $C \neq 0$ and ϕ is a continuously differentiable function on $[-r, 0]$, then there exists a unique function $x : (-\infty, \infty) \to \mathbb{R}$ that coincides with ϕ on $[-r, 0]$, is continuously differentiable and satisfies Equation* (7.1) *except maybe at the points kr, $k = 0, \pm 1, \pm 2, \ldots$. This solution x can have* *no more derivatives than ϕ and is continuously differentiable if and only if Relation* (7.3) *is satisfied.*

For Equation (7.1), one can continue to develop the theory with the preceding definition of a solution. However, if one wishes to consider many delays and more general functional equations, the exceptional set where the derivative is not required to exist becomes impossible to describe. There are many ways to overcome this difficulty and we now discuss one.

Rewrite Equation (7.1) as

$$\frac{d}{dt}\,[x(t) - Cx(t-r)] = Ax(t) + Bx(t-r) + f(t). \tag{7.4}$$

In this form, it is at least meaningful to consider the following initial-value problem. Suppose ϕ is a *continuous* function on $[-r, 0]$. A solution of Equation (7.4) through ϕ is a continuous function defined on $[-r, \infty)$, which coincides with ϕ on $[-r, 0]$ such that the difference $x(t) - Cx(t-r)$ is differentiable and satisfies Equation (7.4) for $t \geq 0$.

Can one obtain such a solution? It is not difficult to prove that a solution exists by letting $x(t) = (\exp At)y(t)$ and observing that y satisfies the equation

$$\frac{d}{dt}\,[y(t) - e^{-Ar}Cy(t-r)] = e^{-Ar}(AC + B)y(t-r) + e^{-At}f(t).$$

This latter equation may now always be integrated by the method of steps in the direction of increasing t and $C \neq 0$, in the direction of decreasing t. These remarks are summarized in the following theorem.

Theorem 7.2. *If ϕ is a continuous function on $[-r, 0]$, then there is a unique solution of Equation* (7.4) *on $[-r, \infty)$ through ϕ. If $C \neq 0$, this solution exists on $(-\infty, \infty)$ and is unique.*

Equation (7.4) can be considered as a generalization of the retarded equations ($C = 0$) as well as a generalization of difference equations $[A = B = f = 0, \phi(0) = C\phi(-r)]$. This is certainly sufficient motivation to consider this class of equations in its own right. Also, we shall see that many generalizations are easily accomplished. Therefore, in the remainder of this section, we will consider only Equation (7.4) with initial data ϕ that is continuous. The corresponding homogeneous equation is

$$\frac{d}{dt}\,[x(t) - Cx(t-r)] = Ax(t) + Bx(t-r). \tag{7.5}$$

As the theory evolves, one cannot keep from observing that retarded equations are very similar to ordinary differential equations and parabolic partial differential equations and that neutral equations resemble difference equations and hyperbolic partial differential equations.

Theorem 7.3. *Let $x(\phi, f)$ be the solution of Equation* (7.4) *given in Theorem* 7.2. *Then there are positive constants a and b such that*

$$|x(\phi, f)(t)| \leq ae^{bt}\big[|\phi| + \int_0^t |f(s)|\, ds\big], \qquad t \geq 0, \tag{7.6}$$

where

$$|\phi| = \sup_{-r\leq\theta\leq 0} |\phi(\theta)|.$$

Proof. Let $\alpha = (1 + 2|C|)$ and $\beta = |A| + |B|$. If x is a solution of Equation (7.4) for $t \geq 0$, then

$$x(t) = \phi(0) - C\phi(-r) + Cx(t-r) + \int_0^t [Ax(s) + Bx(s-r)]\, ds + \int_0^t f(s)\, ds.$$

If $|\phi| = \sup_{-r\leq\theta\leq 0} |\phi(\theta)|$ and $y(t) = \sup_{-r\leq\theta\leq 0} |x(t+\theta)|$, and $0 \leq t \leq r/2$, then

$$|x(t)| \leq \alpha|\phi| + \int_0^t |f(s)|\, ds + \beta \int_0^t y(s)\, ds.$$

Since $\alpha \geq 1$ and $x(t) = \phi(t)$ for $t \leq 0$, it follows that

$$y(t) \leq \alpha|\phi| + \int_0^t |f(s)| ds + \beta \int_0^t y(s)\, ds, \qquad 0 \leq t \leq \frac{r}{2}.$$

An application of Lemma 3.1 yields

$$y(t) \leq \big[\alpha|\phi| + \int_0^t |f(s)|\, ds\big] \exp \beta t, \qquad 0 \leq t \leq \frac{r}{2}.$$

Also, the same argument implies the following estimate

$$y(t) \leq \big[\alpha y(\tau) + \int_\tau^t |f(s)|\, ds\big] \exp \beta(t-\tau), \qquad 0 \leq \tau \leq t \leq \tau + \frac{r}{2}.$$

Let $\gamma > \beta$ be such that $\alpha \exp(\beta - \gamma)r/2 < 1$ and let us prove that

$$y(t) \leq \big[\alpha|\phi| + \int_0^t |f(s)|\, ds\big] \exp \gamma t, \qquad 0 \leq t \leq \frac{kr}{2}$$

where k is an integer. We know this relation is true for $k = 1$. Assume that it is true for some $k \geq 1$. If $r/2 \leq t \leq (k+1)r/2$, then

$$y(t) \leq \big[\alpha y(t - \frac{r}{2}) + \int_{t-r/2}^t |f(s)|\, ds\big] \exp \frac{\beta r}{2}$$

and, therefore, the induction hypothesis implies

$$y(t) \le \left[\alpha\left\{\alpha|\phi| + \int_0^{t-r/2} |f(s)|\,ds\right\}e^{\gamma(t-r/2)} + \int_{t-r/2}^{t} |f(s)|\,ds\right]e^{\beta r/2}$$
$$\le \left[\alpha|\phi| + \int_0^{t-r/2} |f(s)|\,ds\right]e^{\gamma t} + e^{\beta r/2}\int_{t-r/2}^{t} |f(s)|\,ds.$$

Since $t \ge r/2$ and $\gamma > \beta$, we have

$$y(t) \le \left[\alpha|\phi| + \int_0^t |f(s)|\,ds\right] \exp \gamma t$$

and the theorem is proved. □

Our next objective is to discuss the characteristic equation for Equation (7.5). If Equation (7.5) has a solution $e^{\gamma t}$, then λ must satisfy the characteristic equation

$$H(\lambda) \stackrel{\text{def}}{=} \lambda(1 - Ce^{-\lambda r}) - A - Be^{-\lambda r} = 0. \tag{7.7}$$

Lemma 7.1. *There is a real number α such that all solutions of Equation (7.7) satisfy* Re $\lambda < \alpha$. *If $C \ne 0$, then all solutions of Equation (7.7) lie in a vertical strip $\{\beta <$ Re $\lambda < \alpha\}$ in the complex plane. If $C \ne 0$ and there is a sequence $\{\lambda_j\}$ of solutions such that $|\lambda_j| \to \infty$ as $j \to \infty$, then there is a sequence $\{\lambda_j'\}$ of zeros of*

$$1 - Ce^{-\lambda r} = 0 \tag{7.8}$$

such that $\lambda_j - \lambda_j' \to 0$ as $j \to \infty$. Also, one can show there always exists such a sequence $\{\lambda_j\}$ when $C \ne 0$.

Proof. For $\lambda \ne 0$, Equation (7.7) is equivalent to the equation

$$e^{\lambda r}\left(1 - \frac{A}{\lambda}\right) - C - \frac{B}{\lambda} = 0.$$

This relation implies the existence of α as stated for any C. If $C \ne 0$, then the same relation implies λ must lie in a vertical strip in the complex plane. In this strip, $|\exp(\lambda r)|$ is bounded. Therefore, if $|\lambda| \to \infty$, then $e^{\lambda r} - C \to 0$. This implies the statement concerning λ_j and λ_j'. The proof of the last statement is left as an exercise. □

One important observation is immediate from Lemma 7.1. For $C \ne 0$, the roots of Equation (7.8) are given by

$$\lambda = \frac{\ln C}{r} + \frac{2k\pi i}{4}, \qquad k = 0, \pm 1, \pm 2, \ldots. \tag{7.9}$$

If $\ln C > 0$, then Lemma 7.1 implies there are always infinitely many solutions of Equation (7.5) which approach ∞ at an exponential rate and, therefore, one can never have stability of Equation (7.5).

To exploit this further, suppose one attempts to consider Equation (7.5) for r small as some type of perturbation of the ordinary differential equation

$$\frac{d}{dt}(1-C)x(t) = (A+B)x(t), \tag{7.10}$$

and suppose all solutions of this equation approach zero as $t \to \infty$; that is, $(A+B)/(1-C) < 0$. Lemma 7.1 and Formula (7.9) imply there are infinitely many solutions of Equation (7.5) approaching infinity at an exponential rate if $|C| > 1$ and there can be at most a finite number of values λ with Re $\lambda > 0$ if $|C| < 1$. Furthermore, from Equation (7.7), it is easy to see that for any compact set U in the complex plane and any neighborhood V of the complex number $(A+B)/(1-C)$, there is an r_0 sufficiently small so that Equation (7.7) has no solution in $U - V$ and exactly one solution in V for $0 < r \leq r_0$. Therefore, except for the root of Equation (7.7), which approaches $(A+B)/(1-C)$ as $r \to 0$, the moduli of the other roots approach $+\infty$ as $r \to 0$. Thus, for r sufficiently small, $|C| < 1$, Lemma 7.1, and Formula (7.9) imply all roots of Equation (7.7) satisfy Re $\lambda < -\delta < 0$ for some $\delta > 0$. If these roots govern the asymptotic behavior of solutions, then at least asymptotic stability of the ordinary differential equation (7.10) is preserved for Equation (7.5) if $|C| < 1$.

For Equation (7.5) with $C = 0$, that is, for the retarded equation, Lemma 7.1 implies there are *never* more than a finite number of λ with Re $\lambda > 0$. As in the argument for $|C| < 1$, one concludes that all roots of the characteristic equation except the one close to $A + B$ are in the left half-plane for r sufficiently small.

In order to put more emphasis on these important remarks, we summarize them in the following corollary.

Corollary 7.1. *Suppose the supremum of the real parts of the roots of Equation* (7.7) *govern the asymptotic behavior of the solutions of Equation* (7.5). *If there is a $\delta > 0$ such that every solution of Equation* (7.7) *satisfies $x(t)\exp\delta t \to 0$ as $t \to \infty$, then it is necessary that $|C| < 1$. If $|C| < 1$ and every solution of the ordinary differential equation* (7.10) *approaches zero, then there are $\delta > 0$ and $r_0 > 0$ such that every solution satisfies $x(t)\exp\delta t \to 0$ as $t \to \infty$ for $0 \leq r < r_0$.*

To define the fundamental solution $X(t)$ of Equation (7.5), one proceeds in a manner similar to the one used for the retarded case. The initial data for $X(t)$ will be given by Condition (5.1), which is discontinuous at zero. From our experiences so far with neutral equations, it is to be expected that this discontinuity will persist at multiples of r for the solution, if it exists. Therefore, we define a solution X of Equation (7.5) with initial data (5.1) as one for which $X(t) - CX(t-r)$ is continuous and satisfies Equation (7.5) for $t \geq 0$ except at the points kr, $k = 0, 1, 2, \ldots$. With this definition, and in the same manner as before, it is now possible to prove that a unique solution $X(t)$ of Equation (7.5) exists that satisfies initial data (5.1).

The function $X(t)$ will actually have a continuous first derivative on each interval $(kr, (k+1)r)$, $k = 0, 1, 2, \dots$, the right- and left-hand limits of $X(t)$ exist at each of the points kr, $k = 0, 1, 2, \dots$. Therefore, $X(t)$ is of bounded variation on each compact interval and satisfies

$$\dot{X}(t) - C\dot{X}(t-r) = AX(t) + BX(t-r) \tag{7.11}$$

for $t \neq kr$, $k = 0, 1, 2, \dots$. These assertions are proved easily from the fact that $X(t)$ satisfies the integral equation

$$X(t) = 1 + CX(t-r) + \int_0^t [AX(s) + BX(s-r)]\,ds, \qquad t \geq 0.$$

As in the proof of Theorem 7.3, one shows that $X(t)$ satisfies the inequality

$$|X(t)| \leq ae^{bt}, \qquad t \in \mathbb{R}.$$

Since $X(t)$ has an exponential bound one can define the Laplace transform of X. It is easy to verify that $\mathcal{L}(X) = H^{-1}(\lambda)$.

Theorem 7.4. *The solution $X(t)$ of Equation (7.11) with initial data $X(t) = 0$, $t < 0$, $X(0) = 1$, is the fundamental solution of Equation (7.4); that is, $\mathcal{L}(X) = H^{-1}(\lambda)$.*

The same proof as in the proof of Theorem 6.1 is easily modified to show that the solution $x(\phi, f)$ of Equation (7.1) is obtained from the variation-of-constants formula,

$$x(\phi, f)(t) = x(\phi, 0)(t) + \int_0^t X(t-s)f(s)\,ds. \tag{7.12}$$

It is also possible to express the solution $x(\phi, 0)$ in terms of the fundamental solution X. The formula is more complicated than the one for the retarded equation and will now be derived. Following the same proof as in the retarded case, one obtains

$$\begin{aligned} x(\phi,0)(t) = X(t)[\phi(0) - C\phi(-r)] &+ B\int_{-r}^0 X(t-\theta-r)\phi(\theta)\,d\theta \\ &+ C\phi(-r+t)\omega(-r+t) + C\int_{-r}^0 \dot{X}(t-\theta-r)\phi(\theta)\,d\theta \end{aligned}$$

where $\omega(\theta) = 1$ for $\theta < 0$ and $\omega(\theta) = 0$ for $\theta \geq 0$. If $t \geq r$, the term involving ω is not present. If $t < r$, then this term is $C\phi(-r+t)$, which is precisely the value of the Stieltjes integral $-\int_{-r}^{t-r} d[X(t-\theta-r)]\phi(\theta)$.

Therefore, if we make use of the Stieltjes integral, the relation for $x(\phi,0)(t)$ can be written as

$$\begin{aligned} x(\phi,0)(t) = X(t)[\phi(0) - C\phi(-r)] + B\int_{-r}^{0} X(t-\theta-r)\phi(\theta)\,d\theta \\ - C\int_{-r}^{0} d[X(t-\theta-r)]\phi(\theta). \end{aligned} \tag{7.13}$$

The results are summarized in the following theorem.

Theorem 7.5. *If $X(t)$ is the fundamental solution of Equation* (7.5), *then the solution $x(\phi,f)$ of Equation* (7.1) *is obtained from the variation-of-constants formula* (7.12) *with $x(\phi,0)$ as given in Formula* (7.13).

Of course, the usefulness of the representation formulas (7.12) and (7.13) depends on knowing that the asymptotic behavior of the fundamental solution $X(t)$ is determined by the solutions of the characteristic equation. For the retarded equation, these estimates were obtained in a rather straightforward manner knowing that the Laplace transform of X was $h(\lambda)$.

For the neutral Equation (7.5), we know that $\mathcal{L}(X) = H^{-1}(\lambda)$ and so it should be possible to obtain similar estimates.

The analysis is not a trivial modification of the proof of Theorem 5.2.

Theorem 7.6. *If $\alpha_0 = \sup\{\text{Re }\lambda : H(\lambda) = 0\}$, then, for any $\alpha > \alpha_0$, there is a constant $k = k(\alpha)$ such that the fundamental solution X of Equation* (7.5) *satisfies the inequality*

$$|X(t)| \le ke^{\alpha t}, \qquad \text{Var}_{[t-r,t]}\, X \le ke^{\alpha t}, \qquad t \ge 0,$$

where $\text{Var}_{[t-r,t]}\, X$ denotes the total variation of X on $[t-r,t]$.

Proof. Since $\alpha > \alpha_0$, it follows from Lemma 7.1 that $\alpha r > \ln|C|$. Consequently there is an interval I containing α such that $1 - Ce^{-\lambda r}$ is uniformly bounded away from zero in the strip $S = \{\lambda \in C : \text{Re }\lambda \in I\}$. In a manner similar to the proof of Theorem 5.2, one can show that $X(t) = \int_{(\alpha)} e^{\lambda t} H^{-1}(\lambda)\,d\lambda$. To estimate the value of this integral, observe that

$$\frac{1}{H(\lambda)} = \frac{1}{\lambda(1 - Ce^{-\lambda r})} + \frac{A + Be^{-\lambda r}}{\lambda(1 - Ce^{-\lambda r})H(\lambda)}.$$

On the line $\text{Re }\lambda = \alpha$, the integral

$$\int_{(\alpha)} \frac{A + Be^{-\lambda r}}{\lambda(1 - Ce^{-\lambda r})H(\lambda)}\, e^{\lambda t}$$

is absolutely convergent since the kernel is like λ^{-2}. Therefore, this part of the inverse Laplace transform of $H^{-1}(\lambda)$ admits an estimate of the type specified in the theorem.

The term involving $\lambda^{-1}(1 - Ce^{-\lambda r})^{-1}$ is more difficult. In the strip S, the function $(1 - Ce^{-\lambda r})^{-1}$ is analytic and, if $\lambda = \beta + i\omega$, $\beta \in I$, then this function is periodic of period $2\pi/r$. Therefore, this function possesses an absolutely convergent Fourier series

$$(1 - Ce^{-\lambda r})^{-1} = \sum_{k=-\infty}^{\infty} h_k e^{\lambda k r}, \qquad \lambda \in S,$$

$$\sum_{k=-\infty}^{\infty} |h_k| e^{\beta k r} < \infty, \qquad \beta \in I.$$

Consequently, if $t > 0$, $t + kr \neq 0$, then

$$\int_{(\alpha)} \frac{1}{\lambda(1 - Ce^{-\lambda r})} e^{\lambda t} d\lambda = \sum_{k=-\infty}^{\infty} h_k \int_{(\alpha)} e^{\lambda(t+kr)} \lambda^{-1} d\lambda = \sum_{\alpha(t+kr)>0} h_k.$$

Finally,

$$\Big| \sum_{\alpha(t+kr)>0} h_k \Big| \leq e^{\alpha t} \sum_{\alpha(t+kr)>0} |h_k| e^{\alpha k r} \leq e^{\alpha t} \sum_{k=-\infty}^{\infty} |h_k| e^{\alpha k r},$$

and we have shown that $X(t)$ has an estimate of the type stated in the theorem.

To assert that the variation of X has an estimate of the desired type, we use the difference differential equation (7.11) to show that $\dot{X}(t)$ is bounded by a constant times $\exp \alpha t$ except at the points kr, $k = 0, 1, 2, \ldots$. If $y(t) = \dot{X}(t)$, then $y(t) - Cy(t-r) = p(t)$, where $p(t) = AX(t) + BX(t-r)$ satisfies an inequality of the form $|p(t)| \leq (\text{constant}) \exp \alpha t$. If $y(t) = z(t) \exp \alpha t$, $q(t) = p(t) \exp(-\alpha t)$, then

$$z(t) - C' z(t-r) = q(t),$$

where $|C'| = |C \exp(-\alpha r)| < 1$ and $|q(t)|$ is bounded. It is now very easy to show that this equation has all solutions bounded for $t \geq 0$. To compute the variation of $X(t)$, we also must consider the jump in $X(t)$ at kr. We observe that

$$X(kr^+) - X(kr^-) = C\big(X((k-1)r^+) - X((k-1)r^-)\big)$$

and, thus the jumps are bounded by a constant times $e^{\alpha t}$. This proves the theorem. □

An interesting corollary on the asymptotic behavior of the solutions of Equation (7.5) can now be stated.

Corollary 7.2. *If $\alpha_0 = \sup\{\mathrm{Re}\,\lambda : \lambda(1 - Ce^{-\lambda r}) = A + Be^{-\lambda r}\}$ and $x(\phi)$ is the solution of Equation (7.5), which coincides with ϕ on $[-r, 0]$, then, for any $\alpha > \alpha_0$, there is a constant $K = K(\alpha)$ such that*

$$|x(\phi)(t)| \leq Ke^{\alpha t}|\phi|, \qquad t \geq 0, \qquad |\phi| = \sup_{-r\leq\theta\leq 0} |\phi(\theta)|.$$

In particular, if $\alpha_0 < 0$, then all solutions of Equation (7.5) approach zero exponentially.

Proof. This is an immediate consequence of the representation formula (7.13) and Theorem 7.6. □

Perturbation results similar to the ones for retarded equations can also be stated.

1.8 Supplementary remarks

The presentation in this chapter was certainly influenced by the discussion in the first five chapters of Bellman and Cooke [1]. The treatment of the neutral equations uses also the papers by Hale and Meyer [1] and Henry [1].

In this chapter, we have discussed only scalar equations with one delay. It is essentially a matter of notation to generalize the theory to the matrix case and to include any finite number of delays. Of course, any discussion of specific properties of the characteristic equation will be much more difficult since this equation will be of the form

$$a_0(\lambda) + \sum_{j=1}^{N} a_j(\lambda)e^{-\lambda r_j} = 0 \tag{8.1}$$

where the $a_j(\lambda)$ are polynomials of degree $\leq n$ and $a_0(\lambda)$ is a polynomial of degree n.

The theory of this chapter will be generalized even more in subsequent pages with the emphasis being on qualitative and geometric properties of the solutions. An application of this theory to a concrete example will often require very detailed knowledge of the solutions of equations of the form of Equation (8.1). Therefore, it is absolutely essential that one be aware of existing methods for analyzing Equation (8.1). The theory of this equation will not be developed in this book, but there is an excellent presentation in Chapters 12 and 13 of Bellman and Cooke [1]. In the Appendix, we present some of the available theory on determining conditions for the zeros of Equation (8.1) to have negative real parts.

We briefly mentioned the very interesting question of the expansion of solutions of linear differential difference equations as an infinite series of the form

$$\sum_j p_j(t)e^{\lambda_j t}$$

where the λ_j are the roots of the characteristic function and the p_j are polynomials. Some results on this question are contained in Chapter 7. Additional results can be found in Bellman and Cooke [1], Banks and Manitius [1], Gromova and Zverkin [1], Verduyn Lunel [2,3,4], and Zverkin [3].

As remarked earlier, our objective in later chapters of this book is to develop a comprehensive qualitative theory for general functional differential equations of retarded and neutral type. Before entering into this theory, it is instructive to discuss in an intuitive manner the types of function spaces that should be considered as appropriate spaces of initial functions.

Consider the scalar retarded equation

$$\dot{x}(t) = f(t, x(t), x(t-r)) \tag{8.2}$$

where $r > 0$ and $f : \mathbb{R}^3 \to \mathbb{R}$ is continuous. For a given $\sigma \in \mathbb{R}$ and $\phi : [-r, 0] \to \mathbb{R}$ one would certainly want ϕ to satisfy enough smoothness conditions to ensure that finding a solution of Equation (8.2) for $t \geq \sigma$ satisfying $x(\sigma + \theta) = \phi(\theta)$, $-r \leq \theta \leq 0$, would be equivalent to finding a solution of the integral

$$\begin{aligned} x(t) &= \phi(0) + \int_\sigma^t f(s, x(s), x(s-r))\, ds, \qquad t \geq \sigma, \\ x(\sigma + \theta) &= \phi(\theta), \qquad -r \leq \theta \leq 0. \end{aligned} \tag{8.3}$$

There is very little difficulty finding spaces of initial functions that have this property, and the space C of continuous functions certainly will be sufficient. Even if some space other than continuous functions is used for initial data, then the solution lies in C for $t \geq \sigma + r$. Therefore, for the fundamental theory, the space of initial data does not play a role that is too significant. However, in the applications, it is sometimes convenient to take initial functions with fewer or more restrictions. Even though our subsequent discussion will center around the initial space of continuous functions, we will occasionally amplify on this last remark.

Consider the neutral equation

$$\dot{x}(t) = f(t, x(t), x(t-r), \dot{x}(t-r)) \tag{8.4}$$

where $r > 0$ and $f : \mathbb{R}^4 \to \mathbb{R}$ is continuous. Since a specification of x at t and $x, \dot{x}$ at $t - r$ uniquely determines $\dot{x}(t)$, it is natural to specify the following initial-value problem. Suppose σ is a given real number and ϕ is a given function on $[\sigma - r, \sigma]$ that is continuous together with its first derivative. A solution of Equation (8.4) through (σ, ϕ) is a function x defined on an interval $[\sigma - r, \sigma + A)$, $A > 0$, which coincides with ϕ on $[\sigma - r, \sigma]$ has a continuous first derivative except at the points $\sigma + kr$ for all $k = 0, 1, 2, \ldots$ for which $\sigma + kr$ belongs to $[\sigma, \sigma + A)$. A theory of Equation (8.4) along this line is developed in Bellman and Cooke [1].

One shortcoming of this definition of the initial-value problem is that it cannot be generalized to the situation in which there is general dependence of $\dot{x}(t)$ on values of $x(s)$ for $s \leq t$. Even in Equation (8.4), and even more so when r depends on t, there are great difficulties in discussing the dependence of solutions on the initial data (σ, ϕ). Other objections arise if one tries to develop a geometric theory for Equation (8.4) in the same spirit as for ordinary differential equations. There have been many papers devoted to the formulation of the initial-value problem and the reader may consult *Turdy Sem. Teor. Diff. Urav. Otkl. Argumentom*, Vols. 1–8 (1962–1973) and the survey articles of Zverkin, Kamenskii, Norkin, and El'sgol'tz [1], Kamenskii, Norkin, and El'sogl'tz [1], and Mishkis and El'sgol'tz [1].

A very significant contribution to this question was made by Driver [2], who gave a formulation that has also been generalized by Melvin [1, 2]. We illustrate the ideas for Equation (8.4). Suppose ϕ is a given absolutely continuous function on $[\sigma - r, \sigma]$. A solution of Equation (8.4) through (σ, ϕ) is an absolutely continuous function defined on an interval

$$[\sigma - r, \sigma + A), \qquad A > 0,$$

coinciding with ϕ on $[\sigma - r, \sigma]$ and satisfying Equation (8.4) almost everywhere on $[\sigma, \sigma + A)$. Of course, in order for this initial-value problem to make sense, the function f must satisfy the following property: If x is any given absolutely continuous function on $[\sigma - r, \sigma + A)$ and if

$$F(t) = f(t, x(t), x(t-r), \dot{x}(t-r)), \qquad \sigma \leq t < \sigma + A,$$

then the function F must be locally integrable on $[\sigma, \sigma + A)$. A satisfactory function f is

$$f(t, x, y, z) = g(t, x, y)z + h(t, x, y), \tag{8.5}$$

which is linear in z (the term that corresponds to $\dot{x}(t-r)$ in Equation (8.4)). If f satisfies Equation (8.5) and the initial-value problem is defined as earlier, then a theory of existence, uniqueness, and continuous dependence on the initial data is developed in Driver [2].

One can also consider the case where the initial value ϕ is continuous and has a pth-power integrable first derivative with $1 \leq p \leq +\infty$, including $+\infty$. A solution of Equation (8.4) is required to lie in the same class. This theory has been developed by Melvin [1, 2]. For further discussion of topologies for neutral equations, see Driver [5].

For certain classes of neutral equations that occur frequently in the application, it is also possible to obtain a well-posed initial-value problem with the initial space being only the continuous functions. We have already encountered this situation for linear equations in Section 1.7. Other nonlinear equations of a special form can also be discussed in a similar setting. For example, consider the equation

$$\dot{x}(t) = g(t, x(t-r))\dot{x}(t-r) + h(t, x(t), x(t-r)) \tag{8.6}$$

where g and h are continuous functions of their arguments and $g(t,x)$ has a continuous first derivative in t. If

$$G(t,x) = \int_0^x g(t,s)\,ds, \tag{8.7}$$

then Equation (8.6) can be written as

$$\frac{d}{dt}\left[x(t) - G(t,x(t-r))\right] = H(t,x(t),x(t-r)) \tag{8.8}$$

where $H(t,x,y) = h(t,x,y) - \partial G(t,y)/\partial t$. It is now possible to pose the following initial-value problem for Equation (8.8). Suppose ϕ is a given continuous n-vector function on $[\sigma - r, \sigma]$. A solution of Equation (8.8) through (σ, ϕ) is a continuous function x defined on $[\sigma - r, \sigma + A)$, $A > 0$, coinciding with ϕ on $[\sigma - r, \sigma]$ such that the function $x(t) - G(t, x(t-r))$, not $x(t)$, is continuously differentiable on $[\sigma, \sigma + A)$ and satisfies Equation (8.8) on $[\sigma, \sigma + A)$. A theory in this direction was initiated in Hale and Meyer [1] and Cruz and Hale [1]. It will receive more attention later.

The neutral equations introduced later will be modeled after Equation (8.8). The main reason for taking this approach is that retarded equations and difference equations will be included without imposing too many smoothness conditions on the initial data. Furthermore, this class is mathematically simpler than the general equation of neutral type and yet it is sufficiently general to include many applications.

2
Functional differential equations: Basic theory

In this chapter, we introduce a general class of functional differential equations that generalize the differential difference equations of Chapter 1. The basic theory of existence, uniqueness, continuation, and continuous dependence for retarded equations will be developed in the first five sections. In the last two sections, we introduce a fairly general class of neutral differential equations for which one can extend the basic theory.

2.1 Definition of a retarded equation

Suppose $r \geq 0$ is a given real number, $\mathbb{R} = (-\infty, \infty)$, $\mathbb{R}^n$ is an n-dimensional linear vector space over the reals with norm $|\cdot|$, $C([a,b],\mathbb{R}^n)$ is the Banach space of continuous functions mapping the interval $[a,b]$ into $\mathbb{R}^n$ with the topology of uniform convergence. If $[a,b] = [-r,0]$ we let $C = C([-r,0],\mathbb{R}^n)$ and designate the norm of an element ϕ in C by $|\phi| = \sup_{-r\leq\theta\leq 0} |\phi(\theta)|$. Even though single bars are used for norms in different spaces, no confusion should arise. If

$$\sigma \in \mathbb{R},\ A \geq 0, \qquad \text{and} \quad x \in C([-\sigma - r, \sigma + A], \mathbb{R}^n),$$

then for any $t \in [\sigma, \sigma + A]$, we let $x_t \in C$ be defined by $x_t(\theta) = x(t + \theta)$, $-r \leq \theta \leq 0$. If D is a subset of $\mathbb{R} \times C$, $f : D \to \mathbb{R}^n$ is a given function and "$\cdot$" represents the right-hand derivative, we say that the relation

$$\dot{x}(t) = f(t, x_t) \tag{1.1}$$

is a *retarded functional differential equation* on D and will denote this equation by RFDE. If we wish to emphasize that the equation is defined by f, we write the RFDE(f). A function x is said to be a *solution* of Equation (1.1) on $[\sigma - r, \sigma + A)$ if there are $\sigma \in \mathbb{R}$ and $A > 0$ such that $x \in C([\sigma - r, \sigma + A), \mathbb{R}^n)$, $(t, x_t) \in D$ and $x(t)$ satisfies Equation (1.1) for $t \in [\sigma, \sigma + A)$. For given $\sigma \in \mathbb{R}$, $\phi \in C$, we say $x(\sigma, \phi, f)$ is a *solution of* Equation (1.1) *with initial value* ϕ *at* σ or simply a *solution through* (σ, ϕ)

if there is an $A > 0$ such that $x(\sigma, \phi, f)$ is a solution of Equation (1.1) on $[\sigma - r, \sigma + A)$ and $x_\sigma(\sigma, \phi, f) = \phi$.

Equation (1.1) is a very general type of equation and includes ordinary differential equations ($r = 0$)

$$\dot{x}(t) = F(x(t)),$$

differential difference equations

$$\dot{x}(t) = f(t, x(t), x(t - \tau_1(t)), \ldots, x(t - \tau_p(t)))$$

with $0 \le \tau_j(t) \le r$, $j = 1, 2, \ldots, p$, as well as the integro-differential equation

$$\dot{x}(t) = \int_{-r}^{0} g(t, \theta, x(t + \theta))\, d\theta.$$

Much more general equations are also included in Equation (1.1).

We say Equation (1.1) is *linear* if $f(t, \phi) = L(t)\phi + h(t)$, where $L(t)$ is linear; *linear homogeneous* if $h \equiv 0$ and *linear nonhomogeneous* if $h \not\equiv 0$. We say Equation (1.1) is *autonomous* if $f(t, \phi) = g(\phi)$ where g does not depend on t. The proof of the following lemma is obvious, using Lemma 2.1 of the next section.

Lemma 1.1. *If $\sigma \in \mathbb{R}$, $\phi \in C$ are given, and $f(t, \phi)$ is continuous, then finding a solution of Equation* (1.1) *through* (σ, ϕ) *is equivalent to solving the integral equation*

$$\begin{aligned} x_\sigma &= \phi \\ x(t) &= \phi(0) + \int_\sigma^t f(s, x_s)\, ds, \qquad t \ge \sigma. \end{aligned} \tag{1.2}$$

2.2 Existence, uniqueness, and continuous dependence

In this section, we give a basic existence theorem for the initial-value problem of Equation (1.1) assuming that f is continuous. Also, a rather general result on continuous dependence will be given as well as a simple result on uniqueness.

The ideas in this section are very simple, but the notation naturally will involve some complications. To prove the existence of the solution through a point $(\sigma, \phi) \in \mathbb{R} \times C$, we consider an $\alpha > 0$ and all functions x on $[\sigma - r, \sigma + \alpha]$ that are continuous and coincide with ϕ on $[\sigma - r, \sigma]$; that is, $x_\sigma = \phi$. The values of these functions on $[\sigma, \sigma + \alpha]$ are restricted to the class of x such that $|x(t) - \phi(0)| \le \beta$ for $t \in [\sigma, \sigma + \alpha]$. The usual mapping

T obtained from the corresponding integral equation is defined and it is then shown that α and β can be so chosen that T maps this class into itself and is completely continuous. Thus, Schauder's fixed-point theorem implies existence.

Continuous dependence is slightly more difficult because the mapping T depends on parameters and one must discuss the dependence of the fixed points of T on these parameters. If one is attempting to prove continuous dependence at a value λ_0 of the parameter, then the usual procedure is to assume a unique fixed point at λ_0 and then prove that the "compactness" property of T is uniform with respect to compact sets containing λ_0. This is the pattern followed here.

Our first observation is the following lemma.

Lemma 2.1. *If $x \in C([\sigma - r, \sigma + \alpha], \mathbb{R}^n)$ then x_t is a continuous function of t for t in $[\sigma, \sigma + \alpha]$.*

Proof. Since x is continuous on $[\sigma - r, \sigma + \alpha]$, it is uniformly continuous and thus for any $\epsilon > 0$ there is a $\delta > 0$ such that $|x(t) - x(\tau)| < \epsilon$ if $|t - \tau| < \delta$. Consequently, for t in $[\sigma, \sigma + \alpha]$, $|t - \tau| < \delta$, we have

$$|x(t + \theta) - x(\tau + \theta)| < \epsilon$$

for all θ in $[-r, 0]$. This proves the lemma. □

To bring out the ideas in the proof of existence as well as the results of subsequent sections, it is convenient to introduce some notation and to prove a few technical lemmas.

For any $(\sigma, \phi) \in \mathbb{R} \times C$, let $\tilde{\phi} \in C([\sigma - r, \infty), \mathbb{R}^n)$ be defined by

$$\tilde{\phi}_\sigma = \phi, \qquad \tilde{\phi}(t + \sigma) = \phi(0), \qquad t \geq 0. \tag{2.1}$$

Suppose x is a solution of Equation (1.1) through (σ, ϕ). If $x(t + \sigma) = \tilde{\phi}(t + \sigma) + y(t)$, $t \geq -r$, then Lemma 1.1 implies y satisfies

$$\begin{aligned} y_0 &= 0 \\ y(t) &= \int_0^t f(\sigma + s, \tilde{\phi}_{\sigma+s} + y_s)\, ds, \qquad t \geq 0. \end{aligned} \tag{2.2}$$

Conversely, if y is a solution of this equation, then one obtains a solution x of Equation (1.1) by this transformation. Therefore, finding a solution of (1.1) is equivalent to finding an $\alpha > 0$ and a function $y \in C([-r, \alpha], \mathbb{R}^n)$ such that Equation (2.2) is satisfied for $0 \leq t \leq \alpha$.

If V is a subset of $\mathbb{R} \times C$, then $C(V, \mathbb{R}^n)$ is the class of all functions $f : V \to \mathbb{R}^n$ that are continuous and $C^0(V, \mathbb{R}^n) \subseteq C(V, \mathbb{R}^n)$ is the subset of bounded continuous functions from V to $\mathbb{R}^n$. The space $C^0(V, \mathbb{R}^n)$ becomes a Banach space with the norm

$$|f|_V = \sup_{(t,\phi)\in V} |f(t,\phi)|. \tag{2.3}$$

For any real α and β define

$$\begin{aligned} I_\alpha = [0,\alpha], \qquad B_\beta = \{\psi \in C : |\psi| \le \beta\}, \\ \mathcal{A}(\alpha,\beta) = \{y \in C([-r,\alpha],\mathbb{R}^n) : y_0 = 0,\ y_t \in B_\beta,\ t \in I_\alpha\}. \end{aligned} \tag{2.4}$$

Lemma 2.2. *Suppose $\Omega \subseteq \mathbb{R} \times C$ is open, $W \subseteq \Omega$ is compact and $f^0 \in C(\Omega,\mathbb{R}^n)$ is given. Then there exists a neighborhood $V \subseteq \Omega$ of W such that $f^0 \in C^0(V,\mathbb{R}^n)$, there exists a neighborhood $U \subseteq C^0(V,\mathbb{R}^n)$ of f^0 and positive constants M, α, and β such that*

$$|f(\sigma,\phi)| < M \qquad \text{for } (\sigma,\phi) \in V \text{ and } f \in U. \tag{2.5}$$

Also, for any $(\sigma^0,\phi^0) \in W$, we have $(\sigma^0 + t, y_t + \widetilde{\phi}_{\sigma^0+t^0}) \in V$ for $t \in I_\alpha$ and $y \in \mathcal{A}(\alpha,\beta)$.

Proof. Since W is compact and f^0 is continuous, there is a constant M such that $|f^0(\sigma^0,\phi^0)| < M$ for $(\sigma^0,\phi^0) \in W$. Furthermore, for the same reason, there are positive constants $\bar\alpha, \bar\beta$, and ϵ such that

$$|f^0(\sigma^0+t,\phi^0+\psi)| < M - \epsilon \quad \text{for } (\sigma^0,\phi^0) \in W \text{ and } (t,\psi) \in I_{\bar\alpha} \times B_{\bar\beta}.$$

If $V = \{(\sigma^0+t,\phi^0+\psi) : (\sigma^0,\phi^0) \in W,\ (t,\psi) \in I_{\bar\alpha} \times B_{\bar\beta}\}$, then $f^0 \in C^0(V,\mathbb{R}^n)$ and there is a neighborhood $U \subseteq C^0(V,\mathbb{R}^n)$ of f^0 such that Inequality (2.5) is satisfied.

To prove the last assertion of the lemma, suppose $0 < \beta < \bar\beta$ and choose α so that $\alpha < \bar\alpha$ and $|\widetilde{\phi}_{\sigma^0+t^0} - \phi^0| < \bar\beta - \beta$ for all $(\sigma^0,\phi^0) \in W$, $t \in I_\alpha$. Since W is compact, this last choice is possible. Therefore, $|y_t + \widetilde{\phi}_{\sigma^0+t^0} - \phi^0| < \beta + \bar\beta - \beta = \bar\beta$ for $y \in \mathcal{A}(\alpha,\beta)$. From the manner in which V was constructed, the proof of the lemma is complete. □

The next lemma will be used to apply fixed-point theorems for existence and continuous dependence of solutions of Equation (1.1).

Lemma 2.3. *Suppose $\Omega \subseteq \mathbb{R}\times C$ is open, $W \subseteq \Omega$ is compact, $f^0 \in C(\Omega,\mathbb{R}^n)$ is given, and the neighborhoods U and V and constants M, α, and β are the ones obtained from Lemma 2.2. If*

$$T : W \times U \times \mathcal{A}(\alpha,\beta) \to C([-r,\alpha],\mathbb{R}^n)$$

$$T(\sigma,\phi,f,y)(t) = 0, \qquad t \in [-r,0],$$

$$T(\sigma,\phi,f,y)(t) = \int_0^t f(\sigma+s,\widetilde{\phi}_{\sigma+s}+y_s)\,ds, \qquad t \in I_\alpha,$$

then T is continuous and there is a compact set K in $C([-r,\alpha],\mathbb{R}^n)$ such that

$$T : W \times U \times \mathcal{A}(\alpha, \beta) \to K.$$

Furthermore, if $M\alpha \leq \beta$, *then*

$$T : W \times U \times \mathcal{A}(\alpha, \beta) \to \mathcal{A}(\alpha, \beta).$$

Proof. It is clear that $T : W \times U \times \mathcal{A}(\alpha, \beta) \to C([-r, \alpha], \mathbb{R}^n)$. Also, Relation (2.5) implies

$$|T(\sigma, \phi, f, y)(t) - T(\sigma, \phi, f, y)(\tau)| \leq M|t - \tau|$$
$$|T(\sigma, \phi, f, y)(t)| \leq M\alpha$$

for all t, $\tau \in I_\alpha$. If

$$K = \{g \in C([-r, \alpha], \mathbb{R}^n) : |g(t) - g(\tau)| \leq M|t - \tau|,\ |g(t)| \leq M\alpha\},$$

then K is compact, $T : W \times U \times \mathcal{A}(\alpha, \beta) \to K$. If $M\alpha \leq \beta$, then $K \subseteq \mathcal{A}(\alpha, \beta)$.

It remains only to show that T is continuous. Suppose $(\sigma^k, \phi^k, f^k, y^k) \in W \times U \times \mathcal{A}(\alpha, \beta)$ and $(\sigma^k, \phi^k, f^k, y^k) \to (\sigma^0, \phi^0, f^0, y^0) \in W \times U \times \mathcal{A}(\alpha, \beta)$ as $k \to \infty$. Since $T(\sigma^k, \phi^k, f^k, y^k) \in K$ and K is compact, there is a subsequence we designate with the same symbol and a $\gamma \in K$ such that

$$T(\sigma^k, \phi^k, f^k, y^k) \to \gamma \quad \text{as } k \to \infty.$$

Since

$$f^k(\sigma^k + s, \widetilde{\phi}_{\sigma^k + s^k} + y^k_s) \to f^0(\sigma^0 + s, \widetilde{\phi}_{\sigma^0 + s^0} + y^0_s) \tag{2.6}$$

for all $s \in I_\alpha$ and all of these functions are uniformly bounded by Lemma 2.2, the Lebesgue dominated convergence theorem implies

$$\begin{aligned}\gamma(t) &= \lim_{k\to\infty} \int_0^t f^k(\sigma^k + s, \widetilde{\phi}_{\sigma^k + s^k} + y^k_s)\, ds \\ &= \int_0^t f^0(\sigma^0 + s, \widetilde{\phi}_{\sigma^0 + s^0} + y^0_s)\, ds = T(\sigma^0, \phi^0, f^0, y^0)(t)\end{aligned}$$

for all $t \in I_\alpha$. This implies the limit of any convergent subsequence is independent of the subsequence. But since every subsequence has a convergent subsequence, this obviously implies the sequence itself converges. Therefore, T is continuous and the lemma is proved. □

Even with the proof given earlier, the operator T is continuous under weaker conditions than the ones stated in Lemma 2.3. In fact, only Expression (2.6) was used. This implies the convergence of f^k to f^0 was not required in the uniform way as specified by the norm $|\cdot|_V$. This will affect the generality of our later result on continuous dependence. Our purpose is to give only the basic results that are essential for the development to follow in spite of the temptation to develop a more comprehensive theory of existence, etc.

To prove our basic existence theorem, we need the Schauder fixed-point theorem. If U is a subset of a Banach space X and $T : U \to X$, then T is said to be *completely continuous* if T is continuous and for any bounded set $B \subseteq U$, the closure of TB is compact. We now state the Schauder theorem without proof.

Lemma 2.4. (Schauder fixed-point theorem). *If U is a closed bounded convex subset of a Banach space X and $T : U \to U$ is completely continuous, then T has a fixed point in U.*

Theorem 2.1. (Existence). *Suppose Ω is an open subset in $\mathbb{R} \times C$ and $f^0 \in C(\Omega, \mathbb{R}^n)$. If $(\sigma, \phi) \in \Omega$, then there is a solution of the* RFDE(f^0) *passing through (σ, ϕ). More generally, if $W \subseteq \Omega$ is compact and $f^0 \in C(\Omega, \mathbb{R}^n)$ is given, then there is a neighborhood $V \subseteq \Omega$ of W such that $f^0 \in C^0(V, \mathbb{R}^n)$, there is a neighborhood $U \subseteq C^0(V, \mathbb{R}^n)$ of f^0 and an $\alpha > 0$ such that for any $(\sigma, \phi) \in W$, $f \in U$, there is a solution $x(\sigma, \phi, f)$ of the* RFDE(f) *through (σ, ϕ) that exists on $[\sigma - r, \sigma + \alpha]$.*

Proof. For the first part, take $W = \{(\sigma, \phi)\}$, a single point. Lemma 2.3 and Schauder's fixed-point theorem imply that $T(\sigma, \phi, f^0, \cdot)$ has a fixed point in $\mathcal{A}(\alpha, \beta)$ since $\mathcal{A}(\alpha, \beta)$ is a closed bounded convex set of $C([-r, \alpha], \mathbb{R}^n)$. This yields a solution of the RFDE(f^0) by Formula (2.2) with f replaced by f^0. To obtain the last statement of the theorem, simply apply the same reasoning but use the general form of Lemma 2.3. □

Theorem 2.2. (Continuous dependence). *Suppose $\Omega \subseteq \mathbb{R} \times C$ is open, $(\sigma^0, \phi^0) \in \Omega$, $f^0 \in C(\Omega, \mathbb{R}^n)$, and x^0 is a solution of the* RFDE(f^0) *through (σ^0, ϕ^0) which exists and is unique on $[\sigma^0 - r, b]$. Let $W^0 \subseteq \Omega$ be the compact set defined by*

$$W^0 = \{(t, x_t^0) : t \in [\sigma^0, b]\}$$

and let V^0 be a neighborhood of W^0 on which f^0 is bounded. If (σ^k, ϕ^k, f^k), $k = 1, 2, \ldots$ satisfies $\sigma^k \to \sigma^0$, $\phi^k \to \phi^0$, and $|f^k - f^0|_{V^0} \to 0$ as $k \to \infty$, then there is a k^0 such that the RFDE(f^k) *for $k \geq k^0$ is such that each solution $x^k = x^k(\sigma^k, \phi^k, f^k)$ through (σ^k, ϕ^k) exists on $[\sigma^k - r, b]$ and $x^k \to x^0$ uniformly on $[\sigma^0 - r, \beta]$. Since all x^k may not be defined on $[\sigma^0 - r, b]$, by $x^k \to x^0$ uniformly on $[\sigma^0 - r, b]$, we mean that for any $\epsilon > 0$, there is a $k_1(\epsilon)$ such that $x^k(t)$, $k \geq k_1(\epsilon)$, is defined on $[\sigma^0 - r + \epsilon, b]$, and $x^k \to x^0$ uniformly on $[\sigma^0 - r + \epsilon, b]$.*

Proof. The proof is an easy application of Lemma 2.3. In fact, the set $W^0 \cup \{(\sigma^k, \phi^k) : k = 1, 2, \ldots\}$ is compact. By taking

$$W = W^0 \cup \{(\sigma^k, \phi^k) : k \geq k_0\}$$

for k_0 sufficiently large and restricting α and β so that the resulting neighborhood V of Lemma 2.2 belongs to V^0, one may now apply Lemma 2.3 and Theorem 2.1 in the following manner.

From Theorem 2.1, each of the solutions $x^k = x^k(\sigma^k, \phi^k, f^k)$ through (σ^k, ϕ^k) exists on $[\sigma^k - r, \sigma^k + \alpha]$ where α is independent of k. Furthermore, Lemma 2.3 asserts that the $y^k(t) = x^k(\sigma^k + t) - \widetilde{\phi}^k(\sigma^k + t)$ belong to a compact set K of $C([-r, \alpha), \mathbb{R}^n)$. Therefore, there is a subsequence labeled the same way such that y^k converges uniformly to some function y^* on $[-r, \alpha]$. Since $y^k = T(\sigma^k, \phi^k, f^k, y^k)$ and T is continuous by Lemma 2.3, this implies $y^* = T(\sigma^0, \phi^0, f^0, y^0) = y^0$. Since every subsequence of the sequence $\{y^k\}$ has a convergent subsequence that must converge to y^0, it follows that the entire sequence converges to y^0. Translating these remarks back into x^k gives the result stated in the theorem for the interval $[\sigma^0 - r, \sigma^0 + \alpha]$. The proof is completed by successively stepping intervals of length α, which we know is possible by Theorem 2.1. □

Theorem 2.3. *Suppose Ω is an open set in $\mathbb{R} \times C$, $f : \Omega \to \mathbb{R}^n$ is continuous, and $f(t, \phi)$ is Lipschitizian in ϕ in each compact set in Ω. If $(\sigma, \phi) \in \Omega$, then there is a unique solution of Equation* (1.1) *through* (σ, ϕ).

Proof. Define I_α and B_β as in Equation (2.4) and suppose x and y are solutions of Equation (1.1) on $[\sigma - r, \sigma + \alpha]$ with $x_\sigma = \phi = y_\sigma$. Then

$$x(t) - y(t) = \int_\sigma^t [f(s, x_s) - f(s, y_s)]\, ds, \qquad t \geq \sigma,$$

$$x_\sigma - y_\sigma = 0.$$

If k is the Lipschitz constant of $f(t, \phi)$ in any compact set containing the trajectories $\{(t, x_t)\}$, $\{(t, y_t)\}$, $t \in I_\alpha$, then choose $\bar{\alpha}$ so that $k\bar{\alpha} < 1$. Then, for $t \in I_{\bar{\alpha}}$,

$$|x(t) - y(t)| \leq \int_\sigma^t k|x_s - y_s| ds \leq k\bar{\alpha} \sup_{\sigma \leq s \leq t} |x_s - y_s|$$

and this implies $x(t) = y(t)$ for $t \in I_{\bar{\alpha}}$. One completes the proof of the theorem by successively stepping intervals of length $\bar{\alpha}$. □

2.3 Continuation of solutions

Suppose f in Equation (1.1) is continuous. If x is a solution of Equation (1.1) on an interval $[\sigma, a)$, $a > \sigma$, we say $\hat{x}$ is a *continuation of* x if there is a $b > a$ such that $\hat{x}$ is defined on $[\sigma - r, b)$, coincides with x on $[\sigma - r, a)$, and x satisfies Equation (1.1) on $[\sigma, b)$. A solution x is *noncontinuable* if no such continuation exists; that is, the interval $[\sigma, a)$ is the maximal interval of existence of the solution x. The existence of a noncontinuable solution follows from Zorn's lemma. Also, the maximal interval of existence must be open.

Theorem 3.1. *Suppose Ω is an open set in $\mathbb{R} \times C$ and $f \in C(\Omega, \mathbb{R}^n)$. If x is a noncontinuable solution of Equation (1.1) on $[\sigma - r, b)$, then, for any compact set W in Ω, there is a t_W such that $(t, x_t) \notin W$ for $t_W \le t < b$.*

Proof. The case $b = \infty$ is trivial so we suppose b finite. Consider first the case $r = 0$ (an ordinary equation). Since W is compact, Theorem 2.1 implies there is an $\alpha > 0$ such that the equation has a solution through any $(c, y) \in W$ that exists at least on $[c, c + \alpha]$. Now suppose the assertion of the theorem is false; that is, there is a sequence $(t_k, x(t_k)) \in W$, $y \in \mathbb{R}^n$, $(b, y) \in W$ such that $t_k \to b^-$, $x(t_k) \to y$ as $k \to \infty$. Using the fact that f is bounded in a neighborhood of (b, y), the function x is uniformly continuous on $[\sigma, b)$ and $x(t) \to y$ as $t \to b^-$. There is obviously an extension of x to the interval $[\sigma, b + \alpha]$. Since $b + \alpha > b$, this is a contradiction. The proof for the case $r = 0$ is complete.

If the conclusion of the theorem is not true for $r > 0$, then there are a sequence of real numbers $t_k \to b^-$ as $k \to \infty$ and a $\psi \in C$ such that

$$(t_k, x_{t_k}) \in W, \ (b, \psi) \in W, \ (t_k, x_{t_k}) \to (b, \psi)$$

as $k \to \infty$. Thus, for any $\epsilon > 0$,

$$\lim_{k \to \infty} \sup_{\theta \in [-r, -\epsilon]} |x_{t_k}(\theta) - \psi(\theta)| = 0.$$

Since $x_t(\theta) = x(t + \theta)$, $-r \le \theta \le 0$, and $r > 0$, this implies $x(b + \theta) = \psi(\theta)$, $-r \le \theta < 0$. Hence $\lim_{t \to b^-} x(t)$ exists and x can be extended to a continuous function $[\sigma - r, b]$ by defining $x(b) = \psi(0)$. Since $(b, x_b) \in \Omega$, one can find a solution of Equation (1.1) through this point to the right of b. This contradicts the noncontinuability hypothesis on x and proves the theorem. □

Corollary 3.1. *Suppose Ω is an open set in $\mathbb{R} \times C$ and $f \in C(\Omega, \mathbb{R}^n)$. If x is a noncontinuable solution of Equation (1.1) on $[\sigma - r, b)$ and W is the closure of the set $\{(t, x_t) : \sigma \le t < b\}$ in $\mathbb{R} \times C$, then W compact implies there is a sequence $\{t_k\}$ of real numbers, $t_k \to b^-$ as $k \to \infty$ such that (t_k, x_{t_k}) tends to $\partial\Omega$ as $k \to \infty$. If $r > 0$, then there is a $\psi \in C$ such that $(b, \psi) \in \partial\Omega$ and $(t, x_t) \to (b, \psi)$ as $t \to b^-$.*

Proof. Theorem 3.1 implies that W does not belong to Ω and proves the first part of the corollary. If $r > 0$, then the same argument as in the proof of Theorem 3.1 implies $\lim_{t \to b^-} x(t)$ exists and, thus, x can be extended as a continuous function on $[\sigma - r, b]$. Clearly, $(b, x_b) \in \partial\Omega$ and $(t, x_t) \to (b, x_b)$ as $t \to b^-$. □

Theorem 3.2. *Suppose Ω is an open set in $\mathbb{R}\times C$, $f : \Omega \to \mathbb{R}^n$ is completely continuous; that is, f is continuous and takes closed bounded sets of Ω into bounded sets of $\mathbb{R}^n$, and x is a noncontinuable solution of Equation* (1.1) *on $[\sigma - r, b)$. Then, for any closed bounded set U in $\mathbb{R} \times C$, U in Ω, there is a t_U such that $(t, x_t) \notin U$ for $t_U \le t < b$.*

Proof. The case $r = 0$ is contained in Theorem 3.1. Therefore, we suppose $r > 0$ and it is no restriction to take b finite. Suppose the conclusion of the theorem is not true. Then there is a sequence of real numbers $t_k \to b^-$ such that $(t_k, x_{t_k}) \in U$ for all k. Since $r > 0$, this implies that $x(t)$, $\sigma - r \le t < b$ is bounded. Consequently, there is a constant M such that $|f(\tau, \phi)| \le M$ for (τ, ϕ) in the closure of $\{(t, x_t) : \sigma \le t < b\}$. The integral equation for the solutions of Equation (1.1) imply

$$|x(t + \tau) - x(t)| = \Big| \int_t^{t+\tau} f(s, x_s)\, ds \Big| \le M\tau$$

for all t, $t + \tau < b$. Thus, x is uniformly continuous on $[\sigma - r, b)$. This implies $\{(t, x_t) : \sigma \le t < b\}$ belongs to a compact set in Ω. This contradicts Theorem 3.1 and proves the theorem. □

Theorem 3.2 gives conditions under which the trajectory (t, x) in $\mathbb{R}\times C$ of a noncontinuable solution of $[\sigma, b)$ approaches the boundary of Ω as $t \to b^-$. The approach to the boundary of Ω was described by saying that the trajectory must leave and remain outside every closed bounded set in Ω. If the condition that f is completely continuous is not imposed, then it is conceivable that the trajectory $\{(t, x_t) : \sigma \le t < b\}$ itself is a closed bounded subset of Ω; that is, the curve $(t, x(t))$ oscillates so badly as a subset of $\mathbb{R} \times \mathbb{R}^n$ that there are no limit points of (t, x_t) in $\mathbb{R} \times C$ as $t \to b^-$. An explicit example illustrating this fact will now be given.

Let $\Delta(t) = t^2$ and select two sequences $\{a_k\}$, $\{b_k\}$ of negative numbers, $a_1 < a_2 < \cdots$, $b_1 < b_2 < \cdots$, $a_k \to 0$, $b_k \to 0$ as $k \to \infty$ such that

$$a_k = b_k - \Delta(b_k),\ b_k \le a_{k+1} - \Delta(a_{k+1}), k = 1, 2, \dots.$$

For example, choose $b_k = -2^{-k}$, $k = 1, 2, \dots$.

Let $\psi(t)$ be an arbitrary continuous differentiable function satisfying

$$\psi(t) = \begin{cases} +1, & \text{for } t \text{ in } (-\infty, a_1],\quad [b_{2k}, a_{2k+1}],\ k = 1, 2, \dots, \\ -1, & \text{for } t \text{ in } [b_{2k-1}, a_{2k}],\ k = 1, 2, \dots \end{cases}$$

$$\psi'(t) \ne 0, \qquad t \in (a_k, b_k),\ k = 1, 2, \dots.$$

Let H be the set of points (t, x) such that $|x| < 1 - t$. These inequalities are equivalent to

$$\begin{aligned} x + t < 1 \quad & \text{if } x > 0, \\ -x + t < 1 \quad & \text{if } x < 0. \end{aligned}$$

Thus H is the wedge in Fig. 2.1. We now define a function $h(t,x)$ on H. On the graph of the curve $\psi(t)$, let

$$h(t - \Delta(t), \psi(t - \Delta(t))) = \psi'(t),$$

where the prime denotes the derivative, $-\infty < t < 0$. The function h is continuous on the graph of ψ. For any t in (a_k, b_k), $k \geq 2$ (i.e., a point of increase or decrease on the graph), $t - \Delta(t) \in [b_{k-1}, a_k]$. For t in $(-\infty, b_1]$, $t - \Delta(t) \in (-\infty, a_1]$. Therefore, $h = 0$ for any t in $(-\infty, b_1]$, (a_k, b_k), $k \geq 2$, and, in particular, $h = 0$ on all points of increase or decrease of the graph of the curve $\psi(t)$. Now continue the function $h(t,x)$ in any manner whatsoever as long as it remains continuous and is equal to zero in the square $P : |t| + |x| \leq 1$.

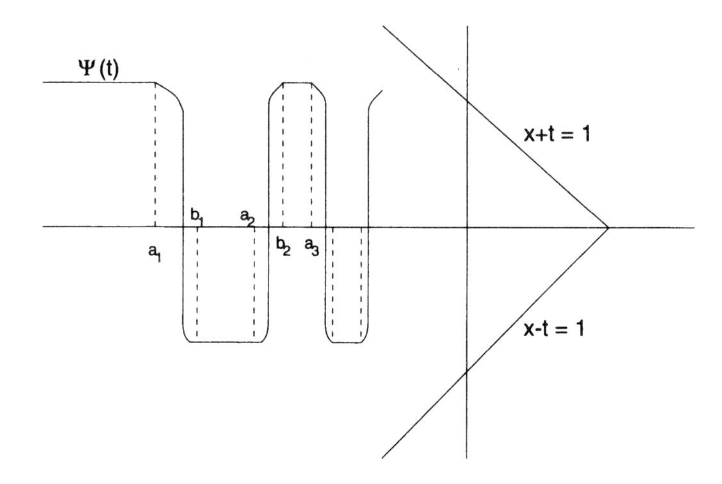

Fig. 2.1.

Now consider the equation

$$\dot{x}(t) = h(t - \Delta(t), x(t - \Delta(t))), \qquad t < 0 \text{ and } \Delta(t) = t^2. \tag{3.1}$$

Choose $\sigma < a_1$ and let $r = \sigma - \min\{(t - t^2) : \sigma \leq t \leq 0\}$. We consider the initial-value problem starting at σ. The function $x(t) = \psi(t)$ is a solution of this equation for $t < 0$ and is a noncontinuable solution on $[\sigma - r, 0)$.

If the right-hand side of Equation (3.1) is denoted by $f(t, x_t)$, $t \in \mathbb{R}$, $x_t \in C([-r,0], \mathbb{R})$, then $f(t, \phi)$ does not map closed bounded sets of $\Omega = \mathbb{R} \times C([-r,0], \mathbb{R})$ into bounded sets. In fact, the set $\{(t, \psi_t) : t \geq 0\}$ is a bounded set and it is closed since there are no sequences $t_k \to 0$ such that ψ_{t_k} converges.

This strange behavior is one of the ways in which the noncompactness of the unit ball in $C([-r,0],\mathbb{R}^n)$, $r>0$ can influence the solution of RFDE. The particular property exhibited in this example is actually a general phenomenon. In fact, one can prove the following result.

Theorem 3.3. *Suppose $x:[-r,\alpha)\to\mathbb{R}^n$, $r>0$, α finite, is an arbitrary bounded continuously differentiable function satisfying the property that $x(t)$ does not approach a limit as $t\to\alpha^-$. Then there exists a continuous function $f:C\to\mathbb{R}^n$ such that x is a noncontinuable solution of the* RFDE(f) *on $[-r,\alpha)$.*

Proof. Suppose x as stated and let $A=\{x_t : t\in[0,\alpha)\}$. By hypothesis, the set A is closed in C. Let $g:A\to\mathbb{R}^n$ be defined by $g(x_t)=\dot{x}(t)$, $t\in[0,\alpha)$. The function g is continuous on A. Therefore, by a generalization of the Tietze extension theorem due to Dugundji [1], there is a continuous extension f of g to all of C. It is clear that x is a noncontinuable solution of the RFDE(f) and the proof is complete. □

2.4 Differentiability of solutions

In Theorem 2.2, sufficient conditions were given to ensure that the solution $x(\sigma,\phi,f)$ on a RFDE(f) depends continuously on (σ,ϕ,f). In this section, some results are given on the differentiability with respect to (σ,ϕ,f).

If Ω is an open set in $\mathbb{R}\times C$, let $C^p(\Omega,\mathbb{R}^n)$, $p\geq 0$, designate the space of functions taking Ω into $\mathbb{R}^n$ that have bounded continuous derivatives up through order p with respect to ϕ on Ω. The space $C^p(\Omega,\mathbb{R}^n)$ becomes a Banach space if the norm is chosen as the supremum norm over all derivatives up through order p. The norm will be designated by $|\cdot|_p$.

For our main theorem, we make use of some basic results on the dependence of fixed points of contraction mappings on parameters. The proofs are omitted.

Definition 4.1. Suppose U is a subset of a Banach space X and $T:U\to X$. The mapping T is said to be a *contraction on U* if there is a λ, $0\leq\lambda<1$, such that

$$|Tx-Ty|\leq\lambda|x-y| \qquad \text{for all } x,y\in U.$$

If V is also a subset of a Banach space Y and $T:U\times V\to X$, then T is said to be a *uniform contraction* if there is a $0\leq\lambda<1$, such that

$$|T(x,v)-T(y,v)|\leq\lambda|x-y| \qquad \text{for all } x,y\in U \text{ and } v\in V.$$

Lemma 4.1. (Contraction mapping principle). *If U is a closed subset of a Banach space X and $T:U\to U$ is a contraction, then T has a unique fixed point in U.*

Lemma 4.2. *If U is a closed subset of a Banach space X, V is a subset of a Banach space Y, $T : U \times V \to U$ is a uniform contraction, and T is continuous, then the unique fixed point $x(v)$ of $T(\cdot, v)$ in U is continuous in v. Furthermore, if U, V are the closures of sets U^0, V^0 and $T(x, v)$ has continuous first derivatives in x, v, then $x(v)$ has a continuous first derivative with respect to v. The same conclusion holds for higher derivatives.*

Theorem 4.1. *If $f \in C^p(\Omega, \mathbb{R}^n)$, $p \geq 1$, then the solution $x(\sigma, \phi, f)(t)$ of the* RFDE(f) *through (σ, ϕ) is unique and continuously differentiable with respect to (ϕ, f) for t in any compact set in the domain of definition of $x(\sigma, \phi, f)$. Furthermore, for each $t \geq \sigma$, the derivative of x with respect to ϕ, $D_\phi x(\sigma, \phi, f)(t)$ is a linear operator from C to $\mathbb{R}^n$, $D_\phi x(\sigma, \phi, f)(\sigma) = I$, the identity, and $D_\phi x(\sigma, \phi, f)\psi(t)$ for each ψ in C satisfies the linear variational equation*

$$\dot{y}(t) = D_\phi f(t, x_t(\sigma, \phi, f))y_t. \tag{4.1}$$

Also, for each $t \geq \sigma$, $D_f x(\sigma, \phi, f)(t)$ is a linear operator from $C^p(\Omega, \mathbb{R}^n)$ into $\mathbb{R}^n$, $D_f x(\sigma, \phi, f)(\sigma) = 0$, and $D_f x(\sigma, \phi, f)g(t)$ for each $g \in C^p(\Omega, \mathbb{R}^n)$ satisfies the nonhomogeneous linear variation equation

$$\dot{z}(t) = D_\phi f(t, x_t(\sigma, \phi, f))z_t + g(t, x_t(\sigma, \phi, f)). \tag{4.2}$$

Proof. Since $p \geq 1$, it follows from Theorem 2.3 that the solution $x = x(\sigma, \phi, f)$ of the RFDE(f) through (σ, ϕ) is unique. Let the maximal interval of existence of x be $[\sigma - r, \sigma + \omega)$ and fix $b < \omega$.

Our first objective is to show that $x(\sigma, \phi, f)(t)$ is continuously differentiable with respect to ϕ on $[\sigma - r, \sigma + b]$. There is an open neighborhood U of ϕ such that $x(\sigma, \psi, f)(t)$, $\psi \in U$, is defined for $t \in [\sigma - r, \sigma + b]$. If $W = \{(t, x_t) : t \in [\sigma, \sigma + b]\}$, then W is compact. Using the notation of Section 2.2, we can determine M, α, β, U, and V as in Lemma 2.2. Choose α so that $M\alpha \leq \beta$ and $k\alpha < 1$, where k is a bound of the derivative of f with respect to ϕ on Ω. If $x(\sigma + t) = \widetilde{\phi}(\sigma + t) + y(t)$, $t \in I_\alpha$, then y is a fixed point of the operator $T(\sigma, \phi, f)$ of Lemma 2.3. On the other hand, the restriction on α, β implies that $T(\sigma, \phi, f)$ takes $\mathcal{A}(\alpha, \beta)$ into itself for each α, β and is a contraction. Furthermore, the contraction constant is independent of $(\sigma, \phi, f) \in V \times U$. Since the mapping $T(\sigma, \phi, f)$ is easily shown to be continuously differentiable in Ω, it follows from Lemma 4.2 that the fixed point $y = y(\sigma, \phi, f)$ is continuously differentiable in Ω. The same proof shows that $x(\sigma, \phi, f)(t)$ is continuously differentiable in f for $t \in [\sigma, \sigma + \alpha]$. Using the fact that the basic interval $[\sigma - r, \sigma + b]$ is compact, one completes the proof of the differentiability.

Knowing that $x(\sigma, \phi, f)$ is continuously differentiable with respect to ϕ, f, one can use the interval equation for x to easily obtain Formulas (4.1) and (4.2). □

The next question of interest is the differentiability of x with respect to σ under the same hypotheses on f as in Theorem 4.1. In general, this derivative will not exist unless the function $f(t, \phi)$ satisfies some additional smoothness properties in t. To see this, consider the following example,

$$\dot{x}(t) = a(t)x(t-1) \tag{4.3}$$

where a is a continuous function. If $x(\sigma, \phi)$ is the solution of Equation (4.3) through (σ, ϕ) and $\sigma < t < \sigma + 1$, $h > 0$, then

$$\begin{aligned} x(\sigma + h, \phi)(t) &= \phi(0) + \int_{\sigma+h}^{t} a(s)x(\sigma + h, \phi)(s-1)\,ds \\ &= \phi(0) + \int_{\sigma+h}^{t} a(s)\phi(s - \sigma - h - 1)\,ds, \\ x(\sigma, \phi)(t) &= \phi(0) + \int_{\sigma}^{t} a(s)x(\sigma, \phi)(s-1)\,ds \\ &= \phi(0) + \int_{\sigma}^{t} a(s)\phi(s - \sigma - 1)\,ds. \end{aligned}$$

Therefore,

$$\begin{aligned} \lim_{h\to 0} \frac{x(\sigma + h, \phi)(t) - x(\sigma, \phi)(t)}{h} &= -a(t)\phi(t - \sigma - 1) \\ &\quad + \lim_{h \to 0} \int_{\sigma}^{t} \frac{a(s+h) - a(s)}{h} \phi(s - \sigma - 1)\,ds \end{aligned}$$

and, without some additional smoothness of a, the expression on the right-hand side will not exist. If a is such that its derivative $\dot{a}$ is integrable, then $\partial x(\sigma, \phi)(t)/\partial\sigma$ exists and satisfies

$$\frac{\partial x(\sigma, \phi)(t)}{\partial \sigma} = -a(t)\phi(t - \sigma - 1) + \int_{\sigma}^{t} \dot{a}(s)\phi(s - \sigma - 1)\,ds.$$

It would be interesting to discuss the general problem of when

$$\partial x(\sigma, \phi, f)/\partial\sigma$$

exists. To carry out this investigation the following formula for $t > \sigma + h$, $h > 0$, is useful:

$$\begin{aligned} x(\sigma + h, \phi, f)(t) - x(\sigma, \phi, f)(t) &= x(\sigma + h, \phi, f)(t) \\ &\quad - x\big(\sigma + h, x_{\sigma+h}(\sigma, \phi, f), f\big)(t). \end{aligned}$$

For the ordinary differential equation ($r = 0$), this last formula shows immediately that

$$\frac{\partial x(\sigma, \phi, f)(t)}{\partial \sigma} = -\frac{\partial x(\sigma, \phi, f)(t)}{\partial \phi} f(\sigma, \phi).$$

Finally, we remark that the pth-derivatives of $x(\sigma, \phi, f)$ with respect to ϕ, f exist under the hypotheses of Theorem 4.1, but the proof of this fact will not be given.

2.5 Backward continuation

In ordinary differential equations with a continuous vector field, one can prove the existence of a solution through a point (σ, x_0) defined on an interval $[\sigma - a, \sigma + a]$, $a > 0$; that is, the solution exists to the right and left of the initial t-value. For an RFDE, this is not necessarily the case. We will state some sufficient conditions for existence to the left of the initial t-value.

Definition 5.1. Suppose $\Omega \subseteq \mathbb{R} \times C$ is open and $f \in C(\Omega, \mathbb{R}^n)$. We say a function $x \in C([\sigma - r - \alpha, \sigma], \mathbb{R}^n)$, $\alpha > 0$, is a *solution of* Equation (1.1) *on* $[\sigma - r - \alpha, \sigma]$ *through* (σ, ϕ) for $(\sigma, \phi) \in \Omega$ if $x_\sigma = \phi$ and for any $\sigma_1 \in [\sigma - \alpha, \sigma]$, $(\sigma_1, x_{\sigma_1}) \in \Omega$ and x is a solution of Equation (1.1) on $[\sigma_1 - r, \sigma]$ through (σ_1, x_{σ_1}). We sometimes refer to such a solution as the *backward continuation of a solution through* (σ, ϕ).

Definition 5.1 is very natural and says only that a function defined on $[\sigma - r - \alpha, \sigma]$ is a solution of Equation (1.1) on this interval if it has the property that it will satisfy the equation in the forward direction of t no matter where the initial time is chosen.

General results on backward continuation are very difficult to prove although the ideas are relatively simple. To motivate the definitions to follow, let us consider a simple example

$$\dot{x}(t) = a(t)x(t-1). \tag{5.1}$$

If $\sigma = 0$ and ϕ is a given function in C and there exists a backward continuation of a solution through $(0, \phi)$, then it is necessary that ϕ be continuously differentiable on a small interval $(-\epsilon, 0]$ and $\dot{\phi}(0) = a(0)\phi(-1)$. Conversely, if this condition is satisfied and $a(t) \neq 0$ for $t \in (-\epsilon, 0]$, then one can define

$$x(t-1) = \frac{1}{a(t)}\,\dot{\phi}(t), \qquad t \in (-\epsilon, 0]$$

and x will be a solution of Equation (5.1) on $(-r - \epsilon, 0]$ with $x_0 = \phi$. Therefore, there is a backward continuation through $(0, \phi)$.

We wish to generalize this idea to a general RFDE(f). The important remark in the example was $a(t) \neq 0$ for $t \in (-\epsilon, 0]$; that is, the evolution of the system $x(t)$ actually used the information specified at $x(t-1)$. In the general case, the manner in which $f(t, \phi)$ varies with $\phi(-r)$ is determined by a "coefficient" that depends on t and ϕ. The precise specification of this coefficient leads to the definition of "atomic at $-r$", given later.

A careful examination of the technical details shows that the proof of existence is also a natural generalization of the preceding remarks.

For the statement of the main result, we need some definitions. For Banach spaces X and Y, $\mathcal{L}(X,Y)$ is the Banach space of bounded linear mappings from X to Y with the operator topology. If $L \in \mathcal{L}(C, \mathbb{R}^n)$, then the Riesz representation theorem implies there is an $n \times n$ matrix function η on $[-r, 0]$ of bounded variation such that

$$L\phi = \int_{-r}^{0} d[\eta(\theta)]\phi(\theta).$$

For any such η, we always understand that we have extended the definition to $\mathbb{R}$ so that $\eta(\theta) = \eta(-r)$ for $\theta \leq -r$, $\eta(\theta) = \eta(0)$ for $\theta \geq 0$.

Definition 5.2. Let Λ be an open subset of a metric space. We say $L : \Lambda \to \mathcal{L}(C, \mathbb{R}^n)$, has *smoothness on the measure* if for any $\beta \in \mathbb{R}$, there is a scalar function $\gamma(\lambda, s)$ continuous for $\lambda \in \Lambda$, $s \in \mathbb{R}$, $\gamma(\lambda, 0) = 0$, such that if $L(\lambda)\phi = \int_{-r}^{0} d\eta[(\lambda, \theta)]\phi(\theta)$, $\lambda \in \Lambda$, $0 < s$, then

$$\Big| \lim_{h \to 0+} \int_{\beta+h}^{\beta+s} + \int_{\beta-s}^{\beta-h} d[\eta(\lambda, \theta)]\phi(\theta) \Big| \leq \gamma(\lambda, s)|\phi|. \tag{5.2}$$

If $\beta \in \mathbb{R}$ and the matrix $A(\lambda; \beta, L) = \eta(\lambda, \beta^+) - \eta(\lambda, \beta^-)$ is nonsingular at $\lambda = \lambda_0$, we say $L(\lambda)$ is *atomic at* β at λ_0. If $A(\lambda : \beta, L)$ is nonsingular on a set $K \subseteq \Lambda$, we say $L(\lambda)$ is *atomic at* β on K.

Lemma 5.1. *If $L \in C(\Lambda, \mathcal{L}(C, \mathbb{R}^n))$, then L has smoothness on the measure.*

Proof. Let elements of Λ be denoted by t. If f is a function of bounded variation on an interval I, let $\mathrm{Var}_I\, f$ denote the total variation of f on I. If $\eta = (\eta_{ij})$ is a matrix of bounded variation on $[-r, 0]$, let

$$\begin{aligned}
\|\eta\| &= \max_{1 \leq i \leq n} \sum_{j=1}^{n} \mathrm{Var}_{[-r,0]}\, \eta_{ij} \\
&= \max_{1 \leq i \leq n} \sum_{j=1}^{n} \sup_{P[-r,0]} \sum_{k=1}^{N} |\eta_{ij}(\tau_k) - \eta_{ij}(\tau_{k-1})|
\end{aligned}$$

where the supremum is taken over all partitions of $[-r, 0]$.

If $L(t)\phi = \int_{-r}^{0} d[\eta(t, \theta)]\phi(\theta)$, then there is a constant $k > 0$ such that $k\|\eta(t, \cdot)\| \leq \|L(t)\| \leq \|\eta(t, \cdot)\|$. Since $L \in C(\Lambda, \mathcal{L}(C, \mathbb{R}^n))$, for any $t \in \Lambda$ and $\epsilon > 0$, there is a $\delta > 0$ such that $\|\eta(t, \cdot) - \eta(\tau, \cdot)\| < \epsilon$ if $|t - \tau| < \delta$. This means for any $[a, b] \subseteq [-r, 0]$ and any $i, j = 1, 2, \ldots, n$,

$$\mathrm{Var}_{[a,b]}\, \big(\eta_{ij}(t, \cdot) - \eta_{ij}(\tau, \cdot)\big) < \epsilon \qquad \text{if} \quad |t - \tau| < \delta. \tag{5.3}$$

For a fixed i and $0 < s \leq r$, let

$$\mu_i(t,s) = \sum_{j=1}^{n} \text{Var}_{[\beta+,\beta+s]\cup[\beta-s,\beta-]}\ \eta_{ij}(t,\cdot).$$

From Inequality (5.3), we have $\mu_i(t,s)$ is continuous in t uniformly with respect to s. Also, $\mu_i(t,s)$ is nondecreasing in s, is uniformly bounded in s and $\mu_i(t,s) \to 0$ as $s \to 0$.

In $\mathbb{R}^2$, consider the set $\{(s,y) : y = \mu_i(t,s),\ s \in (0,\infty)\}$, and its closed convex hull $\Gamma_i(t)$. Let $\gamma_i(t,s) = \sup\{y : (s,y) \in \Gamma_i(t)\}$. Then $\gamma_i(t,s)$ is continuous in t uniformly with respect to s. Also, for each fixed t, it is continuous in s with $\gamma_i(t,s) \to 0$ as $s \to 0$. If we define $\gamma_i(t,0) = 0$, then $\gamma_i(t,s)$ is jointly continuous in t, s. If $\gamma(t,s) = \max_{1\leq i\leq n} \gamma_i(t,s)$, then γ satisfies the conditions of the lemma and the proof is complete. □

In the following our interest lies in the case where $\Lambda = \Omega \subseteq \mathbb{R}\times C$; that is, $L \in C(\Omega, \mathcal{L}(C,\mathbb{R}^n))$. If $D : \Omega \to \mathbb{R}^n$ has a continuous first derivative with respect to ϕ, then Lemma 5.1 implies D_ϕ has smoothness on the measure. This remark justifies the following definition.

Definition 5.3. Suppose $\Omega \subseteq \mathbb{R}\times C$ is open with elements (t,ϕ). A function $D : \Omega \to \mathbb{R}^n$ (not necessarily linear) is said to be *atomic at* β *on* Ω if D is continuous together with its first and second Fréchet derivatives with respect to ϕ; and D_ϕ, the derivative with respect to ϕ, is atomic at β on Ω.

If $D(t,\phi)$ is linear in ϕ and continuous in $(t,\phi) \in \mathbb{R}\times C$,

$$D(t)\phi = \int_{-r}^{0} d[\eta(t,\theta)]\phi(\theta)$$

then $A(t,\phi,\beta) = A(t,\beta)$ is independent of ϕ and

$$A(t,\beta) = \eta(t,\beta+) - \eta(t,\beta-).$$

Thus, $D(t)$ is atomic at β on $\mathbb{R}\times C$ if $\det A(t,\beta) \neq 0$ for all $t \in \mathbb{R}$. In particular, if $\beta \neq 0$, $\beta \in [-r,0]$, $D(t)\phi = \phi(0) + B(t)\phi(\beta)$, then $A(t,\beta) = B(t)$ and $D(t)$ is atomic at β on $\mathbb{R}\times C$ if $\det B(t) \neq 0$ for all $t \in \mathbb{R}$.

For the following results, the smoothness conditions on the function D in the definition of atomic at β are more severe than necessary. What is needed for a given f is the existence of functions $L(t,\phi), g(t,\phi,\psi), \epsilon(t,\phi,s)$, continuous such that $\epsilon(t,\phi,0) = 0$, Relation (5.2) is satisfied, and

$$f(t,\phi+\psi) = f(t,\phi) + L(t,\phi)\psi + g(t,\phi,\psi) \tag{5.4}$$

$$|g(t,\phi,\psi) - g(t,\phi,\xi)| \leq \epsilon(t,\phi,s)|\psi - \xi|, \qquad |\psi|,\ |\xi| \leq s \tag{5.5}$$

for all $(t,\phi) \in \Omega$, $(t,\phi+\psi) \in \Omega$, $(t,\phi+\xi) \in \Omega$, $s \geq 0$. If $A(t,\phi,\beta;L)$ is the matrix defined in Definition 5.2, then for any $(t,\phi) \in \Omega$, there is an $s_0 = s_0(t,\phi)$ such that $A(t,\phi,\beta;L)$ and ϵ are related by

$$|\det A(t,\phi,\beta;L)| - \epsilon(t,\phi,\psi) > 0 \quad \text{for } |\psi| \leq s_0. \tag{5.6}$$

If $|\det A(t,\phi,\beta;L)| \geq a > 0$ for all $(t,\phi) \in \Omega$ and $\epsilon(t,\phi,s) \leq \epsilon_0(s)$ for all $(t,\phi) \in \Omega$, then Relation (5.6) holds uniformly on Ω. We will make use of this remark in the examples.

We can now state

Theorem 5.1. (Backward continuation). *If Ω is an open set in $\mathbb{R} \times C$, $f : \Omega \to \mathbb{R}^n$ is atomic at $-r$ on Ω, $(\sigma,\phi) \in \Omega$, and there is an α, $0 < \alpha < r$, such that $\dot\phi(\theta)$ is continuous for $\theta \in [-\alpha, 0]$, $\dot\phi(0) = f(\sigma,\phi)$, then there is an $\bar\alpha > 0$ and a unique solution x of the* RFDE(f) *on $[\sigma - r - \bar\alpha, \sigma]$ through $[\sigma,\phi)$.*

Proof. A function x is a solution of the RFDE(f) on $[\sigma - r - \alpha, \sigma]$ through (σ,ϕ) if and only if $x_\sigma = \phi$, $(t,x_t) \in \Omega$, $t \in [\sigma - \alpha, \sigma]$, and

$$f(t,x_t) = \dot x(t) = \dot\phi(t-\sigma), \qquad t \in [\sigma-\alpha,\sigma] \tag{5.7}$$

since $0 < \alpha < r$.

For any $\alpha > 0$, let $\widehat\phi : [-r-\alpha, 0] \to \mathbb{R}^n$ be defined by $\widehat\phi(t) = \phi(t)$, $t \in [-r,0]$, $\widehat\phi(t) = \phi(-r)$, $t \in [-r-\alpha,-r]$. If $x(\sigma+t) = \widehat\phi(t) + z(t)$, $t \in [-r-\alpha, 0]$, then x is a solution of Equation (5.7) if and only if z satisfies $z_0 = 0$ and

$$f(\sigma+t,\widehat\phi_t + z_t) = \dot\phi(t), \qquad t \in [-\alpha,0]. \tag{5.8}$$

If $f(t,\phi+\psi) = f(t,\phi) + L(t,\phi)\psi + g(t,\phi,\psi)$, where $L(t,\phi) = f'_\phi(t,\phi)$, then the hypotheses on f imply that $g(t,\phi,\psi)$ is continuous in (t,ϕ,ψ), $g(t,\phi,0) = 0$ and, for any $(t,\phi) \in \Omega$ there is a $\beta = \beta(t,\phi) \geq 0$ and a function $\epsilon(t,\phi,\beta)$ continuous in (t,ϕ,β) such that $\epsilon(t,\phi,0) = 0$ and

$$|g(t,\phi,\psi) - g(t,\phi,\xi)| \leq \epsilon(t,\phi,\beta)|\psi - \xi|, \qquad |\xi| \leq \beta \text{ and } |\psi| \leq \beta.$$

If we make use of this in Equation (5.8), then Equation (5.7) is equivalent to $z_0 = 0$ and

$$L(\sigma+t,\widehat\phi_t)z_t = -f(\sigma+t,\widehat\phi_t) - g(\sigma+t,\widehat\phi_t,z_t) + \dot\phi(t) \qquad t \in [-\alpha,0].$$

Let $A(t,\phi) = A(t,\phi,-r;L)$ be the matrix associated with $L(t,\phi)$ given in Definition 5.2. Then Equation (5.7) is equivalent to $z_0 = 0$ and

$$\begin{aligned} z(t-r) = A^{-1}(\sigma+t,\widehat\phi_t)\Big[\int_{-r^+}^0 & d[\eta(t,\widehat\phi_t,\theta)]z_t(\theta) - f(\sigma+t,\widehat\phi_t) \\ & - g(\sigma+t,\widehat\phi_t,z_t) + \dot\phi(t)\Big], \qquad t \in [-\alpha,0]. \end{aligned} \tag{5.9}$$

For any $\beta > 0$, let $B_\beta = \{\psi \in C : |\psi| \le \beta\}$. For any v, $0 < v < 1/4$, there are $\alpha > 0$, $\beta > 0$, such that $(\sigma + t, \phi + \psi) \in \Omega$,

$$|A^{-1}(\sigma + t, \phi + \psi)|\,\epsilon(\sigma + t, \phi + \psi, \beta) < v$$

$$|A^{-1}(\sigma + t, \phi + \psi)|\,\gamma(\sigma + t, \phi + \psi, \alpha) < v$$

for $t \in [-\alpha, 0]$, $\psi \in B_\beta$, and γ the function in Relation (5.2).

Choose α and β so that these relations are satisfied. For any nonnegative real $\bar{\alpha}$ and $\bar{\beta}$, let $\mathcal{B}(\bar{\alpha}, \bar{\beta})$ be the set defined by

$$\mathcal{B}(\bar{\alpha}, \bar{\beta}) = \{\zeta \in C([-r - \bar{\alpha}, 0], \mathbb{R}^n) : \zeta_0 = 0,\ \zeta_t \in B_{\bar{\beta}},\ t \in [-\bar{\alpha}, 0]\}.$$

For any $0 < \bar{\beta} < \beta$, there is an $\bar{\alpha}$, $0 < \bar{\alpha} < \alpha$, so that $|\widehat{\phi}_t - \phi| < \beta - \bar{\beta}$. Further restrict $\bar{\alpha}$ so that

$$|f(\sigma + t, \widehat{\phi}_t) - f(\sigma, \phi)| \le v\bar{\beta}$$

$$|\dot{\phi}(0) - \dot{\phi}(t)| \le v\bar{\beta}$$

for $t \in [-\bar{\alpha}, 0]$.

For any $\zeta \in \mathcal{B}(\bar{\alpha}, \bar{\beta})$, define the transformation

$$T : \mathcal{B}(\bar{\alpha}, \bar{\beta}) \to C([-r - \bar{\alpha}, 0], \mathbb{R}^n)$$

by the relation $(T\zeta)_0 = 0$ and

$$\begin{aligned}(T\zeta)(t - r) = A^{-1}(\sigma + t, \widehat{\phi}_t)\Big[\int_{-r^+}^{0} d_\theta[\eta(t, \widehat{\phi}_t, \theta)]\zeta_t(\theta) \\ - f(\sigma + t, \widehat{\phi}_t) + f(\sigma, \phi) - g(\sigma + t, \widehat{\phi}_t, \zeta_t) \\ + \dot{\phi}(t) - \dot{\phi}(0)\Big], \qquad t \in [-\bar{\alpha}, 0].\end{aligned} \tag{5.10}$$

By hypothesis $\dot{\phi}(0) = f(\sigma, \phi)$ and therefore the fixed points ζ on T in $\mathcal{B}(\bar{\alpha}, \bar{\beta})$ yield the solution x of Equation (5.7) on $[\sigma - r - a, \sigma]$ with $x(\sigma + t) = \widehat{\phi}(t) + \zeta(t)$, $t \in [-\bar{\alpha}, 0]$.

We now show that T is a contraction on $\mathcal{B}(\bar{\alpha}, \bar{\beta})$. It is clear from Equation (5.10) and the restrictions on $\bar{\alpha}$ and $\bar{\beta}$ that

$$|(T\zeta)(t - r)| \le v\bar{\beta} + v\bar{\beta} + v\bar{\beta} + v\bar{\beta} < \bar{\beta},$$

$$|(T\zeta)(t - r) - (T\xi)(t - r)| \le v|\zeta_t - \xi_t| + v|\zeta_t - \xi_t| < \frac{1}{2}|\zeta_t - \xi_t|$$

for all $t \in [-\alpha, 0]$, ζ, $\xi \in \mathcal{B}(\bar{\alpha}, \bar{\beta})$. Therefore, $T : \mathcal{B}(\bar{\alpha}, \bar{\beta}) \to \mathcal{B}(\bar{\alpha}, \bar{\beta})$ and is a contraction. Thus, there is a unique fixed point in $\mathcal{B}(\alpha, \bar{\beta})$. This proves the theorem. □

Suppose the conditions of Theorem 2.1 are satisfied. For any $(\sigma, \phi) \in \Omega$, there exists a $t_{\sigma,\phi}$ and a function x that is a noncontinuable solution of the RFDE(f) on $[\sigma - r, t_{\sigma,\phi})$ through (σ, ϕ). Let $\Omega_\sigma \subseteq C$ be defined by $\Omega_\sigma = \{\phi \in C : (\sigma, \phi) \in \Omega\}$. If we only speak of noncontinuable solutions, then we can define a map $T(t, \sigma) : \Omega_\sigma \to C$ defined for each $\phi \in \Omega_s$, and $t \in [\sigma, t_{\sigma,\phi})$ by $T(t, \sigma)\phi = x_t(\sigma, \phi)$. This map will be called the *solution map of the* RFDE(f). We shall discuss many properties of this mapping in the following section and content ourselves with the following observation at the present time.

Corollary 5.1. *Suppose Ω is an open set in $\mathbb{R} \times C$, $f : \Omega \to \mathbb{R}^n$ is continuous, and $T(t, \sigma) : \Omega_\sigma \to C$, $t \in [\sigma, t_{\sigma,\phi})$ is defined above. If f is atomic at $-r$ on Ω, then $T(t, \sigma)$ is one-to-one.*

Proof. If the assertion is not true, then there are $\psi \neq \phi$ in C and a $t_1 > \sigma$ such that $x_{t_1}(\sigma, \phi) = x_{t_1}(\sigma, \psi)$, $x_t(\sigma, \phi) \neq x_t(\sigma, \psi)$ for $\sigma \leq t < t_1$. If $x = x(\sigma, \phi)$ and $y = x(\sigma, \psi)$, then $\dot{x}(t) = f(t, x_t)$ and $\dot{y}(t) = f(t, y_t)$ for $\sigma \leq t \leq t_1$. Since f is atomic at $-r$ on Ω there is an $\alpha = \alpha(t_1) > 0$ such that there is a unique solution of the RFDE(f) on $[t_1 - r - \alpha, t_1]$ through $(t_1, x_{t_1}) = (t_1, y_{t_1})$. Thus $(t, x_t) = (t, y_t)$ for $t_1 - \alpha \leq t \leq t_1$. This is a contradiction and proves the corollary. □

Let us now consider some examples. Consider the linear system

$$\dot{x}(t) = \int_{-r}^{0} d[\eta(t, \theta)]x(t + \theta) \stackrel{\text{def}}{=} L(t)x_t \tag{5.11}$$

where $\eta(t, -r^+) - \eta(t, -r) = A(t)$ is continuous and

$$\left| \int_{-r}^{-r+s} d[\eta(t, \theta)]\psi(\theta) - A(t)\psi(-r) \right| \leq \gamma(t, s) \sup_{-r \leq \theta \leq r+s} |\psi(\theta)|$$

for a continuous scalar function $\gamma(t, s)$, $t \in \mathbb{R}$, $s \geq 0$, $\gamma(t, 0) = 0$. If $\det A(t) \neq 0$ for all t, then $L(t)$ is atomic at $-r$ on $\mathbb{R} \times C$ and the map $T(t, \sigma)$ defined by the solutions of Equation (5.11) is one-to-one.

As another example, consider the equation

$$\dot{x}(t) = L(t)x_t + N(t, x_t) \stackrel{\text{def}}{=} F(t, x_t) \tag{5.12}$$

where L is the same function as in Equation (5.11), $N(t, \phi)$ is continuous for $(t, \phi) \in \mathbb{R} \times C$, and the Fréchet derivative $N_\phi(t, \phi)$ of N with respect to ϕ is continuous and $|N_\phi(t, \phi)| \leq \mu(|\phi|)$ for $(t, \phi) \in \mathbb{R} \times C$, where μ is a continuous function with $\mu(0) = 0$. If $|\det A(t)| \geq a > 0$ for $t \in \mathbb{R}$, then $F(t, \phi)$ is atomic at $-r$ on $\mathbb{R} \times U$ where U is a sufficiently small neighborhood of the origin in C. Here, we are using the relations (5.4)–(5.6) after the Definition 5.3 of atomic at $-r$. Consequently, the solution map $T(t, \phi)$ defined by the solutions of Equation (5.12) is one-to-one on its domain of definition.

As a final example, consider the equation

$$\text{(5.13)} \qquad \dot{x}(t) = -\alpha x(t-1)[1 + x(t)], \qquad \alpha > 0.$$

For this case, $f(t,\phi) = -\alpha\phi(-1)[1+\phi(0)]$ and

$$f'_\phi(t,\phi)\psi = -\alpha\psi(-1)[1+\phi(0)] - \alpha\phi(-1)\psi(0)$$

$$A(t,\phi) = -\alpha[1+\phi(0)].$$

As long as $\phi(0) \neq -1$, the function $f(t,\phi)$ is atomic at $-r$ and the mapping $T(t,\sigma)$ will be one-to-one as long as the solution $x(\sigma,\phi)(t) \neq -1$. But, from Equation (5.13), we have

$$x(t) = -1 + [1+\phi(0)]\exp\Big[-\int_\sigma^t \alpha x(s-1)\,ds\Big], \qquad t \geq \sigma.$$

Therefore, any solution with $\phi(0) \neq -1$ will always have $x(t) \neq -1$ and $T(t,\sigma)$ defined by $T(t,\sigma)\phi = x_t(\sigma,\phi)$ is one-to-one on the sets

$$\{\phi \in C : \phi(0) > -1\}, \qquad \text{and} \qquad \{\phi \in C : \phi(0) < -1\}.$$

On the set

$$C_{-1} \stackrel{\text{def}}{=} \{\phi \in C : \phi(0) = -1\}$$

the map $T(t,\sigma)$ is not one-to-one and, in fact, $T(t,\sigma)\phi$ is the constant function one for $t \geq \sigma + 1$ and $\phi \in C_{-1}$.

Let us state more precisely the type of general question that should be discussed in connection with the results in this section and consider only the autonomous equation

$$\text{(5.14)} \qquad \dot{x}(t) = f(x_t).$$

Suppose Γ is an open set in C, $f \in C^1(\Gamma, \mathbb{R}^n)$ with the usual norm, $|f|_\Gamma = \sup_{\psi\in\Gamma}[|f(\psi)| + |D_\phi(\psi)|]$. A *residual set* in $C^1(\Gamma, \mathbb{R}^n)$ is a set that is the countable intersection of open dense sets.

Question. For a given Ω, does there exist a residual set $S \subseteq C^1(\Omega, \mathbb{R}^n)$ such that the solution map $T_f(t) : \Omega \to C$ of Equation (5.14) is one-to-one from Ω to its range?

If we restrict ourselves to the subclass of linear RFDE consisting only of

$$\text{(5.15)} \qquad \dot{x}(t) = Lx_t,$$

then the answer is yes. In fact, for any $\delta > 0$ there is an $n \times n$ matrix B such that $|B| < \delta$ and

$$L\phi + B(-r)$$

is atomic at $-r$ on C. Furthermore, if L is atomic at $-r$ on C, then every continuous linear $I : C \to \mathbb{R}^n$ sufficiently close to L in $\mathcal{L}(C, \mathbb{R}^n)$ is atomic at $-r$. Therefore, we can take S to be the set of continuous linear $L : C \to \mathbb{R}^n$

such that L is atomic at $-r$. The set S is actually open and dense in $\mathcal{L}(C, \mathbb{R}^n)$.

In later sections, the importance of this question to the qualitative theory will be discussed. Also, some further specific results for both the autonomous and nonautonomous equations will be given.

2.6 Caratheodory conditions

In Section 2.1, we defined a functional differential equation for continuous $f : \mathbb{R} \times C \to \mathbb{R}^n$. On the other hand, it was then shown that the initial-value problem was equivalent to

$$
\begin{aligned}
&x_\sigma = \phi \\
(6.1) \qquad &x(t) = \phi(0) + \int_\sigma^t f(s, x_s)\, ds, \qquad t \geq \sigma.
\end{aligned}
$$

Equation (6.1) is certainly meaningful for a more general class of functions f if it is not required that $x(t)$ have a continuous first derivative for $t > \sigma$. We give in this section the appropriate generalization to functional differential equations of the well-known Caratheodory conditions of ordinary differential equations.

Suppose Ω is an open subset of $\mathbb{R} \times C$. A function $f : \Omega \to \mathbb{R}^n$ is said to satisfy the *Caratheodory condition* on Ω if $f(t, \phi)$ is measurable in t for each fixed ϕ, continuous in ϕ for each fixed t and for any fixed $(t, \phi) \in \Omega$, there is a neighborhood $V(t, \phi)$ and a Lebesgue integrable function m such that

$$
(6.2) \qquad |f(s, \psi)| \leq m(s), \qquad (s, \psi) \in V(t, \phi).
$$

If $f : \Omega \to \mathbb{R}^n$ is continuous, it is easy to see that f satisfies the Caratheodory condition on Ω. Therefore, a theory for Equation (6.1) in this more general setting will include the previous theory.

If f satisfies the Caratheodory condition on Ω, $(\sigma, \phi) \in \Omega$, we say a function $x = x(\sigma, \phi, f)$ is a *solution of* Equation (6.1) *through* (σ, ϕ) if there is an $A > 0$ such that $x \in C([\sigma - r, \sigma + A], \mathbb{R}^n)$, $x_\sigma = \phi$ and $x(t)$ is absolutely continuous on $[\sigma, \sigma + A]$ and satisfies $\dot{x}(t) = f(t, x_t)$ almost everywhere on $[\sigma, \sigma + A]$.

Using essentially the same arguments, one can extend all of the previous results to the case where f satisfies a Caratheodory condition on Ω. The most difficult result is the analogue of Theorem 2.2 on continuous dependence. To obtain a result on continuous dependence, it is sufficient to require that all f^k satisfy the Caratheodory condition on Ω, $f^k(s, \psi) \to f^0(s, \phi)$ as $k \to \infty$, $\psi \to \phi$ for almost all s and satisfy the following condition: For any compact set W in Ω, there is an open neighborhood $V(W)$ of W and a Lebesgue integrable function M such that the sequence of functions f_k, $k = 0, 1, 2, \dots$, satisfies

$$|f_k(s, \psi)| \le M(s), \qquad (s, \psi) \in V(W), \quad k = 0, 1, 2, \dots.$$

We remark in passing that more general existence theorems are easily given if the function $f(t, \phi)$ depends on ϕ in some special way. In particular, if for any $\epsilon \ge 0$ we let ϕ^ϵ denote the restriction of ϕ to the interval $[-r, \epsilon]$ and

$$f(t, \phi) = F(t, \phi(0), \phi^\epsilon),$$

then the basic existence theorem can be proved by the process of stepping forward a step of size less than ϵ (if $\epsilon > 0$) under very weak conditions on the dependence of $F(t, x, \psi)$ on ψ.

2.7 Definition of a neutral equation

In this section, we define a neutral functional differential equation (NFDE) and give some examples.

Definition 7.1. Suppose $\Omega \subseteq \mathbb{R} \times C$ is open, $f : \Omega \to \mathbb{R}^n$, $D : \Omega \to \mathbb{R}^n$ are given continuous functions with D atomic at zero (see Definition 5.3 of Section 2.5). The relation

$$\frac{d}{dt} D(t, x_t) = f(t, x_t) \tag{7.1}$$

is called the *neutral functional differential equation* NFDE(D, f). The function D will be called the *difference operator* for the NFDE(D, f).

Definition 7.2. For a given NFDE(D, f), a function x is said to be a *solution of the* NFDE(D, f) if there are $\sigma \in \mathbb{R}$, $A > 0$, such that

$$x \in C([\sigma - r, \sigma + A), \mathbb{R}^n), \qquad (t, x_t) \in \Omega, \quad t \in [\sigma, \sigma + A),$$

$D(t, x_t)$ is continuously differentiable and satisfies Equation (7.1) on $[\sigma, \sigma + A)$. For a given $\sigma \in \mathbb{R}$, $\phi \in C$, and $(\sigma, \phi) \in \Omega$, we say $x(\sigma, \phi, D, f)$ is a *solution of Equation* (7.1) *with initial value* ϕ *at* σ or simply a *solution through* (σ, ϕ) if there is an $A > 0$ such that $x(\sigma, \phi, D, f)$ is a solution of Equation (7.1) on $[\sigma - r, \sigma + A)$ and $x_\sigma(\sigma, \phi, D, f) = \phi$.

Definition 7.3. If $D(t, \phi) = D_0(t)\phi - g(t)$, $f(t, \phi) = L(t)\phi + h(t)$ where $D_0(t)$ and $L(t)$ are linear in ϕ, the NFDE(D, f) is called *linear*. It is *linear homogeneous* if $g \equiv 0$, $h \equiv 0$ and *linear nonhomogeneous* if either $g \not\equiv 0$ or $h \not\equiv 0$. An NFDE(D, f) is called *autonomous* if $D(t, \phi)$ and $f(t, \phi)$ do not depend on t.

Let us now consider some examples of NFDE.

Example 7.1. If $D\phi = \phi(0)$ for all ϕ, then D is atomic at 0. Therefore, for any continuous $f : \Omega \to \mathbb{R}^n$, the pair (D, f) defines an NFDE. Consequently, RFDE are NFDE.

Example 7.2. If $r > 0$, B is an $n \times n$ constant matrix, $D(\phi) = \phi(0) - B\phi(-r)$, and $f : \Omega \to \mathbb{R}^n$ is continuous, then the pair (D, f) defines an NFDE; that is, the equation

$$\frac{d}{dt}[x(t) - Bx(t-r)] = f(t, x_t) \tag{7.2}$$

is an NFDE.

Example 7.3. If $r > 0$, x is a scalar, $D\phi = \phi(0) - \phi^2(-r)$ and $f : \Omega \to \mathbb{R}$ is continuous, then the pair (D, f) defines an NFDE; that is, the equation

$$\frac{d}{dt}[x(t) - x^2(t-r)] = f(t, x_t) \tag{7.3}$$

is an NFDE.

Example 7.4. Suppose $r_j, 0 < r_j \le r, j = 1, 2, \dots, N$, are given real numbers and $g : \mathbb{R}^N \to \mathbb{R}^n$ is a given continuous function. If

$$D(\phi) = \phi(0) - g(\phi(-r_1), \dots, \phi(-r_N))$$

and 0 is the zero function on $\mathbb{R} \times C$, then the pair $(D, 0)$ defines an NFDE on $\mathbb{R} \times C$; that is, the equation

$$\frac{d}{dt}[x(t) - g(x(t-r_1), \dots, x(t-r_N))] = 0 \tag{7.4}$$

is an NFDE. Obviously, the solutions of Equation (7.4) are given by

$$x(t) = g(x(t-r_1), \dots, x(t-r_N)) + c, \tag{7.5}$$

where c is a constant given by $c = \phi(0) - g(\phi(-r_1), \dots, \phi(-r_N))$. Therefore, if one defines the subset U of C by

$$U = \{\phi \in C : \phi(0) = g(\phi(-r_1), \dots, \phi(-r_N))\}$$

then any solution of the NFDE (7.4) with initial data in U is a solution of the difference equation

$$x(t) = g(x(t-r_1), \dots, x(t-r_N)). \tag{7.6}$$

Conversely, any solution of Equation (7.6) is a solution of Equation (7.4). Consequently, NFDE defined as in Definition 7.1 include difference equations.

In the definition of a solution of Equation (7.1), it is only required that $D(t, x_t)$ be differentiable in t. The function $x(t)$ may not be differentiable. The same situation was encountered in Chapter 1. If the function $x(t)$ is continuously differentiable on $[\sigma - r, \sigma + A)$ and the function $D(t, \phi)$ is also differentiable in t, ϕ then the function x satisfies the equation

$$D_\phi(t, x_t)\dot{x}_t = f(t, x_t) - D_t(t, x_t) \tag{7.7}$$

for $\sigma \leq t < \sigma + A$. The symbol $\dot{x}_t$ designates the function in C given by $\dot{x}_t(\theta) = \dot{x}(t + \theta)$, $-r \leq \theta \leq 0$.

As a result of this discussion, we see that smooth solutions of Equation (7.1) satisfy Equation (7.7), which is linear in the derivative of x. This is a basic limitation of the development of the theory using the Definition 1.1 for an NFDE. For the special cases, Equations (7.2) and (7.3), Equation (7.7) is given by

$$\dot{x}(t) - B\dot{x}(t - r) = f(t, x_t), \tag{7.8}$$

$$\dot{x}(t) - 2x(t - r)\dot{x}(t - r) = f(t, x_t). \tag{7.9}$$

Example (7.9) also shows that our theory will not include any general theory of NFDE based only on the assumption that the derivative of x enters in a linear fashion. In fact, the terms involving x that multiply $\dot{x}$ must occur with the same delays.

2.8 Fundamental properties of NFDE

In this section, we consider the questions of existence, uniqueness, and continuous dependence of solutions of NFDE.

Theorem 8.1. (Existence). *If Ω is an open set in $\mathbb{R} \times C$ and $(\sigma, \phi) \in \Omega$, then there exists a solution of the* NFDE(D, f) *through (σ, ϕ).*

Proof. We give an outline of the proof and leave the details of the computations to the reader. If the derivative $D_\phi(t, \phi)$ of $D(t, \phi)$ with respect to ϕ is represented as

$$D_\phi(t, \phi)\psi = A(t, \phi)\psi(0) - \int_{-r}^{0} d[\mu(t, \theta, \phi)]\psi(\theta)$$

then the definition of atomic at 0 implies $\det A(t, \phi) \neq 0$, $A(t, \phi)$ is continuous in t, ϕ, and the linear operator $D_\phi(t, \phi)$ has smoothness on the measure (see Definition 5.2 and Lemma 5.1 of Section 2.5).

Let $\widetilde{\phi} : [-r, \infty) \to \mathbb{R}^n$ be defined by $\widetilde{\phi}_0 = \phi$, $\widetilde{\phi}(t) = \phi(0)$, $t \geq 0$. If x is a solution of Equation (1.1) through (σ, ϕ) on $[\sigma - r, \sigma + \alpha]$ and $x_{t+\sigma} = \widetilde{\phi}_t + z_t$, then z satisfies the equation

$$D(\sigma+t,\widetilde{\phi}_t+z_t)=D(\sigma,\phi)+\int_0^t f(\sigma+s,\widetilde{\phi}_s+z_s)\,ds \tag{8.1}$$

for $0\le t\le\alpha$ and $z_0=0$. Using the preceding notation for $D_\phi(t,\phi)$, we have

$$z=Sz+Uz \tag{8.2}$$

where the operators S and U are defined by $(Sz)(t)=0$ and $(Uz)(t)=0$ for $-r\le t\le 0$, and

$$\begin{aligned}A(t+\sigma,\widetilde{\phi}_t)(Sz)(t)&=\int_{-r}^{0-} d_\theta[\mu(\sigma+t,\theta,\widetilde{\phi}_t)]z(t+\theta)\\&\quad+\big[D(\sigma,\phi)-D(\sigma+t,\widetilde{\phi}_t)\big]-\big[D(\sigma+t,\widetilde{\phi}_t+z_t)\\&\quad-D(\sigma+t,\widetilde{\phi}_t)-D_\phi(\sigma+t,\widetilde{\phi}_t)z_t\big]\\A(t+\sigma,\widetilde{\phi}_t)(Uz)(t)&=\int_0^t f(\sigma+s,\widetilde{\phi}_s+z_s)\,ds\end{aligned}$$

for $0\le t\le\alpha$.

Any $z\in C([-r,\alpha],\mathbb{R}^n)$ satisfying the operator equation (8.2) is a solution of Equation (8.1) and, therefore, $x_{t+\sigma}=\widetilde{\phi}_t+z_t$ is a solution of Equation (7.1). Therefore, the existence of a solution of Equation (7.1) through (σ,ϕ) is equivalent to determining an $\alpha>0$ such that Equation (8.2) has a solution in $C\big([-r,\alpha],\mathbb{R}^n\big)$.

Let $\mathcal{A}(\alpha,\beta)=\{\zeta\in C\big([-r,\alpha],\mathbb{R}^n\big):\zeta_0=0,\ |\zeta_t|\le\beta,\ t\in[0,\alpha]\}$. It is now a relatively simple matter to show that one can choose α and β so that the map $S+U:\mathcal{A}(\alpha,\beta)\to\mathcal{A}(\alpha,\beta)$. Furthermore, S is a contraction on $\mathcal{A}(\alpha,\beta)$ and U is completely continuous on $\mathcal{A}(\alpha,\beta)$. Therefore, $S+U$ is an α-contraction and Darbo's theorem (see Theorem 6.3 of Section 4.6) implies there is a fixed point of $S+U$ in $\mathcal{A}(\alpha,\beta)$. This completes the proof of the theorem. □

Continuous dependence is more difficult. Suppose Λ is a subset of a Banach space and Ω is an open set in $\mathbb{R}\times C$. If $D:\Omega\times\Lambda\to\mathbb{R}^n$ is a given function, we have different possibilities for the manner in which the property of being atomic at 0 depends on the parameter λ. One can assume that $D(t,\phi,\lambda)$ is atomic at each $(t,\phi)\in\Omega$ for each $\lambda\in\Lambda$. One can also assume the stronger hypothesis that $D(t,\phi,\lambda)$ is atomic at zero at each $(t,\phi)\in\Omega$ uniformly with respect to $\lambda\in\Lambda$, with uniformly meaning that all estimates on determinants, measures, etc., in the definition of atomic at 0 are uniform with respect to $\lambda\in\Lambda$. Each of these hypotheses on D will require corresponding hypotheses on $f:\Omega\times\Lambda\to\mathbb{R}^n$ in order to obtain continuous dependence of the solution of the NFDE(D,f).

A useful tool in obtaining results on continuous dependence is the following lemma on continuous dependence of fixed points of condensing mappings.

Lemma 8.1. *Suppose Γ is a closed, bounded, convex set of a Banach space, Λ is a subset of another Banach space, and $T : \Gamma \times \Lambda \to \Gamma$ is a given mapping satisfying the following hypotheses:*

(h_1) *$T(\cdot, \lambda)$ is continuous for each $\lambda \in \Lambda$ and there exists a $\lambda_0 \in \Lambda$ such that $T(x, \lambda)$ is continuous at (x, λ_0) for each $x \in \Gamma$.*

(h_2) *For every $\Gamma' \subseteq \Gamma$, $\alpha(\Gamma') > 0$, there is an open neighborhood $B = B(\Gamma')$ of λ_0 such that for any precompact set $\Lambda' \subseteq \Lambda \cap B$, we have $\alpha(T(\Gamma', \Lambda')) < \alpha(\Gamma')$, where α is the Kuratowskii measure of noncompactness.*

(h_3) *The equation*

$$x = T(x, \lambda) \tag{8.3}$$

for $\lambda = \lambda_0$ has a unique solution $x(\lambda_0)$ in Λ.

Then the solutions $x(\lambda)$ in Γ of Equation (8.3) are continuous at λ at $\lambda = \lambda_0$.

Proof. If $\Gamma' \subseteq \Gamma$ and $\alpha(\Gamma') = 0$, define $B(\Gamma') = \Gamma$. Suppose $\{\lambda_k\} \subseteq \Lambda \cap B(\Gamma)$ is a sequence converging to λ_0 as $k \to \infty$, and let $x(\lambda_k) \in \Gamma$ be a solution of Equation (8.3) for $\lambda = \lambda_k$. If $\Gamma' = \{x(\lambda_k)\}$, choose k so large that $\Lambda' = \{\lambda_k\} \subseteq \Lambda \cap B(\Gamma')$. Since Λ' is precompact and Hypothesis (h_2) is satisfied,

$$\alpha(\Gamma') = \alpha(\{T(x_{\lambda_k}, \lambda_k)\}) \le \alpha(T(\Gamma', \Lambda')) < \alpha(\Gamma')$$

if $\alpha(\Gamma') > 0$. Since this is impossible, $\alpha(\Gamma') = 0$ and Γ' is precompact. Since Γ is closed, there is a subsequence $\{v_k\}$ of $\{\lambda_k\}$ and $z \in \Gamma$ such that $x(v_k) \to z$ as $k \to \infty$. Hypothesis (h_1) implies $z = T(z, \lambda_0)$ and Hypothesis (h_3) implies $z = x(\lambda_0)$. Since every convergent subsequence of the $x(\lambda_k)$ must converge to the same limit, it follows that $x(\lambda_k) \to x(\lambda_0)$ as $k \to \infty$. Since the sequence $\{\lambda_k\}$ was an arbitrary sequence converging to λ_0, the proof of the lemma is complete. □

Corollary 8.1. *Suppose Γ and Λ are as in Lemma* 8.1, *$T : \Gamma \times \Lambda \to \Gamma$ satisfies Hypotheses* (h_1) *and* (h_3), *$T = S + U$, and, for each compact set $\Lambda' \subseteq \Lambda$,*

(h_4) *$S(\cdot, \lambda)$ is a contraction on Γ uniformly with respect to $\lambda \in \Lambda'$.*

(h_5) *$U(\Gamma, \Lambda')$ is precompact.*

Then the solutions $x(\lambda)$ of Equation (8.3) *are continuous at λ_0.*

Proof. To verify that Hypothesis (h_2) in Lemma 8.1 is satisfied, simply observe that for every $\Gamma' \subseteq \Gamma$, $\alpha(T(\Gamma', \Lambda')) = \alpha(S(\Gamma', \Lambda')) \le k\alpha(\Gamma')$ for some $k \in [0, 1)$. □

Corollary 8.2. *Suppose Γ and Λ are as in Lemma* 8.1, *$T : \Gamma \times \Lambda \to \Gamma$ satisfies Hypothesis* (h_3), *$T = S + U$ where U satisfies Hypothesis* (h_5), *and*

(h_6) *$S(\cdot,\lambda)$ is a contraction for $\lambda \in \Lambda$ and $U(\cdot,\lambda)$ is continuous for each $\lambda \in \Lambda'$,*

(h_7) *$S(x,\lambda)$ is continuous at λ_0 uniformly for $x \in \Gamma$.*

Then the solutions $x(\lambda)$ of Equation (8.3) are continuous at λ_0.

Proof. Hypothesis (h_1) is obviously satisfied. To prove Hypothesis (h_2) is satisfied, let $B_\beta = \{\lambda : |\lambda - \lambda_0| < \beta\}$ and, for any $\epsilon > 0$, choose $\beta(\epsilon)$ such that

$$|S(x,\lambda) - S(x,\lambda_0)| < \epsilon \quad \text{if } \lambda \in B_{\beta(\epsilon)},\ x \in \Gamma.$$

This choice is possible by Hypothesis (h_7). Let the contraction constant for $S(\cdot,\lambda_0)$ be $k_0 < 1$. If $\Gamma' \subseteq \Gamma$, then $\alpha((S(\cdot,\lambda) - S(\cdot,\lambda_0))\Gamma') < \epsilon$ if $\lambda \in B_{\beta(\epsilon)}$. If we choose ϵ so that $k_0 + \epsilon < 1$, then

$$\alpha(T(\cdot,\lambda)\Gamma') = \alpha(S(\cdot,\lambda)\Gamma') \le (k_0 + \epsilon)\alpha(\Gamma') < \alpha(\Gamma')$$

for all $\Gamma' \subseteq \Gamma$. Therefore, Hypothesis (h_2) is satisfied and the conclusion of the corollary follows from Lemma 8.1. □

We are now in a position to prove a basic result on continuous dependence.

Theorem 8.2. (Continuous dependence). *Suppose $\Omega \subseteq \mathbb{R} \times C$ is open, Λ is a subset of a Banach space, $D : \Omega \times \Lambda \to \mathbb{R}^n$, $f : \Omega \times \Lambda \to \mathbb{R}^n$ satisfy the following hypotheses:*

(i) *$D(t,\phi,\lambda)$ is atomic at zero for each $(t,\phi) \in \Omega$ uniformly with respect to λ.*

(ii) *$D(t,\phi,\lambda)$ and $f(t,\phi,\lambda)$ are continuous in $(t,\phi) \in \Omega$ for each $\lambda \in \Lambda$ and continuous also at (t,ϕ,λ_0) for $(t,\phi) \in \Omega$.*

(iii) *The* NFDE$(D(\cdot,\lambda_0), f(\cdot,\lambda_0))$ *has a unique solution through $(\sigma,\phi) \in \Omega$ that exists on an interval $[\sigma - r, b]$.*

Then there is a neighborhood $N(\sigma,\phi,\lambda_0)$ of (σ,ϕ,λ_0) such that for any $(\sigma',\phi',\lambda') \in N(\sigma,\phi,\lambda_0)$, all solutions $x(\sigma',\phi',\lambda')$ of the NFDE$(D(\cdot,\lambda')$, $f(\cdot,\lambda'))$ *through (σ',ϕ') exist on $[\sigma' - r, b]$ and $x_t(\sigma',\phi',\lambda')$ is continuous at $(t,\sigma',\phi',\lambda_0)$ for $t \in [\sigma,\sigma + A]$, $(\sigma',\phi',\lambda') \in N(\sigma,\phi,\lambda_0)$.*

Proof. For each $\lambda \in \Lambda$, one can define the operators $S(z,\lambda)$ and $U(z,\lambda)$ as in Equation (8.2) for the solutions of the NFDE$(D(\cdot,\lambda), f(\cdot,\lambda))$ through (σ',ϕ'). It is left as an exercise to verify that the hypotheses of Corollary 8.1 are satisfied for the set Γ chosen as an appropriate set $\mathcal{A}(\alpha,\beta)$ used in the proof of Theorem 8.1. Using the compactness of the set $\{(t, x_t(\sigma,\phi,\lambda_0)) : t \in [\sigma, b]\}$, one can successively step intervals of length α to complete the proof of the theorem. □

Almost exactly as in the proof of Theorem 2.3 of Section 2.2, one obtains the following result on the uniqueness of solutions.

Theorem 8.3. (Uniqueness). *If $\Omega \subseteq \mathbb{R} \times C$ is open and $f : \Omega \to \mathbb{R}^n$ is Lipschitzian in ϕ on compact sets of Ω, then, for any $(\sigma, \phi) \in \Omega$, there exists a unique solution of the* NFDE(D, f) *through (σ, ϕ).*

The problems concerning the behavior of the solutions of an NFDE as one approaches the maximal interval of existence is not as well understood as the ones for RFDE. The proofs of results are also much more technical. The following results are stated without proof with references given in Section 2.9.

Theorem 8.4. (Continuation). *Suppose $\Omega \subseteq \mathbb{R} \times C$ is open, (D, f) defines an* NFDE *on Ω and $W \subseteq \Omega$ is closed and bounded and there is a δ-neighborhood of W in Ω. If f maps W into a bounded set in $\mathbb{R}^n$, $D(t, \phi)$ and $D_\phi(t, \phi)$ are uniformly continuous on W, D is uniformly atomic at* 0 *on W, and x is a noncontinuable solution of the* NFDE(D, f) *on $[\sigma - r, b)$, then there is a $t' \in [\sigma, b)$ such that $(t', x_{t'}) \notin W$.*

If $D(t, \phi)$ is linear in ϕ, we obtain the same conclusion with weaker hypotheses.

Theorem 8.5. (Continuation). *Suppose (D, f) defines an* NFDE *on an open set $\Omega \subseteq \mathbb{R} \times C$ and $D(t)\phi = D(t, \phi)$ is linear in ϕ. If x is a noncontinuable solution of the* NFDE(D, f) *on $[\sigma - r, b)$, $D(t)$ is defined on $[\sigma, b]$, and W is a closed bounded set in Ω for which $f(W)$ is bounded, then there is a $t' \in [\sigma, b)$ such that $(t', x_{t'}) \notin W$.*

2.9 Supplementary remarks

Except for technical details, the methods used in proving the existence and continuous dependence theorems in Section 2.2 are natural generalizations of methods from ordinary differential equations. One can investigate these questions with much weaker conditions on the function f as well as different types of hereditary dependence than the one implied by x_t (see Neustadt [1], Tychonov [1], Jones [1], Cruz and Hale [1], Imaz and Vorel [1]). Coffman and Schäffer [1] have investigated the weakest possible conditions on linear operators $L(t)$ that will permit the development of a general theory. The paper of Stokes [1] contains some results on differentiability of solutions.

By applying differential inequalities, one can obtain very general results on uniqueness of solutions (see Laksmikantham and Leela [1]).

The problem of continuation of solutions has features not encountered in ordinary differential equations as Example (3.1), due to Mishkis [2], demonstrates. Theorem 3.3 is due to Yorke [1] and the extension theorem used in the proof is due to Dugundji [1].

The first general results on backward continuation are due to Hastings [1]. The presentation in the text follows Hale [1] with the improvements made possible by Definition 5.2 and Lemma 5.1 introduced by Hale and Oliva [1]. Lillo [1] discusses the backward continuation of solutions of linear equations that are not necessarily atomic at $-r$. A similar situation is encountered in Hale and Oliva [1] in their discussion of the "size" of the set of linear RDFE for which the solution operator is one-to-one. For some other results on the inverse of the solution operator, see Kamenskii [1].

In the supplementary remarks at the end of Chapter 1, we gave a rather lengthy discussion of and references for the different ways in which one can define a neutral functional differential equation. The approach taken in this chapter corresponds to one of the simplest definitions. The definition of a NFDE in Section 2.7 is due to Hale [1] (see also Hale [19]). A more general definition using only Lipschitz continuous functions D as well as a more general dependence on the past history is contained in Cruz and Hale [1].

The existence Theorem 8.1 is due to Hale [1]. The general result on continuous dependence on parameters of fixed points of condensing maps (Lemma 8.1) is due to Hale [12] (see also Artstein [1]), and generalized earlier results of Melvin [3] and Cruz and Hale [3]. Corollary 8.1 is due to Melvin [3] and Corollary 8.2 is due to Cruz and Hale [1]. The continuation Theorem 8.4 may be found in Hale [1] and the continuation Theorem 8.5 is due to Lopes.

As remarked in Section 2.8 the solution operator for a NFDE(D, f) is a homeomorphism if D is atomic at zero and $-r$. If f is also C^1, then it can be shown that the solution operator is differentiable with respect to the initial data and so it is a diffeomorphism. In this case, as in ordinary equations, the equation defines a group rather than a semigroup.

We have shown how the theory of α-contractions played an important role in the existence theory of Section 2.8. We will see in the next chapter that this same concept plays an important role in the representation formula for the solution operator. Nussbaum [8] also has interesting applications of α-contractions to existence. Other applications of α-contractions to NFDE as well as ordinary differential equations are contained in Hale [16]. For the applications of coincidence degree to NFDE, see Hale and Mawhin [1] and Hetzer [1].

3
Properties of the solution map

In the study of retarded functional differential equations, the space of initial functions is preassigned, but the space in which one considers the trajectories is not. To be more specific, if $x(\sigma, \phi, f)$ is a solution of an RFDE(f) through (σ, ϕ), should the solution map be considered as the map $x(\sigma, \cdot, f)(t) : C \to \mathbb{R}^n$ or the map $T(t, \sigma) : C \to C$ defined in Section 2.5 as $T(t, \sigma)\phi = x_t(\sigma, \phi, f)$?

By the consideration of simple examples, one can see that $x(\sigma, \cdot)(t)$ has some rather undesirable properties. In fact, the scalar equation

$$\dot{x}(t) = -x\left(t - \frac{\pi}{2}\right) \tag{1}$$

has a unique solution through each $(\sigma, \phi) \in \mathbb{R} \times C$, but it also has the solutions $x(t) = \sin t$ and $x(t) = \cos t$. These latter solutions plotted in (x, t) space intersect an infinite number of times on any interval $[\sigma, \infty)$ and yet are not identical on any interval.

If we use the map $T(t, \sigma)$ for this example, then Corollary 5.1 of Section 2.5 implies $T(t, \sigma)$ is one-to-one on all of C. In the general situation, uniqueness implies if there is a $\tau > \sigma$ such that $T(\tau, \sigma)\phi = T(\tau, \sigma)\psi$, then $T(t, \sigma)\phi = T(t, \sigma)\psi$ for all $t \geq \sigma$.

For autonomous equations, it is more natural to consider the orbits of solutions rather than the trajectories, that is, the path traced out by the solution in the phase space X rather than the graph of the solution $\mathbb{R} \times X$. If the phase space for Equation (1) is chosen as $\mathbb{R}$ and the orbits as $\bigcup_{t \geq 0} x(0, \phi)(t)$, then the orbits for the solutions $x(t) = \sin t$ and $x(t) = \cos t$ coincide and are equal to the interval $[-1, 1]$. That the orbits coincide is expected because $\sin(t + (\pi/2)) = \cos t$. Equation (1) is autonomous and therefore, a solution shifted in phase is still a solution. The difficulty encountered by choosing the phase space $\mathbb{R}$ is that the orbit of one solution may completely contain the orbit of another solution and not be related in any way to a phase shift. The orbit of the solution $x = 0$ is contained in the orbit of $\cos t$.

On the other hand, if the phase space is chosen as $C = C([-\pi/2, 0], \mathbb{R})$, then the orbit of the solution $\sin t$ of Equation (1) is the set

$$\Gamma = \{\psi : \psi(\theta) = \sin(t+\theta),\ \frac{-\pi}{2} \le \theta \le 0,\ \text{for } t \in [0,\infty)\},$$

of points in C. The set Γ, as before, is also the orbit of the solution $\cos t$. Furthermore, because of uniqueness of solutions and one-to-oneness of the mapping $T(t,\sigma)$, any solution x of Equation (1) for which there is a τ with $x_\tau \in \Gamma$ must be a phase shift on $\sin t$. Therefore, Γ is determined by phase shifts of a solution. Finally, Γ is a closed curve in C that is intuitively satisfying since $\sin t$ is periodic.

This simple example suggests the geometric theory for Equation (1) will probably be richer if the map $T(t,\sigma)$ is used. However, in some situations, it is very advantageous to know that $T(t,\sigma)\phi$ is determined by taking a restriction over an interval of a function in $\mathbb{R}^n$.

It is the purpose of this chapter to discuss a few of the good and bad properties of the *solution map* $T_f(t,\sigma)$ of an RFDE(f) defined by $T_f(t,\sigma)\phi = x_t(\sigma,\phi,f)$. We will assume, unless otherwise explicitly stated, that f is continuous and there is a unique solution of the RFDE(f) through (σ,ϕ) so that $T_f(t,\sigma)\phi$ is continuous in (t,σ,ϕ,f) by Theorem 2.2 of Section 2.2.

If the RFDE(f) is autonomous, let $T_f(t,0) = T_f(t)$, $t \ge 0$. It follows that

$$\begin{aligned} & T_f(0) = I \\ (2) \qquad & T_f(t)T_f(\tau) = T_f(t+\tau), \qquad t,\tau \ge 0, \\ & T_f(t)\phi \quad \text{is continuous in} \quad (t,\phi,f); \end{aligned}$$

that is, $\{T(t)\}$, $t \ge 0$ is a *strongly continuous semigroup of transformations* on a subset of C. Of course, it is understood here that t and τ are allowed to range over an interval that may depend on $\phi \in C$. Since our interest in this chapter does not concern the dependence of $T_f(t,\sigma)$ on f, we write simply $T(t,\sigma)$.

3.1 Finite- or infinite-dimensional problem?

From the special way in which an RFDE is defined, it is necessary to discuss whether the problem is actually infinite-dimensional or finite-dimensional. More specifically, can the noncompactness of the unit ball in C have any adverse effects on the solutions considered either in $\mathbb{R}^n$ or C? The purpose of this section is to consider this question in some detail.

Theorem 3.2 of Section 2.3 on the continuation of solutions states that a noncontinuable solution of an RFDE(f) must leave every closed bounded set W in the domain of definition Ω of the equation, provided f is completely continuous on Ω. A continuous function on Ω need not be completely continuous on Ω if $r > 0$, that is, if C is infinite-dimensional. Thus, it is natural to ask if this latter condition is necessary. The answer is yes and we state this as

Property 1.1. *The continuation theorem is not valid if f is not a completely continuous map.*

Proof. Equation (3.1) and Theorem 3.3 of Section 2.3 demonstrate this result. □

We say that a mapping from one metric space to another metric space is *bounded* if it takes closed bounded sets into bounded sets. The map is *locally bounded* if it takes some neighborhood of each point into a bounded set.

Property 1.2. *$T(t, \sigma)$ is locally bounded for $t \geq \sigma$.*

Proof. Since $T(t, \sigma)\phi$ is assumed to be continuous in (t, σ, ϕ), it follows that for any $t \geq \sigma$, $\phi \in C$ for which $(\sigma, \phi) \in \Omega$ and $T(t, \sigma)\phi$ is defined, there is a neighborhood $V(t, \sigma, \phi)$ of ϕ in C such that $T(t, \sigma)V(t, \sigma, \phi)$ is bounded. □

Property 1.3. *$T(t, \sigma)$ may not be a bounded map.*

Proof. Suppose $r = \frac{1}{4}$, $C = C([-r, 0], \mathbb{R})$ and consider the equation

$$\dot{x}(t) = f(t, x_t) \overset{\text{def}}{=} x^2(t) - \int_{\min(t-r,0)}^{0} |x(s)|\, ds. \tag{1.1}$$

It is clear that f takes closed bounded sets into bounded sets and is locally Lipschitzian. If $B = \{\phi \in C : |\phi| \leq 1\}$ and $x(b)$, $b \in B$, is the solution of Equation (1.1), then $x(b)$ is always ≥ -1. Also, for $b \neq 0$, $x(b)(0) \leq 1$, $\dot{x}(b)(t) < x^2(t)$ for all t implies $x(t) < y(t)$ for all $0 < t < 1$, where $\dot{y} = y^2(t)$, $y(0) = 1$. Therefore, $x(b)(t)$ exists on $[0, 1)$,

$$x(b)(t) < y(t) = (1 - t)^{-1}$$

and, in particular, $x(b)(r) < (1 - r)^{-1}$ for all $b \in B$. For $t \geq r$, $\dot{x}(b)(t) = x^2(b)(t)$ and the fact that $x(b)(r) < (1 - r)^{-1}$ implies $x(b)(t)$ exists for $-r \leq t \leq 1$.

If we show that for any $\epsilon > 0$, there is a $b \in B$ such that

$$x(b)(r) > (1 - r)^{-1} - \epsilon,$$

then the set $x(B)(1)$ is not bounded since the solution $\dot{y} = y^2(t)$ through $(r, (1 - r)^{-1})$ is unbounded at $t = 1$ and $\dot{x}(t) = x^2(t)$ for $t \geq r$. Therefore, suppose $\epsilon > 0$ is given, $C = |1 - r|^{-1}$.

Choose $b \in B$ so that $b(0) = 1$, $\int_{-r}^{0} |b(t)|\, dt < 2C\epsilon$ and let $y(t) = y(t, 0, 1)$, $y(0, 0, 1) = 1$, be the solution of $\dot{y}(t) = y^2(t)$ and $x(t) = x(b)(t)$. If $\psi(t) = y(t) - x(t)$ for $0 < t < r$, then $\psi(t) \geq 0$ and $\dot{\psi}(t) \leq 2C\psi(t) + 2C\epsilon$. Since $\psi(0) = 0$, one thus obtains $\psi(r) \leq \epsilon$. This shows that

$$x(r) = y(r) - \psi(r) = (1-r)^{-1} - \psi(r)$$
$$\geq (1-r)^{-1} - \epsilon$$

and proves the general assertion made earlier. □

As another illustration of the infinite dimensionality of an RFDE, consider the control problem

$$\dot{x}(t) = Ax(t-r) + Bu \tag{1.2}$$

where A and B are constant matrices, $r > 0$, $x \in \mathbb{R}^n$, $u \in \mathbb{R}^p$, $|u| \leq 1$, and $u = u(t)$ is a locally integrable function. Suppose $\phi \in C$, $t \geq 0$ are given and let $x(\phi, u)$ designate the solution of Equation (1.2) with $x_0(\phi, u) = \phi$. Suppose

$$\mathcal{A}(t,\phi) = \{\psi \in C : \text{there is a locally integrable } u,\ |u| \leq 1, \text{ with } x_t(\phi, u) = \psi\}.$$

The set $\mathcal{A}(t,\phi)$ is the set attainable at time t along solutions of Equation (1.2) using the controls u and starting at $t = 0$ with ϕ. For ordinary differential equations ($r = 0$), it is known that every element of the attainable set at time t can also be reached by using only the bang-bang controls in Equation (1.2); that is, by only using control function $u(\tau)$ with $|u(\tau)| = 1$ for $0 \leq \tau \leq t$.

Property 1.4. *Bang-bang controls are not always possible for* RFDE.

Proof. The following counterexample demonstrates this property. Suppose $\phi = 0$ and consider

$$\dot{x}(t) = x(t-1) + u(t), \qquad |u| \leq 1. \tag{1.3}$$

Then

$$x(0,u)(t) = \int_0^t u(s)\, ds \qquad \text{for } 0 \leq t \leq 1$$

and $\mathcal{A}(1,0)$ contains zero since the control $u(t) = 0$, $0 \leq t \leq 1$, gives $x_1(0,u) = 0$. On the other hand, there is no way to reach zero with a bang-bang control. □

3.2 Equivalence classes of solutions

In the previous chapter, Equation (5.13) of Section 2.5, we gave an example in which the mapping $T(t,\sigma)$ was not one-to-one. To reemphasize this remark and to discuss some more geometry of the solutions, we state this result explicitly and give another example.

Property 2.1. *The map $T(t, \sigma)$ may not be one-to-one.*

Proof. Consider the equation

$$\dot{x}(t) = -x(t-r)[1 - x^2(t)]. \tag{2.1}$$

Equation (2.1) has the solution $x(t) = 1$ for all t in $(-\infty, \infty)$. Furthermore, if $r = 1$, $\sigma = 0$, and $\phi \in C$, then there is a unique solution $x(0, \phi)$ of Equation (2.1) through $(0, \phi)$ that depends continuously on ϕ. If $-1 \leq \phi(0) \leq 1$, these solutions are actually defined on $[-1, \infty)$. On the other hand, if $\phi \in C$, $\phi(0) = 1$, then $x(0, \phi)(t) = 1$ for all $t \geq 0$. Therefore, for all such initial values, $x_t(0, \phi)$, $t \geq 1$, is the constant function 1. A translation of a subspace of C of codimension one is mapped into a point by $T(t, 0)$ for all $t \geq 1$. □

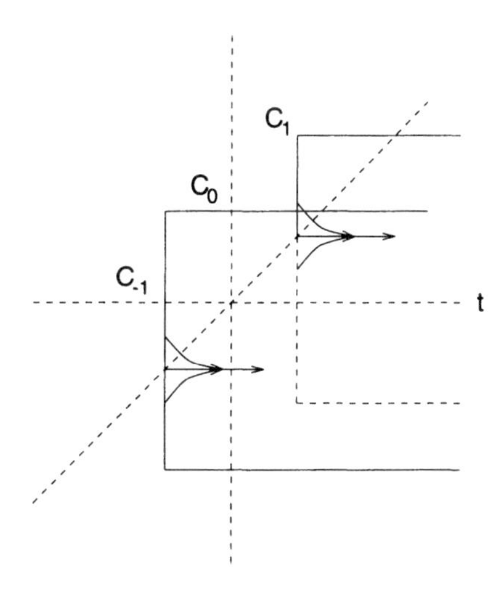

Fig. 3.1.

The function $x(t) = -1$ is also a solution of Equation (2.1) and for any $\phi \in C$, $\phi(0) = -1$, the solution $x(0, \phi)(t)$ is -1 for $t \geq 0$. Therefore, $x_t(0, \phi)$ is the constant function -1 for $t \geq 1$.

For this example, it is interesting to try to depict the trajectories in $\mathbb{R} \times C$. For any constant a let $C_a = \{\phi \in C : \phi(0) = a\}$. The set C_a is the translate of a subspace of C of codimension 1 (a hyperplane) and $\mathbb{R} \times C$ can be represented schematically as in the accompanying diagram. We have put on this diagram the sets $\mathbb{R} \times C_1$ and $\mathbb{R} \times C_{-1}$ as well as the constant functions 1 and -1 and representative trajectories in these planes. Notice that solutions are trapped between these planes if the initial values ϕ satisfy $-1 \leq \phi(0) \leq 1$. Also notice that any solution that oscillates about zero must have a trajectory that crosses the set $\mathbb{R} \times C_0$.

That the solution map $T(t, \sigma)$ may not be one-to-one is an annoying feature of the theory of RFDE. Sufficient conditions for one-to-oneness were given in Corollary 5.1 of Section 2.5. We also posed the general question of whether or not there is a residual set in the set of all RFDE for which the map $T(t, \phi)$ is one-to-one. Even if the answer to this question is affirmative, it does not necessarily take care of applications. It may be that the form of the equation is fixed and one is not allowed to change it too much. In this situation, a better understanding of one-to-oneness is needed.

One way to begin to understand why the map $T(t, \sigma)$ is not one-to-one is to define and study equivalence classes of initial data in the following manner. Suppose $\Omega = \mathbb{R} \times C$ and all solutions $x(\sigma, \phi)$ of the RFDE(f) are defined on $[\sigma - r, \infty)$. We say $(\sigma, \phi) \in \mathbb{R} \times C$ is *equivalent* to $(\sigma, \psi) \in \mathbb{R} \times C$, $(\sigma, \phi) \sim (\sigma, \psi)$, if there is a $\tau \geq \sigma$ such that $x_\tau(\sigma, \phi) = x_\tau(\sigma, \psi)$; that is (σ, ϕ) is equivalent to (σ, ψ) if the trajectories through (σ, ϕ) and (σ, ψ) have a point in common. It is easy to see that $\sim$ is an equivalence relation and the space C is decomposed into equivalence classes $\{V_\alpha\}$ for each fixed σ. If $T(t, \sigma)$ is one-to-one, then each equivalence class consists of a single point; namely, the initial value (σ, ϕ). For each equivalence class V_α, choose a representation element $\phi^{\sigma,\alpha}$ and let

$$W(\sigma) = \bigcup_\alpha \phi^{\sigma,\alpha}. \tag{2.2}$$

From the point of view of the qualitative theory of functional differential equations, the set $W(\sigma)$ is very interesting since it is a maximal set on which the map $T(t, \sigma)$ is one-to-one. However, it seems to be very difficult to say much about the properties of $W(\sigma)$. In fact, without some more precise description of the manner in which $\phi^{\sigma,\alpha}$ is chosen from V_α, one cannot hope to discuss such topological properties of $W(\sigma)$ as connectedness. For example, consider the scalar equation

$$\dot{x}(t) = 0$$

considered as a functional differential equation on C. If $C_a = \{ f \in C : \phi(0) = a\}$, then $\phi \in C_a$ implies $x_t(\sigma, \phi)$ is the constant function a for $t \geq \sigma + r$. Therefore, for each σ, the equivalence classes V_α are the sets C_α, $-\infty < \alpha < \infty$. An arbitrary choice of $\phi^{\sigma,\alpha}$ leads to a very uninteresting set $W(\sigma)$. On the other hand, $W(\sigma)$ consisting of all constant functions is certainly the set that is of interest for the equation.

In a general situation, we know nothing about the "appropriate" choice of $\phi^{\sigma,\alpha}$. On the other hand, in the example discussed in some detail in Property 2.1, the equivalence classes are also very easy to determine. In fact, for any $\phi \in C$, $\phi \notin C_1$, $\phi \notin C_{-1}$, it follows from Corollary 5.1 of Section 2.5 that the equivalence class corresponding to ϕ consists only of the single element ϕ. On the other hand, if $\phi \in C_1$, then the equivalence class corresponding to ϕ is C_1. Similarly, C_{-1} is the equivalence class corresponding to $\phi \in C_{-1}$. A good choice for $W(0)$ in Equation (2.1) in this case would be $C\backslash\{(C_1\backslash\{1\}) \cup (C_{-1}\backslash\{-1\})\}$ where 1 and -1 are constant functions.

We say that an *equivalence class V_α is determined in a finite time* if there exists $\tau > 0$ such that for any $\phi, \psi \in V_\alpha$, $x_{\sigma+t}(\sigma,\phi) = x_{\sigma+t}(\sigma,\psi)$ for $t \geq \tau$. We now give some rather surprising results about this concept.

Property 2.2. *The equivalence classes may not be determined in finite time.*

Proof. Consider the equation

$$\dot{x}(t) = \beta[|x_t| - x(t)]. \tag{2.3}$$

For any $\beta > 0$, we show the equivalence classes for Equation (2.3) are in one-to-one correspondence with the constant functions. Also, we show there is a $\beta > 0$ such that the equivalence classes are not determined in finite time.

For a given ϕ in $C = C([-1,0],\mathbb{R})$, there is a unique solution $x = x(\phi,\beta)(t)$ of this equation through $(0,\phi)$ that is continuous in (ϕ,β,t).

If $\phi(0) \geq 0$, $\phi \neq 0$, then $x(\phi,\beta)(t)$ is a positive constant for $t \geq 1$. In fact, since $\dot{x}(t) \geq 0$, it follows that $|x_t| = x(t)$ for $t \geq 1$ and uniqueness implies $x(t)$ is a constant $\geq \phi(0)$ for $t \geq 1$. Also, if $\phi(0) = 0$, then $\phi \neq 0$ implies $\dot{x}(0) > 0$ and $x(t) > 0$ for $t \geq 1$. Therefore, for any positive constant function, the corresponding equivalence class contains more than one element. Also, the preceding argument and the autonomous nature of the equation show that the equivalence class corresponding to the constant function zero contains only zero.

If $\phi(0) < 0$, then it is clear that $x(\phi,\beta)(t)$ approaches a constant as $t \to \infty$. If $x(\phi,\beta)(t)$, $\phi(0) < 0$, has a zero $z(\phi,\beta)$, it must be simple, and therefore, $z(\phi,\beta)$ is continuous in ϕ,β. For any $\beta > 0$, there is a $\phi \in C$, $\phi(0) < 0$, such that $z(\phi,\beta)$ exists. In fact, let $\phi \in C$, $\phi(0) = -1$, $\phi(\theta) = -\gamma$, $\gamma > 1$, $-1 \leq \theta \leq -\frac{1}{2}$ and let $\phi(\theta)$ be a monotone increasing function for $-\frac{1}{2} \leq \theta \leq 0$. As long as $x(t) \leq 0$ and $0 < t < \frac{1}{2}$, we have $|x_t| = \gamma$ and

$$\dot{x}(t) = \beta[\gamma - x(t)] \geq \beta\gamma.$$

Therefore, $x(t) \geq \beta\gamma t - 1$ if $x(t) \leq 0$ and $0 \leq t \leq \frac{1}{2}$. For $\beta\gamma/2 > 1$, it follows that x must have a zero $z(\phi,\beta) < \frac{1}{2}$.

The closed subset $C_{-1} = \{\phi \in C : \phi(0) = -1\}$ can be written as $C_{-1} = C_{-1^o} \cup C_{-1^n}$ where $C_{-1^o} = \{\phi \in C_{-1} : z(\phi,\beta) \text{ exists}\}$ and $C_{-1^n} = \{\phi \in C_{-1} : z(\phi,\beta) \text{ does not exist}\}$. Since $z(\phi,\beta)$ is continuous, the set C_{-1°} is open and, therefore, C_{-1^n} is closed. If C_{-1^n} is not empty, then there is a sequence $\phi_j \in C_{-1^o}$, $\phi_j \to \phi \in C_{-1^n}$ as $f \to \infty$ and $z(\phi_j,\beta) \to \infty$.

There is a $\beta_0 > 0$ such that C_{-1^n} is not empty. In fact, choose $\beta_0 > 0$ less than or equal to that value β for which the equation $\lambda + \beta = -\beta e^{-\lambda}$ has a real root λ_0 of multiplicity two. For this β_0, the equation $\lambda + \beta = -\beta e^{-\lambda}$

has two real negative roots. If $-\lambda_0$ is one of these roots, then $x(t) = -e^{-\lambda_0 t}$ is a solution of the Equation (2.3) with initial value $\phi_0(\theta) = -e^{-\lambda_0\theta}$, $-1 \leq \theta \leq 0$, $\phi_0 \in C_{-1}$. Therefore, C_{-1^n} is not empty. It follows that

$$\delta(\beta_0) \stackrel{\text{def}}{=} \sup\{z(\phi, \beta_0) : \phi \in C_{-1^0}\} = \infty.$$

Since the original equation is positive homogeneous of degree 1 in x, it follows that, for any positive constants a and t_0, there exists $\phi \in C$, such that $x(\phi, \beta_0)(t) = a$, $t \geq t_0$, and $x(\phi, \beta_0)(t) < a$ for $0 \leq t < t_0$. This proves the assertion. □

3.3 Small solutions for linear equations

A *small solution* x is a solution such that

$$\lim_{t\to\infty} e^{kt}x(t) = 0 \quad \text{for all} \quad k \in \mathbb{R}. \tag{3.1}$$

In this section we study the existence of small solutions of linear autonomous RFDE(L)

$$\begin{cases} \dot{x}(t) = \displaystyle\int_{-r}^{0} d[\eta(\theta)]x(t+\theta) \\ x_0 = \phi. \end{cases} \tag{3.2}$$

The zero solution is always a small solution, and the question becomes whether there are initial conditions $\phi \neq 0$ such that the solution $x(\,\cdot\,, \phi)$ to System (3.2) is a small solution. Such solutions will be called *nontrivial* small solutions. It is easy to see that nontrivial small solutions can exist. For example, consider the system

$$\begin{aligned} \dot{x}_1(t) &= x_2(t-1) \\ \dot{x}_2(t) &= x_1(t). \end{aligned} \tag{3.3}$$

Any initial condition $\phi = (\phi_1, \phi_2)^T$ with $\phi_1(0) = 0$ and $\phi_2 = 0$ yields a small solution.

In this section we present necessary and sufficient conditions for the existence of nontrivial small solutions. The proof is based on a fine analysis of the Laplace transform of System (3.2) and for this we need the notion of the exponential type of an entire function.

An entire function $h : \mathbb{C} \to \mathbb{C}$ is of *order* 1 if and only if

$$\limsup_{r\to\infty} \frac{\log\log M(r)}{\log r} = 1$$

where

$$M(r) = \max_{0 \le \theta \le 2\pi} \{|h(re^{i\theta})|\}.$$

An entire function of order 1 is of *exponential type* if and only if

$$\limsup_{r \to \infty} \frac{\log M(r)}{r} = \mathrm{E}(h)$$

where $0 \le \mathrm{E}(h) < \infty$. In that case, $\mathrm{E}(h)$ is called the exponential type of h. A vector-valued function $h = (h_1, \dots, h_n) : \mathbb{C} \to \mathbb{C}^n$ will be called an entire function of exponential type if and only if the components h_j of h are entire functions of order 1 that are of exponential type. In this case, the exponential type will be defined by

$$\mathrm{E}(h) = \max_{1 \le j \le n} \mathrm{E}(h_j).$$

The following lemma is a special case of the Paley-Wiener theorem, theorem 6.9.1 of Boas [1].

Lemma 3.1. *Let $h : \mathbb{C} \to \mathbb{C}$ be an entire function. If h is uniformly bounded in the closed right half plane $\operatorname{Re} z \ge 0$, then h is of exponential type τ and L^2-integrable along the imaginary axis if and only if*

$$h(z) = \int_0^\tau e^{-zt} \phi(t)\, dt$$

where $\phi \in L^2[0, \tau]$ and ϕ does not vanish a.e. in any neighborhood of τ.

Next we collect some consequences of this lemma.

Lemma 3.2. *For $j = 1, 2$, let α_j be a function of bounded variation on the interval $[0, a_j]$, normalized so that α_j is continuous from the left on $(0, a_j)$ and $\alpha_j(0) = 0$. If α_1 and α_2 are not constant in neighborhoods of a_1 and a_2, respectively, then*

$$\mathrm{E}\Big(\int_0^{a_j} e^{-zt}\, d\alpha_j(t)\Big) = a_j, \quad j = 1, 2 \tag{3.4}$$

and

$$\mathrm{E}\Big(\int_0^{a_1} e^{-zt}\, d\alpha_1(t) \int_0^{a_2} e^{-zt}\, d\alpha_2(t)\Big) = a_1 + a_2. \tag{3.5}$$

Proof. Using partial integration,

$$\int_0^{a_j} e^{-zt}\, d\alpha_j(t) = z \int_0^{a_j} e^{-zt} (\alpha_j(t) - \alpha_j(a_j))\, dt, \quad j = 1, 2$$

and Relation (3.4) follows from Lemma 3.1. From the convolution property, we find

$$\int_0^{a_1} e^{-zt}\, d\alpha_1(t) \int_0^{a_2} e^{-zt}\, d\alpha_2(t) = \int_0^{a_1+a_2} e^{-zt} d\alpha_3(t)$$

where α_3 is defined to be the convolution of α_1 and α_2

$$\alpha_3(t) = \int_0^t d[\alpha_2(\theta)]\alpha_1(t-\theta).$$

Therefore, Relation (3.5) follows from Lemma 3.1 as well. □

In the sequel we call entire functions of the form $l_j(z) = \int_0^\tau e^{-zt} d\alpha(t)$, where $\tau \geq 0$ and α is a function of bounded variation, *finite Laplace-Stieltjes transforms*. From Lemma 3.2, we find the following useful representation for the determinant of the characteristic matrix $\Delta(z)$ for Equation (3.2)

$$\det \Delta(z) = z^n - \sum_{j=1}^{n} l_j(z) z^{n-j}$$

where, for $j = 1, \ldots, n$, the coefficient l_j is a finite Laplace-Stieltjes transform of exponential type at most jr. So $\det \Delta(z)$ is an entire function of exponential type with

$$\mathrm{E}\big(\det \Delta(z)\big) \leq nr.$$

Let $\operatorname{adj} \Delta(z)$ denote the matrix of cofactors of $\Delta(z)$. Since the cofactors $C_{ij}(z)$ are $(n-1)$ by $(n-1)$-subdeterminants of $\Delta(z)$, it follows that

$$C_{ij}(z) = \sum_{j=1}^{n-1} l_j(z) z^{n-j}$$

where, for $j = 1, \ldots, n-1$, the coefficient l_j is a finite Laplace-Stieltjes transform of exponential type at most jr. So the exponential type of the cofactors is less than or equal to $(n-1)r$. After these preliminary results, we return to Equation (3.2). The following result is the first ingredient in the characterization of the small solutions of Equation (3.2).

Theorem 3.1. *Suppose that $x(\,\cdot\,;\phi)$ is a small solution of Equation* (3.2), *then $x(t;\phi) = 0$ for $t \geq nr - E(\det \Delta(\lambda))$.*

Proof. The proof is based on the observation that the Laplace transform of a small solution is an entire function which is L^2-integrable along the imaginary axis. First we compute the Laplace transform of the derivative of the solution of Equation (3.2):

$$\begin{aligned}
\mathcal{L}(\dot{x})(z) &= \int_0^\infty e^{-zs}\dot{x}(s)\,ds \\
&= \int_0^\infty e^{-zs}\int_{-r}^0 d[\eta(\theta)]x(s+\theta)\,ds \\
&= \int_{-r}^0 d[\eta(\theta)]\int_0^\infty e^{-zs}x(s+\theta)\,ds \\
&= \int_{-r}^0 e^{z\theta}d[\eta(\theta)][-\int_{-r}^\theta e^{-z\tau}\phi(\tau)\,d\tau + \int_{-r}^\infty e^{-zs}x(s)\,ds] \\
&= \int_{-r}^0 e^{z\theta}d[\eta(\theta)]\int_{-r}^\theta e^{-z\tau}\phi(\tau)\,d\tau + \int_{-r}^0 e^{z\theta}d[\eta(\theta)]\,\widehat{x}(z).
\end{aligned}$$

Therefore, the Laplace transform of the solution of Equation (3.2) becomes

$$\begin{aligned}
\widehat{x}(z) &= \int_0^\infty e^{-zt}x(t-r)\,dt = e^{-zr}\int_{-r}^\infty e^{-zs}x(s)\,ds \\
&= \frac{e^{-zr}}{z}\Big(\phi(0) + \int_{-r}^0 e^{z\theta}d\eta(\theta)\int_{-r}^\theta e^{-z\tau}\phi(\tau)\,d\tau + z\int_{-r}^0 e^{-zs}\phi(s)\,ds \\
&\qquad + \int_{-r}^0 e^{z\theta}d\eta(\theta)\widehat{x}(z)\Big).
\end{aligned}$$

So, $\widehat{x}$ is explicitly given by the equation

$$\begin{aligned}
\widehat{x}(z) = \Delta(z)^{-1}\Big[e^{-zr}\phi(0) + z\int_{-r}^0 e^{-z(r+s)}\phi(s)\,ds \\
- \int_{-r}^0\int_s^0 d[\eta(\theta)]e^{-z(r+s-\theta)}\phi(s)\,ds\Big].
\end{aligned} \tag{3.6}$$

Since x decays faster than any exponential, the Laplace transform of x converges for every z in $\mathbb{C}$. Further, the Plancherel theorem implies that $\widehat{x}$ is L^2-integrable along the imaginary axis. So $\widehat{x}$ is an entire function that is L^2-integrable along the imaginary axis. From the explicit representation (3.6) for $\widehat{x}$, we next compute the exponential type of $\widehat{x}$. From this representation of the right-hand side of Equation (3.6), it follows that the exponential type of $\widehat{x}$ is at most $nr - \mathrm{E}(\det \Delta(z))$. Thus, the Paley-Wiener theorem (Lemma 3.1) implies the result. □

In contrast to the general nonlinear case, the following result is true for linear systems.

Corollary 3.1. *For linear autonomous systems of* RFDE, *the equivalence classes are determined in a finite time.*

Proof. If ϕ and ψ belong to the same equivalence class of a linear system, then the solution x corresponding to $\phi - \psi$ is a solution and must be identically zero after some finite time. Therefore, this solution approaches zero faster than any exponential and, thus, is a small solution of (3.2). Theorem 3.1 implies that any small solution is identically zero for $t \geq nr - E(\det \Delta(z))$. Therefore, each equivalence class is determined in a finite time and the assertion is proved. □

Observe that in order to study the nontrivial small solutions of Equation (3.2), it suffices to analyze the null space of the solution operator.

Define the *ascent* α of a semigroup $T(t)$ by the value

$$\alpha = \inf \{ t : \text{ for all } \epsilon > 0 : \mathcal{N}(T(t)) = \mathcal{N}(T(t+\epsilon)) \}. \tag{3.7}$$

We shall prove an explicit expression for the ascent of the solution operator for Equation (3.2) solely in terms of the characteristic matrix $\Delta(z)$. For this, we have to introduce the following two numbers: Define ϵ and σ by

$$\mathrm{E}(\det \Delta(z)) = nr - \epsilon, \qquad \max_{1 \leq i,j \leq n} \mathrm{E}(C_{ij}(z)) = (n-1)r - \sigma. \tag{3.8}$$

Theorem 3.2. *The ascent α of the solution operator $T(t)$ associated with Equation* (3.2) *is finite and is given by*

$$\alpha = \epsilon - \sigma. \tag{3.9}$$

Before we sketch a proof of Theorem 3.2, we would like to state the following important consequence.

Theorem 3.3. *The solution operator $T(t)$ associated with Equation* (3.2) *is one-to-one if and only if* $\mathrm{E}(\det \Delta(z)) = nr$.

The following example will illustrate the results.

Example 3.1. Consider the following system of differential difference equations

$$\begin{aligned} \dot{x}_1(t) &= -x_2(t) + x_3(t-1) \\ \dot{x}_2(t) &= x_1(t-1) + x_3(t - \frac{1}{2}) \\ \dot{x}_3(t) &= -x_3(t). \end{aligned} \tag{3.10}$$

The characteristic matrix is given by

$$\Delta(z) = \begin{pmatrix} z & 1 & -e^{-z} \\ -e^{-z} & z & -e^{-\frac{1}{2}z} \\ 0 & 0 & z+1 \end{pmatrix}$$

with determinant

$$\det \Delta(z) = (z+1)(z^2 + e^{-z}).$$

So, $\epsilon = 2$. The cofactor

$$C_{23}(z) = -\begin{vmatrix} z & -e^{-z} \\ -e^{-z} & -e^{-\frac{1}{2}z} \end{vmatrix}$$
$$= ze^{-\frac{1}{2}z} + e^{-2z}$$

has exponential type 2, and hence $\sigma = 0$. Therefore, from Theorem 3.2, the ascent of the System (3.10) equals two.

Sketch of proof of Theorem 3.2. For $t \geq 0$, the solution of Equation (3.2) satisfies the integral equation

$$x(t) = \int_0^t \eta(s-t)x(s)\,ds + F\phi(t)$$

where

$$F\phi(t) = \phi(0) + \int_0^t \int_{-r}^s d[\eta(\theta)]\phi(s+\theta)\,ds.$$

Let $\mathcal{F}$ be the Banach space of continuous functions on the half-line $[0,\infty)$ that are constant on $[r,\infty)$, provided with the supremum norm. The mapping F, considered as $F : C \to \mathcal{F}$ is not onto and it is not easy to characterize the range of F. Therefore, we first consider the auxiliary problem to analyze the small solutions of the integral equation

$$y(t) = \int_0^t \eta(s-t)y(s)\,ds + f(t) \tag{3.11}$$

where f belongs to $\mathcal{F}$. The Laplace transform of y satisfies the equation

$$\int_0^\infty e^{-zt}y(t)\,dt = \Delta(z)^{-1}\big[f(r) + z\int_0^h e^{-zt}\big(f(t) - f(r)\big)\,dt\big]. \tag{3.12}$$

Suppose that the Laplace transform of y is a finite Laplace transform given by $\int_0^\alpha e^{-zt}y(t)\,dt$. This implies that the right-hand side of (3.12) is an entire function. The exponential type of the right-hand side of (3.12) is easily computed to be less than or equal to $\epsilon - \sigma$. So

$$\mathrm{E}(\int_0^\alpha e^{-zt}y(t)\,dt) \leq \epsilon - \sigma.$$

Therefore, if y is a solution of (3.11) that is identically zero after finite time, then $y(t) = 0$ for $t \geq \epsilon - \sigma$. On the other hand, in Verduyn Lunel [1], we constructed a function f in $\mathcal{F}$ such that the right-hand side of (3.12) is a finite Laplace transform of exponential type precisely $\epsilon - \sigma$. So for this function f, we have a solution y of Equation (3.11) such that $y(t) \neq 0$ for t in a neighborhood of $\epsilon - \sigma$. To complete the proof, we claim a one-to-one correspondence between solutions of Equation (3.2) and the integral equation (3.11). For f in $\mathcal{F}$, the solution $y(\,\cdot\,; f)$ of Equation (3.11) is continuous on the interval $[0, r]$ and since f is constant on $[r, \infty)$, $y(\,\cdot\,; f)$ satisfies the differential equation

$$\dot{y}(t) = \int_{-r}^{0} d\eta(\theta) y(t+\theta) \quad \text{on} \quad [r, \infty).$$

So, if we define $\phi(\theta) = y(r+\theta)$, $-r \le \theta \le 0$, then the solution $x(\,\cdot\,; \phi)$ of Equation (3.2) if given by $x(\,\cdot\,; \phi) = y(r + \,\cdot\,)$. On the other hand, given an initial condition ϕ in C, there exists a solution to Equation (3.11) such that $y = \phi(r + \,\cdot\,)$ on the interval $[-r, 0]$; just define f in $\mathcal{F}$ by

$$f = \phi(r + \,\cdot\,) - \int_{0}^{t} \int_{-r}^{-s} d\eta(\theta)\phi(r+s+\theta)\, ds. \tag{3.13}$$

This proves the correspondence between the solutions of the integral equation (3.11) and Equation (3.2) and completes the proof of Theorem 3.2. □

Proof of Theorem 3.3. By Theorem 3.2, it suffices to prove that for all $\epsilon > 0$, $\sigma < \epsilon$. Suppose $\sigma = \epsilon$. We shall calculate $\mathrm{E}\big(\det \operatorname{adj} \Delta(z)\big)$ in two different ways. Since $\sigma = \epsilon$, we have

$$\begin{aligned} \mathrm{E}\big(\det \operatorname{adj} \Delta(z)\big) &\le n\big((n-1)r - \epsilon\big) \\ &= (n-1)(nr - \epsilon) - \epsilon. \end{aligned}$$

On the other hand, by Lemma 3.2, we have

$$\begin{aligned} \mathrm{E}\big(\det \operatorname{adj} \Delta(z)\big) &= \mathrm{E}\big((\det \Delta(z))^{n-1}\big) \\ &= (n-1)(nr - \epsilon). \end{aligned}$$

This implies that

$$(n-1)(nr-\epsilon) \le (n-1)(nr-\epsilon) - \epsilon,$$

which is a contradiction if $\epsilon > 0$. □

In Chapter 1, Section 5, we have seen that the existence of nontrivial small solutions is closely related to the question whether a solution of Equation (3.2) has a convergent series expansion in characteristic solutions $c_j(t)e^{\lambda_j t}$, where p_j is a polynomial and λ_j a root of the characteristic equation

$$\det \Delta(z) = 0. \tag{3.14}$$

The procedure to obtain a series expansion was through the Laplace transform. From Representation (3.6) for the Laplace transform of a solution of Equation (3.2) one finds for $t > 0$,

$$x(t;\phi) = \lim_{\omega\to\infty} \frac{1}{2\pi i} \int_{\gamma-\omega i}^{\gamma+\omega i} e^{zt} H(z,\phi)\, dz \tag{3.15}$$

where

$$\begin{aligned} H(z,\phi) = \Delta(z)^{-1}\big[e^{-zr}\phi(0) + z \int_{-r}^{0} e^{-z(r+s)}\phi(s)\, ds \\ - \int_{-r}^{0}\int_{s}^{0} d[\eta(\theta)] e^{-z(r+s-\theta)}\phi(s)\, ds\big]. \end{aligned} \tag{3.16}$$

By Cauchy's residue theorem, we deduce that

$$\begin{aligned} x(t;\phi) &= \sum_{j=1}^{m} \operatorname*{Res}_{z=\lambda_j} e^{zt} H(z,\phi) + x_{\alpha_m}(t) \\ &= \sum_{j=1}^{m} p_j(t) e^{\lambda_j t} + x_{\alpha_m}(t) \end{aligned} \tag{3.17}$$

where the summation is over the roots of the characteristic equation (3.14) with real part larger than α_m. For the remainder, we have the following integral representation:

$$\begin{aligned} x_\alpha(t) = \lim_{\omega\to\infty} \frac{1}{2\pi i} \int_{\alpha-\omega i}^{\alpha+\omega i} e^{zt} \Delta(z)^{-1}\big[e^{-zr}\phi(0) + z \int_{-r}^{0} e^{-z(r+s)}\phi(s)\, ds \\ - \int_{-r}^{0}\int_{s}^{0} d[\eta(\theta)] e^{-z(r+s-\theta)}\phi(s)\, ds\big]\, dz. \end{aligned}$$

In general, one cannot expect that the series expansion converges to the solution. In particular, small solutions have an entire Laplace transform and hence a series expansion in which all terms are zero. However, if the series expansion converges, we can decompose the solution as a convergent series of characteristic solutions and a small solution. At this point, we encounter two problems. Firstly, when does the series expansion actually converge to the solution, and secondly, can we identify the set of initial values that correspond to a solution with a convergent series expansion. The first question can be answered similarly to the presentation in Chapter 1, Section 1. The second question boils down to the analysis of the closure of the span of the system of eigenvectors and generalized eigenvectors of an unbounded operator. We return to these questions in Chapter 7. In the remaining part of this section, we present the main results and illustrate them with several examples.

To get a better feeling for these questions, let us consider, in detail, the equation

$$\dot{x}(t) = Ax(t-1), \qquad t \ge 0,$$

$$A = \begin{bmatrix} 0 & 1 & 0 \\ 0 & 0 & 1 \\ 0 & 0 & 0 \end{bmatrix}. \tag{3.18}$$

The determinant of the characteristic matrix is a polynomial det $\Delta(z) = z^3$. So the series expansion in characteristic solutions is always finite. As before, let $\widehat{x}(z) = \int_0^\infty e^{-zt} x(t-1)\,dt$. An easy computation yields the following representation for the Laplace transform of a solution

$$\begin{aligned} \widehat{x}_1(z) &= \frac{e^{-z}}{z}\left[\phi_1(0) + \int_{-1}^0 e^{-zs}\phi_1(s)\,ds + \widehat{x}_2(z)\right] \\ \widehat{x}_2(z) &= \frac{e^{-z}}{z}\left[\phi_2(0) + \int_{-1}^0 e^{-zs}\phi_2(s)\,ds + \widehat{x}_3(z)\right] \\ \widehat{x}_3(z) &= \frac{e^{-z}}{z}\left[\phi_3(0) + \int_{-1}^0 e^{-zs}\phi_3(s)\,ds\right]. \end{aligned} \tag{3.19}$$

System (3.19) contains all of the information. First we compute the characteristic solutions at $\lambda = 0$ by computing the residue at $z = 0$. This yields $c_3(t) = c$, $c_2(t) = b + ct$ and $c_1(t) = a + bt + ct^2$, where a, b, and c are arbitrary constants. Define $\mathcal{M}$ to be the subspace spanned by the initial conditions corresponding to characteristic solutions

$$\begin{aligned} \mathcal{M} = \Big\{\phi = (\phi_1, \phi_2, \phi_3) \in C : \phi_3(\theta) = c,\ \phi_2(\theta) = b + c\theta, \\ \phi_1(\theta) = a + b\theta + c\frac{\theta^2}{2}\Big\}. \end{aligned} \tag{3.20}$$

So $\mathcal{M}$ is a subspace of dimension three that is invariant under the solution operator; that is, $T(t)\mathcal{M} \subseteq \mathcal{M}$ for all $t \geq 0$. In fact, if $\phi \in \mathcal{M}$, then $x_3(t) = c$ for $-1 \leq t \leq 0$, and so $x_2(t) = b + \int_0^t c ds$ for $t \geq 0$. Since $x_2(\theta) = b + c\theta$, $-1 \leq \theta \leq 0$, it follows that $x_2(t) = b + ct$ for $t \geq -1$. In the same way, one shows that $x_1(t) = a + bt + (ct^2/2)$ for $t \geq -1$. This proves $x_t(\phi) \in \mathcal{M}$ for all $t \geq 0$ if $\phi \in \mathcal{M}$.

If x is a solution of Equation (3.18) with initial value ϕ that approaches zero faster than any exponential, then $\phi_3(0) = 0$ and $x_3(t) = 0$ for $t \geq 0$. Therefore,

$$x_2(t) = \phi_2(0) + \int_0^t \phi_3(s-1)\,ds, \qquad 0 \leq t \leq 1$$

$$x_2(t) = \phi_2(0) + \int_0^1 \phi_3(s-1)\,ds, \qquad t \geq 1,$$

and one must have $\phi_2(0) = -\int_{-1}^0 \phi_3(\theta)\,d\theta$. In the same way, one must have

$$\phi_1(0) = \int_{-1}^0 [\phi_3(\theta) - \phi_2(\theta)]\,d\theta - \int_{-1}^0\int_{-1}^\theta \phi_3(s)\,ds\,d\theta. \tag{3.21}$$

Therefore, if x is a small solution of Equation (3.18), then the initial value ϕ of x must belong to the set $\mathcal{S}$ defined by

$$
\begin{aligned}
\mathcal{S} = \Big\{ \phi = (\phi_1, \phi_2, \phi_3) \in C : \phi_3(0) = 0,\ \phi_2(0) = -\int_{-1}^{0} \phi_3(\theta)\, d\theta, \\
\phi_1(0) \text{ satisfies Equation (3.21)} \Big\}.
\end{aligned}
\tag{3.22}
$$

The closed subspace $\mathcal{S}$ has codimension three. By Theorem 3.2 $\mathcal{S} = \mathcal{N}(T(3))$, which can be verified directly by choosing ϕ_1 arbitrary, $\phi_1(0) = -3/\pi, \phi_2(\theta) = 2/\pi$, $-1 \le \theta \le 0$, and $\phi_3(\theta) = \sin \pi\theta$, $-1 \le \theta \le 0$. Observe that $\mathcal{M} \cap \mathcal{S} = \{0\}$ and we can decompose C as the direct sum of two closed invariant subspaces

$$
C = \mathcal{S} \oplus \mathcal{M}. \tag{3.23}
$$

For any initial condition ϕ not belonging to $\mathcal{M}$, it follows from System (3.18) that the solution $x(\,\cdot\,; \phi)$ has a component that is a small solution.

What is the form of the original equation (3.18) on the invariant subspace $\mathcal{M}$? Let us first observe that one can extend the definition of $T(t)$ for negative values of t on $\mathcal{M}$ in such a way that $T(t)\mathcal{M} \subseteq U$, $t \in (-\infty, \infty)$. To accomplish this, one simply defines the polynomial functions (x_1, x_2, x_3) as earlier and verifies that Equation (3.18) is satisfied. Choose as a basis for $\mathcal{M}$ the functions $\Phi_1(\theta) = 1$, $\Phi_2(\theta) = \theta$, and $\Phi_3(\theta) = \theta^2/2$, $-1 \le \theta \le 0$. Since $T(t)\mathcal{M} \subseteq \mathcal{M}$, let

$$
T(t)\phi = \Phi_1 y_1(t) + \Phi_2 y_2(t) + \Phi_3 y_3(t) \tag{3.24}
$$

where $y = (y_1, y_2, y_3)$ is a vector in $\mathbb{R}^3$ that depends on t and the initial function ϕ. If $\phi \in \mathcal{M}$ is given as in Expression (3.20), then one observes that $y(0) = (a, b, c)$. Also, a direct calculation shows that y satisfies the ordinary differential equation

$$
\dot{y} = Ay \tag{3.25}
$$

where A is the same matrix as in Equation (3.18). Since A is in Jordan canonical form, it is certainly natural to say that the canonical form for Equation (3.18) is the equation itself.

The small solutions can also be found from System (3.19). Recall that the Laplace transform of a small solution is an entire function. Therefore, an initial value ϕ corresponds to a small solution if and only if the right-hand side of System (3.19) is entire.

This decomposition could be computed directly without invoking the Laplace transform, but we have chosen to illustrate the use of the Laplace transform since this approach also can be used to analyze more complicated examples when the characteristic equation does not reduce to a polynomial. In that case, the space $\mathcal{M}$ becomes infinite dimensional and the question becomes: Does Decomposition (3.23) still hold with $\mathcal{M}$ replaced by $\overline{\mathcal{M}}$ the closure of $\mathcal{M}$.

First we introduce some more notation. Set $H(z,\phi) \stackrel{\text{def}}{=} \int_0^\infty e^{-zt}x(t;\phi)dt$. Then

$$H(z,\phi) = \frac{P(z,\phi)}{\det \Delta(z)}$$

where

$$\begin{aligned} P(z,\phi) = \operatorname{adj} \Delta(z)\big[e^{-zr}\phi(0) + z\int_{-r}^{0} e^{-z(r+s)}\phi(s)\,ds \\ - \int_{-r}^{0}\int_{s}^{0} d[\eta(\theta)]e^{-z(r+s-\theta)}\phi(s)\,ds\big]. \end{aligned} \tag{3.26}$$

From the representation for the Laplace transform of the solution $x(\,\cdot\,;\phi)$ of Equation (3.2), we find the following characterization of the null space of the solution operator.

Corollary 3.2. $\mathcal{N}\big(T(\alpha)\big) = \{\phi \in C : \Delta(z)^{-1}H(z,\phi) \quad \textit{is entire}\,\}$.

In general, convergence results for the series expansion in characteristic solutions are delicate. However, it turns out that $\overline{\mathcal{M}}$, the closure of the set of initial values such that the solution has a convergent series expansion can be characterized.

Theorem 3.4. $\overline{\mathcal{M}} = \big\{\phi \in C : \mathrm{E}(P(z,\phi)) \le \mathrm{E}(\det \Delta(z))\big\}$.

An important consequence of these characterizations is the fact that

$$\overline{\mathcal{M}} \cap \mathcal{N}\big(T(\alpha)\big) = \{0\}.$$

In fact, $\phi \in \overline{\mathcal{M}} \cap \mathcal{N}\big(T(\alpha)\big)$ implies that the Laplace transform $\widehat{x}(z) = \int_0^\infty e^{-zt}x(t-r)\,dt$ of the solution $x = x(\,\cdot\,;\phi)$ is an entire function of zero exponential type, which is L_2-integrable along the imaginary axis. Therefore, the Paley-Wiener theorem (Lemma 3.1) implies that $x(t-r) = 0$ for $t \ge 0$. Hence $\phi = 0$. Another easy consequence of the theorem is that $\overline{\mathcal{M}}$ is invariant under the solution operator.

Furthermore, we have the following "almost" decomposition.

Theorem 3.5. $C = \overline{\overline{\mathcal{M}} \oplus \mathcal{N}\big(T(\alpha)\big)}$.

So the subspace $\overline{\mathcal{M}}$ represents, in a certain sense, the minimal amount of initial data needed to specify a solution of Equation (3.2). If more initial data are specified, then the extra part belongs to $\overline{\mathcal{M}}$ for $t \ge \alpha$.

The proofs of these results are involved and will be discussed in Section 7.8. Here we would like to illustrate the results with an example.

Example 3.2. Consider the following system

$$\begin{aligned} \dot{x}_1(t) &= x_2(t-1) \\ \dot{x}_2(t) &= x_1(t). \end{aligned} \tag{3.27}$$

The characteristic matrix is given by

$$\Delta(z) = \begin{bmatrix} z & -e^{-z} \\ -1 & z \end{bmatrix}, \qquad \det \Delta(z) = z^2 + e^{-z}.$$

So the ascent α of the solution map equals one and it is easy to see that

$$\mathcal{N}\big(T(1)\big) = \big\{(\phi_1, \phi_2) \in C : \phi_1(0) = 0,\ \phi_2 = 0\big\}.$$

Next, we use Theorem 3.4 to compute $\overline{\mathcal{M}}$. Using Equation (3.27), the Laplace transform of the solution satisfies the system of equations

$$\begin{aligned} e^z z\widehat{x}_1(z) &= \phi_1(0) + \widehat{x}_2(z) + z\int_{-1}^{0} e^{-zt}\phi_1(t)\,dt, \\ e^z z\widehat{x}_2(z) &= \phi_2(0) + e^z\widehat{x}_1(z) - \int_{-1}^{0} e^{-zt}\phi_1(t)\,dt + z\int_{-1}^{0} e^{-zt}\phi_2(t)\,dt. \end{aligned}$$

Hence

$$\Delta(z)\begin{bmatrix} \widehat{x}_1(z) \\ \widehat{x}_2(z) \end{bmatrix} = \begin{bmatrix} f_1(z,\phi) \\ f_2(z,\phi) \end{bmatrix}$$

where

$$\begin{aligned} f_1(z,\phi) &= \phi_1(0) + z\int_{-1}^{0} e^{-zt}\phi_1(t)\,dt \\ f_2(z,\phi) &= e^{-z}\phi_2(0) - e^{-z}\int_{-1}^{0} e^{-zt}\phi_1(t)\,dt + ze^{-z}\int_{-1}^{0} e^{-zt}\phi_2(t)\,dt. \end{aligned}$$

Therefore,

$$\begin{bmatrix} \widehat{x}_1(z) \\ \widehat{x}_2(z) \end{bmatrix} = \frac{1}{z^2 + e^{-z}} \begin{bmatrix} zf_1(z,\phi) + e^{-z}f_2(z,\phi) \\ f_1(z,\phi) + zf_2(z,\phi) \end{bmatrix}.$$

The exponential type of both $f_1(z,\phi)$ and $f_2(z,\phi)$ is at most one. By Theorem 3.4

$$\overline{\mathcal{M}} = \big\{\phi \in C : \mathrm{E}(f_2(z,\phi)) = 0\big\}$$

which implies that $\phi_2(t) = \phi_2(0) + \int_0^t \phi_1(s)\,ds$. So

$$\overline{\mathcal{M}} = \big\{\phi \in C : \phi_2(t) = \phi_2(0) + \int_0^t \phi_1(s)\,ds\big\}.$$

Note that indeed $\mathcal{N}\big(T(1)\big) \cap \mathcal{M} = \{0\}$, but that $\mathcal{N}\big(T(1)\big) \oplus \overline{\mathcal{M}}$ is only dense in C.

3.4 Unique backward extensions

In this section, we wish to discuss backward extensions of solutions of an RFDE that exists on $(-\infty, 0]$. The question of interest is: When is this backward extension unique?

Property 4.1. *Backward extension on* $(-\infty, 0]$ *of linear autonomous* RFDE *is unique.*

Proof. Suppose x and y are backward extensions on $(-\infty, 0]$ of $\phi \in C$ for a linear autonomous RFDE. These solutions have unique extensions for $t \geq 0$. If $z = x - y$, then z is a solution of the equation on $(-\infty, \infty)$ with $z_0 = 0$. Fix $\sigma < 0$ and consider the solution z on $[\sigma - nr - r, \infty)$. This is a solution that approaches zero faster than any exponential. Consequently, Theorem 3.1 implies that the solution z_t through $z_{\sigma - nr}$ is equal to zero for $t \geq \sigma - nr + nr = \sigma$. Therefore, $z_\sigma = 0$. Since σ is arbitrary, this implies $z(t) \equiv 0$ for $t \leq 0$ and therefore, the backward extension is unique. □

Property 4.2. *There may exist two distinct backward extensions on* $(-\infty, 0]$ *for an autonomous* RFDE.

Proof. Let $r = 1$, $f(s) = 0$, $0 \leq s \leq 1$, $f(s) = -3(s^{1/3} - 1)^2$, $s > 1$, and consider the equation

$$\dot{x}(t) = f(|x_t|).$$

The function $x \equiv 0$ is a solution of this equation on $(-\infty, \infty)$. Also, the function $x(t) = -t^3$, $t < 0$, $= 0$, $t \geq 0$ is also a solution. In fact, since $x \leq 1$ for $t \geq -1$, it is clear that x satisfies the equation for $t \geq 0$. Since x is monotone decreasing for $t \leq 0$, $|x_t| = x(t-1) = -(t-1)^3$ and $\dot{x}(t) = -3t^2$. It is easy to verify that $-3t^2 = f((1-t)^3)$ for $t < 0$. The proof is complete. □

If $f : C \to \mathbb{R}^n$ has infinitely many bounded derivatives and x is a bounded solution of the RFDE(f) on $(-\infty, 0]$, then x will have infinitely many derivatives. Even in this case, it is possible to give examples for which bounded backward extensions on $(-\infty, 0]$ are not necessarily unique. When can one make a positive statement? One can prove:

Property 4.3. *If* $f : C \to \mathbb{R}^n$ *is analytic, then bounded backward extensions on* $(-\infty, 0]$ *are unique.*

Proof. If one could show that any bounded backward extension x on $(-\infty, 0]$ is an analytic function, then the uniqueness would be immediate. That this is true is a consequence of the following result, which is stated without proof. □

Theorem 4.1. *If* x *is a bounded solution on* $(-\infty, 0]$ *of an analytic* RFDE(f), $f : C \to \mathbb{R}^n$, *then* x *is analytic in a region containing* $(-\infty, 0]$.

3.5 Range in $\mathbb{R}^n$

For ordinary differential equations, the solution map defines a homeomorphism and so the image of a ball under this map contains a ball. For an RFDE, the phenomenon may not be true even if we consider the map $x(\sigma, \cdot)(t) : C \to \mathbb{R}^n$. More precisely, we have

Property 5.1. *For nonautonomous linear equations $x(\sigma, C)(t)$ may be zero dimensional for some $t > \sigma$. In fact, one may have $x(\sigma, C)(t) = 0$ for all t greater than or equal to some τ.*

Proof. Consider the equation

$$\dot{x}(t) = b(t)x\big(t - \frac{3\pi}{2}\big) \tag{5.1}$$

where, for an arbitrary $g \in C([-\infty, 0], \mathbb{R}^n)$, the function b is defined by

$$b(t) = \begin{cases} g(t) & t \le 0, \\ 0 & 0 \le t \le \frac{3\pi}{2}, \\ -\cos t & \frac{3\pi}{2} \le t \le 3\pi, \\ 1 & t \ge 3\pi. \end{cases} \tag{5.2}$$

For $\sigma = 0$ and an arbitrary $\phi \in C([-3\pi/2, 0], \mathbb{R})$, Equation (5.1) has the solution

$$x(0, \phi)(t) = \begin{cases} \phi(0) & 0 \le t \le \frac{3\pi}{2}, \\ (-\sin t)\phi(0) & t \ge \frac{3\pi}{2}. \end{cases} \tag{5.3}$$

Thus, $x(0, \phi)(t) = 0$ for $t = k\pi$, $k = 2, 3, \ldots$ and all $\phi \in C([-3\pi/2, 0], \mathbb{R})$. This proves the first statement of Property 5.1.

To prove the more striking conclusion in the second part, we consider the equation

$$\dot{x}(t) = -\alpha(t)x(t - 1) \tag{5.4}$$

where

$$\alpha(t) = \begin{cases} 2\sin^2(\pi t) & t \in [2n, 2n + 1] \\ 0 & t \in (2n - 1, 2n) \end{cases}$$

for each integer n. For any $\sigma \in \mathbb{R}$, $\phi \in C$, we show $T(t, \sigma)\phi = 0$, $t \ge \sigma + 4$. In fact, if N is the smallest odd integer such that $N \ge \sigma$, then $x(t) = x(N)$, $t \in [N, N + 1]$ and

$$\dot{x}(t) = -\alpha(t)x(N), \qquad t \in [N + 1, N + 2].$$

Thus,

$$x(N + 2) = x(N)\big[1 - 2\int_{N+1}^{N+2} \sin^2(\pi s)\, ds\big] = 0$$

and $x(t) = 0$ for $t \in [N+2, N+3]$ and $x(t) = 0$ for $t \geq N+2$. This completes the proof. □

For autonomous linear systems, one has the following surprising result.

Property 5.2. *For n-dimensional autonomous linear differential difference equations, $x(\sigma, C)(t)$ may have dimension $< n$ for some t.*

Proof. The following example is an autonomous linear equation of dimension 3 for which $x(0, C)(t)$ has dimension 2 for $t \geq$ some t_0. Consider the solutions of

$$\begin{aligned} \dot{x}(t) &= 2y(t) \\ \dot{y}(t) &= -z(t) + x(t-1) \\ \dot{z}(t) &= 2y(t-1) \end{aligned} \tag{5.5}$$

defined on $[-1, \infty)$. For $t \geq 1$, $\ddot{y}(t) = 0$, $\dddot{x}(t) = 0$ and thus $y(t)$ is at most a first-degree polynomial in t and $x(t)$ is at most a second-degree polynomial in t. Therefore

$$\begin{aligned} y(t-1) - y(t) &= -\dot{y}(t) \\ x(t) - x(t-1) &= \frac{1}{2}[\dot{x}(t) + \dot{x}(t-1)] \end{aligned}$$

and

$$\begin{aligned} (x, y, z)(t) \cdot (1, -2, -1) &= x(t) - 2y(t) - z(t) \\ &= x(t-1) + \frac{\dot{x}(t-1) + \dot{x}(t)}{2} - 2y(t) - z(t) \\ &= x(t-1) + y(t-1) + y(t) - 2y(t) - z(t) \\ &= x(t-1) - \dot{y}(t) - z(t) = 0. \end{aligned}$$

Consequently, $(x, y, z)(t)$ is orthogonal to $(1, -2, -1)$ and must lie in a plane for $t \geq 1$.

With one delay as in the preceding example, it is actually possible to prove that dimension 3 is the lowest dimension for which this phenomenon can happen. We do not give the proof. For more than one delay, it can even happen in dimension 2. Suppose $x = (x_1, x_2)$ is a 2-vector and consider the equation

$$\dot{x}(t) = \begin{pmatrix} 1 & -1 \\ 0 & 0 \end{pmatrix} x(t) + \begin{pmatrix} -4 & 3 \\ -4 & 4 \end{pmatrix} x(t - \ln 2) + \begin{pmatrix} 4 & -2 \\ 8 & -4 \end{pmatrix} x(t - \ln 4). \tag{5.6}$$

If $x(t) = c \exp \lambda t$, $c = (c_1, c_2)$ is a solution, then λ must satisfy the characteristic equation $\lambda^2 - \lambda = 0$ and the corresponding exponential solutions are $x(t) = (0, 1)$ and $x(t) = (0, 1)e^t$.

One could prove that $x(0, C)(t)$ is one-dimensional for $t \geq$ some t_0 by showing directly that this set is orthogonal to some particular nonzero vector. However, we choose to give a different proof using Theorem 3.1

and some elementary results on linear systems that will be developed later. Even though this is not the logical way to proceed, the reader should find the proof interesting.

Consider the linear operator $T(t) = T(0,t) : C \to C$ defined by the solution of Equation (5.6). Later we show $T(t)$ is completely continuous for $t \geq \ln 4$ and that the only elements in the spectrum of $T(t)$ are 1, e^t and 0. Furthermore, 1 and e^t are simple eigenvalues with corresponding eigenvectors (0,1) and $(0,1)e^t$. Consequently, any solution with zero projection on the span of these two eigenvectors must be associated with the zero-point of the spectrum of $T(t)$ and thus approach zero faster than any exponential. Finally, Theorem 3.1 implies the corresponding solution is identically zero for $t \geq \beta$, $\beta = (n-1)r - \tau$. This proves the result. □

3.6 Compactness and representation

The previous sections have been devoted mainly to peculiar properties that some RFDE may have. In this section, we state some positive results that are valid for all equations and will be extremely useful in later chapters.

Lemma 6.1. *For any $t \geq \sigma + r$, the map $T(t,\sigma)$ is locally completely continuous; that is, for any $t \geq \sigma + r$, $\phi \in C$, there is a neighborhood $V(t,\sigma,\phi)$ of ϕ such that $T(t,\sigma)V(t,\sigma,\phi)$ is in a compact set of C.*

Proof. Since f is continuous and $T(\tau,\sigma)\psi$ is continuous in (τ,σ,ψ), for any $t \geq \sigma$, there is a neighborhood $V(t,\sigma,\phi)$ of ϕ and a constant M such that $|T(\tau,\sigma)V(t,\sigma,\phi)| \leq M$, $|f(\tau, T(\tau,\sigma)V(t,\sigma,\phi))| \leq M$ for $\sigma \leq \tau \leq t$. Thus $|\dot{x}(\sigma, V(t,\sigma,\phi))(\tau)| \leq M$ for $\sigma \leq \tau \leq t$. This implies the family of functions $\{x_t(\sigma,\psi) : \psi \in V(t,\sigma,\phi)\}$ is precompact for $t \geq \sigma + r$ and proves Lemma 6.1. □

For simplicity in the statement of the global properties of $T(t,\sigma)$, let us assume $\Omega = \mathbb{R} \times C$ and $T(t,\sigma) : C \to C$ is defined for all $t \geq \sigma$ and, of course, $T(t,\sigma)\phi$ is continuous in (t,σ,ϕ). The following definitions are also convenient.

Definition 6.1. Suppose a mapping $A(t,\sigma) : C \to C$ is defined for all $t \geq \sigma$. The mapping $A(t,\sigma)$, $t \geq \sigma$, is said to be *conditionally completely continuous* if $A(t,\sigma)\phi$ is continuous in (t,σ,ϕ) and, for any bounded set $B \subseteq C$, there is a compact set $B^* \subseteq C$ such that $A(\tau,\sigma)\phi \in B$ for $\sigma \leq \tau \leq t$ implies $A(t,\sigma)\phi \in B^*$.

Definition 6.2. If X, Y are metric spaces, then a mapping $A : X \to Y$ is *bounded* if A takes bounded sets of X into bounded sets of Y. If $A(\lambda)$:

$X \to Y$ depends on a parameter in a metric space Λ, then $A(\lambda)$ is said to *be bounded uniformly on compact sets of* Λ if, for any compact set $\Lambda_0 \subseteq \Lambda$ and any bounded set $X_0 \subseteq Y$, there is a bounded set $Y_0 \subseteq Y$ such that $A(\lambda)X_0 \subseteq Y_0$ for all $\lambda \in \Lambda_0$.

The following lemma is obvious from the definitions.

Lemma 6.2. *If $A(t,\sigma) : C \to C$ is defined for $t \geq \sigma$ and $A(t,\sigma)$ is a map bounded uniformly on compact sets of $[\sigma, \infty)$, then $A(t,\sigma)$ conditionally completely continuous implies $A(t,\sigma)$, $t \geq \sigma$ completely continuous.*

Define the operator $S(t) : C \to C$, $t \geq 0$, by the relation

$$S(t)\phi(\theta) = \begin{cases} \phi(t+\theta) - \phi(0) & t+\theta < 0 \\ 0 & t+\theta \geq 0, \ -r \leq \theta \leq 0. \end{cases} \tag{6.1}$$

The operator $S(t)$ is a bounded linear operator on C for each $t \geq 0$ and satisfies

$$\begin{aligned} S(t+\tau) &= S(t)S(\tau) \quad && \text{for } t \geq 0, \ \tau \geq 0 \\ S(t) &= 0 && \text{for } t \geq r. \end{aligned} \tag{6.2}$$

The last relations are obvious consequences of the definition. Therefore, for any $\alpha > 0$, there is a K_α such that

$$|S(t)| \leq K_\alpha e^{-\alpha t}, \qquad t \geq 0. \tag{6.3}$$

In the statement of the next result, we use the concept of an α-contraction, which is described in Definition 3.6 of Chapter 4.

Theorem 6.1. *For a given RFDE(f) for which $f : \mathbb{R} \times C \to \mathbb{R}^n$ is a bounded continuous map, the solution map $T(t,\sigma)$, $t \geq \sigma$, can be written as*

$$T(t,\sigma) = S(t-\sigma) + U(t,\sigma) \tag{6.4}$$

where $S(t)$, $t \geq 0$, is defined in (6.1), satisfies (6.3), and $U(t,\sigma) : C \to C$, $t \geq \sigma$, is conditionally completely continuous. Thus, $T(t,\sigma)$ is an α-contraction for $t > \sigma$ and is conditionally completely continuous for $t \geq \sigma + r$.

Proof. With $S(t)$, $t \geq 0$, defined as in Equation (6.1), it follows that $U(t,\sigma)\phi$ is defined as

$$U(t,\sigma)\phi(\theta) = \begin{cases} \phi(0) & t+\theta < \sigma, \\ \phi(0) + \int_\sigma^{t+\theta} f(s, T(s,\sigma)\phi)\, ds & t+\theta \geq \sigma, \ -r \leq \theta \leq 0. \end{cases}$$

For any bounded set $B \subseteq C$, the fact that $S(t)$ is a bounded linear operator implies there is a bounded set $B_1 \subseteq C$ such that $T(s,\sigma)\phi \in B_1$ for $\sigma \leq s \leq t$ provided that $U(s,\sigma)\phi \in B$ for $\sigma \leq s \leq t$. Since f is a bounded continuous map from $\mathbb{R} \times C$ to $\mathbb{R}^n$, there is an M depending only on B, σ, and t such that $|f(s, T(s,\sigma)\phi)| \leq M$ for $\sigma \leq s \leq t$ if $U(s,\sigma)\phi \in B$ for $\sigma \leq s \leq t$. Therefore, under these suppositions,

$$\int_\alpha^\beta |f(s, T(s,\sigma)\phi)|\, ds \leq M(\beta - \alpha)$$

for $\sigma \leq \alpha \leq \beta \leq t$. Let

$$B^* = \{\phi \in C : \phi \in B,\ |\phi(\theta) - \phi(\zeta)| \leq M|\theta - \zeta|,\ \theta, \zeta \in [-r, 0]\}.$$

These computations show that $U(t,\sigma)\phi \in B^*$ since $U(t,\sigma)\phi(\theta)$ is constant for $t + \theta \leq \sigma$. The proof of the theorem is completed by observing that $S(t) = 0$ for $t \geq r$. □

The following corollaries are immediate.

Corollary 6.1. *If $f : \mathbb{R} \times C \to \mathbb{R}^n$ is a bounded continuous map and $T(t,\sigma) : C \to C$ is a map bounded uniformly on compact sets of $[\sigma, \infty)$, then*

$$T(t,\sigma) = S(t - \sigma) + U(t,\sigma), \qquad t \geq \sigma,$$

where $S(t)$, $t \geq 0$ is defined in Equation (6.1) and $U(t,\sigma)$, $t \geq \sigma$, is completely continuous.

Corollary 6.2. *If $f : \mathbb{R} \times C \to \mathbb{R}^n$ is a bounded continuous map and $T(t,\sigma) : C \to C$ is a map bounded uniformly on compact sets of $[\sigma, \infty)$, then $T(t,\sigma)$, $t \geq \sigma + r$, is completely continuous.*

Corollary 6.3. *For $r > 0$, the solution map $T(t,\sigma)$ for $t > \sigma$ can never be a homeomorphism if $f : \mathbb{R} \times C \to \mathbb{R}^n$ is a bounded continuous map and $T(t,\sigma)$, $t \geq \sigma$, is a bounded map.*

Proof. This follows immediately from the preceding theorem and the fact that the unit ball in $C([-r, 0], \mathbb{R}^n)$ is not compact for any $r > 0$. □

3.7 The solution map for NFDE

For a NFDE(D, f) on Ω and any $(\sigma, \phi) \in \Omega$, there is a $t_{\sigma,\phi}$ and a noncontinuable solution x through (σ, ϕ) defined on $[\sigma - r, t_{\sigma,\phi})$. If we speak of only noncontinuable solutions, then we can define the *solution map* $T(t,\sigma) : \Omega_\sigma \to C$ where $\Omega_\sigma = \{\psi \in C : (\sigma, \phi) \in \Omega\}$ and $T(t,\sigma) = x_t(\sigma, \phi)$.

For RFDE, the solution map $T(t,\sigma)$ smoothes the initial data as t increases. Also, if a solution of an RFDE can be extended in the negative direction through a point (σ, ϕ), then ϕ must satisfy some differentiability conditions. For a large class of NFDE, neither of these properties holds. This large class is the one for which the function D uses the information about the function x at $t-r$ as well as t. In particular, we can state the following result.

Theorem 7.1. *If the* NFDE(D,f) *on an open set* Ω *satisfies the property that* $D(t,\phi)$ *is atomic at* $-r$ *for each* $(\sigma,\phi)\in\Omega$, *then there are* $t_1(\sigma,\phi)<\sigma<t_2(\sigma,\phi)$ *and a noncontinuable solution* x *of the* NFDE(D,f) *on* $t_1(\sigma,\phi)-r<t_2(\sigma,\phi)$; *that is,* $T(t,\phi):\Omega_\sigma\to C$ *is defined for* $t_1(\sigma,\phi)<t<t_2(\sigma,\phi)$. *Furthermore, if* $f(t,\sigma)$ *has a continuous first derivative in* ϕ, *then*

$$T(t,\sigma):\Omega_\sigma\to C$$

is a homeomorphism for each $t_1(\sigma,\phi)<t<t_2(\sigma,\phi)$.

Proof. The same ideas used in the proof of Theorem 8.1 of Chapter 2 are easily adapted to prove the existence of a solution through (σ,ϕ) in the negative direction. The extra smoothness condition on f implies uniqueness and continuous dependence of the solution. The proof is complete. □

A very simple example illustrating the last result is the equation

$$\frac{d}{dt}[x(t)-Ax(t-r)]=f(t,x_t)$$

where A is an $n\times n$ constant matrix with det $A\neq 0$. If $f(t,\phi)$ is continuously differentiable in ϕ, then the solution operator of this equation defines a homeomorphism.

These methods can be used to prove results on NFDE concerning the differentiability with respect to parameters, but we do not dwell on this question. For the case of Theorem 7.1, the implication will be that the solution operator is a diffeomorphism.

The fact that solutions of NFDE can have backward extension for every $(\sigma,\phi)\in\Omega$ allows one to prove the following result for linear systems, which contrasts drastically with RFDE (see Section 3.3).

Theorem 7.2. *Suppose* $\tau\in\mathbb{R}$ *is given,* $\Omega=(\tau,\infty)\times C$, (D,f) *defines an* NFDE *on* Ω, $D(t,\phi)$ *is linear in* ϕ, $f(t,\phi)$ *is linear in* ϕ, *there is a positive constant* k *such that* $|D(t,\phi)|\le k|\phi|$, $|f(t,\phi)|\le k|\phi|$ *for* $(t,\phi)\in\Omega$, $D(t,\phi)$ *is atomic at* 0 *and* $-r$ *uniformly with respect to* $(t,\phi)\in\Omega$. *Then for any* $(\sigma,\phi)\in\Omega$, *there is a unique solution of the* NFDE(D,f) *through* (σ,ϕ) *defined on* (τ,∞) *and, if a solution* x *approaches zero faster than any exponential as* $t\to\infty$, *then* $x(t)=0$ *for all* $t\in(\tau,\infty)$.

Proof. Existence and uniqueness follow by applying these results. Following the same reasoning as in the proof of Theorem 7.3 of Section 1.7, there are positive constants a and b such that for any $\sigma\in(\tau,\infty)$, any solution x satisfies

$$|x_t| \le a e^{b|t-\sigma|}|x_\sigma|, \qquad t \in (\tau, \infty).$$

Now, suppose $x_\sigma \to 0$ faster than any exponential as $\sigma \to \infty$ and there is a $t \in (\tau, \infty)$ such that $|x_t| > 0$. For $\sigma \ge t$, we have

$$|x_t| \exp bt \le a|x_\sigma| \exp b\sigma.$$

This gives a contradiction and proves the theorem. □

Our next objective is to obtain a representation of the solution operator of an NFDE in a form similar to one in Theorem 6.1 for RFDE. We restrict our attention to the case where the equation is autonomous in order to avoid excessive notation.

As preparatory material, we need some notation associated to the homogeneous "difference" equation

$$(7.1) \qquad Dy_t = 0, \qquad t \ge 0,$$

where $D : C \to \mathbb{R}^n$ is linear, bounded, and atomic at 0. Without loss in generality, we may assume $D\phi = \phi(0) - \int_{-r}^{0} d[\mu(\theta)]\phi(\theta)$, where $\mathrm{Var}_{[-s,0]}\, \mu \to 0$ as $s \to 0$.

If

$$(7.2) \qquad C_D = \{\, \phi \in C : D\phi = 0 \,\}$$

then Equation (7.1) defines a semigroup of linear operators $T_D(t)$, $t \ge 0$, on C_D, where $T_D(t)\psi = y_t(\psi)$, $t \ge 0$, $\psi \in C_D$, and $y_t(\psi)$ is the solution of Equation (7.1) through $(0, \psi)$.

Lemma 7.1. *If $D : C \to \mathbb{R}^n$ is linear, bounded and atomic at zero, then there are n functions $\phi_1, \ldots, \phi_n$ such that $D\Phi = I$, where $\Phi = (\phi_1, \ldots, \phi_n)$.*

Proof. For any $s \in [0, r]$, let $\psi \in C([-r, 0], \mathbb{R})$ be defined by

$$\psi(\theta) = 0, \ -r < \theta < -s, \quad \psi(\theta) = 1 + \frac{\theta}{s}, \ -s < \theta \le 0.$$

Then

$$D(\psi I) = I - \int_{-s}^{0} d[\mu(\theta)](1 + \frac{\theta}{s}).$$

If $s > 0$ is sufficiently small, then $\det D(\psi I) \ne 0$ and the columns of the matrix $D(\psi I)$ form a basis for $\mathbb{R}^n$. If we let $\Phi = \psi I(D(\psi I))^{-1}$, then $D\Phi = I$ and the lemma is proved. □

Let $f : C \to \mathbb{R}^n$ be a continuous function and let $T_{D,f}(t)$ be the semigroup defined by the NFDE (D,f),

$$(7.3) \qquad \frac{d}{dt} Dx_t = f(x_t), \qquad t \ge 0.$$

With this notation, we can state the following important representation theorem for the semigroup associated to Equation (7.3).

Theorem 7.3. *Let Φ, $D\Phi = I$, be defined by Lemma 7.1 and let $\Psi = I - \Phi D$. If f is completely continuous and if $T_{D,f}(t)$ is a bounded map for each $t > 0$, then*

$$T_{D,f}(t) = T_D(t)\Psi + U(t)$$

where $U(t) : C \to C$, $t \geq 0$ is completely continuous.

To prove this result, we need the following auxiliary result on the nonhomogeneous "difference" equation

$$Dy_t = h(t), \qquad t \geq 0, \tag{7.4}$$

with $h \in C([0,\infty), \mathbb{R}^n)$.

Lemma 7.2. *If $D : C \to \mathbb{R}^n$ is bounded, linear, and atomic at zero and $y(\psi, h)$ is the solution of Equation (7.6) satisfying $y_0(\psi, h) = \psi$, then there are positive constants a and b such that for $t \geq 0$,*

$$|y(\psi, h)(t)| \leq be^{at}[\,|\psi| + \sup_{0\leq s\leq t} |h(s)|\,].$$

Proof. The proof is technical, but can be supplied by modifying the argument used in the proof of Theorem 7.3 of Section 1.7. See also Theorem 3.4 of Chapter 9. □

Proof of Theorem 7.3. If $U(t)\phi = T_{D,f}(t)\phi - T_D(t)\Psi\phi$, then $U(t)\phi$ satisfies the equation

$$DU(t)\phi = D\phi + \int_0^t f(T_{D,f}(s)\phi)\, ds.$$

Lemma 7.1 and the boundedness assumption on $T_{D,f}(t)$ imply that $U(t)$ takes bounded sets into bounded sets. Also,

$$D[U(t+\tau)\phi - U(t)\phi] = \int_t^{t+\tau} f(T_{D,f}(s)\phi)\, ds.$$

Now suppose that ϕ belongs to a fixed bounded set B. Then $|U(\tau)\phi - \Phi D\phi|$ can be made arbitrarily small by taking τ small. Since f is completely continuous and $\bigcap_{0\leq s\leq t} T_{D,f}(s)B$ is bounded for each t, it follows from Lemma 7.2 that for any fixed $\alpha > 0$, $t \in [0,\alpha]$, the expression $|U(t+\tau)\phi - U(t)\phi|$ can be made arbitrarily small by taking τ small. This shows that $U(t)$ is completely continuous. □

We now assume that the kernel μ in the definition of D has no singular part; that is,

$$(7.5)\qquad \begin{aligned} D\phi &= D_0\phi - \int_{-r}^{0} A(\theta)\phi(\theta)\,d\theta \\ D_0(\phi) &= \phi(0) - \sum_{k=1}^{\infty} A_k\phi(-r_k) \end{aligned}$$

where

$$0 < r_k \le r, \quad \sum_{k=1}^{\infty} |A_k| + \int_{-r}^{0} |A(\theta)|\,d\theta < \infty.$$

Theorem 7.4. *Suppose that D satisfies (7.5), let Φ_0, $D_0\Phi_0 = I$ be defined by Lemma 7.1, and let $\Psi_0 = I - \Phi_0 D_0$. Then*

$$T_D(t) = T_{D_0}(t)\Psi_0 + U(t)$$

where $U(t) : C \to C$ is completely continuous for $t \ge 0$.

Proof. If $U(t)\phi = T_D(t)\phi - T_{D_0}(t)\Psi_0\phi$, then $U(t)\phi$ is linear and continuous in ϕ and satisfies the equation

$$D_0 U(t)\phi = \int_{-r}^{0} A(\theta)\big(T_D(t)\phi\big)(\theta)\,d\theta.$$

For any $\tau \in \mathbb{R}$, $t \ge 0$, we have

$$\begin{aligned} &D_0[U(t+\tau)\phi - U(t)\phi] \\ &\quad = \int_{-r}^{0} A(\theta)\big(T_D(t+\tau)\phi\big)(\theta)\,d\theta - \int_{-r}^{0} A(\theta)\big(T_D(t)\phi\big)(\theta)\,d\theta \\ &\quad = \int_{\tau}^{-r+\tau} A(\theta-\tau)\big(T_D(t)\phi\big)(\theta)\,d\theta - \int_{-r}^{0} A(\theta)\big(T_D(t)\phi\big)(\theta)\,d\theta \\ &\quad = \Big(\int_{-r}^{-r+\tau} + \int_{0}^{\tau}\Big)\big(A(\theta-\tau) - A(\theta)\big)\big(T_D(t)\phi\big)(\theta)\,d\theta \\ &\quad \equiv h(t,\tau,\phi). \end{aligned}$$

Now we proceed as in the proof of Theorem 7.3 using the fact that $\int_a^b |A(s-\tau) - A(s)|\,ds \to 0$ as $\tau \to 0$. □

Corollary 7.1. *Suppose D as in (7.5), let Φ, $D\Phi = I$, Φ_0, $D_0\Phi_0 = I$, be defined by Lemma 7.1, and let $\Psi = I - \Phi D : C \to C_D$, $\Psi_0 = I - \Phi_0 D_0 : C \to C_{D_0}$. If $f : C \to \mathbb{R}^n$ is completely continuous and if $T_{D,f}(t)$ is a bounded map for each $t > 0$, then*

$$(7.6)\qquad T_{D,f}(t) = T_{D_0}(t)\Psi_0\Psi + U(t)$$

where $U(t) : C \to C$, $t \ge 0$, is completely continuous.

Let a_{D_0} be the order of the semigroup $T_{D_0}(t)$; that is,

$$a_{D_0} = \inf\{\, a \in \mathbb{R} : \text{ there is a } k \text{ such that } \|T_{D_0}(t)\| \leq ke^{at}, t \geq 0\,\}. \tag{7.7}$$

Lemma 7.3. *If a_{D_0} is defined as in (7.7), then, for any $a > a_{D_0}$, there is an equivalent norm $|\cdot|_a$ in C such that*

$$|T_{D_0}(t)\psi|_a \leq e^{at}|\psi|, \qquad t \geq 0, \quad \psi \in C_{D_0}. \tag{7.8}$$

Proof. For any $a > a_{D_0}$, there is a constant k such that

$$|T_{D_0}(t)\psi| \leq ke^{at}|\psi|, \qquad t \geq 0, \quad \psi \in C_{D_0}. \tag{7.9}$$

For any $\psi \in C_{D_0}$, we define

$$|\psi|_a = \sup_{s\geq 0} |T_{D_0}(s)\psi|e^{-as}.$$

From (7.9), we have $|\psi| \leq |\psi|_a \leq k|\psi|$ and so $|\cdot|_a$ is an equivalent norm on C_{D_0} and can be extended to an equivalent norm on C. Also,

$$\begin{aligned} |T_{D_0}(t)\psi|_a &= \sup_{s\geq 0} |T_{D_0}(t+s)\psi|e^{-as} \\ &= e^{at} \sup_{s\geq 0} |T_{D_0}(t+s)\psi|e^{-a(t+s)} \\ &\leq e^{at}|\phi|_a, \end{aligned}$$

and the lemma is proved. □

Corollary 7.2. *If D, f satisfy the hypotheses in Corollary 7.1 and $a_{D_0} < 0$, then there is an equivalent norm in C such that in this new norm, the semigroup $T_{D,f}(t)$ defined by (7.4) is the sum of a contraction and a completely continuous map for each $t > 0$.*

Proof. If we choose a so that $a_{D_0} < a < 0$, then this is an immediate consequence of Lemma 7.3 and Equation (7.5). □

In Chapter 9, we will show that the number a_{D_0} is determined by the characteristic equation for the difference equation

$$y(t) - \sum_{k=1}^{\infty} A_k y(t - r_k) = 0. \tag{7.10}$$

In fact,

$$a_{D_0} = \sup\{\, \operatorname{Re}\lambda : \det\,[I - \sum_{k=1}^{\infty} A_k e^{-\lambda r_k}] = 0\,\}. \tag{7.11}$$

3.8 Supplementary remarks

Krasnovskii [1] was the first to emphasize the importance of considering the state of a system defined by a functional differential equation as the element $x_t(\sigma, \phi)$ of C. He made the observation that the converse theorems of Liapunov on stability could not be proved by using a scalar function $V(t, x)$ that depends only on (t, x) in $\mathbb{R} \times \mathbb{R}^n$. In fact, if uniform asymptotic stability of the solution $x = 0$ of

$$\dot{x}(t) = f(x_t)$$

implies the existence of a positive definite function $V(x)$ such that

$$(\partial V/\partial x)f < 0,$$

then the solution $x = 0$ of

$$\dot{x}(t) = kf(x_t)$$

would be uniformly asymptotically stable for any positive k since

$$(\partial V/\partial x)(kf) < 0.$$

On the other hand, the linear equation

$$\dot{x}(t) = -kx(t-1)$$

has all roots of the characteristic equation $\lambda = -ke^{-\lambda}$ with negative real parts if $k < \pi/2$ and some with positive real parts if $k > \pi/2$. Therefore, the converse theorems of Liapunov must make explicit use of the infinite dimensionality of the problem.

Example (1.1) is due to Hannsgen. The example in Property 1.4 is due to Charrier [1]. The same property was observed by Banks and Kent [1]. Example 2.3 and the example in Property 4.2 are due to Hausrath.

The paper of Henry [2] contains Theorem 3.1 and other interesting implications of this result. Theorem 3.2 and 3.3 are from Verduyn Lunel [1]. The results with their proofs carry over verbatim to linear autonomous neutral functional differential equations NFDE(D, L). Related results can be found in Bartosiewicz [1] and Kappel [4]. Theorem 3.1 is false for time-dependent delay equations. The following example is due to Oliva

$$\dot{x}(t) = -2te^{1-2t}x(t-1) \tag{8.1}$$

which has the solution $x(t) = e^{\ t^2}$ on $[-1, \infty)$. From the result of Cooke and Verduyn Lunel [1], it follows that the sign change in the coefficient of Equation (8.1) is necessary in order to have nontrivial small solutions. Also for linear periodic RFDE, Theorem 3.1 is false as follows from the following example due to Stokes [2]

$$\dot{x}(t) = \sin(2\pi t)x(t-1). \tag{8.2}$$

In Section 3 of Chapter 8 we shall further analyze the existence of nontrivial small solutions for periodic RFDE.

Theorem 4.1 is due to Nussbaum [1]. Example (5.1) is due to Zverkin [1], and Example (5.4) to Winston and Yorke [1].

The special systems considered in Property 5.2 belong to the general category of pointwise degenerate systems. Consider a differential difference equation

$$\dot{x}(t) = Ax(t) + \sum_{j=1}^{m} B_j x(t - jr) \tag{8.3}$$

where A, B_j, $j \geq 1$, are $n \times n$ constant matrices. Equation (8.3) is said to be *pointwise degenerate* at t_1 with respect to the $p \times n$, $p < n$, matrix Q if for all continuous initial functions given on $[-mr, 0]$, the corresponding solution $x(t)$ satisfies $Qx(t_1) = 0$. Otherwise, the equation is *pointwise complete.*

If a system is pointwise degenerate at t_1, then it is pointwise degenerate for all $t \geq t_1$. Therefore, the solutions of Equation (8.3) will lie in the subspace M_Q of $\mathbb{R}^n$ defined by $M_Q = \{x \in \mathbb{R}^n : Qx = 0\}$ for $t \geq t_1$. Weiss [1] was the first to conjecture that there might exist pointwise degenerate systems. With Example (5.5), Popov [1] showed the existence of pointwise degenerate systems and gave some interesting properties of these systems. With one delay, one cannot obtain a pointwise degenerate system for $n \leq 2$. Zverkin [2] was the first to show by means of Example (6.6) that one could have a pointwise degenerate system for $n = 2$.

A number of authors have been concerned with necessary and sufficient conditions for System (8.3) to be pointwise degenerate and the construction of pointwise degenerate systems (see, for example, Asner [1], Asner and Halanay [1, 2], Charrier [1], Kappel [1, 2, 3], Popov [1, 2], Zmood and McClamroch [1], and Zverkin [2]).

Let us reinterpret Example (5.5) of Popov [1] in terms of a control problem. Let $q \in \mathbb{R}^n$ be the vector $(1, -2, -1)$. For System (5.5), it was shown that $qx(t) = 0$ for all $t \geq 1$. Therefore, the ordinary differential equation

$$\dot{w} = Aw, \qquad w = \begin{pmatrix} x \\ y \\ z \end{pmatrix}, \qquad A = \begin{bmatrix} 0 & 1 & 0 \\ 0 & 0 & -1 \\ 0 & 0 & 0 \end{bmatrix}$$

can be steered to the plane $M_q = \{x \in \mathbb{R}^3 : qx = 0\}$ in time ≤ 1 and it will remain in this plane for $t \geq 1$ provided that the feedback control is chosen to be

$$Bw(t-1), \qquad B = \begin{bmatrix} 0 & 0 & 0 \\ 1 & 0 & 0 \\ 0 & 2 & 0 \end{bmatrix}.$$

Therefore, the time optimal control problem is meaningful for this situation. This question was discussed in more detail in Popov [2]. Asner and Halanay [3, 4] considered similar questions by allowing the delayed feedback control to involve the derivative $\dot{x}$ of x as well as x itself.

Popov [1] also observed in this first paper on degenerate systems that it is impossible to solve this problem by using delayed feedback of the form $bcw(t-r)$ where b is a column vector and c is a row vector. Asner and Halanay [4] showed the result of Popov remained valid even with several delays.

Definition 6.1 and the representation of the solution operator in Theorem 6.1 are due to Hale and Lopes [1].

Theorem 7.2 is due to Hale [1] and generalizes an earlier result of Wright [3]. The representation theorem 7.3 first occurred in Henry [1]. The importance of the difference operator D_0 has been recognized by Cruz and Hale [4].

4
Autonomous and periodic processes

In this chapter, we discuss some fundamental properties of autonomous and periodic RFDE. Because many of these properties also hold for abstract processes, we will develop the theory in an abstract setting and point out special features that apply to RFDE. The abstract setting also will be sufficiently general to apply to the neutral functional differential equations in Chapter 9.

In Sections 4.1 and 4.2, we define and develop properties of ω-limit sets and invariant sets of discrete and continuous processes. Section 4.3 is devoted to giving conditions on a discrete dynamical system to ensure the existence of maximal compact invariant sets and global attractors. In Section 4.4, we give sufficient conditions for the existence of fixed points. Sections 4.5 and 4.6 deal with the same topics for continuous ω-periodic processes. Section 4.7 deals with convergent systems.

4.1 Processes

In this section, we give the definition of processes, demonstrate how processes arise from RFDE, and give a few elementary properties of orbits of processes.

Definition 1.1. Suppose X is a Banach space, $\mathbb{R}^+ = [0, \infty)$, $u : \mathbb{R} \times X \times \mathbb{R}^+ \to X$ is a given mapping and define $U(\sigma, t) : X \to X$ for $\sigma \in \mathbb{R}$, $t \in \mathbb{R}^+$ by $U(\sigma, t)x = u(\sigma, x, t)$. A *process on* X is a mapping $u : \mathbb{R} \times X \times \mathbb{R}^+ \to X$ satisfying the following properties

(i) u is continuous.

(ii) $U(\sigma, 0) = I$, the identity.

(iii) $U(\sigma + s, t)U(\sigma, s) = U(\sigma, s + t)$.

A process u is said to be an *ω-periodic process* if there is an $\omega > 0$ such that $U(\sigma + \omega, t) = U(\sigma, t)$ for all $\sigma \in \mathbb{R}$, $t \in \mathbb{R}^+$. A process is said to be an *autonomous process* or a *(continuous) dynamical system* if $U(\sigma, t)$ is independent of σ; that is, the family of transformations $T(t) = U(0, t)$ $t \geq 0$, is a C^0-semigroup; $T(t)x$ is continuous for $(t, x) \in \mathbb{R}^+ \times X$,

$$T(0) = I, \qquad T(t + \tau) = T(t)T(\tau), \qquad t, \ \tau \in \mathbb{R}^+.$$

If $S : X \to X$ is a continuous map, the family $\{S^k, k \geq 0\}$ of iterates of S is called a *discrete dynamical system.*

If u is an ω-periodic process and $S = U(0, \omega)$, then the discrete dynamical system $\{S^k : k \geq 0\}$ coincides with the family of mappings $\{U(0, k\omega) : k \geq 0\}$. In fact, $S = U(0, \omega)$, $S^2 = U(0, \omega)U(0, \omega) = U(\omega, \omega)U(0, \omega) = U(0, 2\omega)$, etc. This particular discrete dynamical system associated with the process u will be referred to as the *dynamical system generated by the period map* of the ω-periodic process.

In a process, $u(\sigma, x, t)$ can be considered as the state of a system at time $\sigma + t$ if initially the state at time σ was x.

Let us now discuss how processes arise from RFDE. Suppose $f : \mathbb{R} \times C \to \mathbb{R}^n$ is completely continuous and let $x(\sigma, \phi)$ denote the solution of the RFDE(f)

$$\dot{x}(t) = f(t, x_t) \tag{1.1}$$

through (σ, ϕ) and suppose x is uniquely defined for $t \geq \sigma - r$. From Theorem 2.2 of Section 2.2, $x(\sigma, \phi)(t)$ is continuous in σ, ϕ, t for $\sigma \in \mathbb{R}$, $\phi \in C$, and $t \geq \sigma$. If $u(\sigma, \phi, \tau) = x_{\sigma+\tau}(\sigma, \phi)$ for $(\sigma, \phi, \tau) \in \mathbb{R} \times X \times \mathbb{R}^+$, then u is a process on C. Furthermore, if U is defined as before, then $U(\sigma, \tau) = T(\sigma + \tau, \sigma)$ where $T(t, \sigma)$ is the solution operator for Equation (1.1), $T(t, \sigma)\phi = x_t(\sigma, \phi)$. In the following, we refer to the process obtained from Equation (1.1) in this way as the *process generated by the* RFDE (f).

If there is an $\omega > 0$ such that $f(\sigma + \omega, \phi) = f(\sigma, \phi)$ for all $(\sigma, \phi) \in \mathbb{R} \times C$, then the process generated by the RFDE(f) is an ω-periodic process. If $f(\sigma, \phi)$ is independent of σ, then the process generated by the RFDE(f) is a dynamical system. For the neutral equations of Chapter 2, one obtains a process in the same manner.

Definition 1.2. Suppose u is a process on X. The *trajectory* $\tau^+(\sigma, x)$ *through* $(\sigma, x) \in \mathbb{R} \times X$ is the set in $\mathbb{R} \times X$ defined by

$$\tau^+(\sigma, x) = \{(\sigma + t, U(\sigma, t)x) : t \in \mathbb{R}^+\}.$$

The *orbit* $\gamma^+(\sigma, x)$ *through* (σ, x) is the set in X defined by

$$\gamma^+(\sigma, x) = \{U(\sigma, t)x : t \in \mathbb{R}^+\}.$$

If H is a subset of X, then $\tau^+(\sigma, H) = \bigcup_{x \in H} \tau^+(\sigma, x)$, $\gamma^+(\sigma, H) = \bigcup_{x \in H} \gamma^+(\sigma, x)$. If $\{T^k : k \geq 0\}$ is a discrete dynamical system, the *orbit* $\gamma^+(x)$ *through* $x \in X$ is defined as $\gamma^+(x) = \{T^k x : k \geq 0\}$ and the *orbit* $\gamma^+(H)$ *through a set* $H \subseteq X$ is $\gamma^+(H) = \bigcup_{x \in H} \gamma^+(x)$.

If u is an ω-periodic process on X, then the trajectory $\tau^+(\sigma + k\omega, x)$ for any integer $k \in \mathbb{R}$ is a translate along the reals by a distance $k\omega$ of the trajectory $\tau^+(\sigma, x)$. The orbits satisfy $\gamma^+(\sigma + k\omega, x) = \gamma^+(\sigma, x)$ for any integer $k \in \mathbb{R}$. If u is a dynamical system on X, then $\tau^+(\sigma + s, x)$ is a translate by s of $\tau^+(\sigma, x)$ for any $s \in \mathbb{R}$ and $\gamma^+(\sigma, x) = \gamma^+(0, x)$ for all $\sigma \in \mathbb{R}$. In this latter case, we will simply write $\gamma^+(x)$ for orbits.

A point $c \in X$ is said to be an *equilibrium or critical point of a process* u if there is an $\sigma \in \mathbb{R}$ such that $\gamma^+(\sigma, c) = \{c\}$; that is, $U(\sigma, t)c = c$ for $t \in \mathbb{R}^+$. If there is a $\sigma \in \mathbb{R}$, $p > 0$, $x \in X$ such that $U(\sigma, t+p)x = U(\sigma, t)x$ for all $t \in \mathbb{R}^+$, then the trajectory $\tau^+(\sigma, x)$ is said to be *p-periodic.*

The proof of the following lemma is obvious.

Lemma 1.1. *For a continuous dynamical system, a trajectory is p-periodic if and only if the corresponding orbit is a closed curve. For an ω-periodic process u, a trajectory through (σ, x) is $k\omega$-periodic for some positive integer k if and only if $T^k x = x$ where $T = U(\sigma, \omega)$.*

Lemma 1.2. *Let $\{T(t) : t \geq 0\}$ be a dynamical system on X. If there are sequences $\{x_n\} \subseteq X$, $\{\omega_n\} \subseteq (0, \infty)$ satisfying $T(\omega_n)x_n = x_n$, $\omega_n \to 0$ as $n \to \infty$ and some subsequence of $\{x_n\}$ converges to x_0, then x_0 is an equilibrium point.*

Proof. Changing the notation if necessary, we may assume x_n converges to x_0. Let $k_n(t)$ be the integer defined by $k_n(t)\omega_n \leq t < (k_n(t) + 1)\omega_n$. Then $T(k_n(t)\omega_n)x_n = x_n$ and

$$\begin{aligned} |T(t)x_0 - x_0| \leq{}& |T(t)x_0 - T(k_n(t)\omega_n)x_0| \\ &+ |T(k_n(t)\omega_n)x_0 - T(k_n(t)\omega_n)x_n| + |x_n - x_0|. \end{aligned}$$

Since $k_n(t)\omega_n \to t$, $x_n \to x_0$ as $n \to \infty$, the right-hand side of this expression approaches zero as $n \to \infty$. Therefore, $T(t)x_0 = x_0$ for all $t \geq 0$ and the lemma is proved. □

Lemmas 1.1 and 1.2 give an effective way of changing the problem of finding $k\omega$-periodic trajectories of ω-periodic processes and critical points of dynamical systems into a problem involving fixed points of mappings. We will make explicit use of this fact in the following pages.

Definition 1.3. Suppose u is a process on X. A point $y \in X$ is said to be in the *ω-limit set* $\omega(\sigma, x)$ of an orbit $\gamma^+(\sigma, x)$ if there is a sequence $t_n \to \infty$ as $n \to \infty$ such that $U(\sigma, t_n)x \to y$ as $n \to \infty$, or equivalently, if

$$\omega(\sigma, x) = \bigcap_{t \geq 0} \overline{\bigcup_{\tau \geq t} U(\sigma, \tau)x}. \tag{1.2}$$

A point $y \in X$ is said to be in the *α-limit set* $\alpha(\sigma, x)$ of an orbit $\gamma^-(\sigma, x) = \bigcup_{t \leq 0} U(\sigma, t)x$ if $U(\sigma, t)x$ is defined for $t \leq 0$ and there is a sequence

$t_n \to -\infty$ as $n \to \infty$ such that $U(\sigma, t_n)x \to y$ as n$\to \infty$, or, equivalently, if

$$\alpha(\sigma, x) = \bigcap_{t \le 0} \overline{\bigcup_{\tau \le t} U(\sigma, \tau)x}. \tag{1.3}$$

For any subset $H \subseteq X$, Relations (1.2), (1.3) with x replaced by H can also be used to define $\omega(\sigma, H)$, $\alpha(\sigma, H)$, the *ω- and α-limit sets of a set H*. If the process u is a dynamical system, the dependence of all sets on σ will be deleted. If $\{T^k : k \ge 0\}$ is a discrete dynamical system of X and $H \subseteq X$, then the *ω-limit* set of H is defined as

$$\omega(H) = \bigcap_{j \ge 0} \overline{\bigcup_{n \ge j} T^n H}, \tag{1.4}$$

and the *α-limit set of* H is defined as

$$\alpha(H) = \bigcap_{j \le 0} \overline{\bigcup_{n \le j} T^n H}. \tag{1.5}$$

Lemma 1.3. *Suppose u is a process on X. If $\gamma^+(\sigma, x)$ is the precompact, then $\omega(\sigma, x)$ exists, is nonempty, compact, connected, and*

$$\operatorname{dist}(U(\sigma, t)x, \omega(\sigma, x)) \to 0 \qquad \text{as } t \to \infty.$$

If $\gamma^-(\sigma, x)$ is precompact, then the same is true of $\alpha(\sigma, x)$ with $t \to -\infty$. The same conclusion is valid for any connected subset $H \subseteq X$ for which $\gamma^+(\sigma, H)$ is precompact. If $\{T^k : k \ge 0\}$ is a discrete dynamical system and $\gamma^+(H)$, $H \subseteq X$, is precompact, then $\omega(H)$ is nonempty and compact and $T^k H \to \omega(H)$ as $k \to \infty$.

Proof. The nonemptiness and compactness of $\omega(\sigma, x)$ follow from Expression (1.2). For the remainder of this part of the proof, let $T(t) = U(\sigma, t)$ and let $\omega(x) = \omega(\sigma, x)$. If $\operatorname{dist}(T(t)x, \omega(x)) \not\to 0$ as $t \to \infty$, then there is an $\epsilon > 0$ and a sequence $t_k \to \infty$ as $k \to \infty$ such that $\operatorname{dist}(T(t_k)x, \omega(x)) > \epsilon$ for $k = 1, 2, \ldots$. Since $T(t_k)x$ belongs to a compact set, there must be a convergent subsequence. This limit belongs to $\omega(x)$, which is a contradiction. Therefore, $\operatorname{dist}(T(t)x, \omega(x)) \to 0$ as $t \to \infty$.

If $\omega(x)$ is not connected, then $\omega(x)$ would be the union of two disjoint compact sets that are a distance δ apart. This obviously contradicts the fact that $T(t)x \to \omega(x)$ as $t \to \infty$ and so $\omega(x)$ is connected. The assertions concerning $\alpha(\sigma, x)$ are verified in the same way.

The reader can supply the details for a set $H \subseteq X$ and a process u as well as the assertions concerning a discrete dynamical system to complete the proof of the lemma. □

Lemma 1.4. *For ω-periodic processes generated by ω-periodic* RFDE, *every bounded orbit is precompact. The same conclusion is true for any $H \subseteq X$ for which $\gamma^+(\sigma, H)$ is bounded.*

Proof. If $|U(\sigma,t)\phi| = |x_{t+\sigma}(\sigma,\phi)| \leq m$ for $t \geq 0$, then f is completely continuous and ω-periodic implies there is a constant N such that $|\dot{x}(\sigma,\phi)(t+\sigma)| \leq N$, $t \geq 0$. Ascoli's theorem implies the result. A similar argument applies to arbitrary $H \subseteq X$ for which $\gamma^+(\sigma,H)$ is bounded. □

4.2 Invariance

In this section, we discuss invariant sets for autonomous processes, ω-periodic processes, and discrete dynamical systems.

Definition 2.1. If u is a process on X, then an *integral of the process on* $\mathbb{R}$ is a continuous function $y : \mathbb{R} \to X$ such that for any $\sigma \in \mathbb{R}$, $\tau^+(\sigma, y(\sigma)) = \{(\sigma+t, y(\sigma+t)) : t \geq 0\}$. An integral y is an *integral through* $(\sigma,x) \in \mathbb{R}\times X$ if $y(\sigma) = x$. An *integral set on* $\mathbb{R}$ is a set M in $\mathbb{R} \times X$ such that for any $(\sigma,x) \in M$, there is an integral y on $\mathbb{R}$ through (σ,x) and $(s,y(s)) \in M$ for $s \in \mathbb{R}$. For any $\sigma \in \mathbb{R}$, let $M_\sigma = \{x \in X : (\sigma,x) \in M\}$.

For a discrete dynamical system, an integral is defined in the same manner. In fact, if $\mathbb{Z}$ denotes the integers in $\mathbb{R}$, an *integral* y *of a discrete dynamical system* $\{T^k : k \geq 0\}$ is a function $y : \mathbb{Z} \to X$ such that for any integer $k_0 \in \mathbb{Z}$, $T^k y(k_0) = y(k+k_0)$ for all integers $k \geq 0$. An *integral set* M of a discrete dynamical system is a subset of $\mathbb{Z} \times X$ such that for any $(k_0,x) \in M$, there is an integral y on $\mathbb{Z}$ through (k_0,x) and $(k,y(k)) \in M$ for all $k \in \mathbb{Z}$.

Definition 2.2. If $\{T(t), t \geq 0\}$ is a *dynamical system* on X, then a set $Q \subseteq X$ is said to be an *invariant* set if $T(t)Q = Q$ for $t \geq 0$. This is equivalent to saying that, through every point $x \in Q$, there is an integral y through $(0,x)$ such that $(y(s) \in Q$, $s \in \mathbb{R}$. If $\{T^k : k \geq 0\}$ is a *discrete dynamical system* on X, a set $Q \subseteq X$ is said to be *invariant* if $TQ = Q$. If u is an *ω-periodic process* in $\mathbb{R} \times X$, an integral set $M \subseteq \mathbb{R} \times X$ is an *invariant set* of u if, for each $\sigma \in \mathbb{R}$, the set $M_\sigma = \{x \in X : (\sigma,x) \in M\}$ is an invariant set of the discrete dynamical system $\{U^k(\sigma,\omega) : k \geq 0\}$ and $M_{\sigma+\omega} = M_\sigma$ for all $\sigma \in \mathbb{R}$.

The concept of invariance for discrete dynamical systems in Definition 2.2 is actually equivalent to the following: Q is invariant if and only if it is possible to extend the definition of T^k on Q to negative integers k and $T^k Q \subseteq Q$ for $k \in (-\infty,\infty)$. In fact, if T satisfies the last condition stated then $TQ \subseteq Q$, $T^{-1}Q \subseteq Q$, which implies $TQ = Q$. If $TQ = Q$, then, in particular, $Q \subseteq TQ$. Therefore, we can define T^{-1} on Q and $T^{-n} = (T^{-1})^n$.

Our next objective is to give sufficient conditions for the ω-limit set of an orbit of a process to be invariant.

Lemma 2.1. *If u is a dynamical system on X, and an orbit $\gamma^+(x)$ is precompact, then $\omega(x)$ is nonempty, compact, connected, and invariant. The same conclusion is true for any connected set $H \subseteq X$ for which $\gamma^+(H)$ is precompact. Finally, $T(t)H \to \omega(H)$ as $t \to x$.*

Proof. We only need to prove invariance since the other assertions are contained in Lemma 1.3. To show invariance, let $U(0,t) = T(t)$, $t \geq 0$, suppose $y \in \omega(x)$, $t_k \to \infty$ and $T(t_k)x \to y$ as $k \to \infty$. For any integer $N \geq 0$, there is an integer $k_0(N)$ such that $T(t_k + t)x$ is defined for $-N \leq t < +\infty$ if $k \geq k_0(N)$. Since $\gamma^+(x)$ is precompact, one can, therefore, find a subsequence $\{t_{k,N}\}$ of the $\{t_k\}$ and a continuous function $G : [-N, N] \to \omega(x)$ such that $T(t_{k,N} + t)x \to G(t)$ as $k \to \infty$ uniformly for $t \in [-N, N]$. Using the diagonalization procedure, we can find a subsequence we label again as $\{t_k\}$ and a continuous function $G : (-\infty, \infty) \to \omega(x)$ such that $T(t + t_k)x \to G(t)$ as $k \to \infty$ uniformly on compact sets of $(-\infty, \infty)$.

From the definition of G, we have $G(0) = y$ and $G(t + \tau) = T(t)G(\tau)$ for all $t \geq 0$, $\tau \in (-\infty, \infty)$. Consequently, $G : \mathbb{R} \to X$ is an integral through $(0, y)$ and $\omega(x)$ is invariant. The lemma will be proved when the reader supplies the details for sets $H \subseteq X$. □

Suppose now the autonomous process $\{T(t) : t \geq 0\}$ on C is generated by an RFDE(f) with f completely continuous and $\gamma^+(\phi)$ is a bounded orbit. Lemma 1.4 implies the orbit $\gamma^+(\phi)$ is precompact and Lemma 2.1 implies $\omega(\phi)$ is nonempty, compact, connected, and invariant. Invariance in Definition 2.2 means that a function $P : \mathbb{R} \times M \to M$ can be defined so that $P(\sigma, \phi) \in M$ for all $(\sigma, \phi) \in \mathbb{R} \times M$, $T(t)P(\sigma, \phi) = P(t + \sigma, \phi)$ for all $\sigma \in \mathbb{R}$, $t \geq 0$. This relation can be used to extend the definition of $T(t)$ on M to negative t. How do we know that the extension of the definition of $T(t)$ on M to negative t can be made in such a way that $T(t)\phi = g_t$, $t \in (\infty, \infty)$, for some function $g : (-\infty, \infty) \to \mathbb{R}^n$ that is a solution of the RFDE (1.1) on $(-\infty, \infty)$? To see that this is indeed the case, it is necessary to return to the proof of Lemma 2.1 and use the fact that $T(t)$ is now generated from an RFDE. Using the same type of argument as in the proof of Lemma 2.1, it is possible to construct a continuous function $g : \mathbb{R} \to \mathbb{R}^n$ such that $g_0 = \phi$ and for any $\sigma \in \mathbb{R}$, $T(t)g_\sigma = g_{t+\sigma}$, $t \geq 0$; that is, the function $G(t) = g_t$, $t \in \mathbb{R}$, is an integral through $(0, \phi)$. Let $x_t = T(t)\phi$. For t in any compact set $[-N, N]$, there is a $K(N)$ such that

$$
\begin{aligned}
|g(t+h) - g(t) - hf(g_t)| \leq\ & |g(t+h) - x(t+h+t_k)| \\
&+ |x(t+h+t_k) - x(t+t_k) - hf(x_{t+t_k})| \\
&+ |x(t+t_k) - g(t)| + h|f(x_{t+t_k}) \quad f(g_t)|
\end{aligned}
$$

for all $k \geq K(N)$. Choose $k(h)$ in such a way that $k(h) \to \infty$ as $h \to 0$ and $|g(t) - x(t + t_{k(h)})| = o(|h|)$ as $h \to 0$ for all t in $[-N, N]$. The right-hand side of this inequality is now $o(|h|)$ as $h \to 0$, which proves $\dot{g}(t) = f(g_t)$. Therefore, g is a solution of Equation (1.1) on $(-\infty, \infty)$ and $g_t \in \omega(\phi)$,

$t \in (-\infty, \infty)$. If $H \subseteq X$ is an arbitrary set for which $\gamma^+(H)$ is precompact, one can prove the same result to obtain the following corollary.

Corollary 2.1. *If an autonomous* RFDE(f) *generates a dynamical system and $\gamma^+(\phi)$ is a bounded orbit then $\omega(\phi)$ is nonempty, compact, connected, and invariant, where invariance means for any $\psi \in \omega(\phi)$, there is a solution x of the* RFDE(f) *on $(-\infty, \infty)$ with $x_0 = \psi$ and $x_t \in \omega(\phi)$, $-\infty < t < \infty$. If $H \subseteq C$ is connected and $\gamma^+(H)$ is bounded, the same conclusion is true.*

Lemma 2.2. *If $\{T^k : k \geq 0\}$ is a discrete dynamical system on X, $H \subseteq X$, and $\gamma^+(H)$ is precompact, then $\omega(H)$ is nonempty, compact, invariant, and $T^k H \to \omega(H)$ as $k \to \infty$. Also, H compact, $\gamma^+(H)$ precompact, and $\omega(H) \subseteq H$ implies $\omega(H) = \bigcap_{n \geq 0} T^n H$.*

Proof. The fact that $\omega(H)$ is nonempty and compact and that $T^k H \to \omega(H)$ is contained in Lemma 1.3. The continuity of T implies $T\omega(H) \subseteq \omega(H)$. If $y \in \omega(H)$, then there is a sequence $\{x_j\} \subseteq H$ and a sequence $\{n_j\}$ of integers, $n_j \to \infty$ as $j \to \infty$, such that $T^{n_j} x_j \to y$ as $j \to \infty$. By the precompactness hypothesis, we can select a subsequence that we again label as n_j such that $T^{n_j - 1} x_j \to z$. Then $z \in \omega(H)$ and $Tz = \lim_{j \to \infty} T^{n_j} x_j = y$ by the continuity of T. Therefore, $\omega(H) \subseteq T\omega(H)$, which shows $\omega(H) = T\omega(H)$. The first part of the lemma is proved.

Suppose now H is compact, $\gamma^+(H)$ is precompact, and $\omega(H) \subseteq H$. If $J(H) = \bigcap_{n \geq 0} T^n H$, then clearly $J(H)$ is compact and $J(H) \subseteq \omega(H)$. To prove the converse, suppose $y \in \omega(H)$ and $T^{n_j} x_j \to y$ as $j \to \infty$ where $n_j \to \infty$ as $j \to \infty$ and each $x_j \in H$. Since $\gamma^+(H)$ is precompact, for any integer i, we can find a subsequence of the sequence $T^{n_j - i}$, which we label as before and a $y_i \in \omega(H) \subseteq H$ such that $T^{n_j - 1} x_j \to y_i$ as $j \to \infty$. Thus $T^i Y_i = y$ for all integers i. Therefore $y \in J(H)$. This proves $\omega(H) \subseteq J(H)$ and the proof of the lemma is complete. □

Lemma 2.3. *If u is an ω-periodic process on X and an orbit $\gamma^+(\sigma, x)$ is precompact, then there is an invariant integral set $M(\sigma, x) \in \mathbb{R} \times X$ such that if $M_s(\sigma, x) = \{x : (s, x) \in M(\sigma, x)\}$, then* $\mathrm{dist}(U(\sigma, t)x, M_{\sigma+t}(\sigma, x)) \to 0$ *as $t \to \infty$. The same conclusion is true for any set $H \subseteq X$ such that $\gamma^+(\sigma, H)$ is precompact.*

Proof. If $T = U(\sigma, \omega)$, then $\{T^k : k \geq 0\}$ is a discrete dynamical system with $\gamma^+(x) = \{T^k x : k \geq 0\}$ precompact. Lemma 2.2 implies the ω-limit set Q is an invariant set of this discrete system. For a given integer $k \in (-\infty, \infty)$ and $t \in [0, \omega)$, define the compact set $M_{\sigma - k\omega + t} = U(\sigma, t)Q$. Then M_s is defined for all $s \in \mathbb{R}$ and $M_{s+\omega} = M_s$. Consider the set $M = M(\sigma, x)$ in $\mathbb{R} \times X$ such that for each integer $k \in (-\infty, \infty)$, $t \in [0, \omega)$, $\{(\sigma - k\omega + t, x) : x \in M_{\sigma - k\omega + t}\} = M$. The set M is an invariant set for the ω-periodic process u. In fact, since $U(\sigma - k\omega + t, \omega) = U(\sigma + t, \omega)$ and $U(\sigma, \omega)Q = Q$, we have

$$\begin{aligned}U(\sigma-k\omega+t,\omega)M_{\sigma-k\omega+t} &= U(\sigma+t,\omega)U(\sigma,t)Q = U(\sigma,\omega+t)Q\\ &= U(\sigma+\omega,t)U(\sigma,\omega)Q\\ &= U(\sigma,t)Q = M_{\sigma-k\omega+t}.\end{aligned}$$

To prove $\operatorname{dist}(U(\sigma,s)x, M_{\sigma+s}) \to 0$ as $s \to \infty$, observe first that Lemma 1.3 implies $U(\sigma,k\omega)x \to Q$ as $k \to \infty$. If $s = k\omega+t$, then $M_{\sigma+s} = U(\sigma,t)Q$ and

$$\begin{aligned}\operatorname{dist}(U(\sigma,k\omega+t)x, U(\sigma,t)Q) &= \operatorname{dist}(U(\sigma+k\omega,t)U(\sigma,k\omega)x, U(\sigma,t)Q)\\ &= \operatorname{dist}(U(\sigma,t)U(\sigma,k\omega)x, U(\sigma,t)Q) \to 0\end{aligned}$$

as $k \to \infty$ for each $t > 0$. This proves the first part of the lemma. The reader may easily supply the details for the case of a set $H \subseteq X$ to complete the proof of the lemma. □

One can combine the ideas in the proofs of Lemmas 1.4, 2.3, and Corollary 2.1 to obtain the following result.

Corollary 2.2. *If an ω-periodic* RFDE(f) *generates an ω-periodic process u and $\gamma^+(\sigma,\phi)$ is a bounded orbit, then there is an invariant integral set $M(\sigma,\phi) \subseteq \mathbb{R} \times C$ of the* RFDE(f), *$M_s(\sigma,\phi) = \{\psi \in C : (s,\psi) \in M(\sigma,\phi)\}$ is compact, and* $\operatorname{dist}(x_t(\sigma,\phi), M_{\sigma+t}(\sigma,\phi)) \to 0$ *as $t \to \infty$. That is, there is a set of functions $Y(\sigma,\phi) = \{y : \mathbb{R} \to \mathbb{R}^n\}$ such that each y is a solution of the* RFDE(f) *on $(-\infty,\infty)$, the set $M(\sigma,\phi) = \{(s,y_s) : s \in \mathbb{R}, y \in Y(\sigma,\phi)\}$ is invariant for the ω-periodic process generated by the* RFDE(f) *and* $\operatorname{dist}(x_t(\sigma,\phi), M_{\sigma+t}(\sigma,\phi)) \to 0$ *as $t \to \infty$.*

4.3 Discrete systems—Maximal invariant sets and attractors

In this section, we discuss discrete dynamical systems $\{T^k : k \geq 1\}$ on X and give conditions that ensure there is a maximal compact invariant set that is a stable global attractor. For any set $H \subseteq X$ and $\epsilon > 0$, let $\mathcal{B}(H,\epsilon) = \{x \in X : \operatorname{dist}(H,x) < \epsilon\}$.

Definition 3.1. For a given continuous function $T : X \to X$, we say a set $K \subseteq X$ *attracts a set* $H \subseteq X$ if, for any $\epsilon > 0$, there is an integer $N(H,\epsilon)$ such that $T^n(H) \subseteq \mathcal{B}(K,\epsilon)$ for $n \geq N(H,\epsilon)$. We say K *attracts points of* X if K attracts each point of X. We say K *attracts compact sets of* X if K attracts each compact set of X. We say K *attracts neighborhoods of points* (*compact sets*) of X if, for each point $x \in X$ (each compact set $H \subseteq X$), there is a neighborhood O_x of x (H_0 of H) such that K attracts $O_x(H_0)$. We say K *attracts bounded sets* of X if K attracts each bounded set of

X. We say that K is a *global attractor* if K is invariant and attracts each bounded set of X.

Definition 3.2. A set $M \subseteq X$ is said to be *stable* (with respect to the dynamical system $\{T^k : k \geq 0\}$) if, for any $\epsilon > 0$, there is a $\delta > 0$ such that $x \in \mathcal{B}(M, \delta)$ implies $T^n x \in \mathcal{B}(M, \epsilon)$ for $n \geq 0$. The set M is said to be *asymptotically stable* if it is stable and there is an $\epsilon_0 > 0$ such that M attracts points of $\mathcal{B}(M, \epsilon_0)$. The set M is said to be *uniformly asymptotically stable* if it is asymptotically stable and, for any $\eta > 0$, there is an integer $n_0(\eta, \epsilon_0)$ such that $T^n x \in \mathcal{B}(M, \eta)$ for $n \geq n_0(\eta, \epsilon_0)$ and $x \in \mathcal{B}(M, \epsilon_0)$.

Definition 3.3. If B is a bounded set in a Banach space X, the *Kuratowski measure* $\alpha(B)$ *of noncompactness of* B is defined as

$$\alpha(B) = \inf\{d : B \text{ has a finite cover of diameter } < d\}.$$

Some properties of $\alpha(B)$ that will be needed in the following are now stated without proof.

(i) $\alpha(B) = 0$ if and only if $\overline{B}$ is compact.

(ii) $\alpha(A \cup B) = \max(\alpha(A), \alpha(B))$.

(iii) $\alpha(\overline{\text{co}}B) = \alpha(B)$, where $\overline{\text{co}}B$ is the closed convex hull of B.

Lemma 3.1. *If $T : X \to X$ and there is a compact set $K \subseteq X$ that attracts compact sets of X, then for any compact set $H \subseteq X$, $\gamma^+(H)$ is precompact.*

Proof. Let $\alpha(B)$ be the Kuratowski measure of noncompactness of a bounded set B given in Definition 3.3. For any compact set $H \subseteq X$, the set

$$A \overset{\text{def}}{=} \bigcup_{n \geq 0} T^n H = \gamma^+(H)$$

is bounded since K attracts compact sets. Since $T^j H$ is compact for any j, we have $\alpha(A) = \alpha\big(\bigcup_{n \geq j} T^n H\big)$ for any j. But, for any $\epsilon > 0$, we have $\bigcup_{n \geq j} T^n H \subseteq \mathcal{B}(K, \epsilon)$ for $j \geq N(H, \epsilon)$. Consequently, $\alpha(A) \leq 2\epsilon$ for every $\epsilon > 0$. This implies $\alpha(A) = 0$ and so A is precompact. □

Our next objective is to define a maximal compact invariant set for a discrete dynamical system satisfying the conditions of Lemma 3.1.

Suppose T is continuous and there is a compact set $K \subseteq X$ that attracts compact sets of X. Lemma 3.1 implies $\gamma^+(K)$ is precompact and Lemma 2.2 implies $\omega(K)$ exists. For any $\epsilon > 0$, there is an integer $n_0(K, \epsilon)$ such that $T^n K \subseteq \mathcal{B}(K, \epsilon)$ for $n \geq n_0(K, \epsilon)$. Thus $\omega(K) \subseteq \mathcal{B}(K, \epsilon)$ for every $\epsilon > 0$. Therefore, $\omega(K) \subseteq K$. It follows from Lemma 2.2 that

$$J \overset{\text{def}}{=} \bigcap_{n \geq 0} T^n K = \omega(K) \tag{3.1}$$

is nonempty, compact, and invariant. We claim that J is independent of the set K, which attracts compact sets of X. In fact, if we designate J by $J(K)$ and K_1 is any other compact set that attracts compact sets of X, then there is an integer $n_0(K, K_1, \epsilon)$ such that $T^n J(K) \subseteq \mathcal{B}(K_1, \epsilon)$, $T^n J(K_1) \subseteq \mathcal{B}(K, \epsilon)$, for $n \geq n_0(K, K_1, \epsilon)$. Since $J(K)$ and $J(K_1)$ are invariant, this implies $J(K) \subseteq K_1$, $J(K_1) \subseteq K$ and $J(K) \subseteq T^n K_1$, $J(K_1) \subseteq T^n K$ for all $n \geq 0$. Therefore, $J(K) = J(K_1)$.

Theorem 3.1. *Suppose $T : X \to X$ and there is a compact set $K \subseteq X$ that attracts compact sets of X and let J be defined by Expression* (3.1). *Then the following conclusions hold:*

(i) *$J = \omega(K)$ is independent of K, is a nonempty compact invariant set, and is maximal with respect to this property.*

(ii) *J is connected.*

(iii) *J is stable.*

(iv) *For any compact set $H \subset X$, J attracts H.*

Proof. The remarks preceding the statement of the theorem imply that $J = \omega(K)$ is a nonempty compact invariant set that is independent of K.

To prove that J is maximal, suppose that H is any compact invariant set. Since K attracts H, for any $\epsilon > 0$, there is an integer $n_1(H, \epsilon)$ such that $T^n H \subset \mathcal{B}(K, \epsilon)$ for $n \geq n_1(H, \epsilon)$. From the invariance of H it follows that $H = T^n H$ for all n. So we can choose a sequence $\epsilon_j \to 0$ as $j \to \infty$ to obtain $H \subset K$. Likewise, $H \subset T^n K$ for all $n \geq 0$ and so $H \subset J$.

Now suppose that $x \in X$ is arbitrary. Then $\gamma^+(x)$ is precompact from Lemma 3.1. Thus, $\omega(x)$ is nonempty, compact, and invariant and $\omega(x) \subset J$. Since $T^k x \to \omega(x)$ as $k \to \infty$, it follows that J attracts points of X.

We now prove that J is connected. If $\overline{\text{co}}\ J$ is the closed convex hull of J, then $\overline{\text{co}}\ J$ is compact and connected and J attracts $\overline{\text{co}}\ J$. If J is not connected, then there are open sets U, V with $U \cap J \neq \emptyset$, $V \cap J \neq \emptyset$. By the continuity of T, the set $T^n(\overline{\text{co}}\ J)$ is connected for each $n \geq 0$. Since $J \subset T^n(\text{co}\ J)$ for all $n \geq 0$, we have $U \cap T^n(\overline{\text{co}}\ J) \neq \emptyset$, $V \cap T^n(\overline{\text{co}}\ J) \neq \emptyset$ for all $n \geq 0$. Since $T^n(\overline{\text{co}}\ J)$ is connected, there is an $x_n \in T^n(\overline{\text{co}}\ J) \setminus (U \cup V)$. Since J attracts $\{ x_n, n \geq 0 \}$, this set must be compact and we may assume that $x_n \to x \in J$. Clearly, $x \notin U \cup V$, which is a contradiction.

To prove that J is stable, suppose that this is not the case. Then for some $\epsilon > 0$, which may be chosen as small as desired, there are sequences of integers n_j and $y_j \in X$ such that $n_j \to \infty$, $y_j \to J$ as $j \to \infty$, $T^n y_j \in \mathcal{B}(J, \epsilon)$, $0 \leq n \leq n_j$, and $T^{n_j+1} y_j$ is not in $\mathcal{B}(J, \epsilon)$. Furthermore, J compact implies that we may assume without loss of generality that $y_j \to y \in J$ as $j \to \infty$. The set $H = \{ y, y_j; j \geq 1 \}$ is compact. Lemma 3.1 implies that $\gamma^+(H)$ is precompact. Therefore, Lemma 2.2 implies that $\omega(H)$ is nonempty, compact, and invariant. The first part of the theorem being proved implies that $\omega(H) \subset J$. Also, $\gamma^+(H)$ precompact implies that with no loss in generality we may assume that $T^{n_j} y_j \to z$ as $j \to \infty$. Then $z \in \omega(H) \subset J$. But the choice of n_j, y_j implies that $Tz \notin \mathcal{B}(J, \epsilon)$ and, therefore, $Tz \notin J$. Since J is invariant, this is a contradiction and the proof of assertion (iii) is complete.

To prove (iv), suppose that $\epsilon > 0$ is given and δ is the number associated with ϵ in the definition of stability. For any $x \in X$, there is an integer $N(x)$ such that $T^n x \subset \mathcal{B}(J, \delta)$ for $n \geq N(x)$. Since T is continuous, there is a neighborhood O_x of x such that $T^{N(x)} O_x \subset \mathcal{B}(J, \delta)$. As a consequence, $T^j O_x \subset \mathcal{B}(J, \epsilon)$ for $j \geq N(x)$. If H is an arbitrary compact set in X, then a finite covering argument proves (iv). The proof of the theorem is complete. □

From Theorem 3.1, we see that the existence of a maximal compact invariant set that is stable and attracts compact sets of X is ensured under very weak conditions on the map T. From the discussion in Section 1 dealing with ω-periodic RFDE, a discrete dynamical system is obtained from the solution map $T(t, \sigma)$ by defining $T = T(\omega, 0)$. Furthermore, Lemma 1.1 implies that $k\omega$-periodic solutions of the RFDE are obtained from the fixed points of T^k. As a consequence, it is important to know under what conditions such fixed points exist and, even more importantly, to know when $k = 1$; that is, when does T itself have a fixed point?

Intuitively, one might suspect that the conditions of Theorem 3.1 would imply that T has a fixed point since J is stable and attracts compact sets of X. This problem, however, as well as the problem of whether even some iterate of T has a fixed point, is unsolved. The next section is devoted to the specification of further conditions on T, which will ensure that T or some iterate of T has a fixed point.

However, before developing the theory of fixed points, it is convenient from the point of view of later applications to specify conditions on T, which will imply a type of stability that is stronger than the conclusion in Theorem 3.1(iv). More specifically, we want the same conclusion for any bounded set $H \subset X$. Some additional definitions are required.

Definition 3.4. A continuous map $T : X \to X$ is said to be *point* (*compact*) (*local*) (*locally compact*) (*bounded*) *dissipative* if there is a bounded set $B \subset X$ such that B attracts points (compact sets) (neighborhoods of points) (neighborhoods of compact sets) (bounded sets) of X.

Obviously, bounded dissipative implies locally compact dissipative implies local dissipative implies compact dissipative implies point dissipative.

Definition 3.5. A continuous map $T : X \to X$ is said to be *asymptotically smooth* if, for any bounded set $B \subset X$ for which $TB \subset B$, there is a compact set $B^* \subset X$ such that B^* attracts B.

It is not difficult to show that this definition is equivalent to the following one. A continuous map $T : X \to X$ is asymptotically smooth if, for any bounded set $B \subset X$, there is a compact set $B^* \subset X$ such that for any $\epsilon > 0$, there is an integer $n_0(B, \epsilon)$ with the property that $T^n x \in B$ for $n \geq 0$ implies that $T^n x \in \mathcal{B}(B^*, \epsilon)$ for $n \geq n_0(B, \epsilon)$.

The concept of asymptotically smooth is a result of a detailed investigation of the relationship between asymptotic stability and uniform asymptotic stability of compact invariant sets in a Banach space X. In fact, it is an interesting exercise to prove the following fact: *If T is asymptotically smooth and J is a compact invariant set of T, then J is asymptotically stable if and only if it is uniformly asymptotically stable.* As we will see, the importance of the concept is related to the fact that uniform asymptotically stable sets enjoy some type of persistence under perturbations of the map T, whereas this is not the case when the set is only asymptotically stable.

Lemma 3.2. *If $T : X \to X$ is local dissipative and asymptotically smooth, then there is a compact set $K \subset X$ that attracts neighborhoods of compact sets of X.*

Proof. If B_1 is the bounded set in the definition of local dissipative and $\delta > 0$, let $B = \mathcal{B}(B_1, \delta)$ and let B^* be the corresponding compact set in the definition of asymptotically smooth. If H is an arbitrary compact set and x is an arbitrary point in H, then there is a neighborhood O_x of x and an integer $N(x)$ such that $T^n O_x \subset B$ for $n \geq N(x)$. Selecting from the covering $\{O_x : x \in H\}$ of H a finite subcovering, we see that there is an integer $N(H)$ and an open neighborhood H_1 of H such that $T^n H_1 \subset B$ for $n \geq N(H)$. Since H is compact and T is continuous, one can choose a neighborhood H_0 of H, $H_0 \subset H_1$, such that $T^n H_0$ is bounded for $0 \leq n \leq N(H)$. Therefore, $\{T^n H_0 : n \geq 0\}$ is bounded since $H_0 \subset H_1$. Since T is asymptotically smooth, for any $\epsilon > 0$, this implies that $T^n H_0 \subset \mathcal{B}(B^*, \epsilon)$ for $n \geq N(H) + n_0(B, \epsilon)$. If we let $K = B^*$, the proof of the lemma is complete. □

Theorem 3.2. *If $T : X \to X$ is local dissipative and asymptotically smooth, then the set J in Expression* (3.1) *defined by K in Lemma* 3.2 *satisfies all of the conclusions of Theorem* 3.1 *and*

(v) *For any compact set $H \subset X$, there is a neighborhood H_1 of H such that $\gamma^+(H_1)$ is bounded and J attracts H_1; in particular, J is uniformly asymptotically stable.*

(vi) *If $B \subset X$ is an arbitrary bounded set and $\gamma^+(B)$ is bounded, then J attracts B; in particular, if $\gamma^+(B)$ is bounded for every bounded set $B \subset X$, then J is a compact global attractor.*

Proof. Lemma 3.2 implies that T satisfies the hypotheses of Theorem 3.1 and, therefore, the first part of the theorem is proved. To prove Property (iv), suppose that $B \subset X$ and $\gamma^+(B) \subset U$ for some bounded set $U \subset X$. Let U^* be the corresponding compact set in the definition of asymptotically smooth. The definition of asymptotically smooth implies that for any $\epsilon > 0$, there is an integer $n_0(U, \epsilon)$ such that $T^n B \subset \mathcal{B}(U^*, \epsilon)$ for $n \geq n_0(U, \epsilon)$. Since J attracts a neighborhood of the compact set U^*, we may assume that $\epsilon > 0$ is fixed and so small that J attracts $\mathcal{B}(U^*, \epsilon)$. For any $\eta > 0$, there is an integer $m_0(U^*, \epsilon)$ such that $T\mathcal{B}(U^*, \epsilon) \subset \mathcal{B}(J, \eta)$ for $m \geq m_0(U^*, \eta)$. If we let $N = n_0(U, \epsilon) + m_0(U^*, \eta)$, then

$$T^{n_0(U,\,\epsilon)+m} B \subset T^m \mathcal{B}(U^*, \epsilon) \subset \mathcal{B}(J, \eta) \quad \text{for } m \geq m_0(U^*, \eta)\,,$$

or $T^n B \subset \mathcal{B}(J, \eta)$ for $n \geq N$. This completes the proof of the theorem. □

Corollary 3.1. *If $T : X \to X$ is asymptotically smooth, then the following are equivalent:*

(i) *There is a compact set K that attracts compact sets of X.*

(ii) *There is a compact set K that attracts neighborhoods of compact sets of X.*

It is possible to obtain a result on the existence of a compact global attractor, which is easier to use in the applications than Theorem 3.2. We need the following lemma.

Lemma 3.3. *Let T be asymptotically smooth and point dissipative. If the orbit of any compact set is bounded, then T is locally compact dissipative. If the orbit of any bounded set is bounded, then T is bounded dissipative.*

Proof. Let B be a nonempty closed bounded set that attracts points of X and let $U = \{ x \in B : \gamma^+(x) \subset B \}$. Then U attracts points of X and $\gamma^+(U)$ is bounded. Since $T\gamma^+(U) \subset \gamma^+(U)$ and T is asymptotically smooth, there is a compact set $K \subset \mathrm{Cl}\, U$ such that K attracts U and attracts points of X. The set K also attracts itself and so $\gamma^+(K)$ is precompact. If $J = \omega(K)$, then J is a compact invariant set that attracts points of X.

Now suppose that the orbit of any compact set is bounded. We first show that there is a neighborhood V of J such that $\gamma^+(V)$ is bounded. If this is not the case, then there is a sequence $x_j \in X$ and a sequence of integers $k_j \to \infty$ and a $y \in J$ such that $\lim_{j\to\infty} x_j = y$ and $|T^{k_j} x_j| \to \infty$ as $j \to \infty$. But then $\mathrm{Cl}\{ x_j,\, j \geq 1 \}$ is a compact set with $\gamma_+(\mathrm{Cl}\{ x_j,\, j \geq 1 \})$ unbounded. This is a contradiction. Thus, there is a neighborhood V of J such that $\gamma^+(V)$ is bounded. Since J attracts points of X and T is continuous, for any $x \in X$, there is a neighborhood O_x of x and an integer n_0 such that $T^n O_x \subset \gamma^+(V)$ for $n \geq n_0$; that is, $\gamma^+(V)$ attracts O_x. For any compact set H in X, one can find a finite covering of H and thus an open set H_1 containing H such that $\gamma^+(V)$ attracts H_1. Thus, T is locally compact dissipative.

Now suppose that orbits of bounded sets are bounded. Then orbits of compact sets are bounded and T is locally compact dissipative from the first part of the lemma. We now apply Theorem 3.2 to complete the proof. □

As a consequence of Theorem 3.2 and Lemma 3.3, we have the following result.

Theorem 3.3. *If T is asymptotically smooth, point dissipative, and orbits of bounded sets are bounded, then there exists a connected compact global attractor.*

We now give some sufficient conditions for the hypotheses of this theorem to hold. To do this, some additional definitions are required.

Definition 3.6. Suppose that α is the Kuratowski measure of noncompactness of a set in X and that $T : X \to X$ is a continuous map. The map T is *conditionally condensing* if $\alpha(TB) < \alpha(B)$ for any bounded set $B \subset X$ for which $\alpha(B) > 0$ and TB is bounded. The map T is said to be a *conditionally α-contracting* if there is a constant k, $0 \le k < 1$, such that $\alpha(TB) \le k\alpha(B)$ for any bounded set $B \subset X$ for which TB is bounded. If T takes bounded sets into bounded sets, then a conditionally α-contracting (resp. condensing) map is an *α-contracting* (resp. *condensing*) *map*. The map T is said to be *conditionally completely continuous* if TB is precompact for any bounded set B in X for which TB is bounded. The discrete dynamical system generated by T is said to be *conditionally completely continuous for $k \ge n_0$*, if, for any bounded set B in X, there is a compact set B^* in X such that for any integer $N \ge n_0$ and any $x \in X$ with $T^n X \in B$, $0 \le n \le N$, it follows that $T^n x \in B^*$ for $n_0 \le n \le N$.

Note that, if T is conditionally completely continuous, then T is a conditional α-contraction, and this implies that T is conditionally condensing.

It is also obvious that, if T takes bounded sets into bounded sets and T^{n_0} is completely continuous for some n_0, then the discrete dynamical system generated by T is conditionally completely continuous for $k \ge n_0$.

One of the most important mappings T, which is an α-contraction and is not necessarily completely continuous, is $T = S + U$, where U is a completely continuous mapping and the mapping S is a bounded linear map on X with $\|S\| < 1$. If S is only a bounded linear map with modulus of the spectrum < 1, then the space X can be renormed so that the property is satisfied (see Lemma 4.4).

For RFDE, it was shown in Theorem 6.1 of Section 3.6 that the solution map $T(t, \sigma) = S(t-\sigma)+U(t, \sigma)$, where $U(t, \sigma)$ is conditionally completely continuous for all $t \ge \sigma$ and the bounded linear map $S(t) = 0$ for $t \ge r$. Therefore, $T(t, \sigma)$ is conditionally completely continuous for $t \ge \sigma + r$.

Also, for any given $\omega > 0$, we can renorm the space X so that the map $T \equiv T(\sigma + \omega, \sigma)$ is a conditional α-contraction.

Lemma 3.4. *If $T : X \to X$ is point dissipative and the dynamical system generated by T is conditionally completely continuous for $k \geq n_0$, then there is a compact set K in X such that for any compact set H in X, there is an open neighborhood H_0 of H and an integer $N(H)$ such that $\gamma^+(H_0)$ is bounded and $T^n H_0 \subset K$ for $n \geq N(H)$. In particular, there is a compact set K that attracts compact sets of X.*

Proof. Since T is point dissipative, for any $\epsilon > 0$, $x \in X$, there is an integer $N(x, \epsilon)$ such that $T^n x \in \mathcal{B}(B, \epsilon)$ for $n \geq N(x, \epsilon)$, where B is the bounded set in Definition 3.4. We will fix ϵ and therefore suppress the explicit dependence of constants and sets on ϵ. Let B^* be the compact set of Definition 3.6 corresponding to $\mathcal{B}(B, \epsilon)$. By the continuity of T, there is an open neighborhood O_x of x such that $T^n O_x \subset \mathcal{B}(B, \epsilon)$ for $N(x) \leq n \leq N(x)+n_0$. Therefore, $T^{n(x)} O_x \subset B^*$, where $n(x) = N(x)+n_0$. Suppose that H is an arbitrary compact set of X. The neighborhoods O_x of $x \in H$ form a covering of H. Selecting from this cover a finite subcover $\{ O_{x_i}(H) \}$ and letting $N(H) = \max_i n(x_i)$, we have $T^{n(x_i)} O_{x_i}(H) \subset B^*$ and $0 \leq n(x_i) \leq N(H)$. Let $H_0 = \bigcup_i O_{x_i}(H)$ and let K be the compact set consisting of the union of $B^*, TB^*, \ldots, T^{N(B^*)} B^*$. We claim that $T^n B^* \subset K$ for $n \geq N(B^*)$. In fact, if $x \in B^*$ and $n \geq N(B^*)$, then there is a least integer j, $0 \leq j \leq n$, such that $T^{n-j} x \in B^*$ and $T^{n-k} x \notin B^*$ for $0 \leq k < j$. But $T^{n-j} x \in O_{x_i}(B^*)$ for some i, $T^{n(x_i)} T^{n-j} x \subset B^*$, $0 \leq n(x_i) \leq N(B^*)$. This implies that $0 \leq j \leq N(B^*)$ and so $T^n x \in K$ for $n \geq N(B^*)$. Let $N(H)' = N(H) + N(B^*)$ and suppose that $x \in H_0$, $n \geq N(H)'$. Then $x \in O_{x_i}(H)$ for some i, $0 \leq n(x_i) \leq N(H)$ and $T^{n(x_i)} O_{x_i}(H) = T^{n-n(x_i)} T^{n(x_i)} O_{x_i}(H) \subset K$ for $n \geq N(H)'$. The fact that $\bigcup_{j \geq 0} T^j H_0$ is bounded is obvious, and the lemma is proved. □

If T is a mapping satisfying the conditions of Lemma 3.4, then T satisfies the hypotheses of Theorem 3.1. Therefore, there is a compact set J that is stable and attracts neighborhoods of compact sets. However, more is implied from Lemma 3.4 if T takes bounded sets into bounded sets. In fact, we can prove the following result.

Corollary 3.2. *If $T : X \to X$ takes bounded sets into bounded sets, is point dissipative, and T^{n_0} is completely continuous for some n_0, then for any bounded set $B \subset X$, there is a bounded set $U \subset X$, a compact set $K \subset X$, and an integer $N(B)$ such that $T^j B \subset U$, $j \geq 0$ and $T^j B \subset K$, $j \geq N(B)$. In addition, K is a connected compact global attractor.*

Proof. If B is bounded, then the hypotheses on T imply that $\mathrm{Cl}\, T^{n_0} B = H$ is compact. Therefore, Lemma 3.4 implies that $T^n H \subset K$ for $n \geq N(H)$, where K is a compact set. Thus, $T^n B \subset K$ for $n \geq N(H) + n_0$. Since T

takes bounded sets into bounded sets, the existence of the bounded set U is clear and the corollary is proved. □

Lemma 3.5. *A conditionally condensing map is asymptotically smooth.*

Proof. Let B be a closed bounded set in X such that $B \subset X$, $TB \subset B$. We want to show first that every sequence $\{T^{k_j}x_j\}$ with $k_j \to \infty$ as $j \to \infty$ has compact closure. Let $C = \{\{T^{k_j}x_j\} : \{x_j\} \subset B, \{k_j\}$ integers, $0 \le k_j \to \infty\}$ and $\eta = \sup\{\alpha(h) : h \in C\}$. Since $h \in C \subset B$, we have $\eta \le \alpha(B)$. We claim that there is an $h^* \in C$ such that $\alpha(h^*) = \eta$. Let $\{h_\ell\} \subset C$ be such that $\alpha(h_\ell) \to \eta$ and define $\tilde{h}_\ell = \{T^{k_j}x_j : T^{k_j}x_j \in h_\ell, k_j > \ell\}$. Then $\alpha(\tilde{h}_\ell) = \alpha(h_\ell)$. If $h^* = \bigcup_\ell \tilde{h}_\ell$ ordered in any way, then $h^* \in C$ and $\alpha(h^*) \ge \alpha(h_\ell)$. Therefore, $\alpha(h^*) = \eta$. If $\tilde{h}^* = \{T^{k_j-1}x_j : T^{k_j}x_j \in h^*\}$, then $\tilde{h}^* \in C$ and $\alpha(\tilde{h}^*) \le \eta$. Since $T(\tilde{h}^*) = h^*$ and $\alpha(h^*) = \eta$, we have $\alpha(\tilde{h}^*) \ge \alpha(\tilde{h}^*)$. Since T is conditionally condensing, it follows that $\alpha(\tilde{h}^*) = 0$. Thus, $\tilde{h}^*$ and h^* are precompact and $\eta = 0$. Thus every sequence in C is precompact.

This argument shows that $\omega(B)$ exists and is given by $\omega(B) = \bigcap_{n\ge 0} T^n B$ and thus $\omega(B)$ is invariant. Since T is condensing, the set $\omega(B)$ is invariant. Finally, $\omega(B)$ attracts B. In fact, if this is not the case, then there is a sequence $\{T^{k_j}x_j\}$ and an $\epsilon > 0$ such that $T^{k_j}x_j \notin N_\epsilon(\omega(B))$ for all $j \ge 1$. However, the set $\{T^{k_j}x_j\}$ is precompact by the previous argument and contains a subsequence that converges to a point z outside $N_\epsilon(\omega(B))$. This obviously is a contradiction and completes the proof of the lemma. □

4.4 Fixed points of discrete dissipative processes

In this section, we give sufficient conditions on a continuous map $T : X \to X$, which will ensure that T or some iterate of T has a fixed point. These fixed points are sometimes called periodic points of T.

To motivate the approach to be taken, let us examine some of the implications of Theorem 3.1. If J is the maximal compact invariant set in Theorem 3.1, and $K = \overline{\text{co}}\, J$, the closed convex hull of J, then K is compact and there is a convex neighborhood B of K such that $\gamma^+(B)$ is bounded and J (and, therefore, K) attracts B, i.e., there exist convex subsets $K \subseteq B \subseteq S$ of X with K compact, S closed and bounded, and B open in S such that $\gamma^+(B) \subseteq S$ and K attracts B. For nested sets of this type and even weaker properties of attraction, one can prove the following interesting result.

Lemma 4.1. *Suppose $K \subseteq B \subseteq S$ are convex subsets of X with K compact, S closed and bounded, and B open in S. If $T : S \to X$ is continuous, $\gamma^+(B) \subseteq S$, and K attracts points of B, then there is a closed, bounded, convex subset A of S such that*

$$A = \overline{\text{co}}\ \Big[\bigcup_{j\geq 1} T^j(B\cap A)\Big], \qquad A\cap K \neq \emptyset.$$

Proof. Let $\mathcal{F}$ be the set of convex, closed, bounded subsets L of S such that $T^j(B\cap L) \subseteq L$ for $j \geq 1$ and $L\cap K \neq \emptyset$. The family $\mathcal{F}$ is not empty because $S \in \mathcal{F}$. If $L \in \mathcal{F}$, let

$$L_1 = \overline{\text{co}}\ \Big[\bigcup_{j\geq 1} T^j(B\cap L)\Big].$$

Since K attracts points of B, there is a sequence $x_n \in L_1$ such that $x_n \to K$ as $n \to \infty$. Therefore, $(\text{Cl } L_1)\cap K \neq \emptyset$. But L_1 is closed and so $L_1\cap K \neq \emptyset$. Also, L_1 is convex and $L_1 \subseteq S$. Since $L \in \mathcal{F}$, we have $L \supseteq L_1$, and thus the definition of L_1 implies $L_1 \supseteq T^j(B\cap L) \supseteq T^j(B\cap L_1)$ for $j \geq 1$. Thus, $L_1 \in \mathcal{F}$ and a minimal element A of $\mathcal{F}$ will satisfy the condition of the lemma.

To prove such a minimal element exists, let $(L_\alpha)_{\alpha\in I}$ be a totally ordered family of sets in $\mathcal{F}$. The set $L = \bigcap_{\alpha\in I} L_\alpha$ is closed, convex, and contained in S. Also, $T^j(B\cap L) \subseteq T^j(B\cap L_\alpha) \subseteq L_\alpha$ for any $\alpha \in I$ and $j \geq 1$. Thus, $T^j(B\cap L) \subseteq L$ for $j \geq 1$. If J is any finite subset of I, then by the same reasoning as for the set L_1, we have $K \cap \big(\bigcap_{\alpha\in J} L_\alpha\big) \neq \emptyset$. From compactness of K, it follows that $K \cap \big(\bigcap_{\alpha\in I} L_a\big) \neq \emptyset$. Thus, $L \in \mathcal{F}$ and Zorn's lemma yields the conclusion of the lemma. □

To proceed further, we need the following asymptotic fixed-point theorem of Horn [1], which is stated without proof.

Theorem 4.1. *Let $S_0 \subseteq S_1 \subseteq S_2$ be convex subsets of a Banach space X with S_0, S_2 compact and S_1 open in S_2. Let $T : S_2 \to X$ be a continuous mapping such that for some integer $m > 0$, $T^jS_1 \subseteq S_2$ for $0 \leq j \leq m-1$ and $T^jS_1 \subseteq S_0$ for $m \leq j \leq 2m-1$. Then T has a fixed point.*

Our basic lemma is now derived from this result and Lemma 4.1.

Lemma 4.2. *Suppose $K \subseteq B \subseteq S$ are convex subsets of X with K compact, S closed and bounded, and B open in S. If $T : S \to X$ is continuous, $\gamma^+(B) \subseteq S$, K attracts compact sets of B, and the set A of Lemma 4.1, is compact, then T has a fixed point.*

Proof. Without loss in generality, we may take $B = \mathcal{B}(K,\epsilon)\cap S$ for some $\epsilon > 0$ since K is compact and convex. Let $S_0 = A\cap \text{Cl } \mathcal{B}(K,\epsilon/2)$, $S_1 = A\cap \mathcal{B}(K,\epsilon)$, and $S_2 = A$. Then $S_0 \subseteq S_1 \subseteq S_2$ are convex subsets of X with S_0, S_2 compact and S_1 open in S_2. Also, $T^jS_1 = T^j(A\cap\mathcal{B}(K,\epsilon)) \subseteq A$, $j \geq 1$, $\gamma^+(S_1) \subseteq A$. Therefore. $T^jS_1 \subseteq S_2$ for $j \geq 1$. Also, the fact that K attracts compact sets of B implies there is an integer $n_1 = n_1(K,\epsilon)$ such that $T^jS_1 \subseteq \mathcal{B}(K,\epsilon/2)$ for $j \geq n_1$. If $n_1 \geq 1$, then $T^jS_1 \subseteq A$, for $j \geq n_1$ and so $T^j S_1 \subseteq S_0$ for $j \geq n_1$. We may now apply Theorem 4.1 to the mapping T to complete the proof of the lemma. □

Lemma 4.2 gives a good motivation for determining conditions on the mapping T, which will ensure that the set A is compact. The following lemma deals with this problem.

Lemma 4.3. *If T is conditionally condensing, then the set A in Lemma 4.1 is compact.*

Proof. If $\widetilde{A} = \bigcup_{j\geq 1} T^j(B \cap A)$ then $\widetilde{A} = T(B \cap A) \cup T(\widetilde{A})$ and $\alpha(A) = \alpha(\widetilde{A}) = \max[\alpha(T(B \cap A)), \alpha(T(\widetilde{A}))]$. Since $\alpha(T(\widetilde{A})) < \alpha(\widetilde{A})$ if $\alpha(\widetilde{A}) > 0$, it follows that $\alpha(A) = \alpha(\tilde{A}) = \alpha(T(B \cap A))$. Thus, if $\alpha(B \cap A) > 0$, then $\alpha(A) = \alpha(\widetilde{A}) < \alpha(B \cap A) \leq \alpha(A)$ and this is a contradiction. Thus, $\alpha(B \cap A) = 0$. This implies $\alpha(T(B \cap A)) = 0$ and thus $\alpha(A) = 0$. But this implies A is compact and the lemma is proved. □

Lemmas 4.2 and 4.3 give sufficient conditions for the existence of fixed points of the map T. We now state a result using the theory of Section 4.3 and these lemmas.

Theorem 4.2. *If $T : X \to X$, there is a compact set that attracts compact sets of X and T is conditionally condensing, then T has a fixed point.*

Proof. This is a direct application of Theorem 3.1 and Lemmas 4.2 and 4.3. □

Our next objective is to give other conditions on the map T that imply the hypotheses of Theorem 4.2 and are easier to verify in applications. These conditions involve the concept of dissipativeness introduced in Definition 3.4.

Corollary 4.1. *If $T : X \to X$ is continuous and point dissipative and $\{T^k : k \geq 0\}$ is conditionally completely continuous for $k \geq n_0$, then T^{n_0} has a fixed point.*

Proof. Lemma 3.3 implies the existence of a compact set that attracts compact sets. Since T^{n_0} is conditionally condensing, Theorem 4.2 may be applied to T^{n_0} to yield the assertion. □

Theorem 4.3. *If $T : X \to X$ is conditionally condensing and point dissipative and $\{T^k : k \geq 0\}$ is conditionally completely continuous for $k \geq n_0$, then T has a connected compact global attractor and T has a fixed point.*

Proof. Corollary 3.2 implies the existence of the global attractor. Therefore, Theorem 4.2 implies the existence of a fixed point of T. □

Theorem 4.4. *If T is conditionally condensing and compact dissipative, then T has a fixed point.*

Proof. This is a consequence of Lemma 3.5 and Theorems 3.2 and 4.2. □

Lemma 4.4. *If $S : X \to X$ is a bounded linear operator with spectrum contained in the open unit ball, then there is an equivalent norm, $|\cdot|_1$, in X such that $|S|_1 < 1$.*

Proof. Define $|x|_1 = \sum_{j\geq 0} |S^j x|$. The assumption on the spectrum implies there is an r, $0 \leq r < 1$, such that $|S^n| < r^n$ if n is sufficiently large. Thus, there is a constant $K \geq 1$ such that $|x| \leq |x|_1 \leq K|x|$ for all $x \in X$. Also, for $x \neq 0$,

$$\frac{|Sx|_1}{|x|_1} = 1 - \left[1 + \frac{|Sx|}{|x|} + \frac{|S^2 x|}{|x|} + \cdots\right]^{-1} = 1 - \left[\frac{|x|_1}{|x|}\right]^{-1} \leq 1 - \frac{1}{K}.$$

This proves the lemma. □

Theorem 4.5. *If $T : X \to X$ is continuous, $T = S + U$ where S is linear with spectrum contained in the open unit ball and U is conditionally completely continuous, then*

(i) *T compact dissipative implies T has a fixed point.*

(ii) *T point dissipative implies Y has a fixed point if S^{n_0} is completely continuous for some integer n_0.*

Proof. Changing the norm in the space does not affect the existence of fixed points. Lemma 4.4 implies T is conditionally condensing in an appropriate norm. Theorem 4.4 implies Part (a). In Part (b), $T^{n_0} = S^{n_0} + V$ and one can show that V is conditionally completely continuous. Theorem 4.3 implies the conclusion of Part (b) and the theorem is proved. □

Theorem 3.1 can also serve as a general motivation for the so-called asymptotic fixed-point theorems, one of which is stated in Theorem 4.1. To see this, we first observe the following easy consequence of Theorem 3.1.

Lemma 4.5. *If $T : X \to X$ is continuous and there is a compact set that attracts compact sets of X, then there exists an integer m and convex bounded sets $S_0 \subseteq S_1 \subseteq S_2$ with S_0, S_2 closed and S_1 open, such that $\gamma^+(S_1) \subseteq S_2$ and $\gamma^+(T^m S_1) \subseteq S_0$.*

Proof. Let J be the maximal compact invariant set of Theorem 3.1 and let K be the closed convex hull of J. Then K is compact and convex. Theorem 3.1(iii) guarantees the existence of a neighborhood K_1 of K such that $\gamma^+(K_1)$ is bounded and J attracts K_1. Since $J \subseteq K$, it follows that K attracts K_1. Let S_2 be a closed bounded convex set containing $\gamma^+(K_1)$. Choose $\epsilon > 0$ such that $\mathrm{Cl}\ \mathcal{B}(K, \epsilon) \subseteq K_1$ and let $S_0 = \mathrm{Cl}\ \mathcal{B}(K, \epsilon/2)$,

$S_1 = \mathcal{B}(K, \epsilon)$. Since K attracts K_1, there is an $m = n_0(K, \epsilon)$ such that $T^n K_1 \subseteq S_0$ for $n \geq m$. Therefore, $T^n S_1 \subseteq S_0$ for $n \geq m$. Also, the definition of S_1 and S_2 implies $\gamma^+(S_1) \subseteq S_2$ and the lemma is proved. □

Lemma 4.5 leads in a natural way to the following fixed-point problem: Suppose there exists an integer m, bounded convex subsets $S_0 \subseteq S_1 \subseteq S_2$ of a Banach space X such that S_0 and S_2 are closed and S_1 is open, and a continuous map $T : S_2 \to X$ such that $\gamma^+(S_1) \subseteq S_2$ and $\gamma^+(T^m S_1) \subseteq S_0$. What additional conditions on T imply the existence of a fixed point of T?

The famous asymptotic fixed-point theorem of Browder states that T completely continuous is sufficient to imply the existence of a fixed point of T if there are sets $S_0 \subseteq S_1 \subseteq S_2$ as asserted in the preceding problem. Using the techniques we have been employing in this chapter, it is possible to prove the following interesting extension of this result.

Theorem 4.6. *Suppose $S_0 \subseteq S_1 \subseteq S_2$ are convex bounded subsets of a Banach space X, S_0 and S_2 are closed, and S_1 is open in S_2, and suppose $T : S_2 \to X$ is condensing in the following sense: if Ω and $T(\Omega)$ are contained in S_2 and $\alpha(\Omega) > 0$, then $\alpha(T(\Omega)) < \alpha(\Omega)$. If $\gamma^+(S_1) \subseteq S_2$ and, for any compact set $H \subseteq S_1$, there is a number $N(H)$ such that $\gamma^+(T^{N(H)}H) \subseteq S_0$, then T has a fixed point.*

Proof. Repeating the same argument as in Lemma 3.4, one can show there is a compact set $\widetilde{K} \subseteq S_0$ that attracts compact sets of S_1. Since S_0 is convex, $K = \overline{\mathrm{co}}\, \widetilde{K} \subseteq S_0$. Let $B \subseteq S_1$ be an open convex neighborhood of K. The conditions of Lemma 4.1 are therefore satisfied and Lemma 4.3 implies that A is compact. Thus, the conditions of Lemma 4.2 are satisfied and T has a fixed point. □

4.5 Continuous systems—Maximal invariant sets and attractors

In this section, we consider the same topics as in Section 4.3 for *continuous dynamical systems* or equivalently *C_0-semigroups* of transformations $T(t)$, $t \geq 0$, on a Banach space X; that is, $T(t)x$, $t \geq 0$, $x \in X$, is continuous in t, x and satisfies the properties $T(0) = I$, $T(t+s) = T(t)T(s)$, t, $s \geq 0$.

For any set $B \subset X$, the *positive orbit* $\gamma^+(B)$ is defined as $\gamma^+(B) = \bigcup_{t \geq 0} T(t)B$ and the *ω-limit set* $\omega(B)$ is defined as

$$\omega(B) = \bigcap_{s \geq 0} \overline{\bigcup_{t \geq s} T(t)B}\,.$$

A *negative orbit* through a point x is a function $\phi : (-\infty, 0] \to X$ such that $\phi(0) = x$ and, for any $s \leq 0$, we have $T(t)\phi(s) = \phi(t+s)$ for $0 \leq t \leq -s$.

In an obvious way, we can define the *α-limit* set of a negative orbit. In the following, we do not need the concept of α-limit set of a set.

We say that a set $K \subset X$ *attracts* a set $H \subset X$ if, for any $\epsilon > 0$, there is a $t_0(H, \epsilon)$ such that $T(t)H \subset \mathcal{B}(K, \epsilon)$ for $t \geq 0$. We say that $T(t)$, $t \geq 0$, is *point dissipative* if there is a bounded set $K \subset X$ such that K attracts points of X. We say that $T(t)$, $t \geq 0$, is *bounded dissipative* if there is a bounded set $K \subset X$ such that K attracts bounded sets of X. We say that K is a *global attractor* if K is invariant and attracts bounded sets of X.

A set $M \subset X$ is said to be *stable* if, for any $\epsilon > 0$, there is a $\delta > 0$ such that $x \in \mathcal{B}(M, \delta)$ implies that $T(t)x \in \mathcal{B}(M, \epsilon)$ for $t \geq 0$. A set M is said to be *asymptotically stable* if it is stable and there is an $\epsilon_0 > 0$ such that M attracts the set $\mathcal{B}(M, \epsilon_0)$. A set M is said to be *uniformly asymptotically stable* if it is asymptotically stable and, for any $\eta > 0$, there is a $t_0(\eta, \epsilon_0) > 0$ such that $T(t)x \in \mathcal{B}(M, \eta)$ for $t \geq t_0(\eta, \epsilon_0)$ and $x \in \mathcal{B}(M, \epsilon_0)$.

The semigroup $T(t)$, $t \geq 0$, is said to be *asymptotically smooth* if, for any bounded set $B \subset X$ such that $T(t)B \subset X$, there is a compact set K such that K attracts B. The semigroup $T(t)$, $t \geq 0$, is said to be an *α-contraction* if, for each bounded set $B \subset X$, the set $\{T(s)B, 0 \leq s \leq t\}$ is bounded and there is a continuous function $k : [0, \infty) \to [0, \infty)$ such that $k(t) \to 0$ as $t \to \infty$, and, for each $t > 0$ and for each bounded set $B \subset X$, we have $\alpha(T(t)B) \leq k(t)\alpha(B)$, where α denotes the Kuratowski measure of noncompactness. As in Section 3, we can show that α-contracting semigroups are asymptotically smooth.

All of the results of Section 3 are valid for continuous semigroups with the proofs being essentially the same. However, we are going to state one of the most important ones, which will be useful for our later discussion. An equilibrium point of $T(t)$, $t \geq 0$, is a point $x \in X$ such that $T(t)x = x$ for $t \geq 0$.

Theorem 5.1. *If $T(t), t \geq 0$, is asymptotically smooth, point dissipative and positive orbits of bounded sets are bounded, then there exists a connected compact global attractor and there is an equilibrium point of $T(t)$.*

Theorem 5.2. *If there is a $t_1 \geq 0$ such that $T(t)$ is completely continuous for $t \geq t_1$ and $T(t)$, $t \geq 0$, is point dissipative, then there exists a connected compact global attractor $\mathcal{A}$ and there is an equilibrium point of $T(t)$.*

The only statement that is not verified by the same method as in the previous section is the assertion about the existence of an equilibrium point. This is proved in the following way. Choose a sequence of positive numbers $\omega_n \to \infty$ as $n \to \infty$ and consider the discrete dynamical system defined by the map $T_n \equiv T(\omega_n)$. From the previous results on discrete maps, there is a fixed point x_n of T_n. It is clear that $x_n \in \mathcal{A}$, the global attractor in the preceding theorems. Since $\mathcal{A}$ is compact, we can extract a subsequence and apply Lemma 4.2.

4.6 Stability and maximal invariant sets in processes

In this section, we generalize the results of Section 4.3 to ω-periodic processes.

Definition 6.1. For a given process u on X and a given $\sigma \in \mathbb{R}$, we say that a set $M \subset \mathbb{R} \times X$ *attracts a set* $H \subset X$ *at* σ if, for any $\epsilon > 0$, there is a $t_0(\epsilon, H, \sigma)$ such that $(\sigma + t, U(\sigma, t)H) \subset \mathcal{B}(M, \epsilon)$ for $t \geq t_0(\epsilon, H, \sigma)$. If M attracts a set H at σ for each $\sigma \in \mathbb{R}$, we simply say that M *attracts* H. We say that M *attracts points* of X *(at* σ*)* if M attracts each point of X (at σ). We say that M *attracts neighborhoods of points of* X *(at* σ*)* if, for any $x \in X$, there is a neighborhood O_x of X such that M attracts O_x (at σ). We say that M *attracts bounded sets* if M attracts each bounded set of X. We say that M is a *global attractor* if M is invariant and attracts bounded sets of X. When we use the term *uniform attracts*, we shall mean that the number t_0 can be chosen independently of σ. Similar definitions are given for M attracting other types of sets of X.

Definition 6.2. For a given process u on X and a given $\sigma \in \mathbb{R}$, we say that a set $M \subset \mathbb{R} \times X$ is *stable at* σ if, for any $\epsilon > 0$, there is a $\delta(\epsilon, \sigma) > 0$ such that $(\sigma, x) \in \mathcal{B}(M, \delta(\epsilon, \sigma))$ implies that $(\sigma + t, U(\sigma, t)x) \in \mathcal{B}(M, \epsilon)$ for $t \geq 0$. The set M is said to be stable if it is stable at σ for all $\sigma \in \mathbb{R}$. The set M is *unstable* if it is not stable. The set M is *uniformly stable* if it is stable and the number δ in the definition of stability is independent of σ. The set M is said to be *asymptotically stable at* σ if it is stable at σ and there is an $\epsilon_0(\sigma)$ such that $(\sigma + t, U(\sigma, t)) \to M$ as $t \to \infty$ for $(\sigma, x) \in \mathcal{B}(M, \epsilon_0(\sigma))$. The set M is said to be *uniformly asymptotically stable* if it is uniformly stable and there is an $\epsilon_0 > 0$ such that for any $\eta > 0$, there is a $t_0(\eta, \epsilon_0)$ having the property that $(\sigma + t, U(\sigma, t)) \in \mathcal{B}(M, \eta)$ for $t \geq t_0(\eta, \epsilon_0)$ and all x such that $(\sigma, x) \in \mathcal{B}(M, \epsilon_0)$.

If a process is generated by an ordinary differential equation in $\mathbb{R}^n$, then $M \subset \mathbb{R} \times \mathbb{R}^n$ stable at a fixed $\sigma \in \mathbb{R}$ implies M is stable at every $\sigma \in \mathbb{R}$; that is, M is stable. For general processes, this is no longer true. In fact, it is not even true for RFDE. To see this, consider the process generated by the linear RFDE discussed in Equation (5.1) of Section 3.5. For $\sigma = 0$, the solution is given by Equation (5.3) and, therefore, the solution $x = 0$ (that is, the set $M = \mathbb{R} \times \{0\}$) is stable. On the other hand, for any $\sigma > 3\pi$, the solution $x(\sigma, \phi)$ must satisfy

$$\dot{x}(t) = x\big(t - \frac{3\pi}{2}\big).$$

For any λ satisfying $\lambda = \exp(-3\pi\lambda/2)$, the function $a \exp \lambda t$ is a solution for any $a \in \mathbb{R}$. But this latter equation has a solution $\lambda_0 > 0$ and so the solution $x = 0$ is unstable at $\sigma > 3\pi$.

If u is a process and the set $M = \mathbb{R} \times K$ with K compact is stable at σ, then M is stable at $\zeta < \sigma$. This is an immediate consequence of continuity of $U(\sigma, t)x$ in (σ, t, x). For an ω-periodic process, $U(\sigma + k\omega, t) = U(\sigma, t)$ for all integers k. Thus a set $M = \mathbb{R} \times K$ with K compact is stable at σ if and only if it is stable.

It is difficult to determine when stability at σ is equivalent to stability. From a practical point of view, it does not seem to be significant to consider systems that are stable at σ and not stable. As a consequence, this weaker concept will not be discussed in more detail.

Now suppose that u is an ω-periodic process, $K \subset X$ is compact, and $M \subset \mathbb{R} \times K$ attracts compact sets of X. For any $\sigma \in \mathbb{R}$, consider the discrete dynamical system $\{U^k(\sigma, \omega),\ k \geq 0\}$. For this discrete system, K attracts compact sets of X and we can define, as in Section 4.3,

$$J_\sigma = \bigcap_{n \geq 0} U^n(\sigma, \omega)K, \quad \sigma \in \mathbb{R}. \tag{6.1}$$

It was proved in Section 4.3 that J_σ is independent of K. If $\mathcal{J} \subset \mathbb{R} \times X$ is defined by

$$\mathcal{J} = \bigcup_{\sigma \in \mathbb{R}} (\sigma, J_\sigma), \tag{6.2}$$

then $\mathcal{J}$ is an invariant set for the process u and

$$\mathcal{J}_\sigma \overset{\text{def}}{=} \{x \in X : (\sigma, x) \in \mathcal{J}\} = J_\sigma$$

is compact. Also, $\bigcup_\sigma J_\sigma$ is compact in X.

Theorem 6.1. *Suppose that u is an ω-periodic process on X, there is a compact set $K \subset X$ such that $\mathbb{R} \times K$ attracts compact sets of X, and let $\mathcal{J} \subset \mathbb{R} \times X$ be defined by* (6.2). *Then the following conclusions hold:*

(i) *$\mathcal{J}$ is connected.*

(ii) *$\mathcal{J}$ is independent of K, is a nonempty invariant set with $\mathcal{J}_\sigma$ compact, and $\mathcal{J}$ is maximal with respect to this property.*

(iii) *$\mathcal{J}$ is stable.*

(iv) *For any compact set $H \subset X$, $\mathcal{J}$ attracts H.*

Proof. The proofs of Properties (i) and (ii) are the same as in the proof of Theorem 3.1.

To prove Property (iii), we first observe that Theorem 3.1 implies that the set J_σ in Equation (6.1) is a stable global attractor for the discrete dynamical system defined by $U(\sigma, \omega)$. Therefore, for any $\epsilon > 0$, there is a $\delta(\epsilon, \sigma) > 0$ such that $x \in \mathcal{B}(J_\sigma, \epsilon)$ for $k \geq 0$. Since $U(\sigma, t)$ is continuous in t, we may further restrict δ to be assured that $U(\sigma, t)x \in \mathcal{B}(J_{\sigma+t}, \epsilon)$ for $0 \leq t \leq \omega$ if $x \in \mathcal{B}(J_\sigma, \delta)$. Since $\mathcal{J}_\sigma = J_\sigma$ and $\mathcal{J}$ is invariant, it follows that $\mathcal{J}$ is stable at σ for every $\sigma \in \mathbb{R}$. Since $U(\sigma + k\omega, t) = U(\sigma, t)$ for all integers k, the set $\mathcal{J}$ is stable at $\sigma + k\omega$ with the constant δ independent of k. Therefore, it is only necessary to vary $\sigma \in [0, \omega]$. For $0 \le \sigma \le \omega$ and $t \ge 0$, $U(\sigma, t + \omega - \sigma) = U(\sigma, t)U(\sigma, \omega - \sigma)$. If $\delta(\epsilon, \sigma)$ is the stability constant at σ, then the continuity of u and the compactness of J_σ imply that there is a $0 < \delta_1(\epsilon) \le \delta(\epsilon, \sigma)$, $0 \le \sigma \le \omega$, such that $U(\sigma, \omega - \sigma)\mathcal{B}(J_\sigma, \delta_1) \subset \mathcal{B}(J_\omega, \delta_1)$ for $0 \le \sigma \le \omega$. This shows that uniform stability and Part (iii) of the theorem is proved.

The proof of Part (iv) may be supplied in a similar fashion using the results of Theorem 3.1(iv) to complete the proof of the theorem. □

Definition 6.3. A process u on X is said to be *point* (*compact*) (*local*) (*locally compact*) (*bounded*) *dissipative* if there is a bounded set $B \subset X$ such that $\mathbb{R} \times B$ attracts points (compact sets) (neighborhoods of points) (neighborhoods of compact sets) (bounded sets) of X.

Definition 6.4. A process u on X is said to be *asymptotically smooth* if, for any bounded set $B \subset X$, there is a compact set $B^* \subset X$ such that for any $\epsilon > 0$, there is a $t_0 = t_0(\epsilon, B)$ such that for any $\sigma \in \mathbb{R}$, $U(\sigma, t)x \in B$ for $t \ge 0$ implies that $U(\sigma, t)x \in B^*$ for $t \ge t_0$.

Following arguments similar to the proof of Theorem 3.2 and Theorem 6.1, one can demonstrate the following result.

Theorem 6.2. *Suppose that u is an ω-periodic process on X that is local dissipative and asymptotically smooth. Then there exists a set $\mathcal{J} \subset \mathbb{R} \times X$ that satisfies all of the conclusions of Theorem 6.1 and*

(v) *For any compact set $H \subset X$, there is a neighborhood H_1 of H such that $\gamma^+(H_1)$ is bounded and $\mathcal{J}$ attracts H_1; in particular, $\mathcal{J}$ is uniformly asymptotically stable.*

(vi) *If $B \subset X$ is bounded and $\gamma^+(\sigma, B)$ is bounded, then $\mathcal{J}$ attracts B. In particular, if $\gamma^+(\sigma, B)$ is bounded for every bounded set $B \subset X$, $\sigma \in \mathbb{R}$, then $\mathcal{J}$ is a compact global attractor.*

Corollary 6.1. *If u is an ω-periodic process, then the following are equivalent statements:*

(i) *There is a compact set $K \subset X$ such that $\mathbb{R} \times K$ attracts compact sets of X.*

(ii) *There is a compact set $K \subset X$ such that $\mathbb{R} \times K$ attracts neighborhoods of compact sets of X.*

We also have the analogue of Theorem 3.3.

Theorem 6.3. *If $U(\sigma, t)$ is asymptotically smooth, point dissipative and positive orbits of bounded sets are bounded, then there exists a connected global attractor. Finally there is an ω-periodic trajectory.*

One also can prove the following generalization of Lemma 3.4 and Corollary 3.2.

Theorem 6.4. *Suppose that u is an ω-periodic process on X that is point dissipative and there is a $t_0 > 0$ such that for every $\sigma \in \mathbb{R}$, $U(\sigma, t)$ is conditionally completely continuous for $t \geq t_0$. Then there is a compact set $K \subset X$ such that for any compact set $H \subset X$, there is an open neighborhood H_0 of H and a $t_1 > 0$ such that for any $\sigma \in \mathbb{R}$, $\gamma^+(\sigma, H_0)$ is bounded and $U(\sigma, t)H_0 \subset K$ for $t \geq t_1$. In particular, there is a compact set $K \subset X$ such that $M = \mathbb{R} \times K$ attracts compact sets of X. If, in addition, $U(\sigma, t)$ takes bounded sets into bounded sets, then, for any bounded set $B \subset X$, the set $\gamma^+(\sigma, B)$ is bounded and there is a $t_1 > 0$ such that $U(\sigma, t)B \subset K$ for $t \geq t_1$. In particular, there is a compact, connected, global attractor. Finally, there is an ω-periodic trajectory.*

Using Corollary 6.1 of Section 3.6 and Theorem 6.4, we have the following result for RFDE.

Theorem 6.5. *If an ω-periodic* RFDE(f) *generates an ω-periodic process u on C, $U(\sigma, t)$ is a bounded map for each σ, t and u is point dissipative, then there is a compact, connected, global attractor. Also, there is an ω-periodic solution of the* RFDE(f). *If the* RFDE(f) *is autonomous and the same conditions are satisfied, then there is an equilibrium solution of the* RFDE(f).

For an RFDE(f), Theorem 6.5 gives the existence of a periodic solution under the weak hypothesis of point dissipative. Some results are available for special types of equations even with weaker hypotheses. For example, for linear equations, one can prove the existence of a periodic solution assuming only the existence of a bounded solution. To do this, we need the following result of Darbo, which we state without proof.

Theorem 6.6. *If Γ is a closed, bounded, convex subset of a Banach space and $T : \Gamma \to \Gamma$ is α-condensing, then T has a fixed point in Γ.*

Theorem 6.7. *For an ω-periodic linear nonhomogeneous* RFDE, *the existence of a solution that is bounded for $t \geq 0$ implies the existence of an ω-periodic solution.*

Proof. We need some results from linear systems in Chapter 6, but they are very elementary and will be repeated here. For a linear RFDE

$$\dot{x}(t) = f(t, x_t) + h(t),$$

where $f(t, \phi)$ is linear in ϕ, the solution $x(\sigma, \phi, h)$ through (σ, ϕ) can be written as $x(\sigma, \phi, h) = x(\sigma, \phi, 0) + x(\sigma, 0, h)$. Therefore, the period map taking ϕ into $x_{\sigma+\omega}(\sigma, \phi, h)$ is an affine transformation $T : C \to C$, $T\phi = L\phi + \psi$, where $\phi \in C$, $\psi \in C$ and $L : C \to C$ is a bounded linear map. Corollary 6.1 of Section 3.6 implies that $L = S + U$, where U is completely continuous and the spectrum of S contains only the zero element. Therefore, Lemma 4.4 implies that there is an equivalent norm in C such that S is a contraction with contraction constant k. For any bounded set B in C, $\alpha(LB) = \alpha(SB) \le k\alpha(B)$. Therefore, L is an α-contraction. Since T is only a translation of L, it follows that T is an α-contraction.

Suppose that $x : [-r, \infty) \to \mathbb{R}^n$ is a bounded solution of the RFDE. If $\phi = x_0$, this implies that $\{T^k\phi : k \ge 1\}$ is a bounded set in C. Let $D = \operatorname{co}\{\phi, T\phi, T^2\phi, \dots\}$. If $\eta \in D$, then $\eta = \sum_{i\in J}\alpha_i T^i\phi$, where J is a finite subset of the positive integers, $\alpha_i \ge 0$, $\sum_{i\in J}\alpha_i = 1$. Furthermore,

$$\begin{aligned} T\eta = L\eta + \psi &= \sum_{i\in J}\alpha_i LT^i\phi + \sum_{i\in J}\alpha_i\psi \\ &= \sum_{i\in J}\alpha_i(LT^i\phi + \psi) = \sum_{i\in J}\alpha_i T^{i+1}\phi \in D, \end{aligned}$$

and $TD \subseteq D$. Since T is continuous, $T(\mathrm{Cl}\ D) \subseteq \mathrm{Cl}\ D$. Since $\mathrm{Cl}\ D$ is a closed bounded convex set of C and T is an α-contraction, there is a fixed point of T in D by Theorem 6.6 and, thus, an ω-periodic solution of the RFDE. □

4.7 Convergent systems

In this section, we investigate the simplest possible processes, namely, ones for which all solutions are asymptotically stable and at least one solution is bounded. We first treat discrete dynamical systems in detail and only state the results for ω-periodic processes.

Lemma 7.1. *Suppose $T : X \to X$ is continuous and there is a compact set that attracts compact sets of X. If the set J in Theorem 3.1 consists of a single point, then there is a bounded orbit of T on $(-\infty, \infty)$, every orbit is stable and attracts neighborhoods of points of X.*

Proof. Let $J = \{x_0\}$. Since J is compact and invariant, $T^n x_0$ is defined for $-\infty < n < \infty$, $T^n x_0 = x_0$ for all n and $\gamma = \{T^n x_0 : -\infty < n < \infty\} = J$ is bounded. By Theorem 3.1, γ is stable and so, for any $\epsilon > 0$, there is a $\delta = \delta(\epsilon) > 0$ such that $|x - x_0| < \delta$ implies $|T^n x - x_0| < \epsilon$ for $n \ge 0$ since $\gamma = J$. Also, Theorem 3.1 implies γ attracts neighborhoods of points of X.

Therefore, for any $x \in X$, there is an $n_0(x)$ and a neighborhood O_x of x such that $|T^{n_0(x)}y - x_0| < \delta$ for $y \in O_x$. Stability of γ implies $|T^n y - T^n x| < 2\epsilon$ for $n \geq n_0(x)$, $y \in O_x$. Continuity of T therefore implies we can choose O_x so that $|T^n y - T^n x| < 2\epsilon$ for $0 \leq n \leq n_0(x)$. This implies the orbit $\gamma^+(x)$ is stable. The same type of argument also gives the fact that each orbit attracts neighborhoods of points of X and the proof is complete. □

Lemma 7.2. *If $T : X \to X$ is continuous, if there is a bounded orbit of T on $(-\infty, \infty)$ and every trajectory is uniformly asymptotically stable, then T is local dissipative.*

Proof. Let $\{T^n x_0 : -\infty < n < \infty\} = \gamma$ be a bounded orbit of T. We first show that

$$\lim_{n\to\infty} |T^n x - T^n x_0| = 0 \tag{7.1}$$

for all $x \in X$. Let $G = \{x \in X : \text{Equation (7.1) is satisfied}\}$. The set G is not empty since $x_0 \in G$. The set G is open because of the uniform asymptotic stability. Also ∂G is empty. Otherwise $x^* \in \partial G$ would not be stable. Therefore $G = X$. This proves that T is point dissipative. The hypothesis of uniform asymptotic stability implies the local dissipativeness of T and the lemma is proved. □

Definition 7.1. A continuous map $T : X \to X$ is said to be *convergent* if there is a unique fixed point of T that is stable and attracts neighborhoods of points of X.

Theorem 7.1. *If T is conditionally condensing, then T is convergent if and only if there is a bounded orbit of T on $(-\infty, \infty)$ and every orbit is uniformly asymptotically stable.*

Proof. If T is conditionally condensing and convergent, then T is condensing and local dissipative. Therefore, Lemma 3.4 implies there is a compact set that attracts compact sets of X. The set J in Theorem 3.1 therefore exists and must contain only a single point. Lemma 7.1 implies the result.

Conversely, if there is a bounded orbit of T on $(-\infty, \infty)$ and every orbit is uniformly asymptotically stable, then Lemma 7.2 implies T is local dissipative. We use Lemma 3.4 again and conclude T is convergent directly from Theorem 3.1. The proof is complete. □

It is possible to weaken the definition of *convergent* T and require only that the unique fixed point is stable and a global attractor. An analogue of Theorem 7.1 is then obtained by requiring that T be conditionally condensing and some iterate of T be conditionally completely continuous. One must repeat arguments similar to the ones used in proving Theorem 3.1 and Lemma 3.1.

The analogue of these results for ω-periodic processes are the following.

Definition 7.2. An ω-periodic process u is said to be *convergent* if
(i) there is a unique ω-periodic orbit;
(ii) the trajectory of this orbit is uniformly stable and uniformly attracts neighborhoods of points of $\mathbb{R} \times X$.

Theorem 7.2. *If the ω-periodic process u is conditionally condensing, then u is convergent if and only if there is a bounded orbit on $(-\infty, \infty)$ and every trajectory is uniformly asymptotically stable.*

As for discrete processes, these hypotheses can be modified to allow only asymptotic stability.

4.8 Supplementary remarks

The theory of dissipative processes had its origin in a fundamental paper of Levinson [1] dealing with periodic differential equations in the plane. By considering the period map T, he formulated very clearly the basic problems. First, one should characterize the smallest set J that contains all of the information about the limits of trajectories and then discuss T restricted to J. And Levinson [1] formulated the concept of point dissipative and proved that point dissipative implies the existence of a maximal compact invariant set. From this fact, he was able to prove that some iterate of T must have a fixed point. In their remarkable work on the van der Pol equation,

$$\ddot{x} + k(x^2 - 1)\dot{x} + x = bk\lambda \cos \lambda t,$$

with k large and $b < 2/3$, Cartwright and Littlewood (see Littlewood [1] and Pliss [1] for references) in the late 1940s and 1950s discovered a large number of periodic solutions of very high periods. These results indicated the possible complicated structure of the set J even for rather innocent-looking nonlinear equations. Levinson [2] also noted this same phenomenon.

Over the years, a tremendous literature on this subject has accumulated and one may consult LaSalle [1], Pliss [1], Reissig, Sansone, and Conti [1], and Yoshizawa [1] for references. Continuing in the spirit of Levinson for finite dimensions, Pliss [1] showed that the maximal compact invariant set is globally asymptotically stable. For the special case of RFDE for which the period ω is greater than the delay, Jones [2] and Yoshizawa [1] showed the existence of ω-periodic solutions by using the asymptotic fixed-point theorem of Browder [1]. For a discrete point dissipative dynamical system T on an arbitrary Banach space X, the existence of fixed points of T were proved by Horn [1] and Gerstein and Krasnoselskii [1] when T is completely continuous. Billotti and LaSalle [1] proved the same result when T is completely continuous, but also characterized the maximal compact invariant set and proved it is globally asymptotically stable.

Gerstein [1] considered the case when T is point dissipative and is condensing on balls in X and showed the existence of a maximal compact invariant set, but concluded nothing about stability of this set or the existence of fixed points of T. Hale, LaSalle, and Slemrod [1] and Hale and Lopes [1] proved many of the same results that are contained in Sections 4.3–4.6 although in a different order and sometimes less generally. Nussbaum [2] has also obtained fixed-point theorems similar to Theorem 4.3. Theorem 4.1 is due to Horn [1]. Theorem 4.6 for T completely continuous is due to Browder [1]. Theorem 6.4 is due to Chow [1] with the proof in the text due to Chow and Hale [1]. The latter paper also contains an interesting class of mappings, called strongly limit compact, which were motivated by Lemma 4.1.

In $\mathbb{R}^n$, Jones [3] and Jones and Yorke [1] have proved results on existence of periodic solutions using the concept of a set $S \subset \mathbb{R}^n$ being compactly constrained by a dynamical system.

For dynamical systems on locally compact spaces, Theorem 6.1 was given in a more general form by Bhatia and Hajek [1]. More specifically, they assume that there is a compact set M such that $\omega(x) \cap M \neq \emptyset$ for each $x \in X$. In particular, this weak hypothesis in $\mathbb{R}^n$ implies that there is a compact set that attracts compact sets. This latter property was basic to our entire investigation in this chapter. It is not known how to weaken the hypotheses on Theorem 6.1 for Banach spaces that are not locally compact.

Section 4.7 is a generalization of the corresponding results of Pliss and Krasovskii (see Pliss [1]) in finite-dimensional space. For a detailed discussion of the relationship between stability and the existence of periodic and almost periodic solutions of ordinary and functional differential equations, see Burton [3], Yoshizawa [2], Hino, Murakami, and Naito [1] and the references therein.

The measure of noncompactness of a set was introduced by Kuratowski [1]. The fixed-point theorem stated in Theorem 6.6 is due to Darbo [1]. For an excellent survey of measures of noncompactness and fixed-point theorems, see Sadovskii [1].

These remarks appeared in the original version of the book of Hale [22]. At that time, he was unaware of the very important work of Ladyzenskaya [1] in which she showed the existence of a compact global attractor for the Navier-Stokes equation on a bounded domain in two space dimensions. During the last fifteen years, there has appeared a tremendous amount of literature on global attractors, especially for partial differential equations. The topics center around the existence of attractors, the finiteness of the Hausdorff dimension of the attractor and estimates of this dimension and the structure of the flow on the attractor. It would take us too far afield to attempt to make a survey of these developments. The interested reader should consult the recent books of Babin and Vishik [1], Hale [23], Ladyzenskaya [2], and Temam [1] for a guide to some of the literature.

The processes described in Section 4.1 can be given a much more dynamical interpretation by using the concept of skew product flows. To see this, let W be the set of processes endowed with some metric and, for some $u \in W$, define the translation $\sigma(\tau)$, $\tau \in \mathbb{R}$, of the process as

$$(\sigma(\tau)u)(s,\, x,\, t) = u(\tau + s,\, x,\, t)\,. \tag{8.1}$$

Then $\sigma(0) = I$, $\sigma(t + \tau) = \sigma(t)\sigma(\tau)$ for all t, $\tau \in \mathbb{R}$, and $\sigma(\tau)$, $\tau \in \mathbb{R}$, is a semigroup. Also, let $\alpha : X \times \mathbb{R}^+ \times W \to X$ be defined by

$$\alpha(x,\, t,\, u) = u(0,\, x,\, t)\,. \tag{8.2}$$

With α and σ as in (8.1) and (8.2), define

$$\pi(t) : X \times W \to X \times W, \quad \pi(t)(x,\, u) = (\alpha(x,\, t,\, u),\, \sigma(t)u)\,. \tag{8.3}$$

One now verifies that $\pi(0) = I$, $\pi(t+\tau) = \pi(t)\pi(\tau)$; that is, $\pi(t)$, $t \geq 0$, is a semigroup of transformations. The nonautonomous process has been converted into an autonomous process by enlarging the system! If the map $\pi(t)(x,\, u)$ were continuous in t, x, u, then $\pi(t)$, $t \geq 0$, would be a C^0-semigroup of transformations and all of the previous theory could be applied. This approach has led to many interesting results, especially in the case where the process arises from a system of evolutionary equations for which the vector field is almost periodic in time (see Miller and Sell [1], Sacker and Sell [1], Sell [1, 2], Hale [23], and the references therein).

For the spectral theory of the linearization of such processes with applications to FDE and PDE, see Magalhães [6, 8]. His results apply at least to periodic systems and asymptotically autonomous systems.

5
Stability theory

In this chapter, we discuss methods for determining stability and ultimate boundedness of solutions of RFDE. The method of Liapunov functionals is discussed as well as the method based on the use of functions on $\mathbb{R}^n$ in the spirit of Razumikhin.

5.1 Definitions

Suppose $f : \mathbb{R} \times C \to \mathbb{R}^n$ is continuous and consider the RFDE(f)

$$\dot{x}(t) = f(t, x_t). \tag{1.1}$$

The function f will be supposed to be completely continuous and to satisfy enough additional smoothness conditions to ensure the solution $x(\sigma, \phi)(t)$ through (σ, ϕ) is continuous in (σ, ϕ, t) in the domain of definition of the function. The following theory for $f : (\alpha, \infty) \times \Omega \to \mathbb{R}^n$, where Ω is an open set in C is valid and it is only for notational purposes that the domain of definition of f is taken to be $\mathbb{R} \times C$.

The definition of stability of the solution $x = 0$ was given in Section 4.5 via the process u generated by the RFDE(f). It is convenient to restate these definitions directly in terms of the solution of Equation (1.1).

Definition 1.1. Suppose $f(t, 0) = 0$ for all $t \in \mathbb{R}$. The solution $x = 0$ of Equation (1.1) is said to be *stable* if for any $\sigma \in \mathbb{R}$, $\epsilon > 0$, there is a $\delta = \delta(\epsilon, \sigma)$ such that $\phi \in \mathcal{B}(0, \delta)$ implies $x_t(\sigma, \phi) \in \mathcal{B}(0, \epsilon)$ for $t \geq \sigma$. The solution $x = 0$ of Equation (1.1) is said to be *asymptotically stable* if it is stable and there is a $b_0 = b_0(\sigma) > 0$ such that $\phi \in \mathcal{B}(0, b_0)$ implies $x(\sigma, \phi)(t) \to 0$ as $t \to \infty$. The solution $x = 0$ is said to be *uniformly stable* if the number δ in the definition is independent of σ. The solution $x = 0$ of Equation (1.1) is *uniformly asymptotically stable* if it is uniformly stable and there is $b_0 > 0$ such that for every $\eta > 0$, there is a $t_0(\eta)$ such that $\phi \in \mathcal{B}(0, b_0)$ implies $x_t(\sigma, \phi) \in \mathcal{B}(0, \eta)$ for $t \geq \sigma + t_0(\eta)$ for every $\sigma \in \mathbb{R}$.

If $y(t)$ is any solution of Equation (1.1), then y is said to be *stable* if the solution $z = 0$ of the equation

$$\dot{z}(t) = f(t, z_t + y_t) - f(t, y_t)$$

is stable. The other concepts are defined in a similar manner.

For some RFDE, there is no distinction between stability and uniform stability. In fact, we now prove

Lemma 1.1. *If there is an $\omega > 0$ such that $f(t+\omega, \phi) = f(t, \phi)$ for all $(t, \phi) \in \mathbb{R} \times C$, then the solution $x = 0$ is stable (asymptotically stable) if and only if it is uniformly stable (uniformly asymptotically stable).*

Proof. Suppose the solution $x = 0$ is stable. Since $x_{t+\sigma+k\omega}(\sigma + k\omega, \phi) = x_{t+\sigma}(\sigma, \phi)$ for all $t \geq 0$, $\sigma \in \mathbb{R}$, it is only necessary to show that the number $\delta(\epsilon, \sigma)$ in the definition of stability can be chosen independent of $\sigma \in [0, \omega]$ in order to prove $x = 0$ is uniformly stable. For $0 \leq \sigma \leq \omega$ and $t \geq 0$, $x_{t+\omega}(\sigma, \phi) = x_{t+\omega}(\omega, x_\omega(\sigma, \phi))$. Therefore, the continuity of $x(\sigma, \phi)(t)$ in (σ, ϕ, t) implies there is a $\delta_1 > 0$, $\delta_1 \leq \delta(\epsilon, \sigma)$, $0 \leq \sigma \leq \omega$, such that $x_\omega(\sigma, \phi) \in \mathcal{B}(0, \delta(\epsilon, \omega))$, if $\phi \in \mathcal{B}(0, \delta_1)$ for $0 \leq \sigma \leq \omega$. Since $x_t(\omega, \phi) \in \mathcal{B}(0, \epsilon)$ for $t \geq \omega$ if $\phi \in \mathcal{B}(0, \delta(\epsilon, \omega))$, this proves uniform stability.

The proof of uniform asymptotic stability is more difficult. If $x = 0$ is asymptotically stable, it is stable and therefore, uniformly stable. In addition, if $\phi \in \mathcal{B}(0, b_0)$, then $x_t(\sigma, \phi) \to 0$ as $t \to \infty$, that is, the set $M = \mathbb{R} \times \{0\}$ attracts points of the ball $\mathcal{B}(0, b_0)$. The representation theorem, Theorem 6.1 of Section 3.6, for the solution operator $T(t, \sigma)$ implies there is a $t_0 > 0$, independent of σ such that $T(\sigma + t, \sigma)$ is conditionally completely continuous for $t \geq t_0$. The complete proof of the uniform approach to zero is now supplied in the same way as for the proof of Theorem 5.3 of Section 4.5. □

Definition 1.2. A solution $x(\sigma, \phi)$ of an RFDE(f) is *bounded* if there is a $\beta(\sigma, \phi)$ such that $|x(\sigma, \phi)(t)| < \beta(\sigma, \phi)$ for $t \geq \sigma - r$. The solutions are *uniformly bounded* if for any $\alpha > 0$, there is a $\beta = \beta(\alpha) > 0$ such that for all $\sigma \in \mathbb{R}$, $\phi \in C$, and $|\phi| \leq \alpha$, we have $|x(\sigma, \phi)(t)| \leq \beta(\alpha)$ for all $t \geq \sigma$. The solutions are *ultimately bounded* if there is a constant β such that for any $(\sigma, \phi) \in \mathbb{R} \times C$, there is a constant $t_0(\sigma, \phi)$ such that $|x(\sigma, \phi)(t)| < \beta$ for $t \geq \sigma + t_0(\sigma, \phi)$. The solutions are *uniformly ultimately bounded* if there is a $\beta > 0$ such that for any $\alpha > 0$, there is a constant $t_0(\alpha) > 0$ such that $|x(\sigma, \phi)(t)| \leq \beta$ for $t \geq \sigma + t_0(\alpha)$ for all $\sigma \in \mathbb{R}$, $\phi \in C$, $|\phi| \leq \alpha$.

This definition of ultimate boundedness coincides with the concept of point dissipative for the process generated by the RFDE(f). As a consequence of Theorem 5.2 of Section 4.5, we can state the following result.

Lemma 1.2. *If a periodic* RFDE(f) *is such that the solution map* $T(t,\sigma): C \to C$ *is defined for all* $t \geq \sigma$, $T(t,\sigma)\phi$ *is continuous in* (t,σ,ϕ) *and* $T(t,\sigma)$ *takes bounded sets into bounded sets, then ultimate boundedness is equivalent to uniform ultimate boundedness.*

5.2 The method of Liapunov functionals

In this section, we give sufficient conditions for the stability and instability of the solution $x = 0$ of Equation (1.1) that generalize the second method of Liapunov for ordinary differential equations. The results are illustrated by examples.

If $V : \mathbb{R} \times C \to \mathbb{R}$ is continuous and $x(\sigma,\phi)$ is the solution of Equation (1.1) through (σ,ϕ), we define

$$\dot{V}(t,\phi) = \limsup_{h\to 0^+} \frac{1}{h}\,[V(t+h, x_{t+h}(t,\phi)) - V(t,\phi)].$$

The function $\dot{V}(t,\phi)$ is the upper right-hand derivate of $V(t,\phi)$ along the solution of Equation (1.1). If we wish to emphasize the dependence on Equation (1.1), we write $\dot{V}_{(1.1)}(t,\phi)$.

Theorem 2.1. *Suppose* $f : \mathbb{R} \times C \to \mathbb{R}^n$ *takes* $\mathbb{R} \times$ (*bounded sets of* C) *into bounded sets of* $\mathbb{R}^n$, *and* $u, v, w : \mathbb{R}^+ \to \mathbb{R}^+$ *are continuous nondecreasing functions,* $u(s)$ *and* $v(s)$ *are positive for* $s > 0$, *and* $u(0) = v(0) = 0$. *If there is a continuous function* $V : \mathbb{R} \times C \to \mathbb{R}$ *such that*

$$u(|\phi(0)|) \leq V(t,\phi) \leq v(|\phi|)$$

$$\dot{V}(t,\phi) \leq -w(|\phi(0)|)$$

then the solution $x = 0$ *of Equation* (1.1) *is uniformly stable. If* $u(s) \to \infty$ *as* $s \to \infty$, *the solutions of Equation* (1.1) *are uniformly bounded. If* $w(s) > 0$ *for* $s > 0$, *then the solution* $x = 0$ *is uniformly asymptotically stable.*

Proof. There is no loss of generality in assuming $r > 0$. For any $\epsilon > 0$, there is a $\delta = \delta(\epsilon)$, $0 < \delta < \epsilon$, such that $v(\delta) < u(\epsilon)$. If $\phi \in \mathcal{B}(0,\delta)$, $\sigma \in \mathbb{R}$, then $\dot{V}(t, x_t(\sigma,\phi)) \leq 0$ for all $t \geq 0$ and the inequalities on $V(t,\phi)$ imply

$$u(|x(\sigma,\phi)(t)|) \leq V(t, x_t(\sigma,\phi)) \leq V(\sigma,\phi) \leq v(\delta) < u(\epsilon), \qquad t \geq \sigma.$$

Therefore, $|x(\sigma,\phi)(t)| < \epsilon$, $t \geq \sigma$. Since $|\phi| < \delta < \epsilon$, this proves uniform stability.

If $u(s) \to \infty$ as $s \to \infty$ and $\alpha > 0$ is any given constant, there is a $\beta > 0$ such that $u(\beta) = v(\alpha)$. Consequently, if $|\phi| \leq \alpha$, the solution $x(\sigma,\phi)$ through (σ,ϕ) satisfies $|x(\sigma,\phi)(t)| \leq \beta$ for all $t \geq \sigma$. This obviously implies uniform boundedness.

To prove the assertion of the theorem concerning uniform asymptotic stability, for $\epsilon = 1$, choose $\delta_0 = \delta(1)$ as the constant for uniform stability. For any $\epsilon > 0$, we wish to show there is a $t_0(\delta_0, \epsilon) > 0$ such that any solution $x(\sigma, \phi)$ of Equation (1.1) with $|\phi| < \delta_0$ satisfies $|x_t(\sigma, \phi)| < \epsilon$ for $t \geq \sigma + t_0(\delta_0, \epsilon)$. Let $\delta = \delta(\epsilon)$ be the constant for uniform stability. Suppose that a solution $x = x(\sigma, \phi)$, $|\phi| < \delta_0$ satisfies $|x_t| \geq \delta$, $t \geq \sigma$. Since each interval of length r contains an s such that $|x(s)| \geq \delta$, there exists a sequence $\{t_k\}$ such that

$$\sigma + (2k-1)r \leq t_k \leq \sigma + 2kr, \qquad k = 1, 2, \dots,$$

and

$$|x(t_k)| \geq \delta.$$

By the assumption of f, there exists a constant L such that $|\dot{x}(t)| < L$ for all $t \geq \sigma$. Therefore, on the intervals $t_k - (\delta/2L) \leq t \leq t_k + (\delta/2L)$, we have $|x(t)| > \delta/2$. Therefore,

$$\dot{V}(t, x_t) \leq -w(\frac{\delta}{2}), \qquad t_k - \frac{\delta}{2L} \leq t \leq t_k + \frac{\delta}{2L}.$$

By taking a large L, if necessary, we can assume that these intervals do not overlap, and hence

$$V(t_k, x_{t_k}) - V(\sigma, \phi) \leq -w(\frac{\delta}{2})\frac{\delta}{L}\,(k-1).$$

Let $K(\delta_0, L)$ be the smallest integer $\geq v(\delta_0)/((\delta/L)w(\delta/2))$. If $k > 1 + K(\delta_0, L)$, then

$$V(t_k, x_{t_k}) < v(\delta_0) - w(\frac{\delta}{2})\frac{\delta}{L}\frac{v(\delta_0)}{w(\delta/2)(\delta/L)} \leq 0,$$

which is a contradiction. Therefore, at some $\tau = \tau(\sigma, \phi)$ such that $\sigma < \tau \leq \sigma + 2rK(\delta_0, L)$ we have $|x_\tau| < \delta$ and $|x_t| < \epsilon$ for $t \geq \sigma + 2rK(\delta_0, L)$. This proves the uniform asymptotic stability. The proof of the theorem is therefore complete. □

Let us consider a method of constructing a particular functional satisfying the conditions of Theorem 2.1 for the equation

$$\dot{x}(t) = Ax(t) + Bx(t-r), \qquad r > 0, \tag{2.1}$$

where A and B are constant matrices. Suppose the eigenvalues of A have negative real parts and choose the symmetric matrix C such that $C > 0$ and $A^T C + CA = -D < 0$, where A^T is the transpose of A. If E is a positive definite matrix and

$$V(\phi) = \phi(0)^T C\phi(0) + \int_{-r}^{0} \phi(\theta)^T E\phi(\theta)\, d\theta \tag{2.2}$$

then $v|\phi(0)|^2 \le V(\phi) \le K|\phi|^2$ for some positive v, K. Furthermore,

$$\begin{aligned} \dot{V}(\phi) = &-\phi(0)^T D\phi(0) + 2\phi(0)^T CB\phi(-r) + \phi(0)^T E\phi(0) \\ &- \phi(-r)^T E\phi(-r). \end{aligned} \tag{2.3}$$

If we consider the right-hand side of Equation (2.3) as a quadratic form in $\phi(0), \phi(-r)$ and impose conditions on the matrices A and B to ensure there exist matrices C and E such that this quadratic form is negative definite in $\phi(0), \phi(-r)$, then Theorem 2.1 will imply uniform asymptotic stability of the solution $x = 0$. More specifically, we try to determine A, B, C, and E so that the symmetric matrix

$$\begin{bmatrix} D-E & (CB)^T \\ -CB & E \end{bmatrix} \tag{2.4}$$

is positive where $-D = A^T C + CA$. We already know that D and E are positive. Consequently, Matrix (2.4) for $B = 0$ is positive if $D - E > 0$. As a result of this, this matrix is positive for B sufficiently small and the solution $x = 0$ is asymptotically stable. This latter fact is also an immediate consequence of the results in Chapter 1, where it was shown that asymptotic stability prevailed if the roots of the characteristic equation

$$\det\left[I - A - Be^{-\lambda r}\right] = 0 \tag{2.5}$$

satisfy $\operatorname{Re}\lambda < 0$.

One can actually use this method to obtain explicit estimates on B for which Matrix (2.4) is positive. To be more specific, suppose $E < D$ and

$$x^T(D-E)x \ge \lambda|x|^2, \qquad x^T Ex \ge \mu|x|^2.$$

Then

$$\dot{V}(\phi) \le -\lambda|\phi(0)|^2 + 2\|CB\|\,|\phi(0)|\,|\phi(-r)| - \mu|\phi(-r)|^2.$$

If $\lambda\mu - \|CB\|^2 > 0$, then $\dot{V}(\phi) \le -k(|\phi(0)|^2 + |\phi(-r)|^2)$, $r > 0$, for a suitable positive constant k and Theorem 2.1 implies uniform asymptotic stability.

These estimates are certainly not optimal and the best estimates using this procedure are not easy to obtain. However, there is one important qualitative remark that can be made about proving stability by insisting that Matrix (2.4) is positive. If Matrix (2.4) is positive, then the solution $x = 0$ of Equation (2.1) is *asymptotically stable for every value of the delay* r.

For the scalar equation

$$\dot{x}(t) = -ax(t) - bx(t-r), \tag{2.6}$$

if

$$V(\phi) = \frac{1}{2}\,\phi^2(0) + \mu\int_{-r}^{0} \phi^2(\theta)\,d\theta, \qquad \mu > 0, \tag{2.7}$$

then

$$\dot{V}(\phi) = -(a-\mu)\phi^2(0) - b\phi(0)\phi(-r) - \mu\phi^2(-r).$$

The corresponding Matrix (2.4) is positive if and only if

$$a > \mu > 0, \qquad 4(a-\mu)\mu > b^2. \tag{2.8}$$

The choice of μ that will allow $|b|$ as large as possible is $\mu = a/2$. For this choice of μ, we see that the solution $x = 0$ of Equation (2.6) is asymptotically stable if $|b| < a$. The exact region of asymptotic stability is shown in Figure 5.1 and is obtained by applying the theory in the appendix to the equation $\lambda = -a - b\exp(-\lambda r)$ (see Theorem A.5). The upper boundary of the region of stability is given parametrically by the equation

$$a = -b\cos\zeta r, \qquad b\sin\zeta r = \zeta, \qquad 0 < \zeta < \frac{\pi}{r}.$$

The region $|b| < a$ is precisely the region for which there is asymptotic stability for all $r > 0$ and an estimate of the rate of decay of zero that is independent of the delay r.

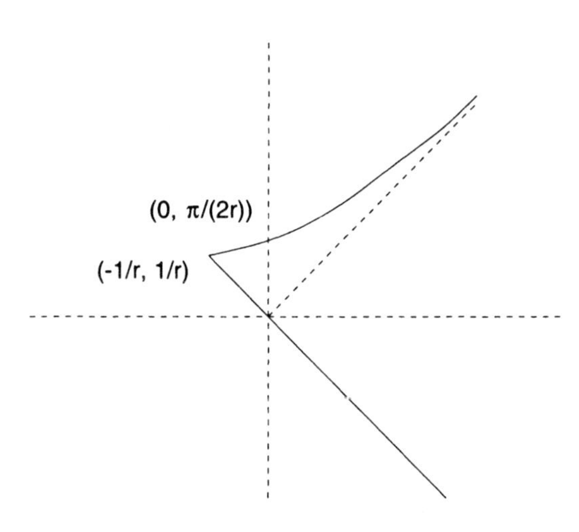

Fig. 5.1.

Even if a and b in Equation (2.6) depend on t, we obtain very interesting information about stability using the same type of functional V in Equation (2.7). In fact, suppose $a(t)$ and $b(t)$ are bounded continuous functions on $\mathbb{R}$ with $a(t) \geq \delta > 0$ for all t and consider the equation

$$\dot{x}(t) = -a(t)x(t) - b(t)x(t-r). \tag{2.9}$$

Using the functional V in Equation (2.7) with $\mu = \delta/2$, one sees that $\dot{V}(\phi)$ is the same expression as before and the solution $x = 0$ of Equation (2.9) is uniformly asymptotically stable if Inequalities (2.8) are satisfied; that is, if $(2a(t) - \delta)\delta > b^2(t)$ uniformly in t. In particular, the conditions are satisfied if there is a constant θ, $0 \le \theta < 1$ such that $|b(t)| \le \theta\delta$ for all $t \in \mathbb{R}$.

If $r = r(t)$ is continuously differentiable and bounded, then using the same argument and the same V as in Equation (2.7) one observes that uniform asymptotic stability prevails if $a(t) \ge \delta > \mu > 0$ and

$$(2a(t) - \mu)(1 - \dot{r}(t))\mu > b^2(t)$$

uniformly in t.

For the autonomous matrix Equation (2.1), the exact region of stability as an explicit function of A, B, and r is not known and probably will never be known. The reason is simple to understand because the characteristic equation (2.5) is so complicated. It is therefore, worthwhile to obtain methods for determining approximations to the region of stability. One possible approach is to make use of Theorem 2.1. If this approach is taken, then the functional $V(\phi)$ must be more complicated than the one in Equation (2.2) since the corresponding stability region is independent of the delay r.

For Equation (2.6), we can give a very simple result on a possible form for the Liapunov functional.

Suppose $\alpha, \beta : [-r, 0] \to \mathbb{R}$, $\gamma : [-r, 0] \times [-r, 0] \to \mathbb{R}$ are continuously differentiable functions and consider the functional V on C defined by the quadratic form

$$\begin{aligned} V(\phi) = \phi^2(0) + 2\phi(0) \int_{-r}^{0} \alpha(\theta)\phi(\theta)\, d\theta + \int_{-r}^{0} \beta(\theta)\phi^2(\theta)\, d\theta \\ + \int_{-r}^{0}\int_{-r}^{0} \phi(\xi)\phi(\eta)\gamma(\xi, \eta)\, d\xi d\eta. \end{aligned} \tag{2.10}$$

A quadratic form H on C is said to be positive if $H(\phi) > 0$ for $\phi \ne 0$. If such an H is positive, we write $H > 0$. To obtain the stability region for Equation (2.6) by means of a functional of the form given in Equation (2.10), we make use of the following result.

Lemma 2.1. *If there exist α, β, γ, and $H > 0$ such that the derivative of V in Equation* (2.10) *along the solutions of Equation* (2.6) *satisfies*

$$\dot{V}(\phi) = -H(\phi)$$

for all $\phi \in C$, then no root of the characteristic equation

$$\lambda = -a - be^{-\lambda r} \tag{2.11}$$

lies on the imaginary axis. Furthermore, $V > 0$ if and only if no roots of Equation (2.11) *have positive real parts.*

Proof. Suppose there exists an H as specified and there is also a purely imaginary root of Equation (2.11). Then there is a periodic solution x of Equation (2.6) such that $x_t \neq 0$ for all t. Thus, $V(x_t)$ is a periodic function and $\dot{V}(x_t) = -H(x_t) < 0$ for all t. This is obviously a contradiction.

To prove the second statement, suppose $V > 0$. If some roots of characteristic equation (2.11) have positive real parts then there is a solution x of Equation (2.6) for which $V(x_t)$ is unbounded. But this contradicts the fact that $0 < V(x_t) \leq V(x_0)$ for $t \geq 0$. Conversely, suppose no roots of Equation (2.11) have positive real parts and V is not positive. If there is a nonzero ϕ such that $V(\phi) \leq 0$, then the solution x through ϕ is such that $x_t \neq 0$ for $0 \leq t < \epsilon$ for some ϵ sufficiently small. Therefore, $V(x_t) < 0$ for $0 < t < \epsilon$. Thus, there is a $\psi \neq 0$ such that $V(\psi) < 0$. The instability Theorem 3.3 of Section 5.3 gives a contradiction and the lemma is proved. □

The problem remains to determine α, β, γ, and H. If we simply proceed to calculate $\dot{V}$ along the solutions of Equation (2.6), we obtain a rather complicated set of equations for which it is difficult to recognize the natural choices for these functions. A more reasonable approach is to try to determine the analogue of the converse theorem of Liapunov for ordinary differential equations. Let us be more specific. For a matrix autonomous ordinary differential equation (Equation (2.1) with $B = 0$), if $V(x) = x^T Cx$ is a given quadratic form, then the derivative along the solutions is given by $\dot{V}(x) = x^T(A^T C + CA)x \equiv -x^T Wx$. If all solutions approach zero as $t \to \infty$, then it is the classical theorem of Liapunov states that, for any positive definite matrix W, there is a positive definite matrix C such that $A^T C + CA = -W$. Conversely, if the latter equation has a solution for a positive definite matrix W, then the solutions of the differential equation approach zero. We now formulate a similar result for RFDE.

Consider the general linear autonomous system

$$\dot{x}(t) = Lx_t = \int_{-r}^{0} d[\eta(\theta)]\, x(t+\theta)\,, \tag{2.12}$$

where η is an $n \times n$ matrix function of bounded variation. Exponential solutions of (2.12) are determined from the characteristic equation

$$\det \Delta(\lambda) = 0, \quad \Delta(\lambda) = \lambda I - \int_{-r}^{0} e^{\lambda\theta} d\eta(\theta)\,. \tag{2.13}$$

If all of the solutions of (2.13) satisfy $\operatorname{Re}\lambda < 0$, then we will see in Chapter 7 that all solutions of (2.12) approach zero as $t \to \infty$.

Theorem 2.2. *If all solutions of* (2.13) *satisfy* $\operatorname{Re}\lambda < 0$, *then for any positive definite* $n \times n$ *matrix* W, *there is a quadratic functional* $V : C \to \mathbb{R}$ *such that for* $\phi \in C$, *the derivative* $\dot{V}(\phi)$ *along the solutions of* (2.12) *satisfies*

$$\dot{V}(\phi) = -\phi(0)^T W \phi(0) \tag{2.14}$$

and there is a constant $k > 0$ and for any $\alpha > 0$ a constant $k_\alpha > 0$ such that for $|\phi| \le \alpha$,

$$k_\alpha |\phi(0)|^3 \le V(\phi) \le k|\phi|^2 . \tag{2.15}$$

We remark that even though $V(\phi)$ is a quadratic form in ϕ, the lower bound on $V(\phi)$ involves $|\phi(0)|^3$ and the lower bound also depends on the upper bound of $|\phi|$. For an equation (2.12) for which there exists a function V satisfying the conditions of Theorem 2.2, we can apply Theorem 2.1 to conclude that the origin is uniformly asymptotically stable. Since the system (2.12) is linear, this implies that all of the solutions approach zero.

We now define the method for constructing the function V. For any given $n \times n$ matrix W, define the matrix $Y(s) \equiv Y_W(s)$, $-r \le s \le r$, by the relation

$$Y(s) = \frac{1}{2\pi} \int_{-\infty}^{\infty} \Delta^{-1}(i\omega)^T W \Delta^{-1}(-i\omega) e^{i\omega s}\, d\omega \tag{2.16}$$

and the function $V(\phi)$ by

$$\begin{aligned} V(\phi) =& \phi(0)^T Y(0)\phi(0) + 2\phi(0)^T \int_{-r}^{0}\int_{u}^{0} Y(-u+\theta)\, d\eta(u)\, \phi(\theta)\, d\theta \\ &+ \int_{-r}^{0}\int_{h}^{0} ds\, \phi(s)^T\, d\eta^T(h) \\ &\quad \Big[\int_{-r}^{0}\int_{u}^{0} Y(-s+h-u+\theta) d\eta(u)\, \phi(\theta)\, d\theta\Big]. \end{aligned} \tag{2.17}$$

It can then be shown that $Y(s)$ is continuously differentiable for $s \ne 0$ and satisfies the properties:

$$\begin{aligned} &Y(0) \text{ is symmetric and } Y(s) = Y^T(-s),\ s \ge 0, \\ &\dot{Y}(s) = \int_{-r}^{0} d[\eta^T(\theta)]\, Y(s+\theta), \quad 0 \le s \le r, \end{aligned} \tag{2.18}$$

where $\dot{Y} = dY/ds$. Also,

$$\begin{aligned} \dot{V}(\phi) &= \phi(0)^T \Big[\int_{-r}^{0} Y(-\theta)\, d[\eta(\theta)] + \int_{-r}^{0} d[\eta^T(\theta)]\, Y(\theta)\Big]\phi(0) \\ &= \phi(0)^T \big[\dot{Y}(0) + \dot{Y}^T(0)\big]\phi(0) \\ &= -\phi(0)^T W \phi(0). \end{aligned} \tag{2.19}$$

If the matrix W is positive definite and the solutions of (2.13) have negative real parts, then the function V satisfies the properties stated. The proofs of these facts are far from trivial and the reader may consult the supplementary remarks for references to the details.

We now give more details on the construction of the functional V for the special equation (2.1). Relation (2.18) implies that

$$\dot{Y}(s) = A^T Y(s) + B^T Y(s-r) = A^T Y(s) + B^T Y^T(r-s).$$

This implies that

$$\ddot{Y}(s) = A^T \dot{Y}(s) - \dot{Y}(s)A + A^T Y(s)A - B^T Y(s)B, \tag{2.20}$$

since

$$\begin{aligned}\ddot{Y}(s) &= A^T \dot{Y}(s) - B^T \dot{Y}^T(r-s) \\ &= A^T \dot{Y}(s) - B^T Y^T(r-s)A - B^T Y^T(-s)B \\ &= A^T \dot{Y}(s) - B^T Y(s-r)A - B^T Y(s)B \\ &= A^T \dot{Y}(s) - \dot{Y}(s)A + A^T Y(s)A - B^T Y(s)B.\end{aligned}$$

From (2.19), we see that not only must Y satisfy the second-order ordinary differential equation, but it must also satisfy the initial condition

$$\dot{Y}(0) + \dot{Y}^T(0) = -W. \tag{2.21}$$

If the eigenvalues of the equation

$$\det\left[\lambda I - A - Be^{-\lambda r}\right] = 0 \tag{2.22}$$

have negative real parts, then Theorem 2.2 asserts that for any positive definite matrix W, the function V defined in (2.17) with Y satisfying (2.20), (2.21), and (2.18) satisfies the conditions (2.14) and (2.15).

Even though it will in general be impossible to give the explicit formula for the functional V, we can use the fact that we know that it must exist and be of the preceding form to obtain approximately the region of stability. In fact, we can make intelligent guesses for the function Y and then verify that the conditions of Theorem 2.1 are satisfied.

It is possible to give the explicit form of V for the scalar equation (2.6) with $r = 1$; that is,

$$\dot{x}(t) = -ax(t) - bx(t-1). \tag{2.23}$$

If we choose $W = 1$, then the equations (2.20) and (2.21) for this case are

$$\ddot{Y}(s) = (a^2 - b^2)Y(s), \quad \dot{Y}(0) = -\frac{1}{2} \tag{2.24}$$

and (2.18) implies that

$$\dot{Y}(s) = -aY(s) - bY(s-1) = -aY(s) - bY(1-s). \tag{2.25}$$

The functional V is given by

$$
\begin{aligned}
V(\phi) =& Y(0)\phi^2(0) - 2b\phi(0)\int_{-1}^{0} Y(\theta+1)\phi(\theta)\,d\theta \\
&+ b^2\int_{-1}^{0}\phi(u)[\int_{-1}^{0} Y(\theta-u)\phi(\theta)\,d\theta]\,du.
\end{aligned}
\tag{2.26}
$$

After some rather lengthy computations, it can be shown that if $\beta + a + be^{-\beta} \neq 0$ (which is always the case if the solutions of the characteristic equation have negative real parts) and

$$
Y(s) = -\frac{1}{2}s + \frac{1+a}{4a}, \quad 0 \le s \le 1, \quad a = b \neq 0
$$

and $Y(-s) = Y(s)$, $0 \le s \le 1$, then

$$
Y(s) = -\frac{be^{\beta}e^{\beta s}}{2\beta(\beta+a+be^{-\beta})} + \frac{(\beta+a)e^{-\beta s}}{2\beta(\beta+a+be^{-\beta})}, \quad 0 \le s \le 1
$$

and $\beta^2 = a^2 - b^2 \neq 0$.

Our next objective is to give sufficient conditions for the instability of the solution $x = 0$ of an RFDE(f).

Theorem 2.3. *Consider an* RFDE(f) *and designate the solution through* (σ, ϕ) *by* $x(\sigma, \phi)$. *Suppose* $V(\phi)$ *is a continuous bounded scalar function on* C. *If there exist a* $\gamma > 0$ *and an open set* U *in* C *such that*

(i) $V(\phi) > 0$ *on* U, $V(\phi) = 0$ *on the boundary of* U,

(ii) 0 *belongs to the closure of* $U \cap \mathcal{B}(0,\gamma)$,

(iii) $V(\phi) \le u(|\phi(0)|)$ *on* $U \cap \mathcal{B}(0,\gamma)$,

(iv) $\dot{V}_-(\phi) \ge w(|\phi(0)|)$ *on* $[0,\infty) \times U \cap \mathcal{B}(0,\gamma)$,

$$
\dot{V}_-(\phi) = \liminf_{h\to 0+} \frac{1}{h}[V(x_{t+h}(t,\phi)) - V(\phi)]
$$

where $u(s), w(s)$ *are continuous, increasing, and positive for* $s > 0$, *then the solution* $x = 0$ *of the* RFDE(f) *is unstable. More specifically, each solution* $x_t(\sigma,\phi)$ *with initial function* ϕ *in* $U \cap \mathcal{B}(0,\gamma)$ *at* σ *must reach the boundary of* $\mathcal{B}(0,\gamma)$ *in a finite time.*

Proof. Suppose $\phi \in U \cap \mathcal{B}(0,\gamma)$, $\sigma \in \mathbb{R}$. Then $V(\phi) > 0$. By Hypothesis (iii), $|\phi(0)| \ge u^{-1}(V(\phi))$ and Hypotheses (iii) and (iv) imply $x_t = x_t(\sigma,\phi)$ satisfies $|x(t)| \ge u^{-1}(V(x_t)) \ge u^{-1}(V(\phi))$ as long as $x_t \in U \cap \mathcal{B}(0,\gamma)$. Consequently,

$$
\dot{V}_-(x_t) \ge w(|x(t)|) \ge w(u^{-1}(V(\phi))) > 0 \qquad \text{if } x_t \in U \cap \mathcal{B}(0,\gamma).
$$

If we let $\eta = w(u^{-1}(V(\phi)))$, then this implies

$$
V(x_t) \ge V(\phi) + \eta(t-\sigma)
$$

as long as $x_t \in U \cap \mathcal{B}(0,\gamma)$. Hypotheses (i) and (iv) imply that x_t cannot leave $U \cap \mathcal{B}(0,\gamma)$ by crossing the boundary of U. Since $V(\phi)$ is bounded on $U \cap \mathcal{B}(0,\gamma)$ this implies there must be a t_1 such that $x_{t_1} \in \partial\mathcal{B}(0,\gamma)$. This proves the last assertion of the theorem. But Hypothesis (ii) implies that in each neighborhood of the origin of C, there are ϕ in $U \cap \mathcal{B}(0,\gamma)$. Instability follows and the proof of the theorem is complete. □

As an example, consider Equation (2.6) with $a+b<0$. We wish to prove by use of Liapunov functions that the solution $x=0$ of Equation (2.6) is unstable, for some $r<r_0(a,b)$. Even though better results were obtained before, this Liapunov function may be used for nonlinear and nonautonomous equations.

If F is any given continuously differentiable function and

$$V(x_t) = \frac{x^2(t)}{2} - \frac{1}{2}\int_{t-r}^{t} F(t-u)[x(u)-x(t)]^2\,du,$$

then it easily seen that

$$\begin{aligned}\dot V_-(x_t) = \dot V(x_t) &= -(a+b)x^2(t) - b[x(t-r)-x(t)]x(t)\\ &\quad + \frac{1}{2}F(r)[x(t-r)-x(t)]^2\\ &\quad - \frac{1}{2}\int_{t-r}^{t} \dot F(t-u)[x(u)-x(t)]^2\,du\\ &\quad + \int_{t-r}^{t} F(t-u)[x(u)-x(t)]\\ &\quad \times \left[-(a+b)x(t) - b\{x(t-r)-x(t)\}\right]du.\end{aligned}$$

If the expression for $\dot V$ is written as an integral from $[t-r,t]$, then the integrand will be a positive definite quadratic form in $x(t), [x(t-r)-x(t)], [x(u)-x(t)]$ if the following inequalities are satisfied

$$a+b<0,$$

$$\Delta_1 \overset{\text{def}}{=} -\frac{(a+b)}{2}\,F(r) - \frac{b^2}{4} > 0,$$

$$-\frac{\Delta_1}{r^2}\left(-\frac{1}{2}\,\dot F(\theta)\right) - \frac{(a+b)^2}{8r}\,F^2(\theta)F(r) > 0, \qquad 0 \le \theta \le r.$$

If $a+b<0$, then one can determine an $r_0(a,b)$ and a continuously differentiable positive function $F(\theta)$, $0 \le \theta \le r < r_0(a,b)$ such that these inequalities are satisfied. Consequently, there exists a positive number q such that

$$\dot V_-(\phi) \ge qr\phi^2(0), \qquad V(\phi) \le \frac{\phi^2(0)}{2},$$

for all ϕ in C. If

$$U = \{\phi \in C : \phi^2(0) > \int_{-r}^{0} F(-\theta)[\phi(\theta) - \phi(0)]^2 d\theta\},$$

then U satisfies Hypotheses (i) and (ii) of Theorem 2.2 and the solution $x = 0$ of Equation (2.6) is unstable if $a + b < 0$ and $r < r_0(a, b)$.

Notice that the same conclusions for this example are valid if a and b are functions of t provided that a and b are bounded and $a(t) + b(t) < \delta < 0$ for all t.

As another example, consider the equation

$$\dot{x}(t) = a(t)x^3(t) + b(t)x^3(t - r) \tag{2.27}$$

where $a(t)$ and $b(t)$ are arbitrary continuous bounded functions with $a(t) \geq \delta > 0$, $|b(t)| < q\delta$, $0 < q < 1$. For

$$V(\phi) = \frac{\phi^4(0)}{4} - \frac{\delta}{2} \int_{-r}^{0} \phi^6(\theta)\, d\theta,$$

we have $V(\phi) \leq \phi^4(0)/4$ and

$$\dot{V}_-(\phi) = \dot{V}(\phi) = \big[a(t) - \frac{\delta}{2}\big]\phi^6(0) + b(t)\phi^3(0)\phi^3(-r) + \frac{\delta}{2}\, \phi^6(-r).$$

This last expression is a positive definite quadratic form in $\phi^3(0), \phi^3(-r)$. If

$$U = \{\phi \in C : \phi^4(0) > \frac{\delta}{2} \int_{-r}^{0} \phi^6(\theta) d\theta\},$$

then the same argument as in the previous example shows that $x = 0$ is an unstable solution of Equation (2.27).

If $a(t) \leq -\delta < 0$, $|b(t)| < q\delta$, one can choose

$$V(\phi) = \frac{\phi^4(0)}{4} + \frac{\delta}{2} \int_{-r}^{0} \phi^6(\theta)\, d\theta$$

and use Theorem 2.1 to prove the zero solution is uniformly asymptotically stable.

Notice that a more refined argument using the same V functionals may be employed to show that the zero solution of

$$\dot{x}(t) = ax^3(t) + b(t)x^4(t - r)$$

is stable or unstable according to whether $a < 0$ or > 0, provided only that $|b(t)|$ is bounded on $\mathbb{R}$. One simply must operate in a sufficiently small neighborhood of the origin.

5.3 Liapunov functionals for autonomous systems

Consider the autonomous equation

$$\dot{x}(t) = f(x_t) \tag{3.1}$$

where $f : C \to \mathbb{R}$ is completely continuous and solutions of Equation (3.1) depend continuously on the initial data. We denote by $x(\phi)$ the solution through $(0, \phi)$.

If $V : C \to \mathbb{R}$ is a continuous function, we define the derivative of V along the solution of Equation (3.1) as in Section 5.2; namely

$$\dot{V}(\phi) = \dot{V}_{(3.1)}(\phi) = \limsup_{h \to 0^+} \frac{1}{h} [V(x_h(\phi)) - V(\phi)].$$

Definition 3.1. We say $V : C \to \mathbb{R}$ is a *Liapunov function* on a set G in C relative to Equation (3.1) if V is continuous on $\overline{G}$ (or Cl G), the closure of G, and $\dot{V} \leq 0$ on G. Let

$$S = \{\phi \in \overline{G} : \dot{V}(\phi) = 0\}$$

$M =$ largest set in S that is invariant with respect to Equation (3.1).

Theorem 3.1. *If V is a Liapunov function on G and $x_t(\phi)$ is a bounded solution of Equation* (3.1) *that remains in G, then $x_t(\phi)$ tends to M as $t \to \infty$.*

Theorem 3.2. *If V is a Liapunov function on $U_l = \{\phi \in C : V(\phi) < l\}$ and there is a constant $K = K(l)$ such that ϕ in U_l implies $|\phi(0)| < K$, then any solution $x_t(\phi)$ of Equation* (3.1) *with ϕ in U_l tends to M as $t \to \infty$.*

Proof of Theorem 3.1. If $|x_t(\phi)| < K$, $x_t(\phi) \in G$, $t \geq 0$, then $\{x_t(\phi)\}$ belongs to a compact set of C and has a nonempty ω limit set, $\omega(\gamma^+(\phi))$. Thus, $V(x_t(\phi))$ is nonincreasing, bounded below, and must approach a limit c as $t \to \infty$. Since V is continuous on Cl G, $V(\psi) = c$ for ψ in $\omega(\gamma^+(\phi))$. Since $\omega(\gamma^+(\phi))$ is invariant $\dot{V}(\psi) = 0$ on $\omega(\gamma^+(\phi))$. Since every solution approaches its ω-limit set, this proves Theorem 3.1. □

Proof of Theorem 3.2. If ϕ is in U_l and $\dot{V} \leq 0$ on U_l, then $x_t(\phi) \in U_l$, $t \geq 0$. Also $|x(\phi)(t)| \leq K$ if $t \geq 0$, which implies $x_t(\phi)$ bounded. Now use Theorem 3.1. □

Corollary 3.1. *Suppose $V : C \to \mathbb{R}$ is continuous and there exist non-negative functions $a(r)$ and $b(r)$ such that $a(r) \to \infty$ as $r \to \infty$*

$$a(|\phi(0)|) \leq V(\phi), \qquad \dot{V}(\phi) \leq -b(|\phi(0)|).$$

Then the solution $x = 0$ of Equation (3.1) is stable and every solution is bounded. If, in addition, $b(r)$ is positive definite, then every solution approaches zero as $t \to \infty$.

Proof. Stability is immediate. The solutions are bounded since $a(r) \to \infty$ and $|x(t)|$ bounded for $t \geq 0$ implies x_t bounded for $t \geq 0$. If b is positive definite, the conditions of Theorem 3.2 are satisfied for any l. Furthermore, $S = \{\phi : \phi(0) = 0\}$, $M = \{0\}$. □

Theorem 3.3. (Instability). *Suppose zero belongs to the closure of an open set U in C and N is an open neighborhood of zero in C. Assume that*

(i) *V is a Liapunov function on $G = N \cap U$.*

(ii) *$M \cap G$ is either the empty set or zero.*

(iii) *$V(\phi) < \eta$ on G when $\phi \neq 0$.*

(iv) *$V(0) = \eta$ and $V(\phi) = \eta$ when $\phi \in \partial G \cap N$.*

If N_0 is a bounded neighborhood of zero properly contained in N, then $\phi \neq 0$ in $G \cap N_0$ implies there exists a τ such that $x_\tau(\phi) \in \partial N_0$.

Proof. If $\phi \in G \cap N_0$, $\phi \neq 0$, then $V(x_t(\phi)) \leq V(\phi) < \eta$ for all $t \geq 0$ as long as $x_t(\phi)$ remains in $N_0 \cap G$. If $x_t(\phi)$ remains in the bounded set $N_0 \cap G$ for all $t \geq 0$, then the ω-limit set $\omega(\gamma^+(\phi)) \subseteq N_0 \cap G$. Also, $\omega(\gamma^+(\phi))$ is invariant. Hypothesis (ii) implies $\omega(\gamma^+(\phi)) = \{0\}$. On the other hand, $V(0) = \eta$. This is a contradiction. Therefore, there is a $\tau > 0$ such that $x_\tau(\phi) \in \partial(N_0 \cap G)$. Condition (iv) implies $x_\tau(\phi) \in \partial N_0$ and the theorem is proved. □

As a first example, we reconsider Equation (2.27) in the autonomous case; that is,

$$\dot{x}(t) = ax^3(t) + bx^3(t-r) \tag{3.2}$$

where a and b are constants, $a \neq 0$. If

$$V(\phi) = -\frac{\phi^4(0)}{2a} + \int_{-r}^{0} \phi^6(\theta)\, d\theta,$$

then

$$\dot{V}(\phi) = -\big[\phi^6(0) + \frac{2b}{a}\, \phi^3(0)\phi^3(-r) + \phi^6(-r)\big].$$

Consequently, V is a Liapunov function on C if $|b| \leq |a|$. If $a < 0$, then $V(\phi) \geq \phi^4(0)/(2|a|)$ and Corollary 3.1 implies the origin is stable and every solution is bounded.

If $a < 0$, $|b| < |a|$, then S in Definition 3.1 is $\{\phi \in C : \phi(0) = \phi(-r) = 0\}$. Obviously, the set $M = \{0\}$ and Corollary 3.1 implies the solution $x = 0$ is globally asymptotically stable.

If $a < 0$, $b = a$, then $S = \{\phi : \phi(0) = -\phi(-r)\}$. Therefore, M must be the set of initial values of solutions satisfying $x(t) = -x(t-r)$ for all $t \in \mathbb{R}$; that is, $\dot{x} = 0$ and $x(t) = c$, a constant. But $x(t) = -x(t-r)$ implies $c = 0$ and again the origin is globally asymptotically stable.

If $a < 0$, $b = -a$, then $S = \{\phi \in C : \phi(0) = \phi(-r)\}$. Using the same type of reasoning, one concludes that

$$M = \{\text{constant functions on } [-r, 0]\}.$$

To obtain information about $\omega(\phi)$ for $\phi \in C$, suppose $V(x_t(\phi)) \to c$ as $t \to \infty$. Then $\omega(\phi) \subseteq V^{-1}(c) \cap M$ and this latter set consists of a finite number of constant functions since $V(\alpha)$ is a polynomial of sixth degree in the indeterminate α. Since $\omega(\phi)$ is connected, this implies $\omega(\phi)$ is a single point and each solution of Equation (3.2) approaches a constant.

If $a > 0$ and $|b| < a$ (or $b = a$), then the set $G = \{\phi : V(\phi) < 0\}$ is nonempty and positively invariant. As before the set $M = \{0\}$. Consequently, Theorem 3.3 implies instability of the solution $x = 0$ and, in fact, every solution starting in G is unbounded.

As a more sophisticated example, suppose $n = 1$ and

$$f(\phi) = -\int_{-r}^{0} a(-\theta) g(\phi(\theta))\, d\theta$$

where

$$G(x) \stackrel{\text{def}}{=} \int_{0}^{x} g(s)\, ds \to \infty \qquad \text{as } |x| \to \infty \tag{3.3}$$

$$a(r) = 0, \qquad a(t) \geq 0, \qquad \dot{a}(t) \leq 0, \qquad \ddot{a}(t) \geq 0, \qquad 0 \leq t \leq r, \tag{3.4}$$

are continuous. We consider the special case of Equation (3.1) given by

$$\dot{x}(t) = -\int_{-r}^{0} a(-\theta) g(x(t+\theta))\, d\theta = -\int_{t-r}^{t} a(t-u) g(x(u))\, du. \tag{3.5}$$

Any solution of Equation (3.5) satisfies

$$\ddot{x}(t) = a(0) g(x(t)) = -\int_{t-r}^{t} \dot{a}(t-u) g(x(u))\, du \tag{3.6}$$

or

$$\begin{aligned} \ddot{x}(t) + a(0) g(x(t)) &= -\dot{a}(r) \int_{-r}^{0} g(x(t+\theta))\, d\theta \\ &\quad + \int_{-r}^{0} \ddot{a}(-\theta) \Big(\int_{\theta}^{0} g(x(t+u))\, du \Big)\, d\theta. \end{aligned} \tag{3.7}$$

Equation (3.6) is the model of a special type of circulating fuel nuclear reactor where x represents the neutron density. It can also serve as a one-dimensional model in viscoelasticity where x is the strain and a is the relaxation function.

If we define $V : C \to \mathbb{R}$ by the relation

$$V(\phi) = G(\phi(0)) - \frac{1}{2}\int_{-r}^{0} \dot{a}(-\theta)\Big[\int_{\theta}^{0} g(\phi(s))\,ds\Big]^2 d\theta,$$

then the derivative of V along solutions of Equation (3.5) is given by

$$\dot{V}(\phi) = \frac{1}{2}\,\dot{a}(r)\Big[\int_{-r}^{0} g(\phi(\theta))\,d\theta\Big]^2 - \frac{1}{2}\int_{-r}^{0} \ddot{a}(-\theta)\Big[\int_{\theta}^{0} g(\phi(s))\,ds\Big]^2 d\theta.$$

Since the hypotheses on a imply that $\dot{V}(\phi) \le 0$, it follows from Corollary 3.1 that all solutions are bounded.

Let us now apply Theorem 3.1 to this equation. If for any $s \in [0, r]$, we let

$$H_s(\phi) = \int_{-s}^{0} g(\phi(\theta))\,d\theta \tag{3.8}$$

then the set S of Theorem 3.1 is

$$S = \{\phi \in C : H_r(\phi) = 0 \text{ if } \dot{a}(r) \neq 0, \quad H_s(\phi) = 0 \text{ if } \ddot{a}(s) \neq 0\}.$$

From Equation (3.7), the largest invariant set M of Equation (3.5) in S is contained in the set in C generated by bounded solutions x of the ordinary differential equation

$$\ddot{x} + a(0)g(x) = 0 \tag{3.9}$$

for which

$$\begin{aligned} H_r(x_t) &= 0, \qquad t \in (-\infty, \infty) \quad \text{if } \dot{a}(r) \neq 0 \\ H_s(x_t) &= 0, \qquad t \in (-\infty, \infty) \quad \text{if } \ddot{a}(s) \neq 0. \end{aligned}$$

If $\dot{a}(r) \neq 0$, x satisfying Equation (3.9) is bounded and $H_r(x_t) = 0$ for $t \in (-\infty, \infty)$, then $\dot{x}(t) = \dot{x}(t-r)$ for all t. Therefore, $x(t) = kt +$ (a periodic function of period r) and boundedness of x implies $x(t) = x(t-r)$ for all t. If there is an s_0, $\ddot{a}(s_0) \neq 0$, then there is an interval I_{s_0} containing s_0 such that $\ddot{a}(s) \neq 0$ for $s \in I_{s_0}$. If x satisfies Equation (3.9), is bounded, and $H_s(x_t) = 0$ for $-\infty < t < \infty$, $s \in I_{s_0}$, then $\dot{x}(t)$ is periodic of period s for every $s \in I_{s_0}$. Therefore, $\dot{x}(t)$ is constant and boundedness of x implies x is constant.

Theorem 3.4. *If System* (3.5) *satisfies Conditions* (3.3) *and* (3.4) *and* g *has isolated zeros, then*

(i) *If there is an* s *such that* $\ddot{a}(s) > 0$, *then, for any* $\phi \in C$, *the* ω*-limit set* $\omega(\phi)$ *of the orbit through* ϕ *is an equilibrium point of Equation* (3.5) ; *that is, a zero of* g.

(ii) *If* $\ddot{a}(s) \equiv 0$, $a \not\equiv 0$ (*that is,* a *linear*), *then for any* $\phi \in C$, *the* ω*-limit set* $\omega(\phi)$ *of the orbit through* ϕ *is a single periodic orbit of period* r *generated by a solution of Equation* (3.9).

Proof. (i) The remarks preceding the theorem imply $\omega(\phi)$ contains only equilibrium points and, thus, only zeros of g. Since $\omega(\phi)$ is connected and the zeros of g are isolated, we have the result in Part (i).

(ii) Suppose $\ddot{a}(s) \equiv 0$ and choose $a(s) = (r - s)/r$. If x is a solution of Equation (3.9) of period r, then

$$\begin{aligned} -\int_{t-r}^{t} \frac{r-(t-u)}{r}\, g(x(u))\, du &= \int_{t-r}^{t} \frac{r-(t-u)}{r}\, \ddot{x}(u)\, du \\ &= \dot{x}(u)\, \frac{r-(t-u)}{r}\Big|_{t-r}^{t} - \int_{t-r}^{t} \frac{1}{r}\, \dot{x}(u)\, du \\ &= \dot{x}(t) \end{aligned}$$

that is, x is a solution of Equation (3.5). From the remarks preceding the theorem this implies M consists of the periodic solutions of Equation (3.9) of period r.

We first prove that if $\omega(\phi)$ contains an equilibrium point c of Equation (3.5), then $\omega(\phi) = c$. We know that $\omega(\phi)$ is a closed connected set and must be the union of r-periodic orbits of Equation (3.9). If c is not a local minimum of $G(x)$, then the nature of the orbits of Equation (3.9) in the $(x, \dot{x})$-plane implies there can be no r-periodic orbits in $\omega(\phi)$ except c. If c is a local minimum of $G(x)$, then

$$V(\phi) - G(c) = G(\phi(0)) - G(c) + \frac{1}{2r}\int_{-r}^{0}\Big[\int_{\theta}^{0} g(\phi(s))\, ds\Big]^2 d\theta > 0$$

for $\phi \not\equiv c$ in a neighborhood of c. Since $V(x_t(\phi)) =$ constant for $\psi \in \omega(\phi)$, it follows that $\omega(\phi) = c$.

Therefore, assume $\omega(\phi)$ contains no constant solutions of Equation (3.9). Since the solutions of Equation (3.9) must lie on the curves

$$\frac{\dot{x}^2}{2} + G(x) = \text{constant},$$

it follows that any periodic orbit must be symmetrical with respect to the x-axis. Let $u(t, \alpha)$ be a nonconstant periodic solution of Equation (3.9) of least period p with $u(0, \alpha) = \alpha$ and $\dot{u}(0, \alpha) = 0$. Then there is an integer m such that $mp = r$. If there is an interval of periodic orbits in $\omega(\phi)$, then

p is independent of α in this interval. In fact, $p = p(\alpha)$ is continuous and, therefore, $m = m(\alpha)$ is continuous. But m is an integer and, therefore, must be independent of α. Also

$$\begin{aligned}
V(u_t(\alpha)) &= V(u_0(\alpha)) \\
&= G(\alpha) + \frac{1}{2r}\int_{-r}^{0}\Big[\int_{\theta}^{0} g(u(s))\,ds\Big]^2 d\theta \\
&= G(\alpha) + \frac{1}{2r}\int_{-r}^{0} \dot{u}^2(\theta,\alpha)\,d\theta \\
&= G(\alpha) + \frac{1}{2mp}\int_{-mp}^{0} \dot{u}^2(\theta,\alpha)\,d\theta \\
&= G(\alpha) + \frac{1}{2p}\int_{-p}^{0} \dot{u}^2(\theta,\alpha)\,d\theta \\
&= G(\alpha) + \frac{1}{p}\int_{0}^{p/2} \dot{u}^2(\theta,\alpha)\,d\theta \\
&= G(\alpha) + \frac{2}{p}\int_{0}^{p/2} [G(\alpha) - G(u(\theta,\alpha))]\,d\theta \\
&= G(\alpha) + \frac{\sqrt{2}}{p}\int_{\alpha}^{\gamma(\alpha)} [G(\alpha) - G(\tau)]^{1/2} d\tau
\end{aligned}$$

where $\gamma(\alpha) = u(\alpha, p/2)$. On the other hand, the derivative of this latter expression with respect to α is not zero. Therefore, $V(u_t(\alpha))$ is not constant for α in an interval. This implies $\omega(\phi)$ is a single orbit and proves the theorem. □

It is also possible to analyze the stability and instability properties of the limiting equilibrium point. If c is an equilibrium point in Equation (3.5), let $\beta = g'(c)$. If $x = y+c$ in Equation (3.5), then the linear variational equation for y is given by

$$\dot{y}(t) = -\int_{-r}^{0} a(-\theta)g'(c)y(t+\theta)\,d\theta. \tag{3.10}$$

If $g'(c) > 0$, then we can use the preceding theorem to conclude that every solution of Equation (3.10) approaches zero as $t \to \infty$ if either

(i) $\ddot{a}(s) \not\equiv 0$, or

(ii) if $a(s) = (r - s)/r$ and

$$\frac{m2\pi}{(g'(c))^{1/2}} \neq r$$

for all integers m.

If $g'(c) < 0$, then using the negative of the function $V(\phi)$ and instability Theorem 3.3, one sees that the solution $x = 0$ of Equation (3.10) is unstable.

Later, we will see that these properties of the linear equation also hold true for the nonlinear equation.

Consider now the system

$$A\ddot{x}(t) + Bx(t) = \int_0^r F(\theta)x(t-\theta)\,d\theta \tag{3.11}$$

where A, B, and F are symmetric $n \times n$ matrices and F is continuously differentiable. Let

$$M = B - \int_0^r F(\theta)\,d\theta$$

and write Equation (3.11) as

$$\begin{aligned} \dot{x}(t) &= y(t) \\ A\dot{y}(t) &= -Mx(t) + \int_0^r F(\theta)[x(t-\theta) - x(t)]\,d\theta. \end{aligned} \tag{3.12}$$

Theorem 3.5.

(i) *If $A > 0$, $M > 0$, $F(\theta) \geq 0$, $\dot{F}(\theta) \leq 0$, and there is a θ_0 in $[0, r]$ such that $\dot{F}(\theta_0) < 0$, then every solution of Equation* (3.12) *approaches zero as $t \to \infty$.*

(ii) *If $A > 0$, $M > 0$, $\dot{F} \equiv 0$, and $F > 0$, then all solutions of Equation* (3.12) *are bounded and the ω-limit set of any solution must be generated by periodic solutions of period r of the ordinary system*

$$\dot{x} = y, \qquad A\dot{y} = -Bx. \tag{3.13}$$

(iii) *If $A > 0$, $M < 0$, $F(\theta) \geq 0$, $0 \leq \theta \leq r$, $\dot{F}(\theta) \leq 0$, $0 \leq \theta \leq r$, and there is a θ_0 in $[0, r]$ such that $\dot{F}(\theta_0) < 0$, then the solution $x = 0$, $y = 0$ of Equation* (3.12) *is unstable.*

Proof. Let ϕ, ψ be the initial values for x, y in Equation (3.12) and define

$$\begin{aligned} V(\phi, \psi) = {} & \frac{1}{2}\,\phi(0)^T M\phi(0) + \frac{1}{2}\,\psi(0)^T A\psi(0) \\ & + \frac{1}{2}\int_0^r [\phi(-\theta) - \phi(0)]^T F(\theta)[\phi(-\theta) - \phi(0)]\,d\theta. \end{aligned}$$

A few simple calculations yield

$$\begin{aligned} \dot{V}(\phi, \psi) = {} & -\frac{1}{2}[\phi(-r) - \phi(0)]^T F(r)[\phi(-r) - \phi(0)] \\ & + \frac{1}{2}\int_0^r [\phi(-\theta) - \phi(0)]^T \dot{F}(\theta)[\phi(-\theta) - \phi(0)]\,d\theta \leq 0 \end{aligned} \tag{3.14}$$

if Condition (i), (ii), or (iii) is satisfied. If either Condition (i) or (iii) is satisfied, then there is an interval I_{θ_0} containing θ_0 such that $\dot{F}(\theta) < 0$ for θ in I_{θ_0}. From Equation (3.14), $\dot{V}(\phi, \psi) = 0$ implies $\phi(-\theta) - \phi(0)$ for

$\theta \in I_{\theta_0}$. For a solution x, y to belong to the largest invariant set where $\dot{V}(\phi, \psi) = 0$, we must, therefore, have $x(t - \theta) = x(t)$ for all t in $(-\infty, \infty)$, θ in I_{θ_0}. Therefore $x(t) = c$, a constant. From Equation (3.12), this implies $y = 0$ and thus $Mc = 0$. But $c = 0$ if $M > 0$ or < 0. Therefore, the largest invariant set in the set where $\dot{V}(\phi, \psi) = 0$ is $\{(0, 0)\}$, the origin. If Condition (i) is satisfied, then V satisfies Theorem 3.1 and Corollary 3.1 and every solution of Equation (3.12) approaches $(0, 0)$ as $t \to \infty$. If Condition (iii) is satisfied, then V satisfies Theorem 3.3 and $(0, 0)$ is unstable.

If Condition (ii) is satisfied, then $\dot{F}(\theta) \equiv 0$ and $\dot{V}(\phi, \psi) = 0$ implies $\phi(-r) = \phi(0)$. Thus the largest invariant set in the set where $\dot{V}(\phi, \psi) = 0$ consists of the r-periodic solutions of Equation (3.12). On the other hand, if $x(t), y(t)$ is an r-periodic solution of Equation (3.12) with $F(\theta) =$ constant, then the fact that $M > 0$ implies the integrals of x, y over the interval $[0, r]$ are zero. Therefore, $x(t), y(t)$ must satisfy Equation (3.13). This proves the theorem. □

As a final example, consider the system of equations

$$
\begin{aligned}
\dot{x}(t) &= y(t) \\
\dot{y}(t) &= -\frac{a}{r}\, y(t) - \frac{b}{r}\, \sin x(t) + \frac{b}{r} \int_{-r}^{0} [\cos x(t+\theta)] y(t+\theta)\, d\theta
\end{aligned} \tag{3.15}
$$

where a, b and r are positive constants. If $t \geq r$, solutions of this equation satisfy

$$
\ddot{x}(t) + \frac{a}{r}\, \dot{x}(t) + \frac{b}{r}\, \sin x(t - r) = 0 \tag{3.16}
$$

which is a special case of the equation mentioned in the introduction for the circummutation of plants.

If

$$
V(x_t, y_t) = \frac{r}{2}\, y^2(t) + b(1 - \cos x(t)) + \frac{a}{2r} \int_{-r}^{0} \int_{s}^{0} y^2(t+u)\, du\, ds
$$

then

$$
\begin{aligned}
\dot{V}(x_t, y_t) &= \int_{-r}^{0} [-\frac{a}{2r}\, y^2(t) + b \cos x(t+\theta) y(t) y(t+\theta) \\
&\qquad - \frac{a}{2r}\, y^2(t+\theta)]\, d\theta \\
&\leq \int_{-r}^{0} [-\frac{a}{2r}\, y^2(t) + b\, \mathrm{sgn}(y(t) y(t+\theta)) y(t) y(t+\theta) \\
&\qquad - \frac{a}{2r}\, y^2(t+\theta)]\, d\theta.
\end{aligned}
$$

If $r < a/b$, then the quadratic form in the integral is negative definite in $y(t)$, $y(t+\theta)$. For any $b > 0$, let

$$U_b = \{\phi_1, \phi_2) : V(\phi_1, \phi_2) < b\}.$$

Then $V(\phi_1, \phi_2)$ is a Liapunov function on U_b and, for any $(\phi_1, \phi_2) \in U_b$, the corresponding solution $x(\phi_1, \phi_2), y(\phi_1, \phi_2)$ of System (3.15) satisfies $|x(\phi_1, \phi_2)(t)| < \pi/2$ and $|y^2(\phi_1, \phi_2)(t)| < 2b/r$. We may therefore apply Theorem 3.2. We know $\dot{V} = 0$ if and only if $\phi_2(0) = 0$. Therefore, any solution that remains in M for all $t \in (-\infty, \infty)$ must satisfy $y(t) = 0$ for all t. But this implies $\dot{x}(t) = 0$ for all t or $x(t) = \text{constant}$ for all t. These constants must be the zeros in $\sin x$. Therefore, we have proved that $r < a/b$ implies every solution of Equation (3.16) approaches one of the constants $k\pi$, $k = 0, \pm 1, \pm 2, \ldots$, if the initial values are in U_b.

5.4 Razumikhin theorems

In the previous section, sufficient conditions for stability of an RFDE were given in terms of the rate of change of functionals along solutions. The use of functionals is a natural generalization of the direct method of Liapunov for ordinary differential equations. On the other hand, functions are much simpler to use and it is natural to explore the possibility of using the rate of change of a function on $\mathbb{R}^n$ to determine sufficient conditions for stability. Results in this direction are generally referred to as theorems of Razumikhin type.

If $v : \mathbb{R}^n \to \mathbb{R}^n$ is a given positive definite continuously differentiable function, then the derivative of v along an RFDE(f) is given by

$$\dot{v}(x(t)) = \frac{\partial v(x(t))}{\partial x} f(x_t). \tag{4.1}$$

In order for $\dot{v}$ to be nonpositive for all initial data, one would be forced to impose very severe restrictions on the function $f(\phi)$. In fact, the point $\phi(0)$ must play a dominant role and, therefore, the results will apply only to equations that are very similar to ordinary differential equations.

A few moments of reflection in the proper direction indicate that it is unnecessary to require that Equation (4.1) be nonpositive for all initial data in order to have stability. In fact, if a solution of the RFDE(f) begins in a ball and is to leave this ball at some time t, then $|x_t| = |x(t)|$; that is, $|x(t+\theta| \leq |x(t)|$ for all $\theta \in [-r, 0]$. Consequently, one need only consider initial data satisfying this latter property. This is the basic idea exploited in this section.

If $V : \mathbb{R} \times \mathbb{R}^n \to \mathbb{R}$ is a continuous function, then $\dot{V}(t, \phi(0))$, the derivative of V along the solutions of an RFDE(f) is defined to be

$$\dot{V}(t, \phi(0)) = \limsup_{h \to 0_+} \frac{1}{h} \left[V(t+h, x(t, \phi)(t+h)) - V(t, \phi(0))\right]$$

where $x(t, \phi)$ is the solution of the RFDE(f) through (t, ϕ).

Theorem 4.1. *Suppose $f : \mathbb{R} \times C \to \mathbb{R}^n$ takes $\mathbb{R} \times$ (bounded sets of C) into bounded sets of $\mathbb{R}^n$ and consider the* RFDE(f). *Suppose $u, v, w : \mathbb{R}^+ \to \mathbb{R}^+$ are continuous, nondecreasing functions, $u(s), v(s)$ positive for $s > 0$, $u(0) = v(0) = 0$, v strictly increasing. If there is a continuous function $V : \mathbb{R} \times \mathbb{R}^n \to \mathbb{R}$ such that*

$$u(|x|) \leq V(t, x) \leq v(|x|), \qquad t \in \mathbb{R},\ x \in \mathbb{R}^n, \tag{4.2}$$

and

$$\dot{V}(t, \phi(0)) \leq -w(|\phi(0)|) \quad \text{if } V(t + \theta, \phi(\theta)) \leq V(t, \phi(0)), \tag{4.3}$$

for $\theta \in [-r, 0]$, then the solution $x = 0$ of the RFDE(f) *is uniformly stable.*

Proof. If

$$\overline{V}(t, \phi) = \sup_{-r \leq \theta \leq 0} V(t + \theta, \phi(\theta))$$

for $t \in \mathbb{R}$, $\phi \in C$, then there is a θ_0 in $[-r, 0]$ such that $\overline{V}(t, \phi) = V(t + \theta_0, \phi(\theta_0))$ and either $\theta_0 = 0$ or $\theta_0 < 0$ and $V(t + \theta, \phi(\theta)) < V(t + \theta_0, \phi(\theta_0))$ if $\theta_0 < \theta \leq 0$. If $\theta_0 < 0$, then for $h > 0$ sufficiently small

$$\overline{V}(t + h, x_{t+h}(t, \phi)) = \overline{V}(t, \phi)$$

and $\dot{\overline{V}} = 0$. If $\theta_0 = 0$, then $\dot{\overline{V}} \leq 0$ by the Condition (4.3). Also, Relation (4.2) implies that $u(|\phi(0)|) \leq \overline{V}(t, \phi) \leq v(|\phi|)$ for $t \in \mathbb{R}$, $\phi \in C$. Theorem 2.1 implies the uniform stability of the solution $x = 0$ of the RFDE(f) and the theorem is proved. □

Theorem 4.2. *Suppose all of the conditions of Theorem 4.1 are satisfied and in addition $w(s) > 0$ if $s > 0$. If there is a continuous nondecreasing function $p(s) > s$ for $s > 0$ such that Condition (4.3) is strengthened to*

$$\dot{V}(t, \phi(0)) \leq -w(|\phi(0)|) \quad \text{if } V(t + \theta, \phi(\theta)) < p(V(t, \phi(0)) \tag{4.4}$$

for $\theta \in [-r, 0]$, then the solution $x = 0$ of the RFDE(f) *is uniformly asymptotically stable. If $u(s) \to \infty$ as $s \to \infty$, then the solution $x = 0$ is also a global attractor for the* RFDE(f).

Proof. Theorem 4.1 implies uniform stability. To complete the proof of the theorem, suppose $\delta > 0$, $H > 0$ are such that $v(\delta) = u(H)$. Such numbers always exist by our hypotheses on u and v. In fact, since $v(0) = 0$ and $0 < u(s) \leq v(s)$ for $s > 0$, one can preassign H and then determine a δ such that the desired relation is satisfied. If $u(s) \to \infty$ as $s \to \infty$, then one can fix δ arbitrarily and determine H such that $v(\delta) = u(H)$. This remark and the reasoning that follows will prove the uniform asymptotic stability of $x = 0$ and the fact that $x = 0$ is a global attractor.

If $v(\delta) = u(H)$, the same argument as in the proof of Theorem 4.1 shows that $|\phi| \leq \delta$ implies $|x_t(t_0, \phi)| \leq H$, $V(t, x(t_0, \phi)(t)) < v(\delta)$ for

$t \geq t_0 - r$. Suppose $0 < \eta \leq H$ is arbitrary. We need to show there is a number $\bar{t} = \bar{t}(\eta, \delta)$ such that for any $t_0 \geq 0$ and $|\phi| \leq \delta$, the solution $x(t_0, \phi)$ of the RFDE(f) satisfies $|x_t(t_0, \phi)| \leq \eta$, $t \geq t_0 + \bar{t} + r$. This will be true if we show that $V(t, x(t_0, \phi)(t)) \leq u(\eta)$, for $t \geq t_0 + \bar{t}$. In the remainder of this proof, we let $x(t) = x(t_0, \phi)(t)$.

From the properties of the function $p(s)$, there is a number $a > 0$ such that $p(s) - s > a$ for $u(\eta) \leq s \leq v(\delta)$. Let N be the first positive integer such that $u(\eta) + Na \geq v(\delta)$, and let $\gamma = \inf_{v^{-1}(u(\eta)) \leq s \leq H} w(s)$ and $T = Nv(\delta)/\gamma$.

We now show that $V(t, x(t)) \leq u(\eta)$ for all $t \geq t_0 + T$. First, we show that $V(t, x(t)) \leq u(\eta) + (N-1)a$ for $t \geq t_0 + (v(\delta)/\gamma)$. If $u(\eta) + (N-1)a < V(t, x(t))$ for $t_0 \leq t < t_0 + (v(\delta)/\gamma)$, then, since $V(t, x(t)) \leq v(\delta)$ for all $t \geq t_0 - r$, it follows that

$$p(V(t, x(t))) > V(t, x(t)) + a \geq u(\eta) + Na \geq v(\delta) \geq V(t + \theta, x(t + \theta)),$$
$$t_0 \leq t \leq t_0 + \frac{v(\delta)}{\gamma}, \qquad \theta \in [-r, 0].$$

Hypothesis (4.4) implies

$$\dot{V}(t, x(t)) \leq -w(|x(t)|) \leq -\gamma$$

for $t_0 \leq t \leq t_0 + (v(\delta)/\gamma)$. Consequently,

$$V(t, x(t)) \leq V(t_0, x(t_0)) - \gamma(t - t_0) \leq v(\delta) - \gamma(t - t_0)$$

on this same interval. The positive property (4.2) of V implies that $V(t, x(t)) \leq u(\eta) + (N-1)a$ at $t_1 = t_0 + v(\delta)/\gamma$. But this implies $V(t, x(t)) \leq u(\eta) + (N-1)a$ for all $t \geq t_0 + (v(\delta)/\gamma)$, since $\dot{V}(t, x(t))$ is negative by Condition (4.4) when $V(t, x(t)) = u(\eta) + (N-1)a$.

Now let $\bar{t}_j = jv(\delta)/\gamma$, $j = 1, 2, \ldots, N$, $\bar{t}_0 = 0$, and assume that for some integer $k \geq 1$, in the interval $\bar{t}_{k-1} - r \leq t - t_0 \leq \bar{t}_k$, we have

$$u(\eta) + (N-k)a \leq V(t, x(t)) \leq u(\eta) + (N-k+1)a.$$

By the same type of reasoning, we have

$$\dot{V}(t, x(t)) \leq -\gamma, \; \bar{t}_{k-1} \leq t - t_0 \leq \bar{t}_k$$

and

$$\begin{aligned} V(t, x(t)) &\leq V\big(t_0 + \bar{t}_{k-1}, x(t_0 + \bar{t}_{k-1})\big) - \gamma(t - t_0 - \bar{t}_{k-1}) \\ &\leq v(\delta) - \gamma(t - t_0 - \bar{t}_{k-1}) \leq 0 \end{aligned}$$

if $t - t_0 - \bar{t}_{k-1} \geq v(\delta)/\gamma$. Consequently,

$$V(t_0 + \bar{t}_k, x(t_0 + \bar{t}_k)) \leq u(\eta) + (N-k)a,$$

and, finally, $V(t, x(t)) \leq u(\eta) + (N-k)a$ for all $t \geq t_0 + \bar{t}_k$. This completes the induction and we have $V(t, x(t)) \leq u(\eta)$ for all $t \geq t_0 + Nv(\delta)/\gamma$. This proves Theorem 4.2. □

As a first example, consider the equation

$$\dot{x}(t) = -a(t)x(t) - b(t)x(t - r_0(t)) \tag{4.5}$$

where a, b, and r_0 are bounded continuous functions on $\mathbb{R}$ with $|b(t)| \le a(t)$, $0 \le r_0(t) \le r$, for all $t \in \mathbb{R}$. If $V(x) = x^2/2$, then

$$\begin{aligned} \dot{V}(x(t)) &= -a(t)x^2(t) - b(t)x(t)x(t - r_0(t)) \\ &\le -a(t)x^2(t) + |b(t)|\,|x(t)|\,|x(t - r_0(t))| \\ &\le -[a(t) - |b(t)|\,]x^2(t) \le 0 \end{aligned}$$

if $|x(t)| \ge |x(t - r_0(t))|$. Since $V(x) = x^2/2$, we have shown that $\dot{V}(x(t)) \le 0$ if $V(x(t)) \ge V\big(x(t - r_0(t))\big)$. Theorem 4.1 implies the solution $x = 0$ of Equation (4.5) is uniformly stable.

If, in addition, $a(t) \ge \delta > 0$, and there is a k, $0 \le k < 1$ such that $|b(t)| \le k\delta$, then the solution $x = 0$ of Equation (4.5) is uniformly asymptotically stable. In fact, choose $p(s) = q^2 s$ for some constant $q > 1$. If $V(x) = x^2/2$ as before, then

$$\dot{V}(x(t)) \le -(1 - qk)\,\delta x^2(t)$$

if $p(V(x(t))) > V(x(t - r_0(t)))$. Since $k < 1$, there is a $q > 1$ such that $1 - qk > 0$ and Theorem 4.2 implies the uniform asymptotic stability of the solution $x = 0$. This is an improvement over the results obtained with functionals for Equation (2.9) since the delay can be an arbitrary bounded continuous function.

If we use the same $V(x)$, then a similar argument shows that the zero solution of

$$\dot{x}(t) = -a(t)x(t) - \sum_{j=1}^{n} b_j(t)x(t - r_j(t))$$

is uniformly asymptotically stable for all bounded continuous functions a, b_j, r_j if $a(t) \ge \delta > 0$, $\sum_{j=1}^{n} |b_j(t)| < k\delta$, $0 < k < 1$, $0 \le r_j(t) \le r$ for all $t \in \mathbb{R}$.

As an example of a nonlinear problem, consider the first-order equation

$$\dot{x}(t) = f(x(t - \gamma(t)), t), \qquad 0 \le \gamma(t) \le r,\ f(0,t) = 0 \tag{4.6}$$

where $\gamma(t)$ is a continuous function of t and $f(x,t)$ is a continuous function of x, t for $t \ge 0$, $-\infty < x < \infty$, has a continuous partial derivative such that $|\partial f(x,t)/\partial x| < L$, $t \ge 0$, $-\infty < x < \infty$. For $t \ge 2r$ we can rewrite Equation (4.6) as

$$\begin{aligned} \dot{x}(t) &= f(x(t),t) - \big[f(x(t),t) - f(x(t - \gamma(t)),t)\big] \\ &= f(x(t),t) - \int_{t-\gamma(t)}^{t} \frac{\partial f}{\partial x}(x(\theta),t) f(x(\theta - \gamma(\theta)),\theta)\, d\theta. \end{aligned} \tag{4.7}$$

For $V(\phi) = \phi^2(0)$, we have

$$\begin{aligned}\dot V(x_t) &= 2x(t)f(x(t),t) - 2\int_{t-\gamma(t)}^{t} x(t)\frac{\partial f}{\partial x}(x(\theta),t)f(x(\theta-\gamma(\theta)),\theta)\,d\theta \\ &\le 2x(t)f(x(t),t) + 2L^2\int_{t-\gamma(t)}^{t} |x(t)x(\theta-\gamma(\theta))|\,d\theta \\ &\le 2x(t)f(x(t),t) + 2L^2\gamma(t)|x(t)|\,|x_{t-\gamma(t)}|, \qquad t\ge 2r.\end{aligned}$$

Consequently, if $q>1$ is fixed and we consider the set of all $x(t)$ such that

$$x^2(t-\xi)\le q^2x^2(t), \qquad 0\le\xi\le 2r,$$

then

$$\dot V(x_t)\le 2\Big[\frac{f(x(t),t)}{x(t)} + L^2\gamma(t)q\Big]x^2(t) < -2\mu x^2(t)$$

if $(f(x,t)/x) + L^2\gamma(t)q < -\mu < 0$. For $\mu>0$, $t\ge 0$, $x\in(-\infty,\infty)$, and $p(s) = q^2 s$, Theorem 4.2 implies the origin is uniformly asymptotically stable and a global attractor.

Let us now consider the linear equation

$$\dot x(t) = Ax(t) + Bx(t-\tau) \tag{4.8}$$

where A and B are matrices (A constant, B may be a bounded continuous matrix function) and $\tau=\tau(t)$, $0\le\tau(t)\le r$, is continuous. If $V(x) = x^TDx$, where D is positive definite, then

$$\dot V = x(t)^T(A^TD + DA)x(t) + 2x(t)^TDBx(t-\tau).$$

If there are constants $q>1$, $\eta>0$ such that

$$V(x(\xi)) < qV(x(t)), \quad t-r\le\xi\le t \quad \text{implies } \dot V\le -\eta|x(t)|^2,$$

then Theorem 4.2 implies the solution $x=0$ of Equation (4.8) is uniformly asymptotically stable.

The difficulty in obtaining results along this line arises from attempting to estimate $\dot V$ for the restricted class of initial curves satisfying $V(x(\xi)) < qV(x(t))$, $t-r\le\xi\le t$. Furthermore, there are numerous directions in which one may proceed. In particular, one may wish to obtain stability conditions that are independent of τ or conditions that depend on τ. In the first case, one must obviously have the zero solution of

$$\dot x(t) = (A+B)x(t)$$

asymptotically stable. By an appropriate change of coordinates, one can take $V(x) = x^Tx$ and be assured that $\dot V$ along the trajectories of this ordinary differential equation is a negative definite function. In these new coordinates, $\dot V$ along solutions of Equation (4.8) is

$$\dot{V} = x(t)^T[(A+B)+(A+B)^T]x(t) + 2x(t)^T Bx(t-\tau) - x(t)^T(B+B^T)x(t)$$

and one can estimate $\dot{V}$ along curves satisfying $V(x(\xi)) < qV(x(t))$ (or equivalently, $|x(\xi)| < q|x(t)|$), $q > 1$, $t - \tau \leq \xi \leq t$ in the following way. Since $(A+B) + (A^T + B^T)$ is negative definite, there is a $\lambda > 0$ such that

$$\begin{aligned}\dot{V} &\leq -\lambda|x(t)|^2 + 2q|B|\,|x(t)|^2 + |B+B^T|\,|x(t)|^2 \\ &= -[\lambda - 2q|B| - |B+B^T|]\,|x(t)|^2.\end{aligned}$$

Consequently, if $2q|B| + |B+B^T| < \lambda$, then the solution $x = 0$ of Equation (4.8) is uniformly asymptotically stable. Razumikhin [1] has carried out this type of procedure for a second-order system

$$\ddot{x}(t) + a\dot{x}(t) + bx(t) + cx(t-\tau) = 0$$

to obtain estimates on a, b, and c ensuring asymptotic stability independent of τ.

To obtain estimates that depend on the delay function τ, one may proceed in the following manner. For simplicity in notation, let us consider the initial time to be zero and let $x(t)$, $t \geq 0$, be the solution of Equation (4.8) through $(0, \phi)$. Since $x(t)$ is continuously differentiable for $t \geq 0$ one can write

$$\begin{aligned}x(t-\tau) &= x(t) - \int_{-\tau}^{0} \dot{x}(t+\theta)\,d\theta \\ &= x(t) - \int_{-\tau}^{0} [Ax(t+\theta) + Bx(t-\tau+\theta)]\,d\theta\end{aligned}$$

for $t \geq \tau$. If we return to Equation (4.8) using this expression for $x(t-\tau)$, we obtain the equation

$$\dot{x}(t) = (A+B)x(t) - B\int_{-\tau}^{0} [Ax(t+\theta) + Bx(t-\tau+\theta)]\,d\theta \tag{4.9}$$

for arbitrary continuous initial data on $[-2r, 0]$. If the zero solution of Equation (4.9) is asymptotically stable, then the zero solution of Equation (4.8) is asymptotically stable since Equation (4.8) is a special case of Equation (4.9) with continuous initial data ψ on $[-2r, 0]$ given by $\psi(s)$ arbitrary for $s \in [-2r, -r - \tau(0)]$, $\psi(s) = \phi(s + \tau(0))$, $-r - \tau(0) \leq s \leq -\tau(0)$, and $\psi(s) = x(t+s)$, $-\tau(0) \leq s \leq 0$, where x is the solution of $\dot{x}(t) = Ax(t) + Bx(t-r)$ through $(0, \phi)$.

As an example, consider the equation

$$\dot{x}(t) = -bx(t-r), \qquad r > 0,$$

and the auxiliary problem on $[-2r, 0]$ given by

$$\dot{x}(t) = -bx(t) - b^2 \int_{t-2r}^{t-r} x(s)\, ds.$$

If $V(x) = x^2/2$, then, for any constant $q > 1$,

$$\begin{aligned}\dot{V} &= -bx^2(t) - b^2 \int_{t-2r}^{t-r} x(t)x(s)\, ds \\ &\leq -b(1 - qbr)x^2(t)\end{aligned}$$

if $V(x(\xi)) < q^2 V(x(t))$, $t - 2r \leq \xi \leq t$. Consequently, if $br < 1$, then there is a $q > 1$ such that $qbr < 1$ and Theorem 4.2 implies asymptotic stability.

There are many ways to extend the ideas of the last computation to obtain more precise information about stability for the RFDE(f). For any integer $k \geq 0$, one can artificially interpret this equation on $C([-r, 0], \mathbb{R}^n)$ as an equation on $C([-(k+1)r, 0], \mathbb{R}^n)$. Of course, arbitrary initial conditions in this larger space need not be considered from the point of view of stability in the original equation. One should restrict consideration to initial data that are related in some way to the original equation. An obvious way is the following. Let $\phi \in C([-r, 0], \mathbb{R}^n)$ and let $x(\phi)$ be the solution of the RFDE(f) through ϕ at initial time zero. For initial data $\psi \in C[-(k+1)r, 0]$ for the artificial problem, take $\psi(s) = x(\phi)(kr + s)$, $-(k+1)r \leq s \leq 0$. Such considerations will take into account integration of the equation over k delay intervals.

One can obviously apply analogues of Theorems 4.1 and 4.2 to obtain information about the equation. Rather than go into detail, let us consider specifically what could be done to obtain stability of solutions.

Suppose there is a continuous function $V : \mathbb{R}^n \to \mathbb{R}$ such that

$$u(|x|) \leq V(x) \leq v(|x|) \tag{4.10}$$

where u and v satisfy the same properties as in Theorem 4.1. If $x(\phi)$ is the solution of the RFDE(f) through $\phi \in C([0, r], \mathbb{R}^n)$, let

$$\overline{V}(\phi) = \sup_{-r \leq s \leq kr} V(x(\phi)(s)). \tag{4.11}$$

If $\dot{\overline{V}}(\phi) \leq 0$ along the solutions of the RFDE(f), then the solution $x = 0$ is stable. Therefore, we need only make $\dot{\overline{V}}(\phi) \leq 0$. Using the same argument as in the proof of Theorem 4.1, this condition will be satisfied if one can show that the set

$$\Phi = \{\phi \in C : \overline{V}(\phi) = V(x(\phi)(kr)) > 0; \dot{V}(x(\phi)(kr)) > 0\} \tag{4.12}$$

is empty.

Let us apply the previous remarks to the scalar equation

$$\dot{x}(t) = -bx(t - r), \qquad b > 0 \tag{4.13}$$

with $k = 2$ and $V(x) = x^2/2$. If Φ in Expression (4.12) is not empty, then there is an $\epsilon > 0$ and a $\phi \in C$ such that $|x(\phi)(2r)| = \epsilon$, $\overline{V}(\phi) = V(x(\phi)(2r))$, and

$$\frac{1}{2}\frac{d}{dt}\, x^2(\phi)(2r) = -bx(\phi)(2r)x(\phi)(r) > 0.$$

Without loss of generality, we may assume $x(\phi)(2r) = \epsilon$ and the inequality then implies $x(\phi)(r) < 0$. Consequently, we will obtain a contradiction if we show that

$$\begin{aligned} P \stackrel{\text{def}}{=} \sup\{x(\phi)(2r) : \phi \in C,\ |x(\phi)(t)| \le \epsilon,& \\ -r \le t \le 2r,\ x(\phi)(r) < 0\} < \epsilon.& \end{aligned} \tag{4.14}$$

The variational problem (4.14) imposes a restriction of the magnitude of the parameter b. Rather than integrate the equation for the solution up to $t = 2r$, we can take advantage of the equation to obtain a variational problem that only involves integration over an interval of length r. In fact if $|\phi| < \epsilon$, then the equation implies $|\dot{x}(\phi)(t)| \le b\epsilon$ for $0 \le t \le r$ and then

$$P \le \sup\{x(\phi)(r) : |x_t(\phi)| \le \epsilon,\ |\dot{x}(\phi)(t)| \le b\epsilon, \quad 0 \le t \le r,\ \phi(0) < 0\}.$$

Let W_1^∞ be the subset of C consisting of those functions that are absolutely continuous and have an essentially bounded derivative on $[-r, 0]$ and define

$$\widetilde{P}(y) = \sup\{x(\xi)(r) : \xi \in W_1^\infty,\ |\xi| \le \epsilon,\ |\dot{\xi}(\theta)| \le b\epsilon \text{ a.e. in } \theta,\ \xi(0) = y\}.$$

It is clear that $P \le \sup\{\widetilde{P}(y) : -\epsilon < y < 0\}$. This latter problem involves integration only over $[0, r]$. For any $y \in \mathbb{R}$, $-\epsilon < y < 0$, there is a function $\overline{\xi}$ such that $x(\overline{\xi}, r) = \widetilde{P}(y)$. In fact one can take $\overline{\xi}$ as the function

$$\overline{\xi}(\theta) = \begin{cases} -\epsilon & \theta \in [-r, -(\epsilon + y)/b\epsilon], \\ y + b\epsilon\theta & \theta \in [-(\epsilon + y)/b\epsilon, 0]. \end{cases}$$

For this $\overline{\xi}$ and $br > 1$,

$$x(\overline{\xi})(r) = y + b\epsilon r - \frac{1}{2\epsilon}\,(\epsilon + y)^2$$

$$P \le \sup\{\widetilde{P}(y) : -\epsilon < y < 0\} = b\epsilon r - \frac{\epsilon}{2} = \epsilon\big(br - \frac{1}{2}\big).$$

For $br < 1$, a similar computation gives $P \le \epsilon(1 - br)$. Therefore, if $0 < b < 3/2r$, $P < \epsilon$ and we have stability of the solution $x = 0$. This is a significant improvement over the estimates obtained before.

Our next result is concerned with uniform ultimate boundedness.

Theorem 4.3. *Suppose* $f : \mathbb{R} \times C \to \mathbb{R}^n$ *takes* $\mathbb{R} \times$ (*bounded sets of* C) *into bounded sets of* $\mathbb{R}^n$ *and consider the* RFDE(f). *Suppose* $u, v, w : \mathbb{R}^+ \to \mathbb{R}^+$ *are continuous nonincreasing functions,* $u(s) \to \infty$ *as* $s \to \infty$. *If there is a continuous* $V : \mathbb{R} \times \mathbb{R}^n \to \mathbb{R}$, *a continuous nondecreasing function* $p : \mathbb{R}^+ \to \mathbb{R}^+$, $p(s) > s$ *for* $s > 0$, *and a constant* $H \geq 0$ *such that*

$$u(|x|) \leq V(t,x) \leq v(|x|), \qquad t \in \mathbb{R},\ x \in \mathbb{R}^n \tag{4.15}$$

and

$$\dot{V}(t,\phi) \leq -w(|\phi(0)|) \tag{4.16}$$

if

$$|\phi(0)| \geq H, \qquad V(t+\theta,\phi(\theta)) < p(V(t,\phi(0)), \quad \theta \in [-r,0],$$

then the solutions of the RFDE(f) *are uniformly ultimately bounded.*

The proof of this result will not be given since it is essentially a repetition of the arguments used in the proof of Theorems 4.1 and 4.2.

In the applications one often needs a generalization of this result. More specifically, one may be able to verify Inequality (4.16) only for some coordinate, say ϕ_1, of the function ϕ; that is, one may be able to verify that

$$\dot{V}(t,\phi) \leq -w(|\phi_1(0)|) \tag{4.17}$$

if

$$|\phi_1(0)| \geq H, \qquad V(t+\theta,\phi_1(\theta),x_2,\dots,x_n) < p(V(t,\phi_1(0),x_2,\dots,x_n))$$

for

$$\theta \in [-r,0], \qquad x_j \in \mathbb{R}.$$

In this case, one can prove that the first coordinate of the solutions of the RFDE(f) is uniformly ultimately bounded. The proof of this result is essentially the same as before.

One can also prove results by replacing $|\phi_1(0)| \geq H$ by $\phi_1(0) \geq H$ with the conclusion being that there is an $\alpha > 0$ such that the first coordinate satisfies $x_1(t) \leq \alpha$ for all $t \geq \sigma$.

As a first example of the application of Theorem 4.3, consider the second-order system

$$\begin{aligned} \dot{x}(t) &= y(t) \\ \dot{y}(t) &= -\Phi(t,y(t)) - f(x(t)) + p(t) + \int_{-r}^{0} g(x(t+\theta))y(t+\theta)\,d\theta. \end{aligned} \tag{4.18}$$

If $g(x) = df(x)/dx$, then System (4.18) includes the second-order scalar equation

$$\ddot{x}(t) + \Phi(t,\dot{x}(t)) + f(x(t-r)) = p(t). \tag{4.19}$$

We make the following assumptions on System (4.18):

(i) $\Phi : \mathbb{R}^2 \to \mathbb{R}$ is continuous, Φ takes $\mathbb{R} \times$ (bounded sets of $\mathbb{R}$) into bounded sets and there are constants $a > 0$, $H > 0$, such that

$$\frac{\Phi(t,y)}{y} > a > 0 \qquad \text{for} \quad |y| \geq H,$$

(ii) $f : \mathbb{R} \to \mathbb{R}$ s continuous and $f(x)$ sgn $x \to \infty$ as $|x| \to \infty$,

(iii) $p : \mathbb{R} \to \mathbb{R}$ is continuous and bounded by k,

(iv) $g : \mathbb{R} \to \mathbb{R}$ is continuous and $|g(x)| \leq L$ for all $x \in \mathbb{R}$,

(v) $Lr < a$.

Of course, it is always assumed that a uniqueness result holds for the solutions of System (4.18).

Under the hypotheses, we will show that the solutions of System (4.18) are uniformly ultimately bounded. If, in addition, there is a $\omega > 0$ such that $\Phi(t+\omega, y) = \Phi(t, y)$, $p(t+\omega) = p(t)$ for all $t \in \mathbb{R}$, then Equation (4.18) has an ω-periodic solution. This latter remark is a consequence of the uniform ultimate boundedness and Theorem 6.2 of Section 4.6.

If $V(x,y) = F(x) + y^2/2$, $F(x) = \int_0^x f(s)\,ds$ and $q > 1$, $qLr < a$, then

$$\begin{aligned}\dot V(x(t), y(t)) &= -y(t)\Phi(t, y(t)) + y(t)p(t) \\ &\quad + y(t)\int_{-r}^{0} g(x(t+\theta))y(t+\theta)\,d\theta \\ &\leq -(a - qLr)y^2(t) + |y(t)|k\end{aligned}$$

if $|y(t)| \geq H$ and $|y(t+\theta)| \leq q|y(t)|$. By choosing $H_1 \geq H$ appropriately, one obtains a positive constant μ such that $\dot V(x(t), y(t)) \leq -\mu y^2(t)$ for $|y(t)| \geq H_1$ and $|y(t+\theta)| \leq q|y(t)|$. Therefore, the remarks following Theorem 4.3 imply the y coordinate of the solutions is uniformly ultimately bounded by a constant c.

If $|y| \leq c$ and $V_1(x,y) = V(x,y) + y$, then there is a constant k_1 such that

$$\begin{aligned}\dot V_1(x(t), y(t)) &= \dot V(x(t), y(t)) - \Phi(t, y(t)) - f(x(t)) + p(t) \\ &\quad + \int_{-r}^{0} g(x(t+\theta))y(t+\theta)\,d\theta \\ &\leq -f(x(t)) + k_1.\end{aligned}$$

One can choose a constant $b > 0$ such that

$$\dot V_1(x(t), y(t)) < -1 \qquad \text{if } x(t) \geq b.$$

Consequently, there is an $\alpha > 0$ such that the x coordinate of the solutions satisfies $x(t) \leq \alpha$, in the strip $|y| \leq c$. Using the function $V_2(x,y) = V(x,y) - y$ one obtains the x coordinate that satisfies $-\alpha \leq x(t)$ in the strip $|y| \le c$. This clearly implies the uniform ultimate boundedness of the solutions of System (4.18).

It is natural to ask if this result is valid without any restriction on the delay. Consider the linear equation

$$\ddot{x}(t) + a\dot{x}(t) + b^2 x(t - 2r) = 0.$$

The characteristic equation is

$$\lambda^2 + a\lambda + b^2 e^{-2\lambda r} = 0.$$

For $a = 0$, it is easy to verify that $2rb > \pi$ implies there are at least two roots with positive real parts. Therefore, for $a > 0$ sufficiently small, there will be unbounded solutions of the linear equation and one does not have uniform ultimate boundedness.

As another example, consider the equation

$$\begin{aligned} \dot{x}(t) &= y(t) \\ \dot{y}(t) &= -ay(t-r) - f(x(t)) + p(t) \end{aligned} \tag{4.20}$$

with $a > 0$, f, p satisfying (ii) and (iii). If $V(x, y) = F(x) + y^2/2$, $F(x) = \int_0^x f(s)\, ds$, $q > 1$, then

$$\begin{aligned} \dot{V}(x(t), y(t)) &\le -ay(t)y(t-r) + y(t)p(t) \\ &\le -aqy^2(t) + |y(t)|k \\ &\le -\mu^2 y(t) \end{aligned}$$

if $|y(t)| \ge H$ and $|y(t-r)| \le q|y(t)|$. Therefore, we obtain uniform ultimate boundedness of the y-coordinate of the solutions of (4.20) without any restriction on the delay. Using arguments on the x-coordinate of the solutions as in the previous example, one obtains uniform ultimate boundedness of the solutions of (4.20) only under the hypotheses $a > 0$ and (ii), (iii).

5.5 Supplementary remarks

The proof of Lemma 1.1 had its origin in the work on dissipative processes of Hale, LaSalle, and Slemrod [1] and Hale and Lopes [1]. The result was independently discovered by Izé [1]. Lemma 1.2 is implicitly contained in Billotti and LaSalle [1] and was independently discovered by Pavel [1] (see also Yoshizawa [2]).

Krasovskii [1, p. 151 ff.] proved the asymptotic stability in Theorem 2.1. The proof in the test is due to Yoshizawa [1]. In Theorem 2.1 (and Theorem 4.1), we have required that f takes $\mathbb{R} \times$ (bounded sets of C) into bounded sets of $\mathbb{R}^n$. Burton [1, 2] has shown that it suffices to require that f is a completely continuous map. However, Makay [1] has shown that the conditions cannot be weakened if one only assumes that the estimates on V and $\dot{V}$ are required to be satisfied along the solutions of the differential equation (this is the only requirement in the proofs).

Lemma 2.1 is due to Repin [1] and Datko [1]. The method of constructing the Liapunov functional in Theorem 2.2 is due to Huang [1]. For the special case of (2.1), Infante and Castelan [1] earlier had proved that a quadratic functional exists as in Theorem 2.2 by approximating the difference differential equation by a system of ordinary differential equations, using the Liapunov theorem for this approximate equation and then taking a limit. Mansurov [1] has also considered difference approximations to obtain stability. A special case of the instability Theorem 2.2 was proved by Shimanov [1]. The material in Section 5.3 on the stability of autonomous systems is based on Hale [2], taking into account the improvements by LaSalle [2] and Onuchic [1]. Much more general results for compact and uniform processes have been given by Dafermos [3].

Example (3.2) is due to LaSalle. Example (3.5) and a special case of Theorem 3.4 was originally given by Levin and Nohel [1] by different methods. Under the assumption (i) in Theorem 3.4, it follows from Theorem 5.2 of Chapter 4 that there is a compact connected global attractor. Furthermore, the system is gradient-like with hyperbolic equilibrium points and the attractor is the union of the unstable manifolds (see Chapter 10 for the definition) of the equilibrium points (see Hale [23], for example). The unstable points correspond to the zeros of g for which the derivative at the point is negative and the dimension of the unstable manifold is one. Hale and Rybakowski [1] have discussed the types of orbit connections between equilibrium points and, surprisingly, it is shown that these connections do not always preserve the natural order of the real numbers. By using the recent results on convergence of Hale and Raugel [1], we remark that the same conclusion as in part (i) of Theorem 3.4 (that is, convergence to a single equilibrium point) can be shown to be valid without the hypothesis that the zeros of g are isolated.

Onuchi [1] has an interesting instability theory for Equation (3.5). Theorem 3.5 was first proved by Hale [2] and motivated by Volterra [2].

The Liapunov functional for Equations (3.15) is due to A. Somolinos. Equation (3.16) often is referred to as the sunflower equation because of its origin in the circummutation of plants. It is shown in the text that, for $r < a/b$, the system is gradient-like. If we consider the flow defined by this equation on the space $X = C([-r, 0], S^1 \times \mathbb{R})$, then there is a compact global attractor from Theorem 5.2 of Chapter 4. Since the equilibrium points are hyperbolic, it follows that the attractor is the union of the unstable manifolds, and it is easy to check that these have dimension 1. Since there is a Liapunov functional, only two equilibrium points, and the attractor is connected, it follows that the attractor is homeomorphic to S^1. For more details and further properties, see Lizano [1]. For some interesting stability problems in car following, see Harband [1].

Theorems of the type given in Theorems 4.1 and 4.2 originated with Razumikhin [1, 2] with versions also being stated in the book of Krasovskii [1, p. 157 ff.]. The proof of uniform asymptotic stability was first given by Driver [2]. The global nature of this theorem was independently discovered by Seifert [1]. Example (4.6) is due to Krasovskii [1, p. 174], the reduction of the stability problem for Equation (4.13) to an optimization problem is due to Barnea [1]. Other interesting examples with several delays are in Barnea [1], Bailey and Williams [1], and Noonburg [1]. An instability result similar to Theorem 4.4 is also contained in Barnea [1]. Theorem 4.3 on ultimate boundedness was first stated by Lopes [1, 2, 3] for more general neutral functional differential equations. Example (4.18) was first considered by Yoshizawa [1, p. 208] with a more restrictive hypotheses on the delay.

As indicated in the main text, it is possible to extend Razumikhin Theorems 4.1 and 4.2 by taking into consideration the value of the solution over a few lag intervals. An even more general extension has been given by J. Kato [1] based on comparison principles (see earlier versions in Laksmikantham and Leela [1]). We now summarize some of the results of Kato [1].

Consider a scalar ordinary differential equation

$$\dot{u}(t) = U(t, u(t)),$$

where $U : \mathbb{R}^2 \to \mathbb{R}$ is continuous and denote by $r(t, s, \alpha)$ the maximal solution of this equation through s, α for $t \geq s$ and by $l(t, s, \alpha)$ the maximal solution through s, α for $t \leq s$.

Theorem 5.1. *If $v : [\sigma - r, \infty) \to \mathbb{R}$ is a continuous function whose upper right-hand derivative $\dot{v}$ satisfies*

$$\dot{v}(t) \leq U(t, v(t)) \quad \textit{for } t \geq \sigma, \quad \textit{if } v(s) \leq l(s, t, v(t)),\ s \in [t - r, t],$$

then

$$v(t) \leq r(t, \sigma, \alpha) \qquad \textit{for } t \geq \sigma$$

where α is chosen so that $v(s) \leq l(s, \sigma, \alpha)$, $s \in [\sigma - r, \sigma]$.

Theorem 5.1 can be shown to include Theorem 4.2 by noting an appropriate function $U(t, u)$ is given by the function

$$\min\{\frac{1}{3r}\, u, p(\frac{2}{3}\, u) - \frac{2}{3}\, u, w(u)\}.$$

Kato gives applications of this more general result to the theory of asymptotic stability, stability with respect to delays, and the following interesting result of Yorke [2].

Theorem 5.2. *Suppose $f : \mathbb{R} \times C \to \mathbb{R}$ is continuous and satisfies*

$$-\alpha M(\phi) \leq f(t, \phi) \leq \alpha M(-\phi) \tag{5.1}$$

for some constant $\alpha \geq 0$, where $M(\phi) = \max\{0, \sup_{-r \leq \theta \leq 0} \phi(\theta)\}$. Then the following conclusions hold:

(i) *If $\alpha r \leq \frac{3}{2}$, the zero solution of the* RFDE(f) *is uniformly stable.*

(ii) *If $\alpha r < \frac{3}{2}$, then the zero solution of the* RFDE(f) *is uniformly asymptotically stable if for any sequence $t_m \to \infty$ and any sequence $\{\phi_m\}$ converging to a nonzero constant function, the sequence $f(t_m, \phi_m)$ does not converge to zero.*

A special case of the last result is the linear equation

$$\dot{x}(t) = -bx(t - \tau(t)) \tag{5.2}$$

where $b \geq 0$ and $\tau(t)$ is a continuous function satisfying $0 \leq \tau(t) \leq r$. If $br \leq \frac{3}{2}$, then the solution $x = 0$ is uniformly stable and if $br < \frac{3}{2}$, the solution $x = 0$ is uniformly asymptotically stable. Notice the constant $\frac{3}{2}$ is the same as the one obtained in the analysis of Equation (4.13) by using the knowledge of the solution over $[0, 2r]$. The original proof of Yorke also uses knowledge of the actual solution, but over the larger interval $[0, 3r]$.

The upper bound $\frac{3}{2}$ for br has been previously shown to be the best possible for Equation (5.2) and it is, therefore, quite remarkable that the same upper bound can be obtained for the nonlinear problem as in Theorem 5.2. In fact, it was shown by Mishkis [1] and Lillo [2] that for $br = \frac{3}{2}$, there are Equations (5.2) that have periodic solutions and if $br > \frac{3}{2}$, there are equations with unbounded solutions. The paper of Lillo [2] also contains further remarkable properties on the asymptotic behavior of general linear scalar systems that satisfy the conditions of Theorem 5.2. For other results along this line, see Halanay and Yorke [1].

Generalizations of Theorem 5.2 to n-dimensions have been given by Grossman [1] (see also Kato [1]).

Grimmer and Seifert [1] have applied Razumikhin arguments to discuss the asymptotic behavior of the solutions of integrodifferential equations of the type

$$\dot{x}(t) = Ax(t) + \int_0^t B(t, s)\{x(s) + g_1(x)(s)\}ds + g_2(x)(t) + f(t)$$

where $g_j(x)(s)$ are certain higher-order functionals of x, depending on values of $x(\theta)$ for $\theta \leq t$.

Seifert [2, 3] has also used arguments in the spirit of Razumikhin to prove the existence of a solution of an RFDE(f) that has the range of f in a closed convex set in $\mathbb{R}^n$.

It is certainly worthwhile to discuss the advantages and disadvantages of Liapunov functionals and Razumikhin arguments. As we have seen by examples, it appears to be easier generally to use the Razumikhin theorems for stability of the trivial solution than to construct Liapunov functionals. On the other hand, if the limit of a solution is more complicated than a point, it is not clear how one can take advantage of arguments in the spirit of Razumikhin. This becomes very apparent if we consider autonomous equations. For example, if we consider a scalar equation that has a stable nontrivial periodic solution, then the trace of this solution in $\mathbb{R}$ will be an interval that could contain a constant solution. How can one possibly detect the asymptotic behavior of a solution by only observing what happens in $\mathbb{R}$? The Razumikhin arguments use mainly $\mathbb{R}$ whereas the Liapunov functionals take advantage of C. This is very apparent in the example of Levin and Nohel (see Equation (3.5)).

In Section 5.2, there is the obvious omission of a theorem on ultimate boundedness using Liapunov functionals. A result in this direction is contained in Yoshizawa [1, p. 206]. The reason for not stating this particular result in the text is that the proof imposes a restriction on the size of the delay, in addition to the usual conditions on Liapunov functionals. To obtain a result that depends only on the rate of change of certain functionals seems to put severe restrictions on the form of the RFDE. On the other hand, Theorem 4.3 on ultimate boundedness does not have such additional restrictions on the delays.

Another area of investigation that uses ideas in the same spirit as the ones in the Razumikhin theorems is the method of guiding functions introduced by Krasnoselskii et al. around 1958. The following definition is due to Mawhin [1, 2].

Definition 5.1. A continuously differential function $V : \mathbb{R}^n \to \mathbb{R}$ is a *guiding function* for an RFDE(f) if there is a $\rho > 0$ such that

$$\frac{\partial V(\phi(0))}{\partial x} f(t, \phi) > 0 \tag{5.3}$$

for every $\phi \in C$ satisfying $|\phi(0)| \geq \rho$ and $|V(\phi(0))| \geq |V(\phi(\theta))|$, for $\theta \in [-r, 0]$.

Krasnoselskii was interested in determining periodic solutions of periodic ordinary and differential difference equations. His original definition required that Inequality (5.3) be satisfied for every $\phi \in C$ such that $|\phi(0)| \geq \rho$. This imposed very severe restrictions on the functional f since it required that $\phi(0)$ be the governing factor for the asymptotic behavior of f. In addition, the proof of the results on the existence of periodic solutions by Krasnoselskii et al. was very complicated and the extension of the proof to general RFDE was not clear. Mawhin [1] found a different proof that was much simpler. With this simpler proof, it became clear that the guiding function could be as general as given in Definition 5.1.

The topological method of Wazewski using ingress and egress points on the boundary of an open set is one of the most important tools in the study of the asymptotic behavior of the solutions of ordinary differential equations. By insisting that the direction of the flow in $\mathbb{R}^n$ at time t is determined in a very strong way by the value of the solution x at the same time t, one can give an extension of Wazewski's principle (see Onuchic [2] and Mikolajska [1]).

The general principle of Wazewski has been given by Rybakowski [1, 2, 3] together with its connection to a homotopy index and the Conley index.

Due to lack of space, many areas of the theory of stability of RFDE have not been mentioned. For example, no converse theorems have been given concerning the existence of Liapunov functionals for stable systems. These results are certainly important and, as remarked earlier, were the original motivation for Krasovskii [1] to begin the vigorous development of the theory of RFDE in the state space C as opposed to $\mathbb{R}^n$. If one understands well the method of obtaining converse theorems for ordinary differential equations, it is not too difficult to extend the results of RFDE. The basic ideas are contained in the books of Halanay [1], Krasovskii [1], and Yoshizawa [1] and the paper of Hale [3]. The converse theorems have applications to the implications of stability; for example, stability under constantly acting disturbances (see Corduneanu [2]), the behavior of the solutions of perturbations of stable systems (see Laksmikantham and Leela [1] and Onuchic [3]), and the absolute stability of systems of the form

$$\begin{aligned} \dot{x}(t) &= Ax(t) + Bx(t-\tau) + Q(\sigma(t)) \\ \sigma &= cx \end{aligned} \tag{5.4}$$

(see Halanay [1], and Somolinos [1]). The book of Razvan [1] contains an excellent bibliography on Equation (5.4). Halanay [2] also has used the converse theorems to discuss Equation (5.4) with nonhomogeneous almost-periodic forcing functions. The book of Martynyuk [1] is devoted entirely to the stability theory of functional differential equations.

For a more complete discussion of the recent developments in stability theory and the general theory of RFDE, see Burton [2, 3] Grippenberg, Londen, and Staffans [1], and Hino, Murakami, and Naito [1].

6
General linear systems

This chapter is devoted to the development of linear RFDE, including the variation-of-constants formula and the formal adjoint of a linear system.

For an arbitrary two-point boundary-value problem, it is then shown that there is a two-point boundary-value problem for the formal adjoint equation that fulfills the conditions of the usual Fredholm alternative. Also, relations between various types of stability for linear systems are given.

The chapter is completed with a discussion about perturbed linear systems.

6.1 Resolvents and exponential estimates

For $(\sigma, \phi) \in \mathbb{R} \times C$, consider the linear system

$$\begin{aligned} \dot{x}(t) &= L(t)x_t + h(t), \qquad t \geq \sigma, \\ x_\sigma &= \phi \end{aligned} \tag{1.1}$$

where $h \in \mathcal{L}_1^{\text{loc}}([\sigma, \infty), \mathbb{R}^n)$, the space of functions from $[\sigma, \infty)$ into $\mathbb{R}^n$ that are Lebesgue integrable on each compact set of $[\sigma, \infty)$. Also, assume that there is an $n \times n$ matrix function $\eta(t, \theta)$, measurable in $(t, \theta) \in \mathbb{R} \times \mathbb{R}$, normalized so that

$$\eta(t, \theta) = 0 \quad \text{for} \quad \theta \geq 0, \qquad \eta(t, \theta) = \eta(t, -r) \quad \text{for} \quad \theta \leq -r,$$

$\eta(t, \theta)$ is continuous from the left in θ on $(-r, 0)$ and has bounded variation in θ on $[-r, 0]$ for each t. Further, there is an $m \in \mathcal{L}_1^{\text{loc}}((-\infty, \infty), \mathbb{R})$ such that

$$\operatorname{Var}_{[-r,0]} \eta(t, \cdot\,) \leq m(t) \tag{1.2}$$

and the linear mapping $L(t) : C \to \mathbb{R}^n$ is given by

$$L(t)\phi = \int_{-r}^{0} d[\eta(t, \theta)]\phi(\theta) \tag{1.3}$$

for all $t \in (-\infty, \infty)$ and $\phi \in C$. Obviously, the norm of $L(t)$ satisfies $|L(t)\phi| \leq m(t)|\phi|$.

Theorem 1.1. *Suppose these conditions on η and μ are satisfied. For any given $\sigma \in \mathbb{R}$, $\phi \in C([-r, 0], \mathbb{R}^n)$, and $h \in \mathcal{L}_1^{loc}([\sigma, \infty), \mathbb{R}^n)$, there exists a unique function $x(\sigma, \phi)$ defined and continuous on $[\sigma - r, \infty)$ that satisfies System (1.1) on $[\sigma, \infty)$.*

Proof. Condition (1.2) implies the Caratheodory conditions are satisfied. Therefore, we have local existence from Chapter 2. Local uniqueness follows as in Chapter 1 since $L(t)$ is Lipschitzian. To prove global existence, let x be a noncontinuable solution of System (1.1) on $[\sigma - r, b)$. Integration of System (1.1) yields

$$|x(t)| \leq |\phi(0)| + \int_\sigma^t m(s)|x_s|\, ds + \int_\sigma^t |h(s)|\, ds$$

for all values of $t \in [\sigma, b)$. Thus,

$$|x_t| \leq |\phi| + \int_\sigma^t m(s)|x_s|\, ds + \int_\sigma^t |h(s)|\, ds$$

for $t \in [\sigma, b)$. The inequality in Lemma 3.1 of Section 1.3 implies

$$|x_t| \leq \left[|\phi| + \int_\sigma^t |h(s)|\, ds\right] \exp\left[\int_\sigma^t m(s)\, ds\right] \tag{1.4}$$

for $t \in [\sigma, b)$. But this relation implies $|\dot{x}(t)|$ is bounded by a function in $\mathcal{L}_1^{loc}$. Following the same proof as in Theorem 3.2 of Section 2.3 for equations with continuous right-hand sides, one shows $b = \infty$ and the theorem is proved. □

The most common type of linear systems with finite lag known to be useful in the applications has the form

$$\dot{x}(t) = \sum_{k=1}^{N} A_k x(t - \omega_k) + \int_{-r}^{0} A(t, \theta) x(t + \theta)\, d\theta + h(t) \tag{1.5}$$

where $0 \leq \omega_1 < \omega_2 < \cdots < \omega_N \leq r$ and $A(t, \theta)$ is integrable in θ for each t and there is a function $a \in \mathcal{L}_1^{\mathrm{loc}}((-\infty, \infty), \mathbb{R})$ such that

$$\left|\int_{-r}^{0} A(t, \theta)\phi(\theta)\, d\theta\right| \leq a(t)|\phi|$$

for all $t \in \mathbb{R}$ and $\phi \in C$.

To derive a representation for the solution, it is useful to rewrite the equation. First we split off the part that explicitly depends on the initial data

$$\begin{aligned}\dot{x}(t) &= \int_{\sigma}^{t} d[\eta(t,\theta-t)]x(\theta) + \int_{-r}^{\sigma-t} d_\theta[\eta(t,\theta)]\phi(t-\sigma+\theta) + h(t) \\ &= -\eta(t,\sigma-t)x(\sigma) - \int_{\sigma}^{t} \eta(t,\theta-t)\dot{x}(\theta)\,d\theta \\ &\quad + \int_{-r}^{\sigma-t} d_\theta[\eta(t,\theta)]\phi(t-\sigma+\theta) + h(t).\end{aligned}$$

This equation is a Volterra equation (of the second kind)

$$y(t) = \int_{\sigma}^{t} k(t,s)y(s)\,ds + g(t), \qquad t \geq \sigma, \tag{1.6}$$

where $y(t) = \dot{x}(t)$, $k(t,s) = \eta(t,s-t)$ and

$$g(t) = -\eta(t,\sigma-t)\phi(0) + \int_{-r}^{\sigma-t} d_\theta[\eta(t,\theta)]\phi(t-\sigma+\theta) + h(t).$$

Let J be an interval. A measurable function $k : J \times J \to \mathbb{R}^{n\times n}$ is called a *Volterra kernel of type* $\mathcal{L}_1$ on J if $k(t,s) = 0$ for $s > t$ and $\|k\|_1 < \infty$, where

$$\|k\|_1 \stackrel{\text{def}}{=} \sup_{|f|_1 \leq 1} \int_J \int_J |k(t,s)f(s)|\,ds\,dt = \operatorname{ess\,sup}_{s\in J} \int_J |k(t,s)|\,dt.$$

The kernel $k(t,s) = \eta(t,s-t)$ is a kernel of type $\mathcal{L}_1$ on $[\sigma,\infty)$.

We call a kernel of type $\mathcal{L}_1$ if $J = \mathbb{R}$ and of type $\mathcal{L}_1^{loc}$ if for every interval $[a,b] \subset \mathbb{R}$, the kernel is of type $\mathcal{L}_1$ on $[a,b]$.

If k is a kernel of type $\mathcal{L}_1$ on $[\sigma,\infty)$, then

$$f \mapsto \int_{\sigma}^{t} k(t,s)f(s)\,ds \tag{1.7}$$

maps $\mathcal{L}_1[\sigma,\infty)$ into itself and

$$\left|\int_{\sigma}^{t} k(t,s)f(s)\,ds\right|_1 \leq \|k\|_1\,|f|_1.$$

A kernel $R(t,s)$, $t \geq s$, of type $\mathcal{L}_1$ is called a *Volterra resolvent of* k if

$$\begin{aligned}R(t,s) &= k(t,s) - \int_{s}^{t} R(t,\alpha)k(\alpha,s)\,d\alpha \\ &= k(t,s) - \int_{s}^{t} k(t,\alpha)R(\alpha,s)\,d\alpha.\end{aligned} \tag{1.8}$$

A simple contraction argument shows that if k is a kernel of type $\mathcal{L}_1$ with $\|k\|_1 < 1$, then k has a resolvent of type $\mathcal{L}_1$. Further for $g \in \mathcal{L}_1$, Equation (1.6) has a unique solution in $\mathcal{L}_1$. This solution is given by the variation-of-constants formula

$$(1.9) \qquad y(t) = g(t) - \int_\sigma^t R(t,s)g(s)\,ds.$$

Lemma 1.1. *If the hypotheses on η are satisfied, then the kernel $k(t,s) = \eta(t, s-t)$, $t \geq s$, has a resolvent of type $\mathcal{L}_1^{loc}$.*

Proof. If we define

$$\widetilde{R}(t,s) = R(t,s)e^{\gamma(t-s)}, \quad \widetilde{\eta}(t,s) = \eta(t,s)e^{-\gamma s}, \quad \tilde{k}(t,s) = \widetilde{\eta}(t,s),$$

then $\widetilde{R}(t,s)$ satisfies the equation

$$\widetilde{R}(t,s) = \widetilde{\eta}(t, s-t) - \int_s^t \widetilde{R}(t,\alpha)\widetilde{\eta}(\alpha, s-\alpha)\,d\alpha, \qquad t \geq s.$$

If we choose γ such that

$$\sup_{s\in[\sigma,\infty)} \int_\sigma^\infty \left|\eta(\alpha, s-\alpha)\right| e^{-\gamma(s-\alpha)}\,d\alpha < 1,$$

then $\|\tilde{k}\|_1 < 1$ and $\widetilde{R}(t,s)$ is a resolvent of type $\mathcal{L}_1$ on $[\sigma,\infty)$. Consequently,

$$R(t,s) = \widetilde{R}(t,s)e^{-\gamma(t-s)}$$

is a resolvent of type $\mathcal{L}_1$ on $[\sigma, T]$ for the kernel $\eta(t, s-t)$, $t \geq s$. This proves the lemma. □

Using the resolvent equation, we can give a representation theorem for the solutions of System (1.1).

Theorem 1.2. *If the hypotheses on η are satisfied, then for any given σ, $\phi \in C$, and $h \in \mathcal{L}_1^{loc}([\sigma,\infty), \mathbb{R}^n)$, there exists a unique solution $x(\,\cdot\,;\sigma,\phi)$ defined and continuous on $[\sigma - r, \infty)$ that satisfies System (1.1) on $[\sigma,\infty)$. Furthermore, this solution is given by*

$$(1.10) \quad x(t;\sigma,\phi) = X(t,\sigma)\phi(0) + \int_\sigma^t X(t,\alpha)\,d_\alpha[F(\alpha,\sigma;\phi,h)], \qquad t \geq \sigma,$$

where

$$(1.11) \qquad X(t,\sigma) = I - \int_\sigma^t R(s,\sigma)\,ds$$

and $F(\,\cdot\,,\sigma;\,\cdot\,,h) : C \to \mathcal{L}_1^{loc}([\sigma,\infty), \mathbb{R}^n)$ is defined by

$$(1.12) \quad F(t,\sigma;\phi,h) = \phi(0) + \int_\sigma^t \int_{-r}^{-s} d_\theta[\eta(s,\theta)]\phi(s+\theta)\,ds + \int_\sigma^t h(\alpha)\,d\alpha.$$

Proof. From Representation (1.9), we find that the derivative of any solution of System (1.1) has the form

$$\dot{x}(t) = g(t) - \int_\sigma^t R(t,\alpha)g(\alpha)\,d\alpha. \tag{1.13}$$

From Representation (1.13), the resolvent equation (1.8), and Fubini's theorem, we deduce that

$$\begin{aligned}
x(t) &= \phi(0) + \int_\sigma^t \dot{x}(s)\,ds \\
&= \phi(0) + \int_\sigma^t g(s)\,ds - \int_\sigma^t \int_\sigma^s R(s,\alpha)g(\alpha)\,d\alpha\,ds \\
&= \phi(0) - \int_\sigma^t \eta(s,\sigma-s)\phi(0)\,ds + \int_\sigma^t \int_\sigma^s R(s,\alpha)\eta(\alpha,\sigma-\alpha)\phi(0)\,d\alpha\,ds \\
&\quad + \int_\sigma^t d_\alpha[F(\alpha,\sigma;\phi,h)] - \int_\sigma^t \int_\sigma^s R(s,\alpha)\,d_\alpha[F(\alpha,\sigma;\phi,h)]\,ds \\
&= \phi(0) - \int_\sigma^t R(s,\sigma)\phi(0)\,ds + \int_\sigma^t d_\alpha[F(\alpha,\sigma;\phi,h)] \\
&\quad - \int_\sigma^t \int_\sigma^s R(s,\alpha)\,d_\alpha[F(\alpha,\sigma;\phi,h)]\,ds \\
&= X(t,\sigma)\phi(0) + \int_\sigma^t d_\alpha[F(\alpha,\sigma;\phi,h)] \\
&\quad - \int_\sigma^t \int_\alpha^t R(s,\alpha)\,ds\,d_\alpha[F(\alpha,\sigma;\phi,h)] \\
&= X(t,\sigma)\phi(0) + \int_\sigma^t X(t,\alpha)\,d_\alpha[F(\alpha,\sigma;\phi,h)].
\end{aligned}$$

Thus, any solution of System (1.1) has the representation (1.10) and the theorem follows from Lemma 1.1. □

The matrix solution $X(t,s)$, $t \geq s$, has a natural interpretation for the homogeneous differential equation

$$\dot{x}(t) = L(t)x_t. \tag{1.14}$$

With respect to the initial data

$$X_0(\theta) = \begin{cases} I, & \text{for } \theta = 0, \\ 0, & \text{for } -r \leq \theta < 0, \end{cases}$$

Equation (1.6) becomes the resolvent equation

$$R(t,s) = \eta(t,s-t) - \int_s^t \eta(t,\alpha-t)R(\alpha,s)\,d\alpha, \qquad t \geq s. \tag{1.15}$$

Therefore the matrix solution $X(t,s) = I - \int_s^t R(\alpha,s)\,d\alpha$, $t \geq s$, can be regarded as the solution of System (1.14) with respect to the discontinuous initial data X_0.

In the sequel, we define $X(t,s) = 0$ for $t < s$ and call $X(t,s)$ the *fundamental matrix solution of System* (1.14).

Corollary 1.1. *The fundamental matrix solution* $X(t,s)$, $t \geq s$, *of System* (1.14) *satisfies the following estimates*

$$|X(t,s)| \leq \exp\Big[\int_s^t m(\alpha)\,d\alpha\Big] \tag{1.16}$$

$$\operatorname{Var}_{[s,t]} X(\,\cdot\,,s) \leq \exp\Big[\int_s^t m(\alpha)\,d\alpha\Big] - 1, \qquad t \geq s. \tag{1.17}$$

Proof. Since $X(t,s)$, $t \geq s$, satisfies the integrated equation, the same estimate as in Theorem 1.1 gives (1.16).

The estimation

$$|R(t,s)| \leq m(t) + \int_s^t |R(t,\alpha)| m(\alpha)\,d\alpha$$

and the inequality in Lemma 3.1 of Chapter 1 yield the a priori bound

$$|R(t,s)| \leq m(t) \exp\Big[\int_s^t m(\alpha)\,d\alpha\Big].$$

Since $X(t,s) = I - \int_s^t R(\tau,s)\,d\tau$, we find

$$\begin{aligned}\operatorname{Var}_{[s,t]} X(\,\cdot\,,s) &\leq \int_s^t |R(\tau,\sigma)|\,d\tau \\ &\leq \int_s^t m(\tau) \exp\Big[\int_s^\tau m(\alpha)\,d\alpha\Big]\,d\tau = \exp\Big[\int_s^t m(\alpha)\,d\alpha\Big] - 1.\end{aligned}$$

This proves estimate (1.17). □

As an illustration, we consider Equation (1.5). It is easy to verify that Representation (1.10) becomes

$$\begin{aligned} x(t) = X(t,\sigma)\phi(0) &+ \sum_{k=1}^{N} \int_{\sigma-\omega_k}^{\sigma} X(t,\alpha+\omega_k)A_k(\alpha+\omega_k)\phi(\alpha-\sigma)\,d\alpha \\ &+ \int_{\sigma-\omega}^{\sigma}\int_{\sigma}^{\sigma+\omega} X(t,s)A(s,\alpha-s)\,ds\,\phi(\alpha-\sigma)\,d\alpha \\ &+ \int_\sigma^t X(t,s)h(s)\,ds. \end{aligned} \tag{1.18}$$

6.2 The variation-of-constants formula

In this section, it is our aim to present an abstract version of Equation (1.10) that holds in the state space C. It will become clear that such a formula has great advantages over Equation (1.10).

A two-parameter family $T(t,\sigma)$, $t \geq \sigma$, of bounded linear operators on a real Banach space $\mathcal{B}$ is called a (forward) *evolutionary system* on $\mathcal{B}$ if

(i) $T(\sigma,\sigma) = I$;

(ii) $T(t,s)T(s,\sigma) = T(t,\sigma), \qquad t \geq s \geq \sigma.$

From the existence and uniqueness for solutions of System (1.1), it follows that translation along the solution defines an evolutionary system on C:

$$T(t,\sigma)\phi = x_t(.;\sigma,\phi), \qquad t \geq \sigma. \tag{2.1}$$

From the variation-of-constants formula given by Equation (1.10), we find that the solution of System (1.1) is given by

$$x_t(\theta;\sigma,\phi,h) = T(t,\sigma)\phi(\theta) + \int_\sigma^{t+\theta} X(t+\theta,\alpha)h(\alpha)\,d\alpha, \quad -r \leq \theta \leq 0, \tag{2.2}$$

where it is understood that the integral is considered as a family of Euclidean space integrals parameterized by θ. It our objective to give an interpretation of Equation (2.2) as a Banach space integral.

In order to do so, we have to introduce a little bit of vector-valued integration. Let $S : \mathcal{L}_1([a,b],\mathbb{R}^n) \to \mathcal{B}$ be a continuous linear operator, and let Σ be the σ-field of Lebesgue measurable subsets of $[a,b]$. Define

$$F : \Sigma \to \mathcal{B}, \qquad E \mapsto S(\chi_E),$$

where χ_E denotes the characteristic function of E. One can easily verify that whenever E_1 and E_2 are disjoint members of Σ then $F(E_1 \cup E_2) = F(E_1) + F(E_2)$. Further, from the fact that there is a $\lambda(E) > 0$ (the Lebesgue measure of E) such that

$$\|F(E)\| \leq \lambda(E)\|S\|, \tag{2.3}$$

one verifies that $F(\bigcup_{j=1}^\infty E_j) = \sum_{j=1}^\infty F(E_j)$ in norm for all sequences of pairwise disjoint members of Σ such that $\bigcup_{j=1}^\infty E_j \in \Sigma$. A function F with this property is called a *countable additive vector measure.* The variation of F is the extended nonnegative function

$$|F| : \Sigma \to [0,\infty], \qquad |F|(E) = \sup_\pi \sum_{A\in\pi} \|F(A)\|, \tag{2.4}$$

where the supremum is taken over all partitions π of E into a finite number of pairwise disjoint members of Σ. From (2.3), it follows that $|F|([a,b]) \leq (b-a)\|S\|$. We call F a *vector measure of bounded variation.*

It is easy to define the integral of an integrable scalar function with respect to the vector measure F. First, define S_F on the space of simple functions

$$S_F\Big(\sum_{j=1}^{n} c_j \chi_{E_j}\Big) = \sum_{j=1}^{n} c_j F(E_j). \tag{2.5}$$

One can show that (2.5) defines a linear operator and

$$|S_F(f)| = |\sum_{j=1}^{n} c_j F(E_j)| \leq |F|([a,b])|f|_1.$$

So if the space of simple functions is given the $\mathcal{L}_1$ norm, then S_F has a continuous linear extension to $\mathcal{L}_1\big([a,b], \mathbb{R}\big)$. Therefore, one can define

$$\int_{[a,b]} f\, dF \overset{\text{def}}{=} S_F(f). \tag{2.6}$$

It is convenient to write the integral in (2.6) as a Stieltjes integral. This can be done as follows: If we define $K : [a,b] \to \mathcal{B}$ by $K(t) = F([a,t])$, then K is of strong bounded variation over $[a,b]$; that is, the strong variation function

$$|K|(t) = \sup_{P(a,t)} \sum_{j=1}^{N} |K(\alpha_j) - K(\alpha_{j-1})|, \tag{2.7}$$

where $P(a,t)$ denotes a partition $a = \alpha_0 < \alpha_1 < \cdots < \alpha_N = t$ of $[a,t]$, is bounded. We write the integral in (2.6) as follows

$$\int_{[a,b]} f\, dF = \int_a^b d[K(\alpha)]f(\alpha). \tag{2.8}$$

If f is continuous, then the integral in (2.8) can be understood as a vector-valued Riemann-Stieltjes integral.

After these preparations, we return to Equation (2.2). The following lemma is the key to a variation-of-constants formula in the space C.

Lemma 2.1. *Fix $t \geq \sigma$ and define $S(t) : \mathcal{L}_1\big([\sigma,t], \mathbb{R}^n\big) \to C$ by*

$$S(t)h(\theta) = \int_\sigma^{t+\theta} X(t+\theta, \alpha)h(\alpha)\, d\alpha \tag{2.9}$$

where $X(t,s)$ denotes the fundamental matrix solution to System (1.1). The linear operator $S(t)$ is completely continuous and can be represented by a vector-valued integral

$$S(t)h = \int_\sigma^t d[K(t,\alpha)]h(\alpha), \tag{2.10}$$

where the kernel $K(t,\,\cdot\,):[\sigma,t]\to C$ *is given by*

$$K(t,s)(\theta)=\int_\sigma^s X(t+\theta,\alpha)\,d\alpha. \tag{2.11}$$

Proof. First we show that $S(t)$ is a completely continuous operator. Set $B_1=\{h\in\mathcal{L}_1([\sigma,t],\mathbb{R}^n):|h|_1\le 1\}$. By the Arzela-Ascoli theorem, we have only to check that $S(t)B_1$ is uniformly bounded and equicontinuous. Fix $t>0$. From Corollary 1.1, we have

$$\big|S(t)h(\theta)\big|\le \exp\Big[\int_\sigma^{t+\theta} m(\alpha)\,d\alpha\Big]\,|h|_1.$$

Since for $\epsilon>0$, there is a $\delta>0$ such that for $|\theta_1-\theta_2|<\delta$, we have $|X(t+\theta_1,\alpha)-X(t+\theta_2,\alpha)|<\epsilon$. Therefore, if $|h|_1\le 1$ and $|\theta_1-\theta_2|<\delta$

$$\big|S(t)h(\theta_1)-S(t)h(\theta_2)\big|\le (t-\sigma)\epsilon.$$

This shows that $S(t)$ is completely continuous. Note that $X(t,s)=0$ for $t<s$. Therefore, Representation (2.10) follows immediately from the definition $K(t,s)=S(t)\chi_{[\sigma,s]}$. □

Corollary 2.1. *The solution* $x_t=x_t(\,\cdot\,;\sigma,\phi,h)$ *of System* (1.1) *in* C *satisfies the following abstract variation-of-constants formula*

$$x_t=T(t,\sigma)\phi+\int_\sigma^t d[K(t,\alpha)]h(\alpha) \tag{2.12}$$

where the kernel $K(t,s)$ *is given by* (2.11). *In the special case that* $L(t)\equiv L$ *is independent of* t *in System* (1.1), *we have*

$$T(t,s)=T(t-s,0)\stackrel{\text{def}}{=}T(t-s),\quad X(t,s)=X(t-s,0)\stackrel{\text{def}}{=}X(t-s),$$

and

$$x_t=T(t-\sigma)\phi+\int_\sigma^t d[K(t,\alpha)]h(\alpha) \tag{2.13}$$

where the kernel $K(t,\,\cdot\,):[\sigma,t]\to C$ *is given by*

$$K(t,s)(\theta)=\int_\sigma^s X(t+\theta-\alpha)\,d\alpha.$$

6.3 The formal adjoint equation

In this section, we consider the general linear system

$$\dot{x}(t) = L(t)x_t = \int_{-r}^{0} d[\eta(t,\theta)]x(t+\theta), \qquad t \geq \sigma, \tag{3.1}$$

with $x_\sigma = \phi$, $\phi \in C$, and where η satisfies the conditions from Section 1. The purpose of this section is to introduce the formal adjoint equation for Equation (3.1). A two-parameter family $V(s,t)$, $s \leq t$, of bounded linear operators on $\mathcal{B}$ is called a *backward evolutionary system* if

(i) $V(t,t) = I$;

(ii) $V(s,\sigma)V(\sigma,t) = V(s,t), \qquad s \leq \sigma \leq t.$

Let $\mathcal{B}^*$ denote the dual space of $\mathcal{B}$ and let $T(t,s)$, $t \geq s$, be a forward evolutionary system on $\mathcal{B}$. For every (t,s), $t \geq s$, we can define the adjoint operator $T(t,s)^*$ on $\mathcal{B}^*$. If we set $T^*(s,t) = T(t,s)^*$, then

$$\begin{aligned} \langle \phi, T^*(s,r)T^*(r,t)\phi^* \rangle &= \langle T(t,r)T(r,s)\phi, \phi^* \rangle \\ &= \langle T(t,s)\phi, \phi^* \rangle \\ &= \langle \phi, T^*(s,t)\phi^* \rangle. \end{aligned}$$

So $T^*(s,t), s \leq t$, is a backward evolutionary system on $\mathcal{B}^*$ and is called the *adjoint system* of $T(t,s)$, $t \geq s$. Note that, in particular, we have that

$$\langle T(s,\sigma)\phi, T^*(s,t)\phi^* \rangle, \qquad \sigma \leq s \leq t, \tag{3.2}$$

is independent of s. This property plays an important role in the Fredholm alternative, when dealing with boundary-value problems. In order to use (3.2) we have to compute the adjoint system of $T(t,s)$, $t \geq s$.

Let B_0 denote the Banach space of row-valued functions $\psi : (-\infty, 0] \to \mathbb{R}^{n*}$ that are constant on $(-\infty, -r]$, of bounded variation on $[-r,0]$, continuous from the left on $(-r,0)$ and vanishing at zero with norm $\operatorname{Var}_{[-r,0]} \psi$. Together with the pairing

$$\langle f, \phi \rangle = \int_{-r}^{0} df(\theta)\phi(\theta), \qquad f \in B_0, \quad \phi \in C, \tag{3.3}$$

the space B_0 becomes a representation for the dual space of C.

So the adjoint system of $T(t,s)$, $t \geq s$, is a backward evolutionary system on B_0, and to compute $T^*(s,t), s \leq t$, we use the pairing (3.3). Before we do so, we would like to have some more information about $T^*(s,t), s \leq t$. An obvious question would be: Is $T^*(s,t), s \leq t$, the evolutionary system for a differential equation?

Define the formal adjoint equation by

$$y(s) + \int_{s}^{t} y(\tau)\,\eta(\tau, s-\tau)\,d\tau = \text{constant}, \qquad s \leq t-r, \tag{3.4}$$

where y is a solution that vanishes on $[t, \infty)$, satisfies Equation (3.4) on $(-\infty, t-r]$, and such that $y(t+\theta) = \psi(\theta)$, $-r \le \theta \le 0$ for ψ in B_0. In general, we cannot differentiate the integral equation to obtain a differential equation, since the solution on the interval $[t-r, t]$ is only in B_0.

Many applications have the form given in Equation (1.5). In this case, the formal adjoint equation can be written as

$$\frac{dy(s)}{ds} = -\sum_{k=1}^{N} y(s+\omega_k)A_k(s+\omega_k) - \int_{-\omega}^{0} y(s-\xi)A(s-\xi,\xi)\,d\xi \tag{3.5}$$

with $y_t = \psi$, $\psi \in B_0$ and $s \le t$.

To clarify the relation between the adjoint evolutionary system $T^*(s,t)$ and the formal adjoint equation (3.4), it turns out to be useful to study the adjoint equation as a forced Volterra equation. If we define g to be

$$g(\theta) = \begin{cases} 0, & \text{for } \theta = 0, \\ \psi(\theta) + \int_\theta^0 \psi(\tau)\eta(t+\tau, s-\tau)\,d\tau, & \text{for } -r \le \theta < 0, \\ g(-r), & \text{for } \theta \le -r, \end{cases} \tag{3.6}$$

then g belongs to B_0 and the solution y of Equation (3.4) satisfies the Volterra equation

$$y(s) + \int_s^t y(\tau)\,\eta(\tau, s-\tau)\,d\tau = g^t(s), \qquad s \le t, \tag{3.7}$$

where $g^t(s) = g(s-t)$. The form of this equation resembles Equation (1.6). Recall that we introduced the formal resolvent equation associated with $\eta(t, s-t)$ by

$$R(t,s) = -\eta(t, s-t) + \int_s^t R(t,\alpha)\eta(\alpha, s-\alpha)\,d\alpha$$

or equivalently,

$$R(t,s) = -\eta(t, s-t) + \int_s^t \eta(t, \alpha-t)R(\alpha,s)\,d\alpha. \tag{3.8}$$

In Lemma 1.1, we showed the existence of a unique solution to this equation. The proof of the next theorem follows the same lines as the proof of Theorem 1.1, using the (equivalent) resolvent equation (3.8).

Theorem 3.1. *If the hypotheses on η from Section 1 are satisfied, then for any given t in $\mathbb{R}$ and g in B_0, there exists a unique solution $y(\cdot\,; t, g)$ defined and locally of bounded variation on $(-\infty, t]$ that satisfies the formal adjoint equation*

$$y(s) + \int_s^t y(\tau)\,\eta(\tau, s-\tau)\,d\tau = g^t(s), \qquad s \le t, \tag{3.9}$$

where $g^t(s) = g(s-t)$. *Furthermore, a representation of the solution is given by*

$$
\begin{aligned}
y(s;t,g) &= g^t(s) - \int_s^t g^t(\alpha)R(\alpha,s)\,d\alpha \\
&= -\int_s^t d[g(\alpha - t)]X(\alpha,s), \qquad s \le t.
\end{aligned}
\tag{3.10}
$$

Translation along a solution of Equation (3.9) induces a two-parameter family of bounded operators $V(s,t),\ s \le t$ on B_0.

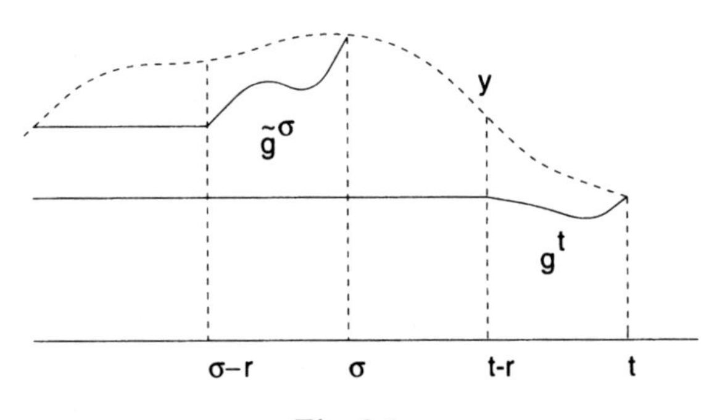

Fig. 6.1.

In fact, the solution y restricted to $(-\infty, \sigma]$ satisfies a Volterra integral equation given by

$$
y(s) + \int_s^\sigma y(\tau)\eta(\tau, s-\tau)\,d\tau = \tilde{g}^\sigma, \qquad s \le \sigma,
$$

where $\tilde{g}$ belongs to B_0 and is given by

$$
\tilde{g}(\theta) = \begin{cases} 0 & \text{for } \theta = 0; \\ y_\sigma(\theta) + \int_\theta^0 y_\sigma(\tau)\eta(\tau+\sigma, \theta-\tau)\,d\tau, & \text{for } -r \le \theta < 0; \\ \tilde{g}(-r), & \text{for } \theta \le -r \end{cases}
$$

(see Fig. 6.1). Define $V(\sigma,t)g = \tilde{g}$, so that $V(\sigma,t)$ maps the forcing function for the solution on $(-\infty,t]$ onto the forcing function for the solution on $(-\infty,\sigma]$. From the uniqueness property for the Volterra equation, it is easy to see that $V(s,t),\ s \le t$, defines a backward evolutionary system on B_0.

We shall see shortly that $V(s,t),\ s \le t$, is the adjoint system of $T(t,s),\ t \ge s$. First we derive a representation for $V(\sigma,t)$, from its definition. The integral equation yields

$$\begin{aligned} g(\sigma+\theta-t) &= y(\sigma+\theta) + \int_{\sigma+\theta}^{t} y(\tau)\eta(\tau,\sigma+\theta-\tau)\,d\tau \\ &= y_\sigma(\theta) + \int_{\theta}^{t-\sigma} y_\sigma(\tau)\eta(\tau+\sigma,\theta-\tau)\,d\tau. \end{aligned} \tag{3.11}$$

Therefore

$$\tilde{g}(\theta) = g(\theta-(t-\sigma)) - \int_0^{t-\sigma} y_\sigma(\tau)\eta(\tau+\sigma,\theta-\tau)\,d\tau$$

and, using Representation (3.10), $V(\sigma,t): B_0 \to B_0$ is given by

$$\begin{aligned} V(\sigma,t)g(\theta) &= g(\theta-(t-\sigma)) - \int_\sigma^t y(\tau)\eta(\tau,\theta-\tau)\,d\tau \\ &= g(\theta-(t-\sigma)) \\ &\quad + \int_\sigma^t \int_\tau^t d[g(-t+\alpha)]X(\alpha,\tau)\eta(\tau,\theta-\tau)\,d\tau. \end{aligned} \tag{3.12}$$

By definition, $V(\sigma,t)g(0)=0$, and using Equation (3.11),

$$V(\sigma,t)g(0-) \stackrel{\text{def}}{=} \lim_{\theta\to 0} V(\sigma,t)(\theta) = y(\sigma). \tag{3.13}$$

Theorem 3.2. *Let $T(t,s),\ t \ge s$ be the evolutionary system associated with System (3.1) on C. If $V(s,t),\ s \le t$ denotes the backward evolutionary system for the adjoint equation defined by (3.12), then $V(s,t),\ s \le t$ is the adjoint system of $T(t,s),\ t \ge s$, that is,*

$$T(t,s)^* = V(s,t).$$

Proof. Compute

$$\begin{aligned} \langle \phi, V(s,t)g \rangle &= \int_{-r}^0 \phi(\theta)\,d\big[V(s,t)g(\theta)\big] \\ &= -\phi(0)y(s) \\ &\quad + \int_{-r}^0 \phi(\theta)\,d[g(\theta-(t-s))] \\ &\quad + \int_{-r}^0 \phi(\theta) \int_s^t \int_\tau^t d[g(-t+\alpha)]X(\alpha,\tau)\,d\eta(\tau,\theta-\tau)\,d\tau. \end{aligned}$$

For the first term, we use the representation for y given in Theorem 3.1,

$$\phi(0)y(s) = -\phi(0)\int_s^t d[g(\alpha - t)]X(\alpha, s)$$
$$= -\int_{s-t}^0 X(\alpha + t, s)\phi(0)\, d[g(\alpha)].$$

Since $\phi(0)\, d[g]$ is a scalar-valued measure, it commutes with X and

$$\phi(0)y(s) = -\int_{-r}^0 X(\alpha + t, s)\phi(0)\, d[g(\alpha)] \tag{3.14}$$

where we used that $X(t, s) = 0$, $t \leq s$. The second integral becomes

$$\int_{-r}^0 \phi(\theta)\, d[g(\theta - (t - s))] = \int_{-r}^0 \phi(\alpha + (t - s))\, d[g(\alpha)]. \tag{3.15}$$

For the third term, integrating first with respect to θ and using that $\phi(0)\, d[g]$ is a scalar-valued measure, yields

$$\int_{-r}^0 \phi(\theta) \int_s^t \int_\tau^t d[g(-t + \alpha)]X(\alpha + \tau)\, d\eta(\tau, \theta - \tau)\, d\tau$$
$$=$$
$$\int_s^t \int_{\tau - t}^0 X(\alpha + t, \tau) \int_{-r}^{-\tau} d_\theta[\eta(\tau, \theta)]\phi(\tau + \theta)\, d[g(\alpha)]\, d\tau$$
$$=$$
$$\int_{-r}^0 \int_s^{\alpha + t} X(\alpha + t, \tau) \int_{-r}^{-\tau} d_\theta[\eta(\tau, \theta)]\phi(\tau + \theta)\, d\tau\, d[g(\alpha)]$$

where in the last identity, we used Fubini's theorem to reverse the order of integration, and the convention that $X(t, s) = 0$, $t \leq s$. Altogether, we find

$$\langle \phi, V(s, t)g \rangle = \int_{-r}^0 X(\alpha + t, s)\phi(0)\, d[g(\alpha)] + \int_{-r}^0 \phi(\alpha + (t - s))\, d[g(\alpha)]$$
$$+ \int_{-r}^0 \int_s^{\alpha + t} X(\alpha + t, \tau) \int_{-r}^{-\tau} d_\theta[\eta(\tau, \theta)]\phi(\tau + \theta)\, d\tau\, d[g(\alpha)]$$
$$= \langle T(t, s)\phi, g \rangle.$$

□

6.4 Boundary-value problems

In this section, we discuss two-point boundary-value problems for the nonhomogeneous equation (1.1) and obtain results of the "Fredholm alternative" type. The notation of the previous section will be employed without explanation.

Suppose V is a real Banach space, $\sigma < \tau$ are given real numbers, $M, N : C \to V$ are linear operators with domain dense in C, and $\gamma \in V$ is fixed. The problem is to find a solution x of

$$\dot{x}(t) = L(t)x_t + h(t) \tag{4.1}$$

subject to the boundary condition

$$Mx_\sigma + Nx_\tau = \gamma. \tag{4.2}$$

Let V^* be the dual space of V, and M^* and N^* the adjoint operators of M and N respectively. The fundamental result is

Theorem 4.1. *In order that Equations* (4.1) *and* (4.2) *have a solution, it is necessary that*

$$\int_\sigma^\tau y(\alpha)h(\alpha)\,d\alpha = -\langle\, \delta, \gamma \,\rangle_V \tag{4.3}$$

for all $\delta \in V^*$ *and solutions* y *of the system of adjoint equations*

$$\begin{cases} y(s) + \displaystyle\int_s^\tau y(\alpha)\eta(\alpha, s-\alpha)\,d\alpha = N^*\delta, & s \le \tau, \\ y(s) + \displaystyle\int_s^\sigma y(\alpha)\eta(\alpha, s-\alpha)\,d\alpha = -M^*\delta, & s \le \sigma. \end{cases} \tag{4.4}$$

If $\mathcal{R}\big(\, M + NT(\tau,\sigma)\,\big)$ *is closed, then the preceding condition is sufficient.*

Proof. Define $S(\tau,\sigma) : \mathcal{L}^1\big([\sigma,\tau]; \mathbb{R}^n\big) \to C$ by

$$S(\tau,\sigma)h(\theta) = \int_\sigma^{\tau+\theta} X(\tau+\theta,\alpha)h(\alpha)\,d\alpha.$$

The solution to Equation (4.1) is given by

$$x_\tau = T(\tau,\sigma)\phi + S(\tau,\sigma)h$$

and Relation (4.2) is equivalent to

$$NS(\tau,\sigma)h - \gamma \in \mathcal{R}\big(\, M + NT(\tau,\sigma)\,\big).$$

Therefore, it is necessary and, under the closure hypothesis, sufficient that

$$NS(\tau,\sigma)h - \gamma \in \operatorname{Cl}\mathcal{R}\big(\, M + NT(\tau,\sigma)\,\big) = \mathcal{N}\big(\, M^* + T^*(\sigma,\tau)N^*\,\big)^\perp;$$

that is, for any $\delta \in V^*$ such that $M^*\delta + T^*(\sigma,\tau)N^*\delta = 0$, we must have $\langle\, \delta, NS(\tau,\sigma)h - \gamma \,\rangle_V = 0$, or

$$
\begin{aligned}
\langle \delta, \gamma \rangle_V &= \langle \delta, NS(\tau,\sigma)h \rangle_V \\
&= \langle N^*\delta, S(\tau,\sigma)h \rangle_V \\
&= \int_{-\tau}^{0} \int_{\sigma}^{\tau+\theta} X(\tau+\theta,\alpha)h(\alpha)\,d\alpha\,d[N^*\delta(\theta)] \qquad (4.5)\\
&= \int_{\sigma}^{\tau} h(\alpha) \int_{\sigma-\tau}^{0} d[N^*\delta(\theta)]X(\tau+\theta,\alpha)\,d\alpha.
\end{aligned}
$$

In the last equality, we used Fubini's theorem to change the order of integration. From Representation (3.10) for the solution of Equation (4.4), we derive that

$$
\begin{aligned}
\int_{\sigma-\tau}^{0} d[N^*\delta(\theta)]X(\tau+\theta,\alpha) &= \int_{\alpha}^{\tau} d_\theta[N^*\delta(\theta-\tau)]X(\theta,\alpha) \\
&= -y(\alpha),
\end{aligned}
$$

where we used that $X(t,s) = 0$ for $t \le s$. Substitution in Equation (4.5) yields Equation (4.3) and proves the theorem. □

As another boundary-value problem, suppose $P, Q : V \to C$ are linear operators with domain dense in V and that p and q are fixed elements of C. The problem is to find a $v \in V$ and a solution of Equation (4.1) on $[\sigma, \tau]$ such that

$$
x_\sigma = Pv + p, \qquad x_\tau = Qv + q. \tag{4.6}
$$

Let P^* and Q^* be the adjoints of P and Q, respectively.

Theorem 4.2. *In order for Equations* (4.1) *and* (4.6) *to have a solution, it is necessary that*

$$
\int_{\sigma}^{\tau} y(\xi)h(\xi)\,d\xi = \langle T^*(\sigma,\tau)g, p \rangle - \langle g, q \rangle \tag{4.7}
$$

for every solution y of the formal adjoint problem

$$
y(s) + \int_{s}^{\tau} y(\alpha)\eta(\alpha, s-\alpha)\,d\alpha = g, \qquad s \le \tau \tag{4.8}
$$

with g in B_0 such that

$$
P^*T^*(\sigma,\tau)g = Q^*g. \tag{4.9}
$$

If $\mathcal{R}\big(Q - T(\tau,\sigma)P\big)$ is closed in C, this condition is both necessary and sufficient.

Proof. Proceeding as in the proof of Theorem 4.1, Relation (4.6) is equivalent to

$$
T(\tau,\sigma)p - q + S(\tau,\sigma)h \in \mathcal{R}\big(Q - T(\tau,\sigma)P\big).
$$

Therefore, it is necessary and, under the closure hypothesis, sufficient that

$$T(\tau,\sigma)p - q + S(\tau,\sigma)h \in \mathcal{N}\big(Q^* - P^*T^*(\sigma,\tau)\big)^{\perp};$$

that is; for any $g \in B_0$ such that $Q^*g - P^*T^*(\sigma,\tau)g = 0$,

$$\begin{aligned} 0 &= \langle g, T(\tau,\sigma)p - q + S(\tau,\sigma)h \rangle \\ &= \langle T^*(\sigma,\tau)g, p \rangle - \langle g, q \rangle + \langle g, S(\tau,\sigma)h \rangle. \end{aligned} \tag{4.10}$$

The same argument as in the proof of Theorem 4.1 shows

$$\langle g, S(\tau,\sigma)h \rangle = -\int_\sigma^\tau y(\xi)h(\xi)\,d\xi,$$

where y is the solution of (4.8). Substitution into (4.10) yields the conditions (4.7)–(4.9) and this proves the theorem. □

Our next objective is to consider Equation (4.1) with $h \in \mathcal{P}_\omega$; that is, h is continuous and $h(t+\omega) = h(t)$ for all $t \geq 0$. Let $\tilde{\pi} : \mathcal{P}_\omega \to \mathcal{P}_\omega$ be a continuous projection of $\mathcal{P}_\omega$ onto the periodic solutions of the homogeneous equation

$$\dot{x}(t) = L(t)x_t \tag{4.11}$$

of period ω. For example, one can define $\tilde{\pi}$ in the following manner. Let $U = (\phi_1, \dots, \phi_d)$ be a basis for the ω-periodic solutions of Equation (4.11) and define

$$\tilde{\pi}h = U\Big[\int_0^\omega U^T(s)U(s)\,ds\Big]^{-1}\int_0^\omega U^T(s)h(s)\,ds. \tag{4.12}$$

We can now state the following result.

Corollary 4.1. *Suppose $h(t)$ and $L(t)$ in Equation* (4.1) *are periodic in t of period $\omega > 0$. The necessary and sufficient condition that there exist ω-periodic solutions of Equation* (4.1) *is*

$$\int_0^\omega y(\alpha)h(\alpha)\,d\alpha = 0 \tag{4.13}$$

for all ω-periodic solutions y of the formal adjoint problem (3.7)*. Furthermore, there is a continuous projection $J : \mathcal{P}_\omega \to \mathcal{P}_\omega$ such that the set of h in $\mathcal{P}_\omega$ satisfying Equation* (4.13) *is $(I - J)\mathcal{P}_\omega$ and there is a continuous linear operator $\mathcal{K} : (I - J)\mathcal{P}_\omega \to (I - \tilde{\pi})\mathcal{P}_\omega$ such that $\mathcal{K}h$ is a solution of Equation* (4.1) *for each $h \in (I - J)\mathcal{P}_\omega$.*

Proof. Take $\sigma = 0$, $\tau = \omega$, $V = C$, $\gamma = 0$, and $M = -N = I$ in Theorem 4.1. Since the solution operator is linear, Corollary 6.1 of Section 3.6 implies that $T(\tau,\sigma) = U(\tau,\sigma) + S(\tau - \sigma)$, where $U(\tau,\sigma)$ is completely continuous and the spectral radius of $S(t-\tau)$ is zero. Therefore, $\mathcal{R}\big(I - T(\omega,0)\big)$ is closed. The corollary now follows from Theorem 4.1.

Let $V = \operatorname{col}(\psi_1, \ldots, \psi_d)$ be a basis for the ω-periodic solutions of Equation (3.7). If we define $J : \mathcal{P}_\omega \to \mathcal{P}_\omega$

$$Jh = V\Big[\int_0^\omega V(s)V^T(s)\,ds\Big]^{-1}\int_0^\omega V(s)h(s)\,ds, \tag{4.14}$$

then $Jh = 0$ if and only if Condition (4.13) is satisfied.

If Condition (4.13) is satisfied, then we have seen there is an ω-periodic solution $x(h)$ of Equation (4.1). If $\mathcal{K}f = (I - \tilde{\pi})x(h)$, then all assertions of the theorem are true and the proof is complete. □

6.5 Stability and boundedness

In this section we consider the homogeneous linear equation (4.1); that is,

$$\dot{x}(t) = L(t)x_t \tag{5.1}$$

where L satisfies the conditions of Section 6.1. Also, $X(t,s)$ denotes the fundamental matrix solution of Equation (5.1) given in Theorem 1.1, and $T(t,\sigma)$ denotes the solution operator of Equation (5.1).

Lemma 5.1. *The following statements are equivalent:*

(i) *The solutions of Equation* (5.1) *are bounded.*

(ii) *The solution $x = 0$ of Equation* (5.1) *is stable.*

(iii) *There is a constant $c(\sigma)$ for each $\sigma \in \mathbb{R}$ such that $|T(t,\sigma)| \le c(\sigma)$, $t \ge \sigma$.*

(iv) *There is a constant $C(s)$ for each $s \in \mathbb{R}$ such that $|X(t,s)| \le C(s)$, $t \ge s$.*

Proof. If each solution of Equation (5.1) is bounded, then, for any $(\sigma,\phi) \in \mathbb{R} \times C$, there is a constant $c(\sigma,\phi)$ such that the bounded linear operator $T(t,\sigma)$ satisfies

$$|T(t,\sigma)\phi| \le c(\sigma,\phi) \qquad \text{for} \quad t \ge \sigma.$$

The principle of uniform boundedness implies that there is a constant $c(\sigma)$ such that $|T(t,\sigma)| \le c(\sigma)$ and so (i) implies (iii).

Since $X(t,s)$ satisfies Equation (1.11), it follows from Lemma 1.1 that

$$|X_t(\cdot\,,s)| \le M\exp\Big[\int_s^t m(u)\,du\Big], \qquad t \ge s. \tag{5.2}$$

Consequently, $|X_t(\cdot\,,s)| \le \gamma(s) = \exp \int_s^{s+r} m(u)\,du$, $s \le t \le s+r$. Since $X(t,s)$ satisfies Equation (5.1) for $t \ge s+r$, it follows that (iii) implies that $|X_t(\cdot\,,s)| \le c(s+r)\gamma(s)$ for $t \ge s$ and so (iii) implies (iv).

Theorem 1.2 shows that (iv) implies (ii). Obviously (ii) implies (i) and the proof is complete. □

One could have avoided the principle of uniform boundedness by proving that (i) implies (iv) implies (iii) implies (ii) implies (i). However, the application of the principle of uniform boundedness can be used to show the equivalence of properties similar to (i) and (iii) for general linear processes.

Lemma 5.2. *If there is a constant m_1 such that*

$$\int_t^{t+r} m(u)\,du \le m_1 \qquad \text{for} \quad t \in \mathbb{R}, \tag{5.3}$$

then the following statements are equivalent:

(i) *The solutions of Equation* (5.1) *are uniformly bounded.*

(ii) *The solution $x = 0$ of Equation* (5.1) *is uniformly stable.*

(iii) *There is a constant c such that for all $\sigma \in \mathbb{R}$, $|T(t,\sigma)| \le c$, $t \ge \sigma$.*

(iv) *There is a constant C such that for all $s \in \mathbb{R}$, $|X(t,s)| \le C$, $t \ge s$.*

Proof. If the solutions of Equation (5.1) are uniformly bounded, then there is a $c > 0$ such that for all $\sigma \in \mathbb{R}$,

$$|x_t(\sigma,\phi)| \le c \qquad \text{for} \quad t \ge \sigma,\ \phi \in C,\ |\phi| \le 1.$$

Therefore, $|T(t,\sigma)| \le c$, $t \ge 0$ and (i) implies (iii). Hypothesis (5.3) and Inequality (5.2) imply that for all $s \in \mathbb{R}$, $|X_t(\cdot\,,s)| \le \gamma$, $\gamma = \exp m_1$, for $s \le t \le s+r$. As in the proof of the previous lemma, for all $s \in \mathbb{R}$, this implies that $|X_t(\cdot\,,s)| \le c\gamma$ for $t \ge s$ if (iii) is satisfied. Thus, (iii) implies (iv). Using Theorem 1.1, one easily sees that (iv) implies (ii). It is obvious that (ii) implies (i) and the lemma is proved. □

Note that statements (i)–(iii) of Lemma 5.2 are equivalent without Hypothesis (5.3).

Lemma 5.3. *If Inequality* (5.3) *is satisfied, then the following statements are equivalent:*

(i) *The solution $x = 0$ of Equation* (5.1) *is uniformly asymptotically stable.*

(ii) *The solution $x = 0$ is exponentially asymptotically stable; that is, there are constants $c > 0$, $\alpha > 0$ such that for all $\sigma \in \mathbb{R}$,*

$$|T(t,\sigma)| \le ce^{-\alpha(t-\sigma)}, \qquad t \ge \sigma.$$

(iii) *There are constants* $C > 0$, $\alpha > 0$ *(α is the same as in (ii)) such that for all* $s \in \mathbb{R}$,

$$|X(t,s)| \leq Ce^{-\alpha(t-s)}, \qquad t \geq s.$$

Proof. If (i) is satisfied, then, for any $\eta > 0$, there is a $\tau = \tau(\eta) > 0$ such that for all $\sigma \in \mathbb{R}$, $|\phi| \leq 1$,

$$|x_t(\sigma,\phi)| < \eta \quad \text{for all} \quad t \geq \sigma + \tau.$$

Consequently, $|T(t,\sigma)| < \eta$ for $t \geq \sigma + \tau$. Choose $\eta < 1$ and let

$$\alpha = -\tau^{-1}\log \eta, \qquad c = c_0 \exp\, \alpha\tau$$

where c_0 is the constant in Lemma 5.2(iii) guaranteed by the uniform stability of the solution $x = 0$ of Equation (5.1). For any $t \geq \sigma$, there is an integer $n \geq 0$ such that $n\tau \leq t - \sigma < (n+1)\tau$. Thus,

$$\begin{aligned}
|T(t,\sigma)| &\leq |T(t,\sigma+n\tau)||T(\sigma+n\tau,\sigma)| \\
&\leq c_0|T(\sigma+n\tau,\sigma)| \\
&\leq c_0\eta|T(\sigma+(n-1)\tau,\sigma)| \\
&\leq c_0\eta^n = c_0\exp(-\alpha n\tau) \\
&= c\exp[-\alpha(n+1)\tau] \leq c\exp[-\alpha(t-\sigma)].
\end{aligned}$$

This proves (ii).

The proofs that (ii) implies (iii) and (iii) implies (i) are supplied in the same manner as in the previous two lemmas. □

Note that Properties (i) and (ii) of Lemma 5.3 are equivalent without Hypothesis (5.3).

6.6 Supplementary remarks

The theory of resolvents used in Section 6.1 is basic classical material and one can consult Grippenberg et al. [1] and Miller [1,2]. Another approach is to integrate Equation (1.1); see Section 9.1 for details. For the theory of vector measures, we refer to Diestel and Uhl [1]. The theory for vector-valued Riemann-Stieltjes integrals can be found in Hille and Phillips [1].

The formal adjoint equation has been used in functional differential equations since 1920. For a complete list of references on its origin and evolution, see Zverkin [3] and Hale [22].

The idea to study a functional differential equation as a Volterra integral equation has been used by Diekmann [1,2,3], Delfour and Manitius [1,2], Staffans [1,2] and Verduyn Lunel [1,2,3].

The interpretation of the adjoint evolutionary system as a backward evolutionary system on a Volterra integral equation was given by Diekmann [2] for autonomous equations. The approach in Section 6.3 has the advantage that it eliminates the use of the "true adjoint" (see Henry [3] and Hale [22]).

Now that we have a good version of the variation-of-constants formula (see Equation (2.12)), we can subject the linear equation to many types of perturbations and use techniques very similar to the ones in ordinary differential equations to obtain properties of solutions. We indicate briefly a few of the results here and suggest that the reader consult the references for more details.

6.6.1 Boundary-value problems

We keep the same notation as in Section 6.4. Also, we make the following hypothesis:

$$\mathcal{R}(M + NT(\tau,\sigma)) = V_E, \quad \mathcal{N}(M + NT(\tau,\sigma)) = C_U \tag{6.1}$$

where $U : C \to C$ and $E : V \to V$ are continuous projections with ranges C_U and V_E, respectively. In particular, this hypothesis implies that these sets are closed. Let $\mathcal{K} : V_E \to C_{I-U}$ be a bounded right inverse of $M + NT(\tau,\sigma)$. If $\gamma \in \mathcal{R}(M+NT(\tau,\sigma))$, then the function $T(t,\sigma)\mathcal{K}\gamma$, $\sigma \le t \le \tau$, is a solution of the linear equation

$$\dot{x}(t) = L(t)x_t \tag{6.2}$$

and the boundary conditions (4.2).

Now suppose that $f : \mathbb{R} \times C \to \mathbb{R}^n$ is a given function that is continuous in (t,ϕ) and continuously differentiable in ϕ, and consider the solutions of the RFDE

$$\dot{x}(t) = L(t)x_t + f(t,x_t), \tag{6.3}$$

which satisfy the boundary conditions (4.2). If we let $x_t(\sigma,\phi)$ be the solution to Equation (6.3) with $x_\sigma(\sigma,\phi) = \phi$ and define

$$W(t,\sigma,\phi) = \int_\sigma^t d[K(t,s)]f(s,x_s(\sigma,\phi)), \tag{6.4}$$

then $x_t(\sigma,\phi)$ is a solution of the boundary-value problem if and only if

$$[M + NT(\tau,\sigma)]\phi = \gamma - NW(\tau,\sigma,\phi). \tag{6.5}$$

Using the projection operators U and E, we observe that Equation (6.5) is equivalent to the system of equations

$$v = \mathcal{K}E[\gamma - NW(\tau,\sigma,u+v)], \tag{6.6a}$$

(6.6b) $$(I-E)[\gamma - NW(\tau,\sigma,u+v)] = 0,$$

where $\phi = u+v$, $u \in C_U = \mathcal{N}(M+NT(\tau,\sigma))$, $v \in C_{I-U}$.

We are in a position now to apply all of the existing methods for the solution of Equation (6.6)—fixed point theorems, the method of Liapunov-Schmidt, or more generally the method of alternative problems (see Cesari [1,2], Hale [7,21] for extensive references) and degree theory (see Mawhin [3], for example).

One of the most elementary, but at the same time very important, observations is that we can easily apply the method of Liapunov-Schmidt to Equations (6.6) if the function f and its derivative $D_\phi f$ with respect to ϕ are small.

Theorem 6.1. *If* (6.1) *is satisfied, then there are constants* $\epsilon > 0$, $\delta_1 > 0$, $\delta_2 > 0$ *such that for any function* f *satisfying* $|f(t,\phi)| < \epsilon$, $|D_\phi f(t,\phi)| < \epsilon$ *for* $t \in [\sigma, \tau]$, $|\phi| < \delta_1$, *there is a unique function* $v^*(u,\phi)$ *satisfying* (6.6a), *having norm* $< \delta_2$, *continuously differentiable in* u, f *and* $v^*(u,0) = \mathcal{K}E\gamma$. *Furthermore, the boundary-value problem* (6.3), (4.2) *has a solution* $x(\sigma,\phi)$ *with norm* $< \delta_1$ *if and only if* $\phi = u + v^*(u,f)$, *where* u *is a solution of the bifurcation equation*

(6.7) $$G(u,\phi) \overset{\text{def}}{=} (I-E)[\gamma - NW(\tau,\sigma,u+v^*(u,\phi))] = 0.$$

The proof of this result is an elementary application of the implicit function theorem.

A trivial consequence of Theorem 6.1 arises when $E = I$; that is, the range of $M + NT(\tau,\sigma)$ is the whole space.

Corollary 6.1. *If the conditions of Theorem* 6.1 *are satisfied and* $E = I$, *then, for any* $u \in C_U$, *there is a solution* $x(\sigma,\phi)$ *of the boundary-value problem* (6.3), (4.2) *with* $\phi = u + v^*(u,f)$ *provided that* $|\phi| < \delta_2$. *If, in addition,* $\mathcal{N}(M+NT(\tau,\sigma)) = \{0\}$ *(that is,* $U = 0$*), then this solution is unique.*

For results related to Corollary 6.1, see Fennell and Waltman [1] and Mosjagin [1].

The most interesting situation is when $E \neq I$ and $U \neq 0$. In this case, the solution of the original boundary-value problem is reduced to the discussion of the solutions of the *bifurcation equation* (6.7).

Let us translate these results to the case where the boundary conditions (4.2) correspond to periodicity conditions (see Perelló [2], Shimanov [5]). More specifically, suppose there is a positive constant ω such that $L(t+\omega) = L(t)$, $f(t+\omega,\phi) = f(t,\phi)$ for all (t,ϕ) and let us determine periodic solutions of Equation (6.3) of period ω. In our previous notation, this is the same as taking $\sigma = 0$, $\tau = \omega$, $V = C$, $M = -N = I$, $\gamma = 0$. The space C_U is the space of ω-periodic solutions of the linear equation (6.2) and the space C_{I-E} is the space of ω-periodic solutions of the formal adjoint equation (3.7). The spaces C_U and C_{I-E} have the same dimension d and $d < \infty$. If we choose a basis Φ for C_U and a basis Ψ for C_{I-E}, and let $u = \Phi a$, where $a \in \mathbb{R}^d$, then there is an n-vector function $B(a, f)$ such that

$$G(\Phi a, f) = \Psi B(a, f), \tag{6.8}$$

and the bifurcation equation (6.7) is equivalent to the equation

$$B(a, f) = 0. \tag{6.9}$$

Therefore, the problem of the existence of an ω-periodic solution of Equation (6.3) is reduced to the discussion of solutions of equation (6.9) in $\mathbb{R}^d$.

Let us state another result of a more global nature.

Theorem 6.2. *If* (6.1) *is satisfied with* $E = I$ *and* $|f(t,\phi)|/|\phi| \to 0$ *as* $|\phi| \to \infty$ *uniformly for* $t \in [\sigma, \tau]$, *then there exists at least one solution of the boundary-value problem* (6.3), (4.2).

We only give an outline of the proof. Since $E = I$, the problem is to find a solution of Equation (6.6a); that is, a solution of the equation $v + S_u(v) = \mathcal{K}\gamma$, where $S_u(v) = \mathcal{K}NW(\tau, \sigma, u + v)$. It is possible to show that the function $S_u(\cdot)$ is completely continuous and $|S_u(v)|/|v| \to 0$ as $|v| \to \infty$. Now, we apply a fixed point theorem of Granas [1].

For a complete proof of Theorem 6.2, see Hale [22]. Theorem 6.2 is a generalization (since it is not assumed that $\tau - \sigma \geq r$) of a result of Waltman and Wong [1].

Related results on boundary-value problems are contained in Kwapisz [1] and Nosov [3,4]. For other results on the existence of periodic solutions, see Burton [3], Perelló [3], Ziegler [1]. For periodic solutions of parabolic equations with delays, see Biroli [1], and Comincioli [1]. Boundary-value problems for special equations is an entire subject in itself and has an extensive literature. See Norkin [1], Nosov [1], Grimm and Schmitt [1,2], Medzitov [1], Sentebova [1], Kobyakov [1], Kovač and Savčenko [1] and de Nevers and Schmitt [1]. For general functional boundary conditions, see Henry [3], Myjak [1], and Kwapisz [2].

6.6.2 Exponential dichotomies, bounded and almost periodic perturbations

In Corollary 6.1 and Theorem 6.2, we have seen that if the range of the solution operator of a homogeneous linear boundary-value problem is the whole space, then the existence of a solution of a perturbed problem can be reduced to the discussion of the existence of a fixed point of some operator. It is possible to extend this idea to more general situations; for example, to the existence of almost periodic solutions.

We say that the linear system (6.2) admits an *exponential dichotomy on* $\mathbb{R}$ (see Henry [5], Hale [28]) if the solution operator $T(t,s)$ satisfies the following property: there exist positive constants K, α, and projection operators $P(s) : C \to C$, $s \in \mathbb{R}$ such that if $Q(s) = I - P(s)$, then

(i) $T(t,s)P(s) = P(t)T(t,s), \quad t \geq s.$

(ii) The restriction $T(t,s)\mathcal{R}(Q(s))$, $t \geq s$, is an isomorphism of $\mathcal{R}(Q(s))$ onto $\mathcal{R}(Q(t))$ and we define $T(s,t)$ as the inverse mapping.

(iii) $|T(t,s)P(s)| \leq Ke^{-\alpha(t-s)}, \quad t \geq s.$

(iv) $|T(t,s)Q(s)| \leq Ke^{-\alpha(t-s)}, \quad s \geq t.$

The operator $P(t)$ is called the *projection operator of the dichotomy.* The solution of System (6.2) with initial value at time s in the range of $P(s)$ (resp. $Q(s)$) approach zero exponentially as $t \to \infty$ (resp. $t \to -\infty$).

Let $BC(\mathbb{R}, Y)$ be the bounded continuous functions from $\mathbb{R}$ to a real Banach space Y. Suppose that $L \in BC(\mathbb{R}, \mathcal{L}(C, \mathbb{R}^n))$ and that System (6.2) admits an exponential dichotomy. For any $h \in BC(\mathbb{R}, \mathbb{R}^n)$, it follows that the equation

$$\dot{x}(t) = L(t)x_t + h(t) \tag{6.10}$$

has a unique solution $\mathcal{K}h \in BC(\mathbb{R}, C)$, and it is given by

$$\mathcal{K}h(t) = \int_{-\infty}^{t} d[P(s)K(t,s)]h(s) - \int_{t}^{\infty} d[Q(s)K(t,s)]h(s). \tag{6.11}$$

If $L(\cdot)$, h are in $\mathcal{AP}$ ($\mathcal{AP}$ is the class of almost periodic functions), then it also can be shown that $\mathcal{K}h$ is $\mathcal{AP}$ and the frequency module is the union of the frequency modules of $L(\cdot)$ and h.

If $f(\cdot, \phi) \in BC(\mathbb{R}, \mathbb{R}^n)$, then the equation

$$\dot{x}(t) = L(t)x_t + f(t, x_t) \tag{6.12}$$

has a solution $t \mapsto x_t \in BC(\mathbb{R}, C)$ if and only if

$$x_t = \mathcal{K}f(t, x_t), \quad t \in \mathbb{R}. \tag{6.13}$$

Using the contraction mapping principle, we can obtain very easily an extension of Corollary 6.1 for the existence of solutions of Equation (6.12) that are bounded on $\mathbb{R}$. If the perturbation function $f(\cdot, \phi)$ is in $\mathcal{AP}$ uniformly with respect to ϕ in bounded sets, then the bounded solution is $\mathcal{AP}$. Special cases of such perturbation results have been given by Halanay [1] and Konovalov [1].

Results on $\mathcal{AP}$-solutions without assumptions of smallness of the perturbed vector field are very difficult to obtain. For some results and references, see Fink [1] for ODE, Yoshizawa [2] for ODE and RFDE and Hino, Murakami, and Naito [1] for RFDE.

If $L \in BC(\mathbb{R}, \mathcal{L}(C, \mathbb{R}^n))$ and, for each $h \in BC(\mathbb{R}, \mathbb{R}^n)$, there is at least one solution of the Equation (6.10) in $BC(\mathbb{R}, \mathbb{R}^n)$, then the homogeneous equation System (6.2) must admit an exponential dichotomy (see Burd and Kolesov [1], and Pecelli [1]). If there is an exponential dichotomy for (6.2), then there is the solution $\mathcal{K}h \in BC(\mathbb{R}, C)$ if $h \in BC(\mathbb{R}, \mathbb{R}^n)$. This suggests the following more general concepts. If X, Y are real Banach spaces, we say that (X, Y) is *admissible for* Equation (6.10) if, for every $h \in X$, there is at least one solution of Equation (6.10) in Y. Coffman and Schäffer [2], Schäffer [1], Pecelli [1] and Corduneanu [1] have discussed admissible pairs for RFDE.

6.6.3 Asymptotic behavior

The variation-of-constants formula in the function space C also is very useful for the study of the asymptotic behavior of solutions of a perturbed system

$$\dot{x}(t) = Lx_t + M(t)x_t \tag{6.14}$$

where $L : C \to C$ is a continuous linear operator, $M(t) : C \to C$ is a continuous linear operator that is "small" at ∞. Since the operator L is independent of t, we would expect to be able to relate the solutions of Equation (6.14) to the autonomous equation

$$\dot{x}(t) = Lx_t. \tag{6.15}$$

More specifically, if we have an eigenvalue μ of Equation (6.15), under what conditions on $M(t)$ will there be a solution of Equation (6.14) that has an asymptotic behavior similar to $e^{\mu t}$? Using the variation-of-constants formula in C, Hale [6,22] has obtained very precise results, which are more general than the ones obtained by Bellman and Cooke [2,3] using the variation-of-constants formula in $\mathbb{R}^n$. Kato [2] has given similar results. For other results on asymptotic behavior, see Kato [3] and Onuchic [3]. For a differential difference equation,

$$\dot{x}(t) = Ax(t) + Bx(t - r(t)) + Cx(t - s(t)), \tag{6.16}$$

it also is of interest to determine the asymptotic behavior of the solutions if $r(t), s(t)$ are "close" to constants for t large. In this case, the space C is too large in general. Cooke [2] has shown that the theory of Chapter 7 and the variation-of-constants formula in the function space $W^{1,\infty}$ can be effectively used to discuss this more general situation (see also Bellman and Cooke [4]). The space $W^{1,\infty}$ or $W^{1,p}$ for some p, $1 < p < \infty$, has also been used for the existence of periodic solutions (see Ginzburg [2], Ruiz-Clayessen [1], and Stephan [1,2]).

Perturbation problems that involve a small parameter in the highest derivative have received much attention and probably should be reexamined in the light of the preceding remarks. For some of the literature, see the nine volumes of *Trudy Sem. Teorii Differentialnije Uravneniyija c Otklonyayushimcya Argumentom*, published by Univ. Patrisa Lumumba and the papers of Cooke [3], Cooke and Meyer [1], Habets [1], and Magalhães [1,2,3,4]. A special singular perturbation problem is considered in some detail in Chapter 12.

7
Linear autonomous equations

A linear autonomous RFDE has the form

$$(1) \qquad \dot{x}(t) = Lx_t$$

where L is a continuous linear mapping from C into $\mathbb{R}^n$. This hypothesis implies that there exists an $n \times n$ matrix $\eta(\theta)$, $-r \leq \theta \leq 0$, whose elements are of bounded variation, normalized so that η is continuous from the left on $(-r, 0)$ and $\eta(0) = 0$, such that,

$$(2) \qquad L\phi = \int_{-r}^{0} d[\eta(\theta)]\phi(\theta), \qquad \phi \in C.$$

The goal is to understand the geometric behavior of the solutions of Equation (1) when they are interpreted in C.

7.1 Strongly continuous semigroups

Let $\mathcal{B}$ be a real Banach space. A strongly continuous semigroup of linear operators, in short a $\mathcal{C}_0$-semigroup, is a one-parameter family $T(t) : \mathcal{B} \to \mathcal{B}$, $t \geq 0$, of bounded linear operators that satisfy the properties

(i) $T(0) = I$;

(ii) $T(t_1 + t_2) = T(t_1)T(t_2), \qquad t_1, t_2 \geq 0$;

(iii) $\lim_{t\downarrow 0} \|T(t)\phi - \phi\| = 0, \qquad \phi \in \mathcal{B}$.

To every $\mathcal{C}_0$-semigroup $T(t)$, we can associate an *infinitesimal generator* $A : \mathcal{D}(A) \to \mathcal{B}$ defined by

$$A\phi = \lim_{t\downarrow 0} \frac{1}{t}\big[T(t)\phi - \phi\big], \qquad \phi \in \mathcal{D}(A),$$

that is, for all $\phi \in \mathcal{B}$ for which the limit exists in the norm topology of $\mathcal{B}$. The following lemma is standard.

Lemma 1.1. *If $T(t)$ is a $\mathcal{C}_0$-semigroup on $\mathcal{B}$, then*

(i) *for every* ϕ *in* $\mathcal{B}$, $t \mapsto T(t)\phi$ *is a continuous mapping from* $\mathbb{R}_+$ *into* $\mathcal{B}$;

(ii) A *is a closed densely defined operator;*

(iii) *for every* $\phi \in \mathcal{D}(A)$, $t \mapsto T(t)\phi$ *satisfies the differential equation*

$$\frac{d}{dt}T(t)\phi = AT(t)\phi = T(t)A\phi.$$

Next, we investigate the abstract properties of the solution operator of Equation (1). Let ϕ be a given function in C. If $x(\,\cdot\,;\phi)$ is the unique solution of Equation (1) with initial function ϕ at zero, then the solution operator $T(t) : C \to C$ is defined by the relation

$$x_t(\,\cdot\,;\phi) = T(t)\phi. \tag{1.1}$$

Lemma 1.2. *The solution operator* $T(t)$, $t \geq 0$, *defined by Relation* (1.1), *is a* $\mathcal{C}_0$-*semigroup with infinitesimal generator*

$$\begin{aligned} \mathcal{D}(A) &= \{\phi \in C : \ \frac{d\phi}{d\theta} \in C,\ \frac{d\phi}{d\theta}(0) = \int_{-r}^{0} d[\eta(\theta)]\phi(\theta)\}, \\ A\phi &= \frac{d\phi}{d\theta}. \end{aligned} \tag{1.2}$$

Furthermore, $T(t)$ *is completely continuous for* $t \geq r$; *that is,* $T(t)$, $t \geq r$, *is continuous and maps bounded sets into relatively compact sets.*

Proof. From the uniqueness of solutions of Equation (1), it is obvious that $T(t)$ is a linear transformation that satisfies the semigroup property. By definition, $T(0) = I$. Since $L : C \to \mathbb{R}^n$ is continuous and linear, it follows that there is a constant γ such that $|L\phi| \leq \gamma|\phi|$ for all ϕ in C. From the definition of $T(t)$, we have, for any fixed $t \geq 0$ and $-r \leq \theta \leq 0$,

$$T(t)\phi(\theta) = \begin{cases} \phi(t+\theta), & t+\theta \leq 0, \\ \phi(0) + \int_0^{t+\theta} LT(s)\phi\, ds, & t+\theta > 0. \end{cases} \tag{1.3}$$

It follows that $|T(t)\phi| \leq |\phi| + \gamma\int_0^t |T(s)\phi|\, ds$. The inequality in Lemma 3.1 of Section 1.3 then implies that

$$|T(t)\phi| \leq e^{\gamma t}|\phi|, \qquad t \geq 0, \quad \phi \in C \tag{1.4}$$

and thus, $T(t)$ is bounded. From (1.3) it is readily seen that $T(t)$ is strongly continuous at zero. So $T(t)$ is a $\mathcal{C}_0$-semigroup. Let $R > 0$. If $S = \{\phi \in C : |\phi| < R\}$, then for any ψ in $T(t)S$, $t \geq r$, Relation (1.4) implies $\|\psi\| \leq e^{\gamma t}R$, and Equation (1) implies $|\dot{\psi}| \leq \gamma e^{\gamma t}R$. Since these functions are uniformly bounded with a uniform Lipschitz constant, $T(t)S$, $t \geq r$, belongs to a compact subset of C. To finish the proof, we compute the infinitesimal generator. From Lemma 1.1 and the strong continuity of $T(t)$, it follows that, for every ϕ in $\mathcal{D}(A)$, we have

$$A\phi = \frac{d\phi}{d\theta} \qquad \text{and} \quad \mathcal{D}(A) \subseteq \{\phi \in C \mid \frac{d\phi}{d\theta} \in C\}.$$

On the other hand, if ϕ is continuously differentiable, the limit

$$\lim_{t\downarrow 0} \frac{1}{t}\left[T(t)\phi(\theta) - \phi(\theta)\right]$$

exists for $\theta > 0$. For $\theta = 0$, we find

$$\frac{d\phi}{d\theta}(0) = \lim_{t\downarrow 0} \frac{1}{t}\int_0^t LT(s)\, ds = LT(0)\phi = L\phi.$$

This proves the lemma. □

We remark that more information on the properties of the semigroup $T(t)$ for $t \in [0, r]$ is contained in Theorem 6.1 of Section 3.6.

From the adjoint theory developed in Section 6.3, we know that the adjoint semigroup $T^*(s) = T(s)^*$ corresponds to the Volterra equation

$$y(s) + \int_s^0 y(\tau)\eta(s-\tau)\, d\tau = g(s), \qquad s \le 0, \tag{1.5}$$

with $g \in B_0$. Recall that B_0 denotes the Banach space of row-valued functions $\psi : (-\infty, 0] \to \mathbb{R}^{n*}$ that are constant on $(-\infty, -r]$, of bounded variation on $[-r, 0]$, continuous from the left on $(-r, 0)$, and vanishing at zero with norm $\mathrm{Var}_{[-r,0]}\, \psi$.

The adjoint semigroup $T^*(s)$ has the following explicit representation

$$T^*(s)g(\theta) = \begin{cases} g(\theta - s) \\ \quad - \int_{-s}^0 \int_\tau^0 d[g(\alpha)]X(\alpha-\tau)\eta(\theta-\tau)\, d\tau & \text{if } \theta < 0; \\ 0 & \text{if } \theta = 0. \end{cases} \tag{1.6}$$

Here X denotes the fundamental matrix solution to System (1). So it is clear that $T^*(t)$ is not a C_0-continuous semigroup. Later, in the decomposition theory, the spectral properties of both A and its adjoint A^* have to be studied. The adjoint A^* of A is defined by $f \in \mathcal{D}(A^*)$ if and only if $g \in B_0$ exists such that

$$\langle f, A\phi \rangle = \langle g, \phi \rangle$$

for all $\phi \in \mathcal{D}(A)$ and in that case $A^* f = g$.

Lemma 1.3. *The adjoint operator $A^* : \mathcal{D}(A^*) \to B_0$ is given by*

$$\begin{aligned} &\mathcal{D}(A^*) = \{f \in B_0 : \frac{df}{d\theta} \in B_0\}, \\ &A^* f(\theta) = f(0-)\eta(\theta) - \frac{df}{d\theta}(\theta), \qquad -r \le \theta \le 0. \end{aligned} \tag{1.7}$$

Proof. It is obvious that given the action of A^*, the domain of definition of A^* cannot be larger than the subspace $\{f \in B_0 : df/d\theta \in B_0\}$. So assume that $g \in B_0$ with $g(-r) = 0$, and

$$f(\theta) = f(0-) - \int_\theta^0 g(s)\,ds, \qquad \text{for} \quad \theta < 0.$$

From Relation (1.2)

$$\begin{aligned}
\langle f, A\phi \rangle &= \int_{-r}^0 A\phi(\theta)\,df(\theta) \\
&= L\phi f(0-) + \int_{-r}^0 \frac{d\phi}{d\theta}(\theta) g(\theta)\,d\theta \\
&= \int_{-r}^0 d\eta(\theta)\phi(\theta) f(0-) - \int_{-r}^0 \phi(\theta)\,dg(\theta) \\
&= \int_{-r}^0 \phi(\theta)\,d[f(0-)\eta(\theta)] - \int_{-r}^0 \phi(\theta)\,dg(\theta) \\
&= \langle A^* f, \phi \rangle.
\end{aligned}$$

This proves the lemma. □

In this chapter, we shall also study the semigroup associated with the transposed equation. Define $C' = C([0,r], \mathbb{R}^{n*})$. For each $s \in [0,\infty)$ let y^s designate the element in C' defined by $y^s(\xi) = y(-s+\xi)$, $0 \le \xi \le r$. The *transpose* of System (1.1) is defined to be

$$\begin{aligned}
\dot{y}(s) &= \int_{-r}^0 y(s-\theta)\,d[\eta(\theta)], \qquad s \le 0, \\
y^0 &= \psi, \qquad \psi \in C'.
\end{aligned} \tag{1.8}$$

Let y be a solution of Equation (1.8) on an interval $(-\infty, r]$. Following the ideas in this section, we can associate a C_0-semigroup $T^T(s)$ with Equation (1.8), the *transposed* semigroup, defined by

$$T^T(s)\psi = y^s(\,\cdot\,;\psi) \tag{1.9}$$

where y is the solution of Equation (1.8). In precisely the same way as we proved Lemma 1.2, one can prove the following result.

Lemma 1.4. *The solution operator $T^T(s)$, $s \ge 0$, defined by Relation* (1.9) *is a C_0-semigroup with infinitesimal generator*

$$\begin{aligned}
\mathcal{D}(A^T) &= \{\psi \in C' : \frac{d\psi}{d\xi} \in C',\ \frac{d\psi}{d\xi}(0) = -\int_{-r}^0 \psi(-\theta)\,d[\eta(\theta)]\}, \\
A^T\psi &= -\frac{d\psi}{d\xi}.
\end{aligned} \tag{1.10}$$

Furthermore $T^T(s)$ is completely continuous (compact) for $s \geq r$.

By integration, Equation (1.8) can be written as a Volterra integral equation

$$y(s) + \int_s^0 y(\tau)\eta(s-\tau)\,d\tau = F^T\psi(s), \qquad s \leq 0,$$

where $F^T : C' \to B_0$ is given by

$$(F^T\psi)(s) = \psi(0) - \int_s^0 \int_{-r}^{\alpha} \psi(\alpha - \theta)\,d[\eta(\theta)]\,d\alpha. \tag{1.11}$$

Since F^T cannot be an invertible mapping from C' onto B_0, the transposed equation (1.8) is not the same as the formal adjoint equation (1.5), but the equations are closely related. From the definitions, it is not difficult to verify that

$$F^T T^T(s)\psi = T^*(s)F^T\psi. \tag{1.12}$$

7.2 Spectrum of the generator—Decomposition of C

It is our objective to determine the nature of $\sigma(T(t))$ and $\sigma(A)$ for the solution operator arising from Equation (1). To introduce the spectra, we have to work with complex Banach spaces. Let $\mathcal{B}_{\mathbb{C}}$ be a complexified real Banach space, and let $B_{\mathbb{C}} : \mathcal{D}(B_{\mathbb{C}}) \to \mathcal{B}_{\mathbb{C}}$ be a complexified linear operator. By this we mean that there exist a real Banach space $\mathcal{B}$ and an operator $B : \mathcal{D}(B) \to \mathcal{B}$ such that $\mathcal{B}_{\mathbb{C}} = \mathcal{B} \oplus i\mathcal{B}$ and $B_{\mathbb{C}}(b_1 + ib_2) = Bb_1 + iBb_2$ for $b_1, b_2 \in \mathcal{D}(B)$. For $\mathcal{B}_{\mathbb{C}}$, we can define the complex conjugate, denoted by an overbar, and given by $\overline{b_1 + ib_2} = b_1 - ib_2$. Whenever there is no confusion, we shall write, by abuse of notation, B for $B_{\mathbb{C}}$ and $\mathcal{B}$ for $\mathcal{B}_{\mathbb{C}}$.

The resolvent set $\rho(B)$ of B is the set of values in the complex plane for which the operator $\lambda I - B$ has a bounded inverse with domain dense in $\mathcal{B}$. This inverse will be denoted by $R(\lambda, B)$, $\lambda \in \rho(B)$, and is called the *resolvent* of B. The complement of $\rho(B)$ in the complex plane is called the spectrum of B and is denoted by $\sigma(B)$. The spectrum of an operator may consist of three different types of points, namely, the *residual spectrum* $\mathrm{R}\sigma(B)$, the *continuous spectrum* $\mathrm{C}\sigma(B)$, and the *point spectrum* $\mathrm{P}\sigma(B)$. The residual spectrum consists of those λ in $\sigma(B)$ for which $R(\lambda, B)$ exists but $\mathcal{D}(R(\lambda, B))$ is not dense in $\mathcal{B}$. The continuous spectrum consists of those λ in $\sigma(B)$ for which $\lambda I - B$ has an unbounded inverse with dense domain. The point spectrum consists of those λ in $\sigma(B)$ for which $\lambda I - B$ does not have an inverse.

For $\lambda \in \mathrm{P}\sigma(B)$, the *generalized eigenspace* of λ will be denoted by $\mathcal{M}_\lambda(B)$ and is defined to be the smallest subspace of $\mathcal{B}$ containing all elements of $\mathcal{B}$ that belong to $\mathcal{N}\big((\lambda I - B)^k\big)$, $k = 1, 2, \ldots$. The dimension of $\mathcal{M}_\lambda(B)$ is called the *algebraic multiplicity* of λ and the smallest integer k such that $\mathcal{N}\big((\lambda I - B)^k\big) = \mathcal{N}\big((\lambda I - B)^{k+1}\big)$ is called the *ascent* of λ. Points of the point spectrum of A are called *eigenvalues*. Isolated points of the point spectrum of A with finite-dimensional generalized eigenspace are called *eigenvalues of finite type* or *normal* eigenvalues.

Lemma 2.1. *If A is defined by Equation (1.2), then $\sigma(A) = \mathrm{P}\sigma(A)$ and λ is in $\sigma(A)$ if and only if λ satisfies the characteristic equation*

$$\det \Delta(\lambda) = 0, \qquad \Delta(\lambda) = \lambda I - \int_{-r}^{0} e^{\lambda\theta}\, d\eta(\theta). \tag{2.1}$$

For any λ in $\sigma(A)$, the generalized eigenspace $\mathcal{M}_\lambda(A)$ is finite dimensional and there is an integer k such that $\mathcal{M}_\lambda(A) = \mathcal{N}\big((\lambda I - A)^k\big)$ and we have the direct sum decomposition:

$$C = \mathcal{N}\big((\lambda I - A)^k\big) \oplus \mathcal{R}\big((\lambda I - A)^k\big).$$

Proof. Let $\psi = R(\lambda, A)\phi$. From Relation (1.2), it follows that $(\lambda I - A)\psi = \phi$ if and only if ψ satisfies the differential equation

$$\lambda\psi - \dot{\psi} = \phi$$

with boundary condition $\lambda\psi(0) - L\psi = 0$. Define

$$\psi(\theta) = e^{\lambda\theta}\psi(0) + \int_{\theta}^{0} e^{\lambda(\theta - s)}\phi(s)ds, \qquad -r \le \theta \le 0.$$

Then ψ satisfies the differential equation and the boundary condition becomes

$$\Delta(\lambda)\psi(0) = c + \int_{-r}^{0} d[\eta(\theta)] \int_{0}^{-\theta} e^{-\lambda s}\phi(s + \theta)\, ds.$$

If $\det \Delta(\lambda) \neq 0$ this equation can be solved. So

$$\{\lambda \in \mathbb{C} : \det \Delta(\lambda) \neq 0\} \subset \rho(A).$$

To prove the reverse inclusion, choose $\lambda \in \mathbb{C}$ such that $\det \Delta(\lambda) = 0$ and define

$$\phi(t) = e^{\lambda t}\phi^0 \qquad \text{for} \quad -h \le t \le 0,$$

where $\phi^0 \neq 0$ is an element of the null space of $\Delta(\lambda)$. Then

$$A\phi = \dot{\phi} = \lambda\phi.$$

Therefore, we conclude that $\lambda \in \mathrm{P}\sigma(A)$. This proves the first part of the lemma.

The characteristic function $\det \Delta$ is an entire function and therefore has zeros of finite order. So the representation for the resolvent implies that the resolvent $R(z, A)$ has a pole of order k, $k \leq k_0$ at λ_0 if λ_0 is a zero of $\det \Delta$ of order k_0. Since A is a closed operator, it follows from Hille and Phillips [1, p.306] that $\mathcal{M}_\lambda(A)$ is finite dimensional and has the properties stated in the lemma. □

From Lemma 2.1, we know that λ in $\sigma(A)$ implies that $\mathcal{M}_\lambda$ is finite dimensional and $\mathcal{M}_\lambda(A) = \mathcal{N}\big((\lambda I - A)^k\big)$ for some integer k. The subspace $\mathcal{M}_\lambda(A)$ satisfies $A\mathcal{M}_\lambda(A) \subseteq \mathcal{M}_\lambda(A)$ since A commutes with $R(\lambda, A)$. Let $\mathcal{M}_\lambda(A)$ have dimension d, let $\phi_1^\lambda, \dots, \phi_d^\lambda$ be a basis for $\mathcal{M}_\lambda(A)$ and let $\Phi_\lambda = \{\phi_1^\lambda, \dots, \phi_d^\lambda\}$. Since $A\mathcal{M}_\lambda(A) \subseteq \mathcal{M}_\lambda(A)$, there is a $d \times d$ constant matrix B_λ such that $A\Phi_\lambda = \Phi_\lambda B_\lambda$.

Lemma 2.2. *The only eigenvalue of B_λ is λ.*

Proof. For any d-vector a, $(\lambda I - A)^k \Phi_\lambda a = 0$ and so $\Phi_\lambda(\lambda I - B_\lambda)^k a = 0$ for all d-vectors a. Therefore, $(\lambda I - B_\lambda)^k a = 0$ for all d-vectors a. But this implies $(\lambda I - B_\lambda)^k = 0$ and the result follows. □

From the definition of A in Expression (1.2), the relation $A\Phi_\lambda = \Phi_\lambda B_\lambda$ implies that

$$\Phi_\lambda(\theta) = \Phi_\lambda(0)e^{B_\lambda \theta}, \qquad -r \leq \theta \leq 0.$$

From Lemma 1.1(iii), one also obtains

$$T(t)\Phi_\lambda = \Phi_\lambda e^{B_\lambda t}$$

for $t \geq 0$, which together with the expression for Φ_λ, implies that

$$[T(t)\Phi_\lambda](\theta) = \Phi_\lambda(0)e^{B_\lambda(t+\theta)}, \qquad -r \leq \theta \leq 0.$$

This relation permits one to define $T(t)$ on $\mathcal{M}_\lambda(A)$ for all values of t in $(-\infty, \infty)$. Therefore, on the generalized eigenspace of an eigenvalue of Equation (1), that is, an element of $\sigma(A)$, the differential equation has the same structure as an ordinary differential equation.

From Lemma 1.1, we also know that $T(t)A\phi = AT(t)\phi$ for all ϕ in $\mathcal{D}(A)$. This implies that $\mathcal{R}\big((\lambda I - A)^k\big)$ is also invariant under $T(t)$. By a repeated application of the preceding process we obtain

Theorem 2.1. *Suppose Λ is a finite set $\{\lambda_1, \dots, \lambda_p\}$ of eigenvalues of Equation (1) and let $\Phi_\Lambda = \{\Phi_{\lambda_1}, \dots, \Phi_{\lambda_p}\}$, $B_\Lambda = diag(B_{\lambda_1}, \dots, B_{\lambda_p})$, where Φ_{λ_j} is a basis for the generalized eigenspace of λ_j and B_{λ_j} is the matrix defined by $A\Phi_{\lambda_j} = \Phi_{\lambda_j} B_{\lambda_j}$, $j = 1, 2, \dots, p$. Then the only eigenvalue of B_{λ_j} is λ_j and, for any vector a of the same dimension as Φ_Λ, the solution $T(t)\Phi_\Lambda a$ with initial value $\Phi_\Lambda a$ at $t = 0$ may be defined on $(-\infty, \infty)$ by the relation*

$$
\begin{aligned}
&T(t)\Phi_\Lambda a = \Phi_\Lambda e^{B_\Lambda t}a, \\
&\Phi_\Lambda(\theta) = \Phi_\Lambda(0)e^{B_\Lambda \theta}, \qquad -r \le \theta \le 0.
\end{aligned} \tag{2.2}
$$

Furthermore, there exists a subspace Q_Λ of C such that $T(t)Q_\Lambda \subseteq Q_\Lambda$ for all $t \ge 0$ and

$$C = P_\Lambda \oplus Q_\Lambda,$$

where $P_\Lambda = \{\phi \in C \mid \phi = \Phi_\Lambda a,$ for some vector $a\}$.

Theorem 2.1 gives a very clear picture of the behavior of the solutions of Equation (1). In fact, on generalized eigenspaces, Equation (1) behaves essentially as an ordinary differential equation and the decomposition of C into two subspaces invariant under A and $T(t)$ tells us that we can separate out the behavior on the eigenspaces from the other type of behavior. The decomposition of C allows one to introduce a direct sum decomposition of C, which plays the same role as the Jordan canonical form in ordinary differential equations. As we know in ordinary differential equations, this is very important for studying systems that are close to linear.

The decomposition of C will be complete provided that we can explicitly characterize the projection operator defined by this decomposition. We shall also need bounds for $T(t)$ on the complementary subspace Q_Λ in order to apply the results to perturbed linear systems. In the next two sections, we shall address these questions for a more general class of generators than given by Relation (1.2). The reason for this extension becomes clear when we study neutral and periodic functional differential equations.

7.3 Characteristic matrices and equivalence

In the last section we saw that the spectrum of the infinitesimal generator A defined by (1.2) is precisely given by the roots of the characteristic equation

$$\det \Delta(z) = 0, \qquad \Delta(z) = zI - \int_{-r}^{0} e^{\lambda\theta}\, d\eta(\theta).$$

We call $\Delta(z)$ the *characteristic matrix* for A. If λ_0 is an eigenvalue of A, then λ_0 is a *characteristic value* of $\Delta(z)$; that is, the matrix $\Delta(\lambda_0)$ is singular. In this section, we shall prove that the null space of $\Delta(\lambda_0)$ describes the complete geometric structure of the generalized eigenspace $\mathcal{M}_\lambda(A)$.

A naive approach to prove such a theorem would be to try to find an equivalence

$$\Delta(z) = F(z)(zI - A)E(z), \quad z \in \mathbb{C},$$

where $E(z)$ and $F(z)$ are invertible operators that depend analytically on the parameter z. But, of course, this would imply that C is finite dimensional, which is a contradiction. So it is obvious that we have to associate with $\Delta(z)$ a simple operator acting on an infinite-dimensional space.

Set $\widehat{C} = \mathbb{R}^n \times C$ endowed with the product norm topology

$$|(c, \phi)| = \left(|c|^2 + |\phi|^2\right)^{\frac{1}{2}}.$$

Define the embedding $j : C \to \widehat{C}$ by $\phi \mapsto (\phi(0), \phi)$. The solution operator $T(t) : C \to C$ induces a solution operator on $jC \subset \widehat{C}$ by

$$jT(t)\phi = T(t)j\phi.$$

It is not possible to extend the solution operator on jC to a solution operator on $\widehat{C}$ since $x_t(\,\cdot\,; c, \phi)$ has a discontinuity at $-t$, $0 \le t \le r$. Indeed, the operator $\widehat{A} : \mathcal{D}(\widehat{A}) \to \widehat{C}$

$$\begin{aligned} \mathcal{D}(\widehat{A}) &= \{(c, \phi) \in \widehat{C} : \frac{d\phi}{d\theta} \in C,\ c = \phi(0)\}, \\ \widehat{A}(c, \phi) &= \left(L\phi, \frac{d\phi}{d\theta}\right) \end{aligned} \tag{3.1}$$

is a closed unbounded operator, but the domain of $\widehat{A}$ is not dense in $\widehat{C}$, and hence, $\widehat{A}$ is not the generator of a semigroup. The closure of the domain is precisely given by jC and the part of $\widehat{A}$ in jC is given by Relation (1.2).

For the complexified operators, it follows that the spectrum of A and $\widehat{A}$ are the same and $j\mathcal{M}_\lambda(A) = \mathcal{M}_\lambda(\widehat{A})$, where $jC \to \widehat{C}$ denotes the embedding $\phi \mapsto (\phi(0), \phi)$. So the spectral analysis of A and $\widehat{A}$ are one and the same.

We shall prove that the operator $\widehat{A}$ is equivalent to

$$\begin{pmatrix} \Delta(z) & 0 \\ 0 & I \end{pmatrix} : \widehat{C} \to \widehat{C},$$

where I denotes the identity on C. This result solely depends on the structure of the operator $\widehat{A}$. For this reason we present the results in this section for a more general class of operators that includes the infinitesimal generators associated with neutral functional differential equations.

There is a general scheme to construct characteristic matrices for a rather general class of unbounded operators. For this purpose, we need auxiliary operators D, L, and M.

The operator $M : \mathcal{D}(M) \to \mathcal{B}$ is a closed linear operator acting in a complex Banach space $\mathcal{B}$ and M is assumed to satisfy the following two conditions:

(H1) $\mathcal{N} := \mathcal{N}(M)$ is finite dimensional and $\mathcal{N} \neq \{0\}$;

(H2) The operator M has a restriction $M_0 : \mathcal{D}(M_0) \to \mathcal{B}$ such that
(i) $\mathcal{D}(M) = \mathcal{N} \oplus \mathcal{D}(M_0)$,
(ii) $\Omega := \rho(M_0) \neq \emptyset$.

Apart from M we need two bounded linear operators

$$D : \mathcal{B} \to \mathbb{C}^n, \qquad \text{and} \quad L : \mathcal{B} \to \mathbb{C}^n,$$

where $n = \dim \mathcal{N}$. One may think about M as a maximal operator and about D and L as generalized boundary-value operators.

With D, L, and M as earlier, we associate two operators $A : \mathcal{D}(A) \to \mathcal{B}$ and $\widehat{A} : \mathcal{D}(\widehat{A}) \to \widehat{\mathcal{B}}$, where $\widehat{\mathcal{B}} = \mathbb{C}^n \times \mathcal{B}$. The definitions are as follows

$$\begin{aligned} &\mathcal{D}(A) = \{\phi \in C : \phi \in \mathcal{D}(M),\ DM\phi = L\phi\}, \qquad A\phi = M\phi \\ &\mathcal{D}(\widehat{A}) = \{(c, \phi) \in \widehat{C} : \phi \in \mathcal{D}(M),\ c = D\phi\}, \\ &\widehat{A}(c, \phi) = (L\phi, M\phi). \end{aligned} \tag{3.2}$$

The operators A and $\widehat{A}$ are well-defined closed linear operators and are closely related. In fact, if $j : \mathcal{B} \to \widehat{\mathcal{B}}$ denotes the embedding $\phi \mapsto (D\phi, \phi)$, then $jA = \widehat{A}j$. We shall refer to A and $\widehat{A}$, respectively, as the *first and second operator associated with* D, L, and M.

Next, we define the candidate for the characteristic matrix function. Let $l : \mathbb{C}^n \to \mathcal{N}$ be some isomorphism, and set

$$\Delta(z) = -(zD - L)M_0(zI - M_0)^{-1}l, \quad z \in \Omega. \tag{3.3}$$

Here M_0 is the operator appearing in Hypothesis (H2) on M and Ω is as defined in (H2).

Let $\widehat{\mathcal{B}}(\widehat{A})$ denote the domain of $\widehat{A}$ provided with the *graph norm*

$$\| \cdot \|_{\widehat{A}} \stackrel{\text{def}}{=} \| \cdot \| + \|\widehat{A} \cdot \|.$$

Since $\widehat{A}$ is closed, $\widehat{\mathcal{B}}(\widehat{A})$ becomes a Banach space and $\widehat{A} : \widehat{\mathcal{B}}(\widehat{A}) \to \widehat{\mathcal{B}}$ a bounded operator.

We then have the following theorem.

Theorem 3.1. *Suppose that $\widehat{A} : \mathcal{D}(\widehat{A}) \to \widehat{\mathcal{B}}$ is the second operator associated with D, L, and M. Then the matrix function $\Delta(z)$ defined in (3.3) is a characteristic matrix for $\widehat{A}$ and the equivalence is given by*

$$\begin{pmatrix} \Delta(z) & 0 \\ 0 & I_{\mathcal{B}} \end{pmatrix} = F(z)(zI - \widehat{A})E(z), \quad z \in \Omega, \tag{3.4}$$

where $E(z) : \widehat{\mathcal{B}} \to \widehat{\mathcal{B}}(\widehat{A})$ and $F(z) : \widehat{\mathcal{B}} \to \widehat{\mathcal{B}}$ are bijective mappings that depend analytically on z in Ω. Furthermore, these operators have the following representation

$$\begin{aligned} E(z)\begin{pmatrix} c \\ \phi \end{pmatrix} &= \begin{pmatrix} -DM_0(zI - M_0)^{-1}lc + D(zI - M_0)^{-1}\phi \\ -M_0(zI - M_0)^{-1}lc + (zI - M_0)^{-1}\phi \end{pmatrix}, \\ F(z)\begin{pmatrix} c \\ \phi \end{pmatrix} &= \begin{pmatrix} c - zD(zI - M_0)^{-1}\phi + L(zI - M_0)^{-1}\phi \\ \phi \end{pmatrix}. \end{aligned} \tag{3.5}$$

Given the formulas for $E(z)$ and $F(z)$, the theorem above is easy to verify directly. With Equivalence (3.4), the problem to determine the structure of the generalized eigenspace $\mathcal{M}_\lambda(\widehat{A})$, $\lambda \in \sigma(\widehat{A})$, has been reduced to the structure of the null space of $\Delta(\lambda)$ when $\det \Delta(\lambda) = 0$.

The following corollary holds for analytic matrix functions and therefore, using (3.4), for $\widehat{A}$ and A.

Corollary 3.1. *Let $\widehat{A}$ satisfy the assumptions in Theorem* 3.1. *If* $\det \Delta(z) \not\equiv 0$, *then*

(i) *The set $\sigma(A) \cap \Omega$ consists of eigenvalues and*

$$\sigma(A) \cap \Omega = \{z \in \Omega : \det \Delta(z) = 0\};$$

(ii) *For $\lambda_0 \in \sigma(A) \cap \Omega$,*

$$\dim \mathcal{M}_{\lambda_0}(A) = m,$$

where $m = m(\lambda_0, \Delta)$, the order of λ_0 as a zero of $\det \Delta$;

(iii) *For $\lambda_0 \in \sigma(A) \cap \Omega$, the ascent of λ_0 equals k; that is,*

$$\mathcal{M}_{\lambda_0}(A) = \mathcal{N}\big((\lambda_0 I - A)^k \big),$$

where $k = k(\lambda_0, \Delta)$, the order of λ_0 as a pole of Δ^{-1}.

To proceed further, we must analyze the null space of $\Delta(\lambda)$. Let $\mathcal{B}_1$ and $\mathcal{B}_2$ be complex Banach spaces and let $K(z) : \mathcal{B}_1 \to \mathcal{B}_2$ be an operator-valued function that depends analytically on z in Ω. For example,

$$K(z) : \mathbb{C}^n \to \mathbb{C}^n, \quad K(z) = \Delta(z),$$

or

$$K(z) : \mathcal{D}(\widehat{A}) \to \widehat{C}, \quad K(z) = z - \widehat{A}$$

with $\Omega = \mathbb{C}$ and $\widehat{A}$ defined by (3.1). A point λ_0 is called a *characteristic value* of $K(z)$ if there exists a vector $x_0 \in \mathcal{B}_1$, $x_0 \neq 0$, such that,

$$K(\lambda_0) x_0 = 0. \tag{3.6}$$

An ordered set $(x_0, x_1, \ldots, x_{k-1})$ of vectors in $\mathcal{B}_1$ is called a *Jordan chain* for $K(\lambda_0)$ if $x_0 \neq 0$ and

$$K(z)[x_0 + (z - \lambda_0)x_1 + \cdots + (z - \lambda_0)^{k-1} x_{k-1}] = O((z - \lambda_0)^k). \tag{3.7}$$

The number k is called the *length* of the chain and the maximal length of the chain starting with x_0 is called the *rank* of x_0. The function

$$\sum_{l=0}^{k-1} (z - \lambda_0)^l x_l$$

in (3.7) is called a *root function* of K corresponding to λ_0. We remark that if $K(z) = z - A$ and if $(x_0, x_1, \ldots, x_{k-1})$ is a Jordan chain for $z - A$ at λ_0, then $x_j \in \mathcal{N}\big((\lambda_0 I - A)^{j+1}\big)$, $j = 0, 1, \ldots, k-1$.

Let $\Delta(z) : \mathbb{C}^n \to \mathbb{C}^n$ be defined by Equation (2.1). If λ_0 is an isolated characteristic value of $\Delta(z)$, then the Jordan chains for $\Delta(\lambda_0)$ have finite rank and we can organize the chains according to the procedure described by Gohberg and Sigal [1]. Choose an eigenvector, say $x_{1,0}$, with maximal rank, say r_1. Next, choose a Jordan chain $(x_{1,0}, \ldots, x_{1,r_1-1})$ of length r_1 and let N_1 be the complement in $\mathcal{N}\big(\Delta(\lambda_0)\big)$ of the subspace spanned by $x_{1,0}$. In N_1 we choose an eigenvector $x_{2,0}$ of maximal rank, say r_2, and let $(x_{2,0}, \ldots, x_{2,r_2-1})$ be a corresponding Jordan chain of length r_2. We continue as follows: let N_2 be the complement in N_1 of the subspace spanned by $x_{2,0}$ and replace N_1 by N_2 in the described procedure.

In this way, we obtain a basis $\{x_{1,0}, \ldots, x_{p,0}\}$ of $\mathcal{N}\big(\Delta(\lambda_0)\big)$ and a corresponding *canonical system* of Jordan chains

$$x_{1,0}, \ldots, x_{1,r_1-1}, \ldots, x_{p,0}, \ldots, x_{p,r_p-1} \tag{3.8}$$

for $\Delta(\lambda_0)$.

Lemma 3.1. *Let $\widehat{A}$ satisfy the assumptions in Theorem* 3.1. *If* $\det \Delta(\lambda_0) = 0$, *then there is a one-to-one correspondence between the Jordan chains of $z - A$ and $\Delta(z)$ at λ_0.*

Proof. Since relation (3.4) implies that the null spaces $\mathcal{N}\big(\Delta(\lambda_0)\big)$ and $\mathcal{N}\big(\lambda_0 - A\big)$ are isomorphic, it suffices to show that there is a one-to-one correspondence between the Jordan chains of length k, $k \geq 1$, of $z - A$ and Δ at λ_0.

Let $(x_0, \ldots, x_{k-1})$ be a Jordan chain for $z - A$ at λ_0 of length k. The equivalence relation (3.4) implies that

$$\Delta(z)E(z)^{-1}\sum_{l=0}^{k-1}(z-\lambda_0)^l x_l = O((z-\lambda_0)^k) \tag{3.9}$$

for $|z - \lambda_0| \to 0$. If $\sum_{l=0}^{k-1}(z-\lambda_0)^l y_l$ denotes the Taylor expansion of order k around $z = \lambda_0$ for the holomorphic function

$$E(z)^{-1}\sum_{l=0}^{k-1}(z-\lambda_0)^l x_l,$$

then

$$\Delta(z)\sum_{l=0}^{k-1}(z-\lambda_0)^l y_l = O((z-\lambda_0)^k) \qquad \text{for} \quad |z-\lambda_0| \to 0$$

and $(y_0, \ldots, y_{k-1})$ is a Jordan chain for Δ at λ_0 of length k. So we proved that a Jordan chain for $z - A$ at λ_0 of length k induces a Jordan chain for

Δ at λ_0 of length k. Since the roles of $z - A$ and Δ can be interchanged, the proof is complete. □

7.4 The generalized eigenspace for RFDE

As a first application, we apply the result from Section 3 to retarded functional differential equations.

Lemma 4.1. *Let $M : \mathcal{D}(M) \to C$ be the operator defined by*

$$\mathcal{D}(M) = \{\phi \in C : \frac{d\phi}{d\theta} \in C\}, \quad M\phi = \frac{d\phi}{d\theta}.$$

Then M satisfies hypotheses (H1) *and* (H2) *in Section* 7.3 *with $M_0 : \mathcal{D}(M_0) \to C$ defined by*

$$\mathcal{D}(M_0) = \{\phi \in \mathcal{D}(M) : \phi(0) = 0\}, \quad M_0\phi = M\phi$$

and $\Omega = \mathbb{C}$. Furthermore, the infinitesimal generator A defined by (1.2) *and the infinitesimal generator $\widehat{A}$ defined by* (3.1) *are the first and second operator associated with D, L, and M, where $D\phi = \phi(0)$.*

Proof. Clearly, the kernel of M consists of the constant functions. It follows that $\mathcal{N} = \mathcal{N}(M)$ has dimension n. We have $\mathcal{D}(M) = \mathcal{N} \oplus \mathcal{D}(M_0)$, and for each $z \in \mathbb{C}$ the operator $z - M_0$ is invertible and the resolvent of M_0 is given by

$$((zI - M_0)^{-1}\phi)(\theta) = \int_\theta^0 e^{(\theta-\sigma)z}\phi(\sigma)\,d\sigma, \quad -r \le \theta \le 0. \tag{4.1}$$

Thus M satisfies (H1) and (H2) with $\Omega = \mathbb{C}$. Furthermore, if we set $D\phi = \phi(0)$, then A is given by

$$\mathcal{D}(A) = \{\phi \in C : \phi \in \mathcal{D}(M),\ DM\phi = L\phi\}, \qquad A\phi = M\phi.$$

So A is the first operator associated with D, L, and M. Finally, it is clear that $\widehat{A}$ defined by (3.1) is the second operator associated with D, L, and M. □

Theorem 4.1. *The matrix function*

$$\Delta(z) = zI - \int_{-r}^0 e^{z\theta}d[\eta(\theta)] \tag{4.2}$$

is a characteristic matrix for the infinitesimal generator $\widehat{A}$ defined by Relation (3.1)*. The equivalence is given by*

$$\begin{pmatrix} \Delta(z) & 0 \\ 0 & I \end{pmatrix} = F(z)(zI - \widehat{A})E(z), \qquad z \in \mathbb{C} \tag{4.3}$$

with $E(z) : \widehat{C} \to \mathcal{D}(\widehat{A})$

$$E(z)(c, \phi)(\theta) = \left(c, e^{\theta z}c + \int_\theta^0 e^{(\theta-\sigma)z}\phi(\sigma)d\sigma\right) \tag{4.4}$$

and $F(z) : \widehat{C} \to \widehat{C}$

$$F(z)(c, \phi) = (c + L(zI - M_0)^{-1}\phi, \phi). \tag{4.5}$$

Proof. Let M and M_0 be as in the statement of Lemma 4.1. Define $l : \mathbb{R}^n \to \mathcal{N}$, $\mathcal{N} = \mathcal{N}(M)$, by

$$(lc)(\theta) = c, \quad -r \le \theta \le 0.$$

From Theorem 3.1, we know that

$$\Delta(z) = -(zD - L)M_0(zI - M_0)^{-1}l, \quad z \in \mathbb{C},$$

is a characteristic matrix for A. To verify the concrete representation (4.2) we use the resolvent formula (4.1) for M_0 and calculate

$$\begin{aligned} -(zD - L)M_0(zI - M_0)^{-1}lc(\theta) &= \big(lc - z(zI - M_0)^{-1}lc\big)(\theta) \\ &= c + z\int_0^\theta e^{(\theta-\sigma)z}c\,d\sigma \\ &= c - e^{(\theta-\sigma)z}c\big|_0^\theta \\ &= e^{\theta z}c, \quad -r \le \theta \le 0. \end{aligned}$$

Finally, the concrete representations for $E(z)$ and $F(z)$ are verified in a similar way. □

From the equivalence relation (4.3), we have the following representation for the resolvent of A.

Corollary 4.1. *Let* $A : \mathcal{D}(A) \to C$ *be the generator defined by Relation* (1.2). *The resolvent of* A *has the following representation*

$$\big(R(z, A)\phi\big)(\theta) = e^{z\theta}\left[\Delta^{-1}(z)K(z)\phi + \int_\theta^0 e^{-z\tau}\phi(\tau)d\tau\right] \tag{4.6}$$

for $-r \le \theta \le 0$, *where*

$$K(z)\phi = \phi(0) + \int_{-r}^0 d[\eta(\tau)]e^{z\tau}\int_\tau^0 e^{-zs}\phi(s)\,ds. \tag{4.7}$$

Since the spectral analysis of A and $\widehat{A}$ are one and the same, we have actually characterized $\mathcal{N}\big((\lambda I - A)^k\big)$ in a manner that is convenient for computations.

Theorem 4.2. *The spectrum of the infinitesimal generator A defined by* (1.2) *consists of eigenvalues of finite type only,*

$$\sigma(A) = \{\lambda : \det \Delta(\lambda) = 0\}. \tag{4.8}$$

For $\lambda \in \sigma(A)$, the algebraic multiplicity of the eigenvalue λ equals the order of λ as a zero of $\det \Delta$, *the ascent of λ equals the order of λ as a pole of Δ^{-1}. Furthermore, a canonical basis of eigenvectors and generalized eigenvectors for A at λ may be obtained in the following way: If*

$$\{(\gamma_{i,0}, \ldots, \gamma_{i,k_i-1}) : i = 1, \ldots, p\}$$

is a canonical system of Jordan chains for Δ at λ, then

$$\phi_{i,0}, \ldots, \phi_{i,k_i-1}, \qquad i = 1, \ldots, p,$$

where

$$\phi_{i,\nu}(\theta) = e^{\lambda\theta} \sum_{l=0}^{\nu} \gamma_{i,\nu-l} \frac{\theta^l}{l!} \tag{4.9}$$

yields a canonical basis for A at λ.

Proof. In order to apply Corollary 3.1, we first show that $\det \Delta \not\equiv 0$. From the representation (4.2) for $\Delta(z)$, it follows that it suffices to prove

$$|z^{-1} \int_{-r}^{0} e^{zt} d\eta(t)| \to 0$$

as $\operatorname{Re} z \to \infty$. But this is obvious since η is of bounded variation.

Next we prove the representation for the canonical basis for A at λ. Let $(\gamma_0, \ldots, \gamma_{k-1})$ be a Jordan chain for Δ at λ of length k. Define $\Gamma(\lambda) = \gamma_0 + \gamma_1(z-\lambda) + \cdots + \gamma_{k-1}(z-\lambda)^{k-1}$. Since

$$E(z)(\Gamma(\lambda), 0) = \big(\Gamma(\lambda), e^{z\,\cdot}\,\Gamma(\lambda)\big),$$

we derive from the equivalence that

$$(zI - \widehat{A})e^{z\,\cdot}\,\Gamma(\lambda) = O((z-\lambda)^k) \qquad \text{for} \quad |z-\lambda| \to 0.$$

From Lemma 3.1, it follows that there is a one-to-one correspondence between the Jordan chains of $\Delta(z)$ at λ and the Jordan chains of $z - A$ at λ. So we have to expand $e^{z\,\cdot}$ up to order k in a neighborhood of λ. Since

$$e^{z\theta} = e^{\lambda\theta}\big[1 + \theta(z-\lambda) + \cdots + \frac{\theta^{k-1}}{(k-1)!}(z-\lambda)^{k-1} + O\big((z-\lambda)^k\big)\big],$$

the Jordan chain for $(zI - A)$ at λ becomes $\{\phi_0, \ldots, \phi_{k-1}\}$, where

$$\phi_i(\theta) = e^{\lambda\theta} \sum_{l=0}^{i} \gamma_{i-l} \frac{\theta^l}{l!}$$

and it becomes clear that this procedure yields a canonical basis for A at λ from a canonical system of Jordan chains for $\Delta(z)$ at λ. $\square$

From Theorem 4.2, we see, in particular, that there is a one-to-one correspondence between the Jordan chains of $\Delta(z)$ at λ of length k and the solutions of the equation

$$(\lambda I - A)^k \phi = 0.$$

The Jordan chains of length 1 are precisely the vectors in the null space of $\Delta(\lambda)$ and already in Lemma 2.1 we saw that if $b \in \mathcal{N}\big(\Delta(\lambda)\big)$, then $e^{\lambda\theta} b$ satisfies $(\lambda I - A)e^{\lambda\theta} b = 0$. The higher-order Jordan chains can also be characterized by vectors in the null space of a certain matrix. Define

$$P_{j+1} = P_{j+1}(z) = \frac{\Delta^{(j)}(z)}{j!}, \qquad \Delta^{(j)}(z) = \frac{d^j \Delta(z)}{dz^j}, \quad j = 0, 1, \ldots$$

and the matrices A_k of dimension $(kn) \times (kn)$

$$A_k = \begin{pmatrix} P_1 & 0 & \cdots & 0 \\ P_2 & P_1 & \cdots & 0 \\ \vdots & & \ddots & \vdots \\ P_k & P_{k-1} & \cdots & P_1 \end{pmatrix}, \qquad k = 1, 2, \ldots.$$

Then $(\gamma_0, \ldots, \gamma_{k-1})$ is a Jordan chain of length k if and only if

$$A_k (\gamma_0 \ldots \gamma_{k-1})^T = 0.$$

7.5 Decomposing C with the adjoint equation

In the previous section, in which we proved Theorem 4.2, we have seen that we can characterize the generalized eigenspace of A corresponding to an eigenvalue λ. In this section, it is our goal to compute the corresponding projection onto this generalized eigenspace. For this, we use that

$$\mathcal{R}\big((\lambda I - A)^k\big) = \mathcal{N}\big((\lambda I - A^*)^k\big)^{\perp}, \qquad k = 1, 2, \ldots.$$

Therefore, we start this section with the spectral analysis of the adjoint operator $A^* : \mathcal{D}(A^*) \to B_0$. Let $\widehat{B}_0 = \mathbb{R}^{n*} \times B_0$ be the dual space of $\widehat{C}$, with the pairing

$$\text{(5.1)} \qquad \langle (\alpha, f), (c, \phi) \rangle = \alpha c + \int_{-r}^{0} d[f(\theta)]\, \phi(\theta)$$

where $(\alpha, f) \in \widehat{\mathcal{B}}_0$, $(c, \phi) \in \widehat{C}$. As we observed in the previous section, the operator A is similar to the part of $\widehat{A}$ in jC. This implies that the operator $A^* : \mathcal{D}(A^*) \to B_0$ is similar to the part of the operator $\widehat{A}^* : \widehat{B}_0 \to \widehat{C}(\widehat{A})^*$ to $j^*\widehat{B}_0$. Here $\widehat{C}(\widehat{A})$ denotes the domain of $\widehat{A}$ provided with the graph norm. In particular, the spectral analysis of A^* and $\widehat{A}^*$ are one and the same.

Together with the properties of the adjoint operation, the equivalence (4.3) for A implies the following equivalence relation for $\widehat{A}^*$:

$$\text{(5.2)} \quad (\alpha \;\, f) \begin{pmatrix} \Delta(z) & 0 \\ 0 & I \end{pmatrix} = E(z)^*(zI - \widehat{A}^*)F(z)^*(\alpha, f), \quad (\alpha, f) \in \widehat{B}_0$$

where $E(z)^* : \widehat{C}(\widehat{A})^* \to \widehat{B}_0$ and $F(z)^* : \widehat{B}_0 \to \widehat{B}_0$ are bijective mappings that depend analytically on z, $z \in \mathbb{C}$.

In particular, we find that the Jordan chains of length k of Δ^T at $z = \lambda$ are in one-to-one correspondence with the solutions of

$$\text{(5.3)} \qquad (\lambda I - A^*)^k \psi = 0.$$

More precisely, we have an explicit representation for the eigenfunctions and generalized eigenfunctions of A^*. This is the contents of Theorem 5.1. First we compute $F(z)^* : \widehat{B}_0 \to \widehat{B}_0$.

Lemma 5.1. *The mapping $F(z)^* : \widehat{B}_0 \to \widehat{B}_0$ is bijective, depends analytically on z, $z \in \mathbb{C}$, and is given by $F(z)^*(\alpha, f) = (\alpha, g)$, where*

$$\text{(5.4)} \qquad g(\theta) = \int_{\theta}^{0} \int_{-r}^{\sigma} d[\alpha\eta(\tau)] e^{-(\sigma-\tau)z}\, d\sigma + f(\theta).$$

Proof. Since $F(z) : \widehat{C} \to \widehat{C}$ is bijective and depends analytically on z, $z \in \mathbb{C}$, it follows that $F(z)^* : \widehat{B}_0 \to \widehat{B}_0$ has the same properties. It remains to compute $F(z)^*$. From the definition

$$\text{(5.5)} \qquad \begin{aligned} \langle F(z)^*(\alpha, f), (c, \phi) \rangle &= \langle (\alpha, f), F(z)(c, \phi) \rangle \\ &= \alpha\big[c + L(zI - M_0)^{-1}\phi\big] + \int_{-r}^{0} \phi(\theta)\, d[f(\theta)]. \end{aligned}$$

Using Fubini's theorem to reverse the order of integration, we have the following identity

$$\begin{aligned} \alpha L(zI - M_0)^{-1}\phi &= \int_{-r}^{0} \alpha d[\eta(\theta)] \int_{\theta}^{0} e^{-(\sigma-\theta)z} \phi(\sigma)\, d\sigma \\ &= \int_{-r}^{0} \phi(\sigma) \int_{-r}^{\sigma} \alpha d[\eta(\theta)] e^{-(\sigma-\theta)z}\, d\sigma, \qquad \alpha \in \mathbb{R}^{n*}. \end{aligned}$$

Substitution of this identity into (5.5) yields

$$\langle F(z)^*(\alpha, f), (c, \phi) \rangle = \alpha c + \int_{-r}^{0} \phi(\theta)\, d[g(\theta)], \tag{5.6}$$

where g is given by (5.4). This proves the lemma. □

Theorem 5.1. *The spectrum of the infinitesimal generator A^* defined by* (3.1) *consists of eigenvalues of finite type only,*

$$\sigma(A^*) = \{\lambda : \det \Delta(\lambda) = 0\}. \tag{5.7}$$

For $\lambda \in \sigma(A^*)$, the algebraic multiplicity of the eigenvalue λ equals the order of λ as a zero of $\det \Delta$, the ascent of λ equals the order of λ as a pole of Δ^{-1}. Furthermore, a canonical basis of eigenvectors and generalized eigenvectors for A^* at λ may be obtained in the following way: If

$$\{(\beta_{i,0}^T, \ldots, \beta_{i,k_i-1}^T) : i = 1, \ldots, p\}$$

is a canonical system of Jordan chains for Δ^T at $\lambda \in \mathbb{C}$, then

$$\chi_{i,0}, \ldots, \chi_{i,k_i-1}, \qquad i = 1, \ldots, p,$$

where $\chi_{i,\nu} = F^T \psi_{i,\nu}$ with

$$\psi_{i,\nu}(\xi) = e^{-\lambda\xi} \sum_{l=0}^{\nu} \beta_{i,\nu-l} \frac{(-\xi)^l}{l!},$$

yields a canonical basis for A^* at λ. Here $F^T : C[0, r] \to B_0$ is the mapping from an initial condition into a forcing function of the integrated equation introduced in Section 7.1

$$(F^T\psi)(\theta) = \psi(0) + \int_{\theta}^{0} \int_{-r}^{\sigma} \psi(\sigma - \tau)\, d[\eta(\tau)]. \tag{5.8}$$

Proof. The first part of the proof is the same as the proof of Theorem 4.2. Let $(\beta_0^T, \ldots, \beta_{k-1}^T)$ be a Jordan chain for Δ^T at λ of length k and let $\alpha(z)^T = \beta_0^T + \beta_1^T(z - \lambda) + \cdots + \beta_{k-1}^T(z - \lambda)^{k-1}$ denote the corresponding root function. From the equivalence relation (5.2), we have $F(z)^*(\alpha(z), 0)$ is a root function for $\widehat{A}^*$. Therefore,

$$\Big(\alpha(z), \int_{\theta}^{0} \int_{-r}^{\sigma} e^{-(\sigma-\tau)z} \alpha(z)\, d[\eta(\tau)]\, d\sigma\Big)$$

is a root function for $z - \widehat{A}^*$ at λ. From Lemma 3.1, it follows that there is a one-to-one correspondence between the Jordan chains of $\Delta(z)^T$ at λ and the Jordan chains of $z - \widehat{A}^*$ at λ. So we have to expand

$$z \mapsto e^{-(\sigma-\tau)z}\alpha(z)$$

up to order k in a neighborhood of λ. A similar expansion, as in the proof of Theorem 4.2, shows that

$$\sum_{l=0}^{k-1}\Big(\beta_l(z-\lambda)^l, \int_\theta^0\int_{-r}^\sigma e^{-\lambda(\sigma-\tau)}\sum_{j=0}^{l}\beta_{l-j}\frac{(-(\sigma-\tau))^j}{j!}\,d[\eta(\tau)]\,d\sigma(z-\lambda)^l\Big)$$

is a root function for $z-\widehat{A}^*$. Since $A^*:\mathcal{D}(A^*)\to B_0$ is similar to the part of $\widehat{A}^*$ in $j^*\widehat{B}_0$, it remains to compute the adjoint of $j: C\to\widehat{C}$, $j\phi=(\phi(0),\phi)$. An easy computation shows that $j^*:\widehat{B}_0\to B_0$ is given by

$$j^*(\alpha,f)(\theta)=\begin{cases}0 & \text{for } \theta=0,\\ \alpha+f & \text{for } -r\le\theta<0.\end{cases}$$

So the corresponding Jordan chain for $z-A^*$ at λ becomes $\{\chi_0,\ldots,\chi_k\}$, where

$$\chi_i(\theta)=\beta_i+\int_\theta^0\int_{-r}^\sigma\sum_{j=0}^{i}e^{-\lambda(\sigma-\tau)}\beta_{i-j}\frac{(-(\sigma-\tau))^j}{j!}\,d[\eta(\tau)]\,d\sigma,$$

and it becomes clear that this procedure yields a canonical basis for A^* at λ from the canonical system of Jordan chains for $\Delta(z)^T$ at λ. This proves the theorem. □

From the preceding theorem, we conclude that the (generalized) eigenfunctions of A^* are precisely given by the images under the operator F^T of the (generalized) eigenfunctions of the transposed generator A^T. This is to be expected from the adjoint theory in Section 6.3. In particular, $T^*(t)F^T=F^TT^T(t)$, and hence

$$A^*F^T\psi=F^TA^T\psi,\qquad \psi\in\mathcal{D}(A^T).$$

In general, the mapping F^T is not one-to-one (see Section 3.3), but, on the generalized eigenspace of A^T, it is. Hence, if ψ is a (generalized) eigenfunction of A^T, then $F^T\psi$ is a (generalized) eigenfunction of A^*. This motivates the introduction of the following bilinear form (see also Hale [22]). For $\psi\in C'$ and $\phi\in C$ define

$$\begin{aligned}(\psi,\phi)&\stackrel{\text{def}}{=}\langle F^T\psi,\phi\rangle=\int_{-r}^0 d[F^T\psi(\theta)]\,\phi(\theta)\\ &=\psi(0)\phi(0)-\int_{-r}^0\int_r^\theta\psi(\theta-\tau)\,d[\eta(\tau)]\phi(\theta)\,d\theta\\ &=\psi(0)\phi(0)-\int_{-r}^0\int_0^\theta\psi(\theta-\tau)\,d[\eta(\tau)]\phi(\theta)\,d\theta\end{aligned}\tag{5.9}$$

between C and C'. With respect to this bilinear form, the transposed operator A^T satisfies

$$(\psi, A\phi) = (A^T\psi, \phi),$$

and we have proved the following lemma.

Lemma 5.2. *For λ in $\sigma(A)$, let $\Psi_\lambda = \mathrm{col}\,(\psi_1, \dots, \psi_p)$ and $\Phi_\lambda = (\phi_1, \dots, \phi_p)$ be bases for $\mathcal{M}_\lambda(A^T)$ and $\mathcal{M}_\lambda(A)$, respectively, and let $(\Psi_\lambda, \Phi_\lambda) = (\psi_i, \phi_j)$, $i, j = 1, 2, \dots, p$. Then $(\Psi_\lambda, \Phi_\lambda)$ is nonsingular and thus may be taken as the identity. The decomposition of C given by Lemma 2.1 may be written explicitly as*

$$\begin{aligned}
&\phi = \phi^{P_\lambda} + \phi^{Q_\lambda}, \qquad \phi^{P_\lambda} \textit{ in } P_\lambda, \phi^{Q_\lambda} \textit{ in } Q_\lambda,\\
&P_\lambda = \mathcal{M}_\lambda(A) = \{\phi \in C : \phi = \Phi_\lambda b \textit{ for some p-vector } b\},\\
&Q_\lambda = \{\phi \in C : (\Psi_\lambda, \phi) = 0\},\\
&\phi^{P_\lambda} = \Phi_\lambda b, \quad b = (\Psi_\lambda, \phi), \qquad \phi^{Q_\lambda} = \phi - \phi^{P_\lambda}.
\end{aligned}$$

It is also interesting to note that $(\Psi_\lambda, \Phi_\lambda) = I$, and $A^T\Psi_\lambda = B_\lambda^*\Psi_\lambda$ and $A\Phi_\lambda = \Phi_\lambda B_\lambda$ implies $B_\lambda^* = B_\lambda$. In fact,

$$\begin{aligned}
(\Psi_\lambda, A\Phi_\lambda) &= (\Psi_\lambda, \Phi_\lambda B_\lambda) = (\Psi_\lambda, \Phi_\lambda)B_\lambda = B_\lambda\\
&= (A^T\Psi_\lambda, \Phi_\lambda) = (B_\lambda^*\Psi_\lambda, \Phi_\lambda) = B_\lambda^*(\Psi_\lambda, \Phi_\lambda) = B_\lambda^*.
\end{aligned}$$

We have already defined the generalized eigenspace of a characteristic value of Equation (1) as the set $\mathcal{M}_\lambda(A)$. If $\Lambda = \{\lambda_1, \dots, \lambda_p\}$ is a finite set of characteristic value of Equation (1), we let $P = P_\Lambda$ be the linear extension of the $\mathcal{M}_{\lambda_j}(A)$, $\lambda_j \in \Lambda$, and refer to this as the *generalized eigenspace of Equation* (1) *associated with* Λ. In a similar manner, we can define $P^T = P_\Lambda^T$ to be the *generalized eigenspace of the transposed equation* (1.8) *associated with* Λ. If Φ and Ψ are bases for P and P_Λ^T, respectively, $(\Psi, \Phi) = I$, the identity, then

$$\begin{aligned}
C &= P_\Lambda \oplus Q_\Lambda\\
P_\Lambda &= \{\Phi \in C : \Phi = \Phi b \text{ for some vector } b\}\\
Q_\Lambda &= \{\Phi \in C : (\Psi, \phi) = 0\}
\end{aligned} \tag{5.10}$$

and, therefore, for any ϕ in C,

$$\begin{aligned}
\phi &= \phi^{P_\Lambda} + \phi^{Q_\Lambda}\\
\phi^{P_\Lambda} &= \Phi(\Psi, \phi).
\end{aligned} \tag{5.11}$$

When this particular decomposition of C is used, we shall briefly express this by saying that C is *decomposed by* Λ.

7.6 Estimates on the complementary subspace

If C is decomposed by Λ, we know, from Theorem 2.1, that there is a constant matrix $B = B_\Lambda$ whose eigenvalues coincide with Λ such that

$$T(t)\phi^{P_\Lambda} = \Phi e^{Bt} a, \quad \text{where } \phi^{P_\Lambda} = \Phi a.$$

For the application of the theory of linear systems, we need to have an estimate for the solutions on the complementary subspace Q_Λ. Such an estimate requires detailed knowledge of the spectrum of $T(t)$. In particular, we need to know the spectral radius of the semigroup $T(t)$ restricted to Q_Λ. A first step in this direction is answered by the following result:

Lemma 6.1. *If the semigroup $T(t)$ is strongly continuous on $[0,\infty)$ with infinitesimal generator A, then $P\sigma(T(t)) = e^{tP\sigma(A)}$ plus possibly $\{0\}$. More specifically, if $\mu = \mu(t) \neq 0$ is in $P\sigma(T(t))$ for some fixed t, then there is a point λ in $P\sigma(A)$ such that $e^{\lambda t} = \mu$. Furthermore, if λ_n consists of all distinct points in $P\sigma(A)$ such that $e^{\lambda_n t} = \mu$, then $\mathcal{N}\big((\mu I - T(t))^k\big)$ is the closed linear extension of the linearly independent manifolds $\mathcal{N}\big((\mu I - A)^k\big)$.*

Proof. Lemma 4.1 is a special case of Theorem 16.7.2, p. 467 of Hille and Phillips [1] for $k = 1$. The reader may complete the proof for arbitrary k. □

The *spectral radius* ρ of an operator T mapping a Banach space into itself is the smallest disk centered at the origin of the complex plane that contains $\sigma(T)$.

We also need the following

Lemma 6.2. *If $T(t)$ is a strongly continuous semigroup of operators of a Banach space $\mathcal{B}$ into itself, if for some $r > 0$, the spectral radius $\rho = \rho_{T(r)} \neq 0$ and if β is defined by $\beta r = \log \rho$, then for any $\gamma > 0$, there is a constant $K(\gamma) \geq 1$ such that*

$$\|T(t)\phi\| \leq K(\gamma) e^{(\beta+\gamma)t} \|\phi\|, \quad \textit{for all} \quad t \geq 0,\ \phi \in \mathcal{B}.$$

Proof. Since $T(t)$ is strongly continuous, it is certainly bounded for each t and, in particular, $T(r)$ is bounded. It then follows that

$$\rho = e^{\beta r} = \lim_{n\to\infty} \|T^n(r)\|^{1/n}.$$

Therefore, for any $\gamma > 0$,

$$e^{-\gamma r} = \lim_{n\to\infty} e^{-(\beta+\gamma)r} \|T^n(r)\|^{1/n}$$

and there is a number N such that

$$e^{-(\beta+\gamma)nr}\|T^n(r)\| = (e^{-\gamma r} + \epsilon_n)^n,$$

where $e^{-\gamma r} + \epsilon_n \le L < 1$ for all $n \ge N$. Therefore,

$$e^{-(\beta+\gamma)nr}\|T^n(r)\| \to 0 \qquad \text{as} \quad n \to \infty.$$

Since $T(t)$ is strongly continuous, there is a constant B such that $\|T(t)\| \le B$ for $0 \le t \le r$. Define $K(\gamma)$ for any γ to be

$$K(\gamma) = Be^{|\beta+\gamma|r} \max_{n\ge 0} e^{-(\beta+\gamma)nr}\|T^n(r)\|.$$

If $0 \le t \le r$, then, for any ϕ in $\mathcal{B}$,

$$\|T(t)\phi\| \le \|T(t)\| \cdot \|\phi\| \le B\|\phi\| \le K(\gamma)e^{(\beta+\gamma)t}\|\phi\|.$$

If $t \ge r$, then there is an integer n such that $nr \le t < (n+1)r$ and, for all ϕ in $\mathcal{B}$,

$$\begin{aligned}
\|T(t)\phi\| &= \|T(t-nr)T(nr)\phi\| \le B\|T^n(r)\| \cdot \|\phi\| \\
&= \big[Be^{-(\beta+\gamma)(t-nr)}e^{-(\beta+\gamma)nr}\|T^n(r)\|\big]e^{(\beta+\gamma)t}\|\phi\| \\
&\le K(\gamma)e^{(\beta+\gamma)t}\|\phi\|.
\end{aligned}$$

This completes the proof of the lemma. □

If we now turn to our original problem posed before the statement of Lemma 6.1, we obtain the following information. Since $T(t)$ is completely continuous for $t \ge r$, it follows that for any μ in $\sigma(T(r))$, $\mu \ne 0$ is an element of $P\sigma(T(r))$ and that the only possible accumulation point in $\sigma(T(r))$ is zero. Furthermore, if $\mu \ne 0$ is in $P\sigma(T(r))$, then $\mathcal{N}\big((\mu I - T(r))^k\big)$ is of finite dimension for every k and $\mathcal{M}_\mu(T(r))$ is finite dimensional. These are well-known properties of completely continuous operators that can be found in Taylor [1, pp. 180–182]. Lemmas 6.1 and 2.1 imply there are only a finite number of λ in $\sigma(A)$ such that $\operatorname{Re}\lambda > \beta$ for any real number β. Consequently, if $\Lambda = \Lambda(\beta) = \{\lambda \in \sigma(A) : \operatorname{Re}\lambda > \beta\}$ and C is decomposed by Λ, then there are constants $\gamma > 0$ and $K = K(\gamma) > 0$, such that

$$\|T(t)\phi^{Q_\Lambda}\| \le Ke^{(\beta+\gamma)t}\|\phi^{Q_\Lambda}\|, \qquad t \ge 0.$$

We summarize these results in

Theorem 6.1. *For any real number β, let $\Lambda = \Lambda(\beta) = \{\lambda \in \sigma(A) : \operatorname{Re}\lambda > \beta\}$. If C is decomposed by Λ as $C = P_\Lambda \oplus Q_\Lambda$, then there exist positive constants γ and $K = K(\gamma)$ such that*

$$\begin{aligned}
\|T(t)\phi^{P_\Lambda}\| &\le Ke^{(\beta+\gamma)t}\|\phi^{P_\Lambda}\|, \qquad t \le 0, \\
\|T(t)\phi^{Q_\Lambda}\| &\le Ke^{(\beta+\gamma)t}\|\phi^{Q_\Lambda}\|, \qquad t \ge 0.
\end{aligned} \tag{6.1}$$

The first of Inequalities (6.1) follows from Theorem 2.1 since we know that $T(t)$ can be defined on P_Λ for $-\infty < t < \infty$ and the eigenvalues of the corresponding matrix B_Λ associated with P_Λ coincides with the set Λ.

An important corollary of Theorem 6.2 concerning exponential asymptotic stability is

Corollary 6.1. *If all of the roots of the characteristic equation* (2.1) *of Equation* (1) *have negative real parts, then there exist positive constants K and δ such that*

$$\|T(t)\phi\| \le Ke^{-\delta t}\|\phi\|, \qquad t \ge 0,$$

for all ϕ in C.

Proof. The proof is obvious since, by choosing $\beta < 0$ in Theorem 6.1 sufficiently close to 0, the set Λ is empty. □

Theorem 6.1 can also be obtained without such sophisticated theory. In fact, the same approach as in Chapter 1 applies (see Chapter 9).

However, Theorems 2.1 and 6.1 give a very clear picture of the behavior of the solutions of an autonomous linear RFDE in C. In particular, one can choose $\beta = 0$ in Theorem 6.1, separate out the eigenvalues with real parts ≥ 0, and then be assured that the finite-dimensional subspace P_Λ attracts all solutions of the differential equation exponentially. Furthermore, the flow on P_Λ is equivalent to an ordinary differential equation. In fact, if Φ is a basis for P_Λ and $T(t)\phi^\Lambda = \Phi y(t)$, then $y(t) = (\exp Bt)b$, where $b = (\Psi, \phi^\Lambda)$.

7.7 An example

Consider the scalar equation

$$\dot{x}(t) = -\frac{\pi}{2}x(t-1) = \int_{-1}^{0} d[\eta(\theta)]x(t+\theta) \tag{7.1}$$

where

$$\eta(\theta) = \begin{cases} 0 & \theta = -1, \\ \frac{\pi}{2} & -1 < \theta \le 0, \end{cases}$$

and the transposed system

$$\dot{y}(s) = \frac{\pi}{2}y(s+1). \tag{7.2}$$

The bilincar form is

$$(\psi, \phi) = \psi(0)\phi(0) - \frac{\pi}{2}\int_{-1}^{0} \psi(\tau+1)\phi(\tau)\,d\tau, \tag{7.3}$$

and the operators A, A^T are given by

$$\mathcal{D}(A) = \{\phi \in C : \frac{d\phi}{d\theta} \in C,\ \dot{\phi}(0) = -\frac{\pi}{2}\phi(-1)\}, \qquad A\phi = \frac{d\phi}{d\theta},$$
$$\mathcal{D}(A^T) = \{\phi \in C' : \frac{d\psi}{ds} \in C',\ \dot{\phi}(0) = -\frac{\pi}{2}\phi(1)\}, \qquad A^T\psi = -\frac{d\psi}{ds}.$$

Moreover, ϕ is in $\mathcal{N}(\,\lambda I - A\,)$ if and only if $\phi(\theta) = e^{\lambda\theta}b$, $-r \le \theta \le 0$, where b is a constant and λ satisfies the characteristic equation

$$\lambda + \frac{\pi}{2}e^{-\lambda} = 0. \tag{7.4}$$

Also, ψ belongs to $\mathcal{N}(\,\lambda I - A^T\,)$ if and only if $\psi(\tau) = e^{-\lambda\tau}c$, $0 \le \tau \le r$, where c is a constant and λ satisfies Equation (7.4).

It is easy to prove (using the Appendix) that Equation (7.4) has two simple roots $\pm i\frac{\pi}{2}$ and the remaining roots have negative real parts.

If $\Lambda = \{i\frac{\pi}{2}, -i\frac{\pi}{2}\}$, then it is obvious that

$$\Phi = (\phi_1, \phi_2), \quad \phi_1(\theta) = \sin\frac{\pi}{2}\theta, \quad \phi_2(\theta) = \cos\frac{\pi}{2}\theta, \quad -1 \le \theta \le 0, \tag{7.5}$$

is a basis for the generalized eigenspace $P = P_\Lambda$ of Equation (7.1) associated with Λ and that

$$\Psi^T = \mathrm{col}\,(\psi_1^T, \psi_2^T), \quad \psi_1^T(\tau) = \sin\frac{\pi}{2}\tau, \quad \psi_2^T(\tau) = \cos\frac{\pi}{2}\tau, \quad 0 \le \tau \le 1,$$

is a basis for the generalized eigenspace $P^T = P_\Lambda^T$ of Equation (7.2) associated with Λ. We wish to decompose C by Λ. In addition, we have seen that the transformations are simpler if $(\Psi^T, \Phi) = (\psi_j^T, \phi_k)$, $j, k = 1, 2$, is the identity matrix, and (ψ, ϕ) is defined in Equation (7.3). However, if we compute this matrix, we see that it is not the identity. Therefore, we define a new basis Ψ for P_Λ^T by

$$\Psi = (\Psi^T, \Phi)^{-1}\Psi^T$$

and then $(\Psi, \Phi) = I$. The explicit expression for the basis Ψ is

$$\begin{aligned}
&\Psi = \mathrm{col}(\psi_1, \psi_2),\\
&\psi_1(\tau) = 2\mu[\sin(\frac{\pi}{2}\tau) + \frac{\pi}{2}\cos(\frac{\pi}{2}\tau)],\\
&\psi_2(\tau) = 2\mu[-\frac{\pi}{2}\sin(\frac{\pi}{2}\tau) + \cos(\frac{\pi}{2}\tau)], \qquad \mu = \frac{1}{1 + \pi^2/4}.
\end{aligned} \tag{7.6}$$

If we now decompose C by Λ and let $Q = Q_\Lambda$ for simplicity in notation, then any ϕ in C can be written as

$$\begin{aligned}
&\phi = \phi^P + \phi^Q,\\
&\phi^P = \Phi b, \qquad b = \mathrm{col}\,(b_1, b_2) = (\Psi, \phi),\\
&b_1 = \mu\pi\phi(0) - \mu\pi\int_{-1}^{0}[\cos(\frac{\pi}{2}s) - \frac{\pi}{2}\sin(\frac{\pi}{2}s)]\phi(s)\,ds,\\
&b_2 = 2\mu\phi(0) + \mu\pi\int_{-1}^{0}[\sin(\frac{\pi}{2}s) + \frac{\pi}{2}\cos(\frac{\pi}{2}s)]\phi(s)\,ds.
\end{aligned} \tag{7.7}$$

The explicit expressions for b_1 and b_2 are obtained by simply substituting the expressions for Ψ into Equation (7.3).

From Theorem 6.1, we know that there are positive constants K and γ such that

$$\|T(t)\phi^Q\| \leq Ke^{-\gamma t}\|\phi^Q\|, \qquad t \geq 0. \tag{7.8}$$

Consequently, the subspace P of C is asymptotically stable. More specifically, with A and Φ defined as earlier, we have

$$A\Phi = \Phi B, \qquad B = \begin{pmatrix} 0 & -\frac{\pi}{2} \\ \frac{\pi}{2} & 0 \end{pmatrix} \tag{7.9}$$

and therefore, $T(t)\Phi = \Phi e^{Bt}$. Since $\phi^Q = \phi - \phi^P$, $\phi^P = \Phi b$, $b = (\Psi, \phi)$, it follows from estimate (7.8) that

$$\|T(t)\phi - \Phi e^{Bt}b\| \to 0$$

exponentially as $t \to \infty$ for every $\phi \in C$, where $b = (\Psi, \phi)$ is given explicitly in Equation (7.7). That is, any solution of Equation (7.1) approaches a periodic function of t given by $b_1 \sin(\pi t/2) + b_2 \cos(\pi t/2)$ where b_1 and b_2 satisfy Equations (7.7).

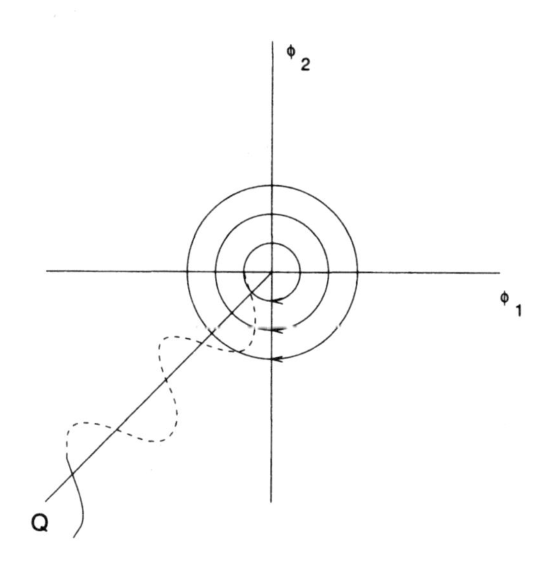

Fig. 7.1.

In the (x, t)-space, it is very difficult to visualize this picture, but in C, everything is very clear. In the subspace P, $T(t)\phi = \Phi e^{Bt}b$, the elements ϕ_1 and ϕ_2 of Φ serve as a coordinate system in P and for any initial value

Φb in P, we have $T(t+4)\Phi b = T(t)\Phi b$ since $\exp[B(t+4)] = \exp Bt$, and in particular, $T(4)\Phi b = \Phi b$; that is, the trajectories in C on P are closed curves. We can, therefore, symbolically represent the trajectories in C as in Figure 7.1. The pictorial representation of P by a (ϕ_1, ϕ_2)-plane is precise, but it should always be kept in mind that Q is an infinite-dimensional space.

7.8 Spectral decomposition according to all eigenvalues

If C is decomposed by $\Lambda_0 = \Lambda(\beta_0) = \{\lambda \in \sigma(A) : \operatorname{Re}\lambda > \beta_0\}$, then we know that there is a constant matrix B_0 whose eigenvalues coincide with Λ_0 such that

$$T(t)\phi^{P_0} = \Phi_0 e^{B_0 t} a, \quad \phi^{P_0} = \Phi_0 a,$$
$$\|T(t)\phi^{Q_0}\| \le M e^{\beta_0 t}, \qquad t \ge 0.$$

The projection $P_0 = P_{\Lambda_0}$ is given by

$$P_0\phi = \Phi_0(\Psi_0, \phi) \tag{8.1}$$

where Φ_0 is a basis of eigenvectors and generalized eigenvectors of A at Λ_0 and Ψ_0 is a basis of eigenvectors and generalized eigenvectors of A^T at Λ_0 such that $(\Psi_0, \Phi_0) = I$.

Suppose we decrease β and cross another pair of eigenvalues of A, say $\Lambda_1 = \Lambda(\beta_1) = \Lambda_0 \cup \{\lambda_k, \bar{\lambda}_k\}$. It is our objective to compute the new projection $P_1 = P_{\Lambda_1}$ from P_0. First of all, observe that if $\lambda_1 \neq \lambda_2$, $\lambda_1 \neq \bar{\lambda}_2$ and if $\phi_1 \in \mathcal{M}_{\lambda_1}(A)$, $\psi_2 \in \mathcal{M}_{\lambda_2}(A^T)$, then $(\psi_2, \phi_1) = 0$. So eigenvectors of A and A^T corresponding to different pairs of eigenvalues are "perpendicular" with respect to the bilinear form $(\cdot\,, \cdot)$. Therefore, adding eigenvalues to Λ does not affect the matrix representation for the projection P_0 and

$$P_1\phi = P_0\phi + \Phi_{\lambda_k,\bar{\lambda}_k}(\Psi_{\lambda_k,\bar{\lambda}_k}, \phi),$$

where the eigenvectors and generalized eigenvectors are normalized according to

$$(\Psi_{\Lambda_1}, \Phi_{\Lambda_1}) = I. \tag{8.2}$$

Next, it is our aim to analyze the projection P_m as β_m decreases to $-\infty$. In order to do so, we need an abstract expression for the coefficients

$$(\psi_\lambda, \phi), \qquad \phi \in C, \quad \psi_\lambda \in \mathcal{M}_\lambda(A^T)$$

so that we can analyze the decay rate as $\operatorname{Re}\lambda$ tends to $-\infty$. Since we have explicit expressions for the eigenfunctions and generalized eigenfunctions of A and A^T, we can compute (ψ_λ, ϕ) explicitly.

Lemma 8.1. *For $\lambda \in \sigma(A)$, let $\psi_1, \dots, \psi_m$ and $\phi_1, \dots, \phi_m$ be bases for $\mathcal{M}_\lambda(A^T)$ and $\mathcal{M}_\lambda(A)$, respectively, such that $(\psi_j, \phi_j) = 1$, for $j = 1, \dots, m$. The projection P_λ onto $\mathcal{M}_\lambda(A)$ is given by*

$$P_\lambda \phi = \sum_{j=1}^{m} (\psi_j, \phi)\phi_j$$

and can be written explicitly as

$$(8.3) \quad (P_\lambda \phi)(\theta) = \operatorname*{Res}_{z=\lambda} e^{z\theta} \Delta(z)^{-1} \big[\phi(0) + \int_{-r}^{0} \int_{-r}^{s} d[\eta(\tau)] e^{-z(s-\tau)} \phi(s)\, ds\big].$$

In particular, if λ is a simple eigenvalue, then

$$(8.4) \quad (\psi_1, \phi)\phi_1 = e^{\lambda \cdot} \frac{d}{dz} \Delta(\lambda)^{-1} \big[\phi(0) + \int_{-r}^{0} \int_{-r}^{s} d[\eta(\tau)] e^{-\lambda(s-\tau)} \phi(s)\, ds\big].$$

Proof. To avoid technical difficulties, we restrict ourselves to the case that λ is a simple eigenvalue. The general case follows by integration by parts.

Since λ is simple, we know from Section 7.4 that

$$\begin{aligned} \psi_1(\xi) &= d_1 e^{-\lambda s}, & 0 \le s \le r, & \quad d_1 \Delta(\lambda) = 0, \\ \phi_1(\theta) &= c_1 e^{\lambda\theta}, & -r \le \theta \le 0, & \quad \Delta(\lambda) c_1 = 0. \end{aligned}$$

Further, from $\Delta(\lambda)^T d_1^T = 0$ it follows that

$$d_1^T = \operatorname*{Res}_{z=\lambda} \, [\Delta(z)^{-1}]^T f, \qquad f \in \mathbb{C}^n.$$

So

$$\begin{aligned} (\psi_1, \phi) &= \int_{-r}^{0} d[F^T \psi_1(\theta)] \phi(\theta) \\ &= d_1 \big[\phi(0) + \int_{-r}^{0} \int_{-r}^{\theta} d[\eta(\tau)] e^{-\lambda(\theta-\tau)} \phi(\theta)\, d\theta\big] \\ &= f^T \operatorname*{Res}_{z=\lambda} \Delta(z)^{-1} \big[\phi(0) + \int_{-r}^{0} \int_{-r}^{\theta} d[\eta(\tau)] e^{-\lambda(\theta-\tau)} \phi(\theta)\, d\theta\big]. \end{aligned}$$

From the normalization $(\psi_1, \phi_1) = 1$

$$\begin{aligned} (\psi_1, \phi_1) &= f^T \operatorname*{Res}_{z=\lambda} \Delta(z)^{-1} \big[c_1 + \int_{-r}^{0} \int_{-r}^{\theta} d[\eta(\tau)] e^{\lambda\tau} c_1\, d\theta\big] \\ &= f^T \frac{d}{dz} \Delta(\lambda)^{-1} \big[I + \int_{-r}^{0} \int_{\tau}^{0} e^{\lambda\tau}\, d\theta\big] c_1 \\ &= f^T c_1 = 1. \end{aligned}$$

Thus

$$(\psi_1, \phi)\phi_1 = e^{\lambda \cdot} \frac{d}{dz} \Delta(\lambda)^{-1} \big[\phi(0) + \int_{-r}^{0} \int_{-r}^{\theta} d[\eta(\tau)] e^{-\lambda(\theta-\tau)} \phi(\theta)\, d\theta \big]$$

and this completes the proof of the lemma. □

As a corollary to the lemma, the spectral projection P_Λ corresponding to a set $\Lambda = \{\lambda_1, \dots, \lambda_m\}$ of eigenvalues of A can be written as

$$(P_\Lambda \phi)(\theta) = \sum_{j=1}^{m} \operatorname*{Res}_{z=\lambda_j} \frac{P(z,\phi)(\theta)}{\det \Delta(z)} \tag{8.5}$$

where

$$P(z,\phi)(\theta) \stackrel{\text{def}}{=} e^{z\theta} \operatorname{adj} \Delta(z) \big[\phi(0) + \int_{-r}^{0} \int_{-r}^{s} d[\eta(\tau)] e^{-z(s-\tau)} \phi(s)\, ds \big]. \tag{8.6}$$

Observe that the residues correspond precisely to the residues in Section 3.3. In particular, the spectral projection P_Λ projects ϕ onto the initial value of a characteristic solution corresponding to Λ.

To analyze the behavior of the residues, we invoke Cauchy's theorem and rewrite the sum of residues as a contour integral. Let Γ_Λ be a simple closed smooth curve. If Γ_Λ encloses Λ, but no other eigenvalues of A, then

$$P_\Lambda \phi = \frac{1}{2\pi i} \int_{\Gamma_\Lambda} \frac{P(z,\phi)}{\det \Delta(z)}\, dz.$$

Note that from Representation (4.6) for the resolvent of A, it follows that P_Λ as defined is the Riesz projection

$$P_\Lambda \phi = \frac{1}{2\pi i} \int_{\Gamma_\Lambda} R(z, A)\phi\, dz$$

onto the generalized eigenspace of A corresponding to Λ.

Define the *generalized eigenspace* of A to be the linear subspace $\mathcal{M}(A)$ generated by all $\mathcal{M}_\lambda(A)$, i.e.,

$$\mathcal{M}(A) = \bigoplus_{\lambda \in \sigma(A)} \mathcal{M}_\lambda(A). \tag{8.7}$$

The system of eigenfunctions and generalized eigenfunctions is called *complete* if $\mathcal{M}(A)$ is dense in C, i.e., $\overline{\mathcal{M}(A)} = C$. The space $\mathcal{M}(A)$ is precisely the space of initial functions such that the corresponding solution has a finite expansion in characteristic solutions. In fact, the solution through $\phi \in \mathcal{M}(A)$ exists for all time and the semigroup $T(t)$ extends to a flow on $\mathcal{M}(A)$. In general, however, the space $\mathcal{M}(A)$ can be too small to be interesting, can even be finite dimensional (see the example in Section 3.3). Further, if it is infinite dimensional it is difficult to characterize and not closed. Therefore, we turn to the closure of $\mathcal{M}(A)$. In Section 3.3, we claimed that this is an invariant subspace on which the semigroup is one-to-one and which contains all information about the equation.

First we analyze the range of the semigroup $T(t)$ using duality. From the results in this chapter, we know that the adjoint semigroup $T^*(t) : B_0 \to B_0$ corresponds to the Volterra integral equation

$$y(s) + \int_s^0 y(\tau)\eta(s-\tau)\,d\tau = g(s), \qquad s \le 0,$$

with $g \in B_0$. Now note that the numbers ϵ and σ introduced in Section 3.3 are invariant under transpose; that is,

$$\epsilon[\Delta(z)] = \epsilon[\Delta(z)^T], \qquad \sigma[\Delta(z)] = \sigma[\Delta(z)^T].$$

Therefore, an argument similar to the proof of Theorem 3.2 of Section 3.3 implies that the ascent of $T^*(t)$ equals $\epsilon - \sigma$ as well. Since

$$\overline{\mathcal{R}\big(T(t)\big)} = \mathcal{N}\big(T^*(t)\big)^{\perp},$$

we conclude that the closure of the range of $T(t)$ becomes independent of t for $t \ge \epsilon - \sigma$. (The range itself becomes smaller with increasing t due to the fact that the solutions become more smooth.) By definition, $\mathcal{M}(A) \subset \mathcal{R}\big(T(t)\big)$ for $t \ge 0$. Hence

$$\overline{\mathcal{M}(A)} \subseteq \overline{\mathcal{R}\big(T(t)\big)}.$$

Theorem 8.1. $\overline{\mathcal{M}(A)} = \overline{\mathcal{R}\big(T(\epsilon - \sigma)\big)}$.

Sketch of the proof. The orthogonal complement of $\overline{\mathcal{M}(A)}$ is given by

$$\overline{\mathcal{M}(A)}^{\perp} = \bigcap_{\lambda \in P\sigma(A)} \mathcal{N}\big((\lambda I - A^*)^{m_\lambda}\big).$$

Therefore, from the Neumann expansion for the resolvent, we have that for any $\psi \in \overline{\mathcal{M}(A)}^{\perp}$, the function

$$z \mapsto R(z, A^*)\psi \tag{8.8}$$

is an entire function. But the resolvent is the Laplace transform of the semigroup, so that $t \mapsto T^*(t)\psi(0)$ is a solution of the adjoint equation which decays faster than any exponential. So an argument similar to the proof of Theorem 3.1 of Section 3.3 implies that $t \mapsto T^*(t)\psi(0)$ is identically zero after finite time and $\psi \subset \mathcal{N}\big(T^*(\epsilon - \sigma)\big)$. The opposite inequality is clear and we have shown that

$$\overline{\mathcal{M}(A)}^{\perp} = \mathcal{N}\big(T^*(\epsilon - \sigma)\big).$$

This proves the theorem. □

A simple observation yields the following important corollary.

Corollary 8.1. *The system of eigenfunctions and generalized eigenfunctions of the generator of Equation* (1) *is complete; that is,* $\overline{\mathcal{M}(A)} = C$, *if and only if*

$$\mathrm{E}(\det \Delta(z)) = nr.$$

Or, equivalently, if and only if there are no small solutions to System (1).

Let Γ_N be a simple closed smooth curve in $\mathbb{C}$ and let Λ_N be the corresponding set of eigenvalues of A enclosed by Γ_N. Assume that $\overline{\mathcal{M}(A)} = C$. To analyze whether the series expansion $\lim_{N\to\infty} P_{\Lambda_N}\phi$ converges to ϕ, we must estimate the resolvent of A outside small circles centered around the eigenvalues of A. From the representation

$$R(z, A)\phi = \frac{P(z,\phi)}{\det \Delta(z)}$$

where $P(z,\phi)$ is given by Formula (8.6), it follows that it suffices to have good estimates for

$$\Delta(z)^{-1} = \frac{\operatorname{adj} \Delta(z)}{\det \Delta(z)}.$$

As a consequence of general properties of entire functions, the zeros of $\det \Delta(z)$ cannot accumulate. So there exists a sequence of simple closed smooth curves Γ_N such that

(i) There is a complex function $\alpha_N : [0, 2\pi] \to \mathbb{C}$ that is differentiable on $[0, 2\pi]$ such that $\Gamma_N = \{z = \alpha_N(t), t \in [0, 2\pi]\}$, for $t \in [0, 2\pi]$ $|\alpha_N(t)| \to \infty$ as $N \to \infty$ and $|\alpha_N'(t)| \le |\alpha_N(t)|$;

(ii) Γ_N encloses Λ_N but no other eigenvalues of $\sigma(A)$;

(iii) there exists an $\epsilon > 0$ such that $\operatorname{dist}(\Gamma_N, \sigma(A)) > \epsilon$ for all $N = 0, 1, \ldots$.

In order to control the behavior of $|\Delta^{-1}(z)|$ as $|z| \to \infty$, we make the following assumption on the kernel η:

(J) The entries η_{ij} of η have an atom before they become constant, i.e., there is a t_{ij} with $\eta_{ij}(t) = \eta_{ij}(t_{ij}+)$ for $t \ge t_{ij}$ and $\eta_{ij}(t_{ij}-) \ne \eta_{ij}(t_{ij}+)$.

For example η can be a step function.

Theorem 8.2. *Suppose that* η *satisfies* (J) *and* $\overline{\mathcal{M}(A)} = C$. *Then for* $\phi \in \mathcal{D}(A^3)$

$$\phi = \lim_{N\to\infty} P_{\Lambda_N}\phi$$

where Λ_N *denotes the set of eigenvalues enclosed by the contour* Γ_N.

Sketch of the proof. From Cauchy's theorem, it suffices to show that

$$\phi = \lim_{N\to\infty} \frac{1}{2\pi i} \int_{\Gamma_N} R(z,A)\phi \, dz.$$

Using the Newton polygon, one can show that if η satisfies (J) and the exponential type of $\det \Delta(z)$ is equal to nr, then for every $\phi \in \mathcal{D}(A)$ there exists a positive constant M such that

$$(8.9) \qquad |R(z,A)\phi| \leq M \qquad \text{for} \quad z \in \Gamma_N, \ N = 0, 1, \ldots.$$

For $\phi \in \mathcal{D}(A^3)$, it follows that

$$R(z,A)\phi = \frac{1}{z}\phi + \frac{1}{z^2}A\phi + \frac{1}{z^2}R(z,A)A^2\phi.$$

Clearly,

$$\lim_{N\to\infty} \frac{1}{2\pi i} \int_{\Gamma_N} \phi \frac{dz}{z} = \phi, \qquad \lim_{N\to\infty} \frac{1}{2\pi i} \int_{\Gamma_N} A\phi \frac{dz}{z^2} = 0,$$

and

$$\Big|\frac{1}{2\pi i} \int_{\Gamma_N} R(z,A)A^2\phi \frac{dz}{z^2}\Big| = \Big|\frac{1}{2\pi} \int_0^{2\pi} R(\alpha_N(t),A)A^2\phi \frac{\alpha_N'}{\alpha_N(t)^2} \, dt\Big|$$
$$\leq \max_{0\leq t\leq 2\pi} \frac{C}{|\alpha_N(t)|}$$

where we used the properties of the contours Γ_N and Estimate (8.9).

Therefore

$$\lim_{N\to\infty} \frac{1}{2\pi i} \int_{\Gamma_N} R(z,A)A^2\phi \frac{dz}{z^2} = 0$$

and this proves the theorem. □

Pointwise expansion for the solution is possible under weaker conditions.

Theorem 8.3. *Suppose that η satisfies* (J) *and $\overline{\mathcal{M}(A)} = C$. Then for $\phi \in C$ the solution $x(\,\cdot\,;\phi)$ to System* (1) *has the following expansion*

$$x(t;\phi) = \lim_{N\to\infty} \sum_{\lambda\in\Lambda_N} e^{\lambda t}\big(P_\lambda\phi\big)(0)$$
$$= \lim_{N\to\infty} \frac{1}{2\pi i} \int_{\Gamma_N} e^{zt}\big(R(z,A)\phi\big)(0) \, dz, \qquad t > 0.$$

Sketch of the proof. It is easy to see that for $\lambda \in \sigma(A)$, the solution of System (1) corresponding to initial condition $P_\lambda\phi$ is given by $x_t(\,\cdot\,;P_\lambda\phi) = e^{\lambda t}P_\lambda\phi$. In particular,

$$x(t;P_\lambda\phi) = e^{\lambda t}\big(P_\lambda\phi\big)(0) = \big(P_\lambda e^{\lambda t}\phi\big)(0)$$
$$= \frac{1}{2\pi i} \int_{\Gamma_\lambda} e^{zt}\big(R(z,A)\phi\big)(0) \, dz$$

where Γ_λ is a small circle surrounding λ. Since the spectrum of A is contained in a left half plane $\operatorname{Re} z < \gamma$, we can modify the contours Γ_N by replacing the arc in $\operatorname{Re} z \geq \gamma$ by the corresponding line segment $\operatorname{Re} z = \gamma$. From the Laplace inversion formula (in $\mathbb{R}^n$), it follows that

$$x(t;\phi) = \lim_{k\to\infty} \frac{1}{2\pi i} \int_{\gamma-ik}^{\gamma+ik} e^{zt}\big(R(z,A)\phi\big)(0)\,dz \qquad \text{for } t > 0.$$

Let C_N denote the arc of the contour Γ_N contained in the half plane $\operatorname{Re} z < \gamma$. In order to prove that

$$x(t;\phi) = \lim_{N\to\infty} \frac{1}{2\pi i} \int_{\Gamma_N} e^{zt}\big(R(z,A)\phi\big)(0)\,dz, \qquad t > 0,$$

it suffices to prove

$$\lim_{N\to\infty} \frac{1}{2\pi i} \int_{C_N} e^{zt}\big(R(z,A)\phi\big)(0)\,dz = 0 \qquad \text{for } \quad t > 0. \tag{8.10}$$

Since we evaluate the resolvent of A at 0, we can improve the estimate for $R(z,A)$ on C_N. Indeed, using the Newton polygon, it follows that

$$\big|\big(R(z,A)\phi\big)(0)\big| \leq M \min\{\frac{|e^{-hz}|}{|z|}, 1\}, \qquad z \in C_N,\ N \to \infty.$$

The limit in Equation (8.10) now follows from a standard result, see for example, the proof of Lemma 4.2 of Bellman and Cooke [1]. □

Without the assumption (J) the results do not hold. However, using similar techniques one can prove that if $\overline{\mathcal{M}(A)} = C$, then for $\phi \in \mathcal{D}\big(A^\infty\big)$

$$\phi = \lim_{N\to\infty} P_{\Lambda_N}\phi.$$

Since A is the generator of a $\mathcal{C}_0$-semigroup, the domain $\mathcal{D}\big(A^\infty\big)$ is still dense, but too small to imply a convergent series expansion for the solution for $t > 0$.

We end this section with some remarks about what to do if $\mathrm{E}(\det \Delta(z))$ is less than nr; that is, there are small solutions and $\overline{\mathcal{M}(A)} \neq C$.

Define the following subspace of C:

$$\mathcal{E} = \big\{\phi \in C \mid \mathrm{E}(P(z,\phi)(\theta)) \leq \mathrm{E}(\det \Delta(z)),\ -r \leq \theta \leq 0\big\}. \tag{8.11}$$

Note that, $\mathcal{E} \neq C$, if and only if the ascent of $T(t)$ is strictly positive. Furthermore, we have the following lemma.

Lemma 8.2. *The restriction of the solution operator $T(t)$ to $\mathcal{E}$ is one-to-one.*

Proof. Suppose that there exists a $\phi \neq 0$ in C such that $T(\sigma)\phi = 0$ for some $\sigma > 0$. Then

$$R(z, A)\phi = \int_0^\infty e^{-zt}T(t)\phi \, dt = \int_0^\sigma e^{-zt}T(t)\phi \, dt. \tag{8.12}$$

On the other hand,

$$R(z, A)\phi = \frac{P(z, \phi)}{\det \Delta(z)}. \tag{8.13}$$

A combination of Equations (8.12) and (8.13) implies that $\mathrm{E}(P(z, \phi)) > \mathrm{E}(\det \Delta(z))$. So $\phi \notin \mathcal{E}$. □

The following two theorems are stated without proof.

Theorem 8.4. *The subspace $\mathcal{E}$ defined by* (8.11) *is a closed subspace of C that is invariant under the solution operator $T(t)$ defined by* (1.2).

Theorem 8.5. *Let A be the infinitesimal generator given by Relation* (1.2). *If $\mathcal{E}$ is the subspace defined by* (8.11), *then*

$$\overline{\mathcal{M}(A)} = \overline{\mathcal{R}\big(T(\epsilon - \sigma)\big)} = \mathcal{E}.$$

Furthermore, the space C can be decomposed as

$$C = \overline{\overline{\mathcal{M}(A)} \oplus \mathcal{N}\big(T(\epsilon - \sigma)\big)} \tag{8.14}$$

and for $\phi \in \mathcal{D}(A^\infty) \cap \overline{\mathcal{M}(A)}$, $\phi = \lim_{N\to\infty} P_{\Lambda_N}\phi$.

7.9 The decomposition in the variation-of-constants formula

Consider the equation

$$\dot{x}(t) = Lx_t, \qquad t \geq \sigma, \tag{9.1}$$

and the inhomogeneous system

$$\dot{x}(t) = Lx_t + f(t) \tag{9.2}$$

where $f \in \mathcal{L}_1^{\mathrm{loc}}\big([\sigma, \infty), \mathbb{R}^n\big)$. From our general results on linear systems, the solution of Equation (9.2) with initial value ϕ at σ is

$$x_t = T(t - \sigma)\phi + \int_\sigma^t d[K(t, s)]f(s) \tag{9.3}$$

where $T(t)$ is the solution semigroup of Equation (9.1),

$$K(t,s)(\theta) = \int_\sigma^s X(t+\theta-\alpha)\,d\alpha, \qquad t \geq \sigma \tag{9.4}$$

and $X(\cdot)$ denotes the fundamental matrix solution to System (9.1). We now wish to obtain a decomposition of this integral relation according to the results on the decomposition of the homogeneous equation (9.1).

Suppose that $\Lambda = \{\lambda_1, \ldots, \lambda_p\}$ where each λ_j belongs to $\sigma(A)$ and that C is decomposed by Λ as $C = P \oplus Q$, Φ is a basis for the generalized eigenspace P of Λ, Ψ is a basis for the generalized eigenspace of the transposed equation associated with Λ, $(\Psi, \Phi) = I$, where $(\cdot\,,\,\cdot)$ is given by Equation (5.9) and the matrix B is defined by the relations

$$A\Phi = \Phi B, \qquad A^T\Psi = B\Psi.$$

If the decomposition of any element ϕ of C is written as $\phi = \phi^P + \phi^Q$, ϕ^P in P, ϕ^Q in Q, then $\phi^P = \Phi(\Psi, \phi)$. Suppose that x is the solution of Equation (9.2) with initial value ϕ at σ, $x_t = x_t^P + x_t^Q$, and let us compute x_t^P directly from the preceding definition. To do this, we observe that y is a solution of the transposed equation on $(-\infty, \infty)$ and x is a solution of Equation (9.2) for $t \geq 0$, then

$$(y^t, x_t) = \int_\sigma^t y(s)f(s)\,ds + (y^\sigma, x_\sigma) \tag{9.5}$$

for all $t \geq 0$. Each row of the matrix $e^{-Bt}\Psi$, $\Psi(\tau) = e^{-B\tau}\Psi(0)$, $0 \leq \theta \leq r$, is a solution of the transposed equation on $(-\infty, \infty)$, and therefore,

$$(e^{-Bt}\Psi, x_t) = \int_\sigma^t e^{-Bs}\Psi(0)f(s)\,ds + (e^{-B\sigma}\Psi, \phi)$$

and

$$\begin{aligned} x_t^P &= \Phi(\Psi, x_t) \\ &= \int_\sigma^t \Phi e^{B(t-s)}\Psi(0)f(s)\,ds + \Phi e^{B(t-\sigma)}(\Psi, \phi) \\ &= \int_\sigma^t T(t-s)\Phi\Psi(0)f(s) + T(t-\sigma)\Phi(\Psi, \phi) \\ &= \int_\sigma^t T(t-s)X_0^P f(s) + T(t-\sigma)\phi^P, \end{aligned}$$

where $X_0^P = \Phi\Psi(0)$. Note that, although $X_0 \notin C$, $X_0^P = \Phi\Psi(0)$ can be considered as the projection of X_0 onto P.

This formula also shows another interesting property: Namely,

$$(\Psi, x_t) = \int_\sigma^t e^{B(t-s)}\Psi(0)f(s)\,ds + e^{B(t-\sigma)}(\Psi, \phi).$$

Therefore, if we let $y(t) = (\Psi, x_t)$ then

$$y(t) = e^{B(t-\sigma)}y(\sigma) + \int_\sigma^t e^{B(t-s)}\Psi(0)f(s)\,ds.$$

With x_t^P given as earlier, define $K(t,s)^Q = K(t,s) - \Phi(\Psi, K(t,s))$. From the variation-of-constants formula (9.3), we find

$$x_t^Q = x_t - x_t^P = T(t-\sigma)\phi^Q + \int_\sigma^t d[K(t,s)^Q]f(s).$$

If $\Lambda = \{\lambda \in \sigma(A) : \operatorname{Re}\lambda > \gamma\}$, then the estimate on the complementary space Q (see Section 7.6) implies the exponential estimate for $T(t-\sigma)\phi^Q$. To estimate the kernel $K(t,s)^Q$, we need a lemma.

Lemma 9.1. *Let Φ is a basis for the generalized eigenspace P of Λ, Ψ is a basis for the generalized eigenspace of the transposed equation associated with Λ, $(\Psi, \Phi) = I$, and $\Lambda = \{\lambda \in \sigma(A) : \operatorname{Re}\lambda > \gamma\}$. If $K(t,s)$ denotes the kernel in the variation-of-constants formula* (9.3) *and $K(t,s)^Q = K(t,s) - \Phi(\Psi, K(t,s))$, then for any $\beta > 0$, there are constants $M_1, M_2 > 0$ such that*

$$\begin{aligned} |K(t,s)^Q| &\le M_1 e^{(\gamma+\beta)t}, \qquad t \ge s \\ \operatorname{Var}_{[\sigma,t]} K(t,\cdot)^Q &\le M_2 e^{(\gamma+\beta)t}. \end{aligned} \tag{9.6}$$

Proof. Without loss of generality assume that $\sigma = 0$. First consider the kernel $K(t,s)$ for $t - s \ge r$. In this case we can write

$$K(t,s)(\theta) = \int_{t-r-s}^{t-r} X(r+\beta+\theta)\,d\beta.$$

So for $t-s \ge r$, the columns of $X(r+\beta+\cdot)$ belong to C and $X(r+\beta+\cdot) = T(\beta)X_r$. Therefore, we can compute the projection

$$\Phi(\Psi, K(t,s)) = \int_{t-r-s}^{t-r} T(\beta)\Phi(\Psi, X_r)\,d\beta. \tag{9.7}$$

Since $X_r - \Phi(\Psi, X_r)$ belongs to Q, the exponential estimates in (6.1) imply

$$\int_{t-r-s}^{t-r} \big|T(\beta)\big(X_r - \Phi(\Psi, X_r)\big)\big|\,d\beta \le Me^{(\beta+\gamma)t}. \tag{9.8}$$

On the other hand, if $t - s \le r$, we write

$$K(t,s) = \int_0^{t-r} T(\beta)X_r\,d\beta + \int_{-r}^{s-t} X_{-\alpha}\,d\alpha \tag{9.9}$$

and $K(t,s)$ can be estimated by a constant independently of t. The projection $K(t,s)^Q$ has the following representation

(9.10)

$$K(t,s)^Q = \begin{cases} \int_{t-r-s}^{t-r} T(\beta)\big(X_r - \Phi(\Psi, X_r)\big)\,d\beta & \text{if } t-s \ge r; \\ \int_0^{t-r} T(\beta)\big(X_r - \Phi(\Psi, X_r)\big)\,d\beta & \\ \quad + \int_{-r}^0 X_{-\alpha}\,d\alpha - \Phi(\Psi, \int_{-r}^0 X_{-\alpha}\,d\alpha) & \text{if } t-s < r. \end{cases}$$

So the exponential estimates (9.6) now follow immediately from Estimate (9.8). □

These results are summarized in

Theorem 9.1. *Let Φ be a basis for the generalized eigenspace P of Λ, Ψ is a basis for the generalized eigenspace of the transposed equation associated with Λ, $(\Psi, \Phi) = I$. If C is decomposed by Λ as $P \oplus Q$, then the solution $x(\sigma, \phi)$ of Equation* (9.2) *satisfies*

(9.11)

$$\begin{aligned} x_t^P &= T(t-\sigma)\phi^P + \int_\sigma^t T(t-s)X_0^P f(s)\,ds \\ x_t^Q &= T(t-\sigma)\phi^Q + \int_\sigma^t d[K(t,s)^Q]f(s), \qquad t \ge \sigma, \end{aligned}$$

where $X_0^P = \Phi\Psi(0)$ and $K(t,s)^Q = K(t,s) - \Phi(\Psi, K(t,s))$ is given by (9.10). *Also, if $x_t^P = \Phi y(t)$ then*

$$\dot{y}(t) = By(t) + \Psi(0)f(t). \tag{9.12}$$

Furthermore, if $\Lambda = \{\lambda \in \sigma(A) : \operatorname{Re}\lambda > \gamma\}$, then for any $\beta > 0$, there is an $M > 0$ such that

$$|T(t)\phi^Q| \le Me^{(\gamma+\beta)t}|\phi^Q|, \qquad \phi^Q \in Q \tag{9.13a}$$

$$|K(t,s)^Q| \le Me^{(\gamma+\beta)t} \tag{9.13b}$$

$$\operatorname{Var}_{[\sigma,t]} K(t,\cdot) \le Me^{(\gamma+\beta)t}. \tag{9.13c}$$

As an example, consider

$$\dot{x}(t) = -\frac{\pi}{2}x(t-1) + f(t) \tag{9.14}$$

and choose $\Lambda = \{+i\frac{\pi}{2}, -i\frac{\pi}{2}\}$. If we let $P = P(\Lambda)$, $Q = Q(\Lambda)$, $\phi^P = \Phi(\Psi, \phi)$ for any $\phi \in C$, where Φ is defined in Equations (7.5), Ψ in Equations (7.6), $\phi^Q = \phi - \phi^P$, and $x_t = x_t^P + x_t^Q = \Phi y(t) + x_t^Q$, then

(9.15)

$$\begin{aligned} \dot{y}(t) &= By(t) + \Psi(0)f(t) \\ x_t^Q &= T(t-\sigma)\phi^Q + \int_\sigma^t d[K(t,s)^Q]f(s) \end{aligned}$$

where

$$\Psi(0) = \operatorname{col}(\pi\mu, 2\mu),$$

$$\mu = \frac{1}{1 + \pi^2/4}, \qquad B = \begin{pmatrix} 0 & -\frac{\pi}{2} \\ \frac{\pi}{2} & 0 \end{pmatrix}. \tag{9.16}$$

If we let the components of y be y_1 and y_2, then the first of Equations (9.15) is given explicitly as

$$\dot{y}_1 = -\frac{\pi}{2} y_2 + \pi\mu f, \qquad \dot{y}_2 = \frac{\pi}{2} y_1 + 2\mu f.$$

If we let

$$\begin{aligned} z_1 &= \frac{\pi}{2} y_1 + y_2, & y_1 &= \mu\big(\frac{\pi}{2} z_1 + z_2\big) \\ z_2 &= y_1 - \frac{\pi}{2} y_2, & y_2 &= \mu\big(z_1 - \frac{\pi}{2} z_2\big) \end{aligned} \tag{9.17}$$

then

$$\dot{z}_1 = -\frac{\pi}{2} z_2, \qquad \dot{z}_2 = \frac{\pi}{2} z_1 + 2f \tag{9.18}$$

and

$$\ddot{z}_1 + \big(\frac{\pi}{2}\big)^2 z_1 = -\pi f. \tag{9.19}$$

It is interesting to see this second-order equation (9.19) for a special type of forcing term, $f = f\big(x(t), x(t-1)\big)$. From the definition of Φ and Relations (9.17) and (9.18), we have

$$x_t(0) - x_t^Q(0) = \Phi(0) y(t) = y_2(t) = \mu\big(z_1(t) + \dot{z}_1(t)\big)$$

and

$$x_t(-1) - x_t^Q(-1) = \Phi(-1) y(t) = -y_1(t) = -\frac{2}{\pi}\mu\big(\frac{\pi^2}{4} z_1(t) - \dot{z}_1(t)\big).$$

Consequently, if we neglect the terms $x_t^Q(0)$ and $x_t^Q(-1)$, the second-order equation (9.18) becomes

$$\ddot{z}_1 + \big(\frac{\pi}{2}\big)^2 z_1 = -\pi f\big[\mu\big(z_1(t) + \dot{z}_1(t)\big), -\frac{2}{\pi}\mu\big(\frac{\pi^2}{4} z_1(t) - \dot{z}_1(t)\big)\big]. \tag{9.20}$$

In the applications to nonlinear oscillations, this equation plays a fundamental role.

7.10 Parameter dependence

In this section, we present some simple results on the dependence of simple eigenvalues on a parameter in the equation.

For any $\alpha \in \mathbb{R}$, suppose $L(\alpha) : C \to \mathbb{R}^n$ is a continuous linear operator that is continuous, together with its first derivative in α. Suppose λ_0 is a characteristic root of the RFDE$[L(0)]$. It is important in applications to know the dependence on α of the characteristic roots of the RFDE$[L(\alpha)]$, which reduce to λ_0 for $\alpha = 0$. We do not discuss the general problem, but prove only the following simple application of the implicit function theorem.

Lemma 10.1. *Suppose that the family of bounded linear operators $\{L(\alpha) : \alpha \in \mathbb{R}\}$ from C to $\mathbb{R}^n$ is continuous, together with the first derivative in α. If λ_0 is a simple characteristic root of the RFDE$[L(0)]$, then there is an $\alpha_0 > 0$ and a simple characteristic root $\lambda(\alpha)$ of the RFDE$[L(\alpha)]$ that is continuous, together with its first derivative for $|\alpha| < \alpha_0$, $\lambda(0) = \lambda_0$. Furthermore, if C is decomposed by $\lambda(\alpha)$ as $C = P_\alpha \oplus Q_\alpha$ and π_α is the corresponding projection with range P_α, then π_α is continuous together with its first derivative.*

Proof. Consider the equation

$$\Delta(\lambda; \alpha)b = 0, \tag{10.1}$$

where $\Delta(z;\alpha)$ is the characteristic equation $\Delta(z;\alpha) = zI - L(\alpha)e^{z\cdot}$ corresponding to the RFDE$[L(\alpha)]$.

Since we have assumed λ_0 is a simple eigenvalue of the RFDE$[L(0)]$, the null space $\mathcal{N}(\Delta(\lambda_0;0))$ has dimension one and $\mathbb{C}^n = \mathcal{N}(\Delta(\lambda_0;0)) \oplus \mathcal{R}(\Delta(\lambda_0;0))$. Let p, $I-p$ be projection operators defined by this decomposition. The existence of a smooth eigenvalue follows easily from the implicit function theorem. We give another proof, which will produce the eigenvalue as well. Rewrite Equation (10.1) as

$$\Delta(\lambda_0;0)b = [\Delta(\lambda_0;0) - \Delta(\lambda;\alpha)]b. \tag{10.2}$$

Fix a complex eigenvector b_0 of $\Delta(\lambda_0;0)$. If $b = b_0 + d$, $d \in (I-p)\mathbb{C}^n$, then (10.2) is equivalent to

$$\Delta(\lambda_0;0)d = (I-p)[\Delta(\lambda_0;0) - \Delta(\lambda;\alpha)](b_0 + d) \tag{10.3a}$$

$$p[\Delta(\lambda_0;0) - \Delta(\lambda;\alpha)](b_0 + d) = 0. \tag{10.3b}$$

Since $\Delta(\lambda_0;0)$ is an isomorphism on $(I-p)\mathbb{C}^n$, the implicit function theorem implies that there is a $\delta > 0$ and a unique solution $d = d^*(\alpha,\lambda,b_0)$ of Equation (10.3a) for $|\lambda - \lambda_0| < \delta$, $|\alpha| < \delta$, $|b_0 + d^*(\alpha,\lambda,b_0)| \neq 0$, and the function $d^*(\alpha,\lambda,b_0)$ is continuously differentiable in α, λ, b_0 and $d^*(0,\lambda,b_0) = 0$. Also, the linearity of (10.3a) implies that $d^*(\alpha,\lambda,b_0) =$

$D^*(\alpha, \lambda)b_0$ where $D^*(\alpha, \lambda)$ is an $n \times n$-matrix. Therefore, if there is a solution b of $\Delta(\lambda; \alpha)b = 0$, with $b = b_0$ for $\alpha = 0$, $\lambda = \lambda_0$, then b must be defined by $b = [b_0 + D^*(\alpha, \lambda)b_0]$ for $|\alpha| < \delta$, $|\lambda - \lambda_0| < \delta$ and satisfy

$$f(\alpha, \lambda_0) \stackrel{\text{def}}{=} p[\Delta(\lambda_0; 0) - \Delta(\lambda; \alpha)][I + D^*(\alpha, \lambda)]b_0 = 0. \tag{10.4}$$

Therefore, the existence of an eigenvalue near λ_0 for α near zero is equivalent to the existence of a solution of Equation (10.4). Since $\frac{\partial D^*(0,\lambda_0)}{\partial \lambda}$ satisfies the equation

$$\Delta(0, \lambda_0)\frac{\partial D^*(0, \lambda_0)}{\partial \lambda} = -(I - p)\frac{\partial \Delta(\lambda_0; 0)}{\partial \lambda},$$

it follows that

$$\frac{\partial f(0, \lambda)}{\partial \lambda} b_0 = p\frac{\partial \Delta(0, \lambda)}{\partial \lambda} b_0.$$

By choosing basis vectors so that $\Delta(0, \lambda_0) = \text{diag}(0, B)$, where B is an $(n-1) \times (n-1)$ nonsingular matrix, one can choose b_0 as $\text{col}(1, 0, \ldots, 0)$ and the projection p can be identified with $p = (1, 0, \ldots, 0)$. Then $p\Delta(\lambda; 0)b_0 = \alpha(\lambda)$ where λ is the term in the left-hand corner of the matrix $\Delta(\lambda; 0)$. An easy computation now shows that

$$p[\frac{\partial \Delta(\lambda_0; 0)}{\partial \lambda}]b_0 = (\det B)^{-1}\frac{\partial \Delta(\lambda_0; 0)}{\partial \lambda}.$$

Thus, $\frac{\partial f(0,\lambda_0)}{\partial \lambda} \neq 0$. Since $f(0, \lambda_0) = 0$, the implicit function theorem implies there is a $\delta > 0$ (which can be taken as the same δ as before) and a continuously differentiable function $\lambda(\alpha)$, $\lambda(0) = \lambda_0$, such that $f(\alpha, \lambda(\alpha)) = 0$ for $|\alpha| < \delta$. The corresponding eigenvector is then $b(\alpha) = b_0 + D^*(\alpha, \lambda)b_0$.

The same remarks apply to the construction of left eigenvectors $c(\alpha)$ of $\Delta(\lambda(\alpha); \alpha)$. The remainder of the proof of the lemma follows directly from the equivalence in Section 7.4. □

Later, we need an explicit formula for the derivative of the eigenvalue $\lambda(\alpha)$ with respect to α.

Lemma 10.2. *Suppose that the conditions of Lemma* 10.1 *are satisfied and let* $\phi_\alpha(\theta) = b(\alpha)exp(\lambda_0(\alpha)\theta)$, $-r \leq \theta \leq 0$, $\psi_0(s) = a(\alpha)exp(-\lambda_0(\alpha)s)$, $0 \leq s \leq r$, *are bases for* $\mathcal{N}(A(\alpha) - \lambda_0(\alpha))$, $\mathcal{N}(A^T(\alpha) - \lambda_0(\alpha))$, *respectively, for the simple eigenvalue* $\lambda(\alpha)$ *of the RFDE*$[L(\alpha)]$. *If* $(\psi_\alpha, \phi_\alpha) = 1$, *then*

$$\lambda'(\alpha) = -a(\alpha)L'(\alpha)e^{\lambda(\alpha)}b(\alpha)$$

for all $\alpha \in \mathbb{R}$, *where prime denotes differentiation with respect to* α.

Proof. From the definition of $a(\alpha)$, $b(\alpha)$, we have

$$\begin{gathered} a(\alpha)\Delta(\lambda(\alpha); \alpha) = 0, \quad \Delta(\lambda(\alpha); \alpha)b(\alpha) = 0 \\ a(\alpha)\Delta(\lambda(\alpha); \alpha)b(\alpha) = 0 \end{gathered} \tag{10.5}$$

for all $\alpha \in \mathbb{R}$. Let us first observe that

$$a(\alpha)\frac{\partial \Delta(\lambda(\alpha);\alpha)}{\partial \alpha}b(\alpha) = (\psi_\alpha, \phi_\alpha) \tag{10.6}$$

(see the proof of Lemma 8.1) Differentiating the latter expression in (10.5) with respect to α and using Equation (10.6), we have

$$\lambda'(\alpha) = -a(\alpha)\frac{\partial \Delta(\lambda(\alpha);\alpha)}{\partial \alpha}b(\alpha)$$

for all $\alpha \in \mathbb{R}$. From the definition of $\Delta(\lambda(\alpha);\alpha)$, one observes that this latter expression is the same as the one in the statement of the lemma. Thus, the lemma is proved. □

Lemma 10.3. *Suppose that the conditions of Lemma* 10.1 *are satisfied and* $\Lambda(\alpha) = \{\lambda_1(\alpha), \dots, \lambda_p(\alpha)\}$, $\lambda_{2j-1}(\alpha) = \bar{\lambda}_{2j}(\alpha)$, $j = 1.2, \dots, k$, $\lambda_j(\alpha)$ *real* $j = 2k+1, \dots, p$, *is a set of simple eigenvalues of the* RFDE$[L(\alpha)]$. *Let* Φ_α, Ψ_α, $(\Phi_\alpha, \Psi_\alpha) = I$, *be real bases for* $\mathcal{M}_{\Lambda(\alpha)}$, $\mathcal{M}^*_{\Lambda(\alpha)}$, *respectively,* $A(\alpha)\Phi_\alpha = \Phi_\alpha B(\alpha)$. *Then*

$$B'(\alpha) = \Psi_\alpha(0)L'(\alpha)\Phi_\alpha(0)e^{B(\alpha)}$$

for all $\alpha \in \mathbb{R}$.

Proof. If $\lambda_{2j-1}(\alpha) = \mu_j(\alpha) + i\sigma_j(\alpha)$, $\mu_j(\alpha)$, $\sigma_j(\alpha)$ real, $j = 1, 2, \dots, k$, then there is no loss in generality in assuming that $B(\alpha)$ is in real canonical form. Then there is a further change of basis taking Φ_α into $\Phi_\alpha M$ where M is a matrix independent of α such that the corresponding $B(\alpha)$ is diagonal. Therefore, we may assume $B(\alpha)$ is diagonal. In this case, $\Psi = \mathrm{col}(\psi_{1\alpha}, \dots, \psi_{p\alpha})$, $\Phi_\alpha = (\phi_{1\alpha}, \dots, \phi_{p\alpha})$ satisfy the relations

$$\begin{aligned} \psi_{j\alpha}(s) &= a_j(\alpha)e^{-\lambda_j(\alpha)s}, \quad 0 \le s \le r, \\ \phi_{j\alpha}(\theta) &= b_j(\theta)e^{\lambda_j(\alpha)\theta}, \quad -r \le \theta \le 0 \end{aligned}$$

for $j = 1, 2, \dots, p$. From Lemma 10.2 one observes that

$$a_j(\alpha)L'(\alpha)e^{\lambda_k(\alpha)}b_k(\alpha) = \begin{cases} 0, & j \ne k, \\ -\lambda'_j(\alpha), & j = k. \end{cases}$$

Since $\Psi_\alpha(0) = \mathrm{col}(a_1(\alpha), \dots, a_p(\alpha))$, $\Phi_\alpha(0) = (b_1(\alpha), \dots, b_p(\alpha))$, one obtains the result stated in the lemma. □

7.11 Supplementary remarks

For basic theory on semigroups of transformations, see Hille and Phillips [1], Yoshida [1], or Pazy [1]. The idea of treating a linear autonomous functional differential equation as a semigroup of transformations on C originated with Krasovskii [1] in his studies on stability. Shimanov [2] continued this investigation and actually used the decomposition in the space C in a very special case. The general theory was first given by Hale [4] and Shimanov [3].

An abstract perturbation theory for delay equations is a nontrivial problem. The reason is the following. If one writes System (1) as an abstract evolutionary system in the Banach space C

$$\dot{u}(t) = Au, \qquad u(0) = u_0 \in C,$$

then A is given by Relation (1.2). Observe that the action of A is independent of Equation (1), the equation only enters in the definition of the domain of A. This means, in particular, that perturbing Equation (1) is the same as perturbing the domain of an infinitesimal generator. This makes perturbations hard to handle.

An important idea to solve this problem is the notion of the part of an operator. If $\widehat{M} : \mathcal{D}(\widehat{M}) \to Z$ is an unbounded operator and $X \subset Z$, then M, the *part of* $\widehat{M}$ in X, is defined to be

$$\mathcal{D}(M) = \{\phi \in \mathcal{D}(\widehat{M}) : \widehat{M}\phi \in X\}, \qquad M\phi = \widehat{M}\phi.$$

It is an immediate consequence of the general theory that if $\widehat{M} : \mathcal{D}(\widehat{M}) \to Z$ is the generator of a semigroup $\widehat{T}(t)$, then the part of $\widehat{M}$ in $X \subset Z$ is a generator of a semigroup $T(t)$ if and only if $\widehat{T}(t)X \subseteq X$ and, in that case, $T(t) = \widehat{T}(t)|_X$.

Now the first question becomes: Can one find a space $Z \supset X$ and an infinitesimal generator $\widehat{A}$ such that A, defined by Relation (1.2), is the part of $\widehat{A}$ in C.

In the text, we have seen that for $Z = \mathbb{R}^n \times C$, $\widehat{A}(c, \phi) = (L\phi, d\phi/d\theta)$, the generator A is indeed the part of $\widehat{A}$ in C. However, $\widehat{A}$ is not the generator of a semigroup on $Z = \mathbb{R}^n \times C$. In fact, one can prove that $\widehat{A}$ is the generator of a once integrated semigroup (see Thieme [1]).

Another choice would be $Z = \mathbb{R}^n \times L_\infty([-r, 0], \mathbb{R}^n)$, where the generator A is still the part of $\widehat{A}$ in C and in this case, $\widehat{A}$ is the generator of a semigroup on $Z = \mathbb{R}^n \times L_\infty([-r, 0], \mathbb{R}^n)$. Unfortunately, $\widehat{T}(t)$ is not a strongly continuous semigroup, but only a weak* continuous semigroup. This is where duality enters the problem. A general method is the (sun star) $\odot *$-framework developed by Clément et al. [1].

The idea is the following. Start with a C_0-semigroup $T_0(t)$ on a Banach space X and consider the adjoint semigroup $T_0^*(t)$ on the dual space X^*. In general, the adjoint semigroup is only weak* continuous and not strongly continuous (see Section 7.1). Therefore, one can restrict to the largest $T_0^*(t)$-invariant subspace, denoted by $X^\odot$, of X^* on which $T_0^*(t)$ is strongly continuous. It is a general result due to Hille and Phillips [1] that

$X^{\odot} = \overline{\mathcal{D}(A^*)}$, the norm closure of the domain of A^* in X^*. From the remarks made earlier, the restriction $T_0^{\odot}(t)$ of $T_0^*(t)$ to $X^{\odot}$ is generated by the part of A^* in $X^{\odot}$.

Now one can repeat the procedure and find a space $X^{\odot *}$ and a weak* continuous semigroup $T^{\odot *}(t)$. Since $X^{\odot}$ is a subspace of the dual space of X, X is embedded in $X^{\odot *}$. Furthermore, from the construction, it follows that $T^{\odot *}(t)|_X = T(t)$. So the largest invariant subspace $X^{\odot\odot}$ on which $T^{\odot *}(t)$ is strongly continuous contains X. When $X \cong X^{\odot\odot}$ one calls X *sun-reflexive* with respect to $T_0(t)$.

Let $X = C$ and $T_0(t)$ be the C_0-semigroup corresponding to

$$\dot{u}(t) = A_0 u(t), \qquad u(0) = u_0 \in X,$$

where

$$\mathcal{D}(A_0) = \{\phi \in C : \dot{\phi}(0) = 0\}, \qquad A_0\phi = \frac{d\phi}{d\theta},$$

that is, A_0 is the generator corresponding to $\dot{x}(t) = 0$ considered as a delay equation. It can be shown that C with respect to $T_0(t)$ is sun-reflexive. System (1) on $X^{\odot *} = \mathbb{R}^n \times L_\infty([-r, 0], \mathbb{R}^n)$ corresponds to the evolutionary system

$$\dot{u}(t) = A_0^{\odot *} u + Bu,$$

where $B : C \to X^{\odot *}$, $\phi \mapsto (L\phi, 0)$ and

$$\mathcal{D}(A_0^{\odot *}) = \{(c, \phi) \in Z : \frac{d\phi}{d\theta} \in L_\infty,\ \dot{\phi}(0) = 0\}, \quad A_0^{\odot *}(c, \phi) = (0, \frac{d\phi}{d\theta}).$$

On $X^{\odot *}$, we have a good perturbation theory and the core of the theory is the following abstract variation-of-constants formula for $T(t)$, the perturbed semigroup on C,

$$T(t) = T_0(t) + \int_0^t T_0^{\odot *}(t - s) B T(s)\, ds$$

where the integral has to be interpreted as a weak* integral. Thus, even though the map B takes C into the larger space $X^{\odot *}$, the convolution integral is in C and one obtains solutions in C.

So although the initial investment is large, the result yields a powerful abstract variation-of-constants formula for a larger class of perturbations than those bounded from C into C. The method not only applies to functional differential equations, but also to different types of problems, for example, structured population dynamics. See Diekmann, van Gils, Verduyn Lunel, and Walther [1].

In this chapter, we have chosen an approach that minimizes the use of functional analysis. Our variation-of-constants formula, which is an abstract integral equation in the state space C, allows us to handle perturbations of System (1). The only disadvantage might be that we do not have a perturbation theory at the generator level.

The text in Section 7.3 follows Kaashoek and Verduyn Lunel [1]. We have included the abstract result since it also can be applied to neutral and periodic functional differential equations as will be illustrated in later chapters. Theorem 4.2 is called a "folk theorem in functional differential equations" and was first proved, using a different approach, by Levinger [1]. See also Kappel and Wimmer [1]

The applications of the decomposition theory of this chapter are numerous. Some applications to perturbed linear systems and behavior near constant and periodic solutions of nonlinear autonomous equations will be given in later chapters.

The theory of existence of small solutions has important applications in the theory of control for linear functional differential equations. See the papers by Delfour and Manitius [1,2], and Manitius [1].

Early work on the closure of $\mathcal{M}(A)$ and, in particular, the convergence of the infinite series representation of a solution in terms of eigenfunctions is contained in Bellman and Cooke [1], Pitt [1] and Banks and Manitius [1].

Estimates for $\Delta^{-1}(z)$ using the Newton polygon are given in Bellman and Cooke [1], Banks and Manitius [1] and Verduyn Lunel [2,3]. The results presented in Section 7.8 hold for neutral equations as well and can be found in Verduyn Lunel [4,5,6]. Convergence in norm rather than pointwise requires delicate estimates and Theorem 8.2 is not optimal. Under assumption (J) the spectral projections $P_{\Lambda_k}\phi$ converge pointwise to ϕ when summation in the sense of Cesàro is used. See Verduyn Lunel [7] for the details.

8
Periodic systems

The purpose of this chapter is to develop the theory of a linear periodic RFDE that is analogous to the Floquet theory for ordinary differential equations. It is shown by example that a complete Floquet theory does not exist. However, it is possible to define characteristic multipliers and exploit the compactness of the solution operator to show that a Floquet representation exists on the generalized eigenspace of a characteristic multiplier. The decomposition of the space C by a characteristic multiplier is also applied to the variation-of-constants formula. The case of periodic delay equations with integer lags is considered in detail.

8.1 General theory

Suppose $L : \mathbb{R} \to \mathcal{L}(C, \mathbb{R}^n)$ satisfies the conditions of Section 6.1 and suppose there is an $\omega > 0$ such that $L(t+\omega) = L(t)$ for all t. In this section, we consider the system

$$\dot{x}(t) = L(t)x_t \tag{1.1}$$

and the extent to which there is a Floquet theory.

For any $s \in \mathbb{R}$, $\phi \in C$, there is a solution $x = x(s, \phi)$ of Equation (1.1) defined on $[s, \infty)$ and $x_t(s, \phi)$ is continuous in t, s, and ϕ. As usual, let $T(t,s)\phi = x_t(s, \phi)$ for all $t \geq s$ and $\phi \in C$. The operator $T(t,s)$ always satisfies $T(t,s)T(s,\tau) = T(t,\tau)$ for all $t \geq s \geq \tau$ and periodicity of Equation (1.1) implies $T(t+\omega, s) = T(t,s)T(s+\omega, s)$ for all $t \geq s$. Let $U : C \to C$ be defined by

$$U\phi = T(\omega, 0)\phi.$$

Since $\omega > 0$, there is an integer $m > 0$ such that $m\omega \geq r$. Therefore, $U^m = T(m\omega, 0)$ is completely continuous. The polynomial spectral theorem and the theory of completely continuous operators imply the spectrum of $\sigma(U)$ of U is at most countable, a compact set of the complex plane with the only possible accumulation point being zero. Also, if $\mu \neq 0$ is in $\sigma(U)$, then μ is in the point spectrum $P\sigma(U)$ of U; that is, there is a $\phi \neq 0$ in C

such that $U\phi = \mu\phi$. Any $\mu \neq 0$ in $P\sigma(U)$ is called a *characteristic multiplier* or *Floquet multiplier* of Equation (1.1) and for λ for which $\mu = e^{\lambda\omega}$ is called a *characteristic exponent* of Equation (1.1).

Lemma 1.1. *$\mu = e^{\lambda\omega}$ is a characteristic multiplier of Equation (1.1) if and only if there is a $\phi \neq 0$ in C such that $T(t+\omega,0)\phi = \mu T(t,0)\phi$ for all $t \geq 0$.*

Proof. If $\mu \in P\sigma(U)$, then there is a $\phi \neq 0$ in C such that $U\phi = \mu\phi$. Periodicity of Equation (1.1) implies $T(t+\omega,0)\phi = \mu T(t,0)\phi$ for all t. The converse is trivial. □

Since U^m is completely continuous, for any characteristic multiplier μ of Equation (1.1), there are two closed subspaces E_μ and Q_μ of C such that the following properties hold:

(i) E_μ is finite dimensional;

(ii) $E_\mu \oplus Q_\mu = C$;

(iii) $UE_\mu \subseteq E_\mu$, $UQ_\mu \subseteq Q_\mu$;

(iv) $\sigma(U|E_\mu) = \{\mu\}$, $\sigma(U|Q_\mu) = \sigma(U) \setminus \{\mu\}$.

The dimension of E_μ is called the *multiplicity* of the multiplier μ.

Let $\phi_1, \dots, \phi_{d_\mu}$ be a basis for E_μ, $\Phi = \{\phi_1, \dots, \phi_{d_\mu}\}$. Since $UE_\mu \subseteq E_\mu$, there is a $d_\mu \times d_\mu$ matrix $M = M_\mu$ such that $U\Phi = \Phi M$ and Property (iv) implies that the only eigenvalue of M is $\mu \neq 0$. Therefore, there is a $d_\mu \times d_\mu$ matrix $B = B_\mu$ such that $B = \omega^{-1}\log M$. Define the vector $P(t)$ with elements in C by $P(t) = T(t,0)\Phi e^{-Bt}$. Then, for $t \geq 0$,

$$\begin{aligned} P(t+\omega) &= T(t+\omega,0)\Phi e^{-B(t+\omega)} = T(t,0)T(\omega,0)\Phi e^{-B\omega}e^{-Bt} \\ &= T(t,0)U\Phi e^{-B\omega}e^{-Bt} = T(t,0)\Phi M e^{-B\omega}e^{-Bt} = P(t); \end{aligned}$$

that is, $P(t)$ is periodic of period ω. Thus, $T(t,0)\Phi = P(t)e^{Bt}$, $t \geq 0$. Extend the definition of $P(t)$ for t in $(-\infty,\infty)$ in the following way. If $t < 0$, there is an integer k such that $t + k\omega > 0$ and let $P(t) = P(t+k\omega)$. The function $x_t(0,\Phi) = T(t,0)\Phi = P(t)e^{Bt}$ is well defined for $-\infty < t < \infty$ and it is easily seen that each column of this matrix is a solution of Equation (1.1) on $(-\infty,\infty)$. We therefore have

Lemma 1.2. *If μ is a characteristic multiplier of Equation (1.1) and Φ is a basis for E_μ of dimension d_μ, there is a $d_\mu \times d_\mu$ matrix B, $\sigma(e^{B\omega}) = \{\mu\}$ and an $n \times d_\mu$-matrix function $P(t)$ with each column in C, $P(t+\omega) = P(t)$, $t \in (-\infty,\infty)$ such that if $\phi = \Phi b$, then $x_t(\phi)$ is defined for $t \in (-\infty,\infty)$ and*

$$x_t(0,\phi) = P(t)e^{Bt}b, \qquad t \in (-\infty,\infty).$$

Therefore, in particular, $\mu = e^{\lambda\omega}$ is a characteristic multiplier of Equation (1.1) if and only if there is a nonzero solution of Equation (1.1) of the form

$$x(t) = p(t)e^{\lambda t}$$

where $p(t+\omega) = p(t)$.

Since $x_t(0,\phi)(\theta) = x(0,\phi)(t+\theta) = x_{t+\theta}(0,\phi)(0)$, $-r \le \theta \le 0$, and $\phi \in E_\mu$, it follows that $P(t)(\theta) = P(t+\theta)(0)e^{B\theta}$, $-r \le \theta \le 0$. Therefore, if we let $\widetilde{P}(t+\theta) = P(t+\theta)(0)$, then $\Phi(\theta) = \widetilde{P}(\theta)e^{B\theta}$ and

$$x(0,\phi)(t) = \widetilde{P}(t)e^{Bt}b, \qquad t \in (-\infty,\infty),\ \phi = \Phi b.$$

Therefore, the solutions of Equation (1.1) with initial value in E_μ are of the Floquet type; namely, if $\mu = e^{\lambda\omega}$, the solutions are of the form $e^{\lambda t}$ times a polynomial in t with coefficients periodic in t of period ω.

We also need the following remark: If $T(t,0)\Phi b = 0$ for any t, then $b = 0$. In fact, if there is a t such that $T(t,0)\Phi b = 0$ and $m\omega \ge t$, $m \ge 1$, then

$$0 = T(m\omega,0)\Phi b = U^m\Phi b = \Phi M^n b.$$

Since the eigenvalue of M^m is $\mu^m \ne 0$, the result follows immediately.

We have defined the characteristic multipliers of Equation (1.1) in terms of the period map U of Equation (1.1) starting with the initial time 0. To justify the terminology, it is necessary to show that the multipliers do not depend on the starting time. We prove much more than this.

For any s in $\mathbb{R}$, let $U(s) = T(s+\omega,s)$. As before, for any $\mu \ne 0$, $\mu \in \sigma(U(s))$, there exist two closed subspaces $E_\mu(s)$ and $Q_\mu(s)$ of C such that Properties (i)–(iv) hold with the appropriate change in notation. Let $\Phi(s)$ be a basis for $E_\mu(s)$, $U(s)\Phi(s) = \Phi(s)M(s)$, $\sigma(M(s)) = \{\mu\}$. As for the case $s = 0$, one can define $T(t,s)\Phi(s)$ for all $t \in (-\infty,\infty)$. For any real number τ

$$\begin{aligned} U(\tau)T(\tau,s)\Phi(s) &= T(\tau+\omega,\tau)T(\tau,s)\Phi(s) \\ &= T(\tau+\omega,s)\Phi(s) \\ &= T(\tau,s)T(s+\omega,s)\Phi(s) \\ &= T(\tau,s)\Phi(s)M(s). \end{aligned}$$

If $M = \mu I - N$, then N is nilpotent and

$$[\mu I - U(\tau)]T(\tau,s)\Phi(s) = T(\tau,s)\Phi(s)N.$$

Since $T(t,s)\Phi(s)b = 0$ implies $b = 0$, it follows that μ is in $\sigma(U(\tau))$ and the dimension of $E_\mu(\tau)$ is at least as large as the dimension of $E_\mu(s)$. Since one can reverse the role of s and τ, we obtain the following

Lemma 1.3. *The characteristic multipliers of Equation* (1.1) *are independent of the starting time and if* $\Phi(s)$ *is a basis for* $E_\mu(s)$*, then* $T(t,s)\Phi(s)$ *is a basis for* $E_\mu(t)$ *for any* t *in* $\mathbb{R}$.

Lemma 1.3 shows in particular that the sets $E_\mu(s)$ and $E_\mu(t)$ are diffeomorphic for all s and t. Similarly the sets $Q_\mu(s)$ and $Q_\mu(t)$ are homeomorphic. In fact, let $\pi_t : C \to Q_\mu(t)$ and $I - \pi_t : C \to E_\mu(t)$ be projections defined by the decomposition $C = E_\mu(t) \oplus Q_\mu(t)$. Since Equation (1.1) is periodic in t, the mapping π_t is uniformly continuous in t. Thus, there is a $\delta > 0$ such that for any $s \in \mathbb{R}$, $|\alpha| < \delta$, the linear mapping $\pi_{s+\alpha}|_{Q_\mu(s)} : Q_\mu(s) \to Q_\mu(s+\alpha)$ is an isomorphism. Therefore, each $Q_\mu(s)$ is homeomorphic to each $Q_\mu(t)$.

There is more information in Lemma 1.3. In fact, let Φ be a basis for $E_\mu(0)$. Then $\Phi(\theta) = P(\theta)e^{B\theta}$, $-r \le \theta \le 0$, where $P(\theta+\omega) = P(\theta)$ and B is a constant matrix. Lemma 1.3 implies Φ_t, $\Phi_t(\theta) = P(t+\theta)e^{B(t+\theta)}$, $-r \le \theta \le 0$, is a basis for $E_\mu(t)$. Therefore, the mapping $h(t) : E_\mu(0) \to E_\mu(t)$ defined by $h(t)\phi = \Phi_t b$, where $\phi = \Phi b$ is differentiable in t. This implies that the set $\bigcup_{t\in\mathbb{R}}(t, E_\mu(t)) \subseteq \mathbb{R}\times C$ is diffeomorphic to $\mathbb{R}\times E_\mu(0)$ through the mapping $(t,\phi) \to (t, h(t)\phi)$.

It is not known if the remarks in the preceding paragraph hold for $Q_\mu(t)$. Our homeomorphism taking $Q_\mu(t)$ into $Q_\mu(s)$ was constructed through the projections π_t. In general, these mappings seem to be only continuous in t. This mapping is differentiable if the function $L(t)\phi$ is continuously differentiable in t. We now prove this fact.

Let $L(t)$ be as in Equation (1.1) and assume that the derivative $\partial L(t)\phi/\partial t$ is continuous. For any $\alpha \in \mathbb{R}$, consider the equation

$$\dot{x}(t) = L(t+\alpha)x_t. \tag{1.2}$$

If μ is a characteristic multiplier of Equation (1.1), then μ is a characteristic multiplier of Equation (1.2). Furthermore, if $x_t(\sigma,\phi)$ is a solution of Equation (1.2) through (σ,ϕ), and $t+\alpha = s$, $x(\sigma,\phi)(t) = z(s)$, then z satisfies Equation (1.1) and $z_{\sigma+\alpha} = \phi$. Therefore, if $T(t,\sigma,\alpha)$, $T(\sigma,\sigma,\alpha) = I$, is the solution operator of Equation (1.2), then $T(t,\sigma,\alpha) = T(t+\alpha,\sigma+\alpha,0)$. Now take $\sigma = 0$, $\alpha = s$. Then the decomposition $C = E_\mu(s) \oplus Q_\mu(s)$ according to the multiplier μ is obtained from the mapping

$$U(s) \stackrel{\text{def}}{=} T(\omega,0,s) = T(\omega+s,s,0).$$

Consequently, if $\partial L(t,\phi)/\partial t$ is continuous, then $\partial U(s)/\partial s$ is continuous from Equation (1.2) and the mapping π_s will be differentiable in s.

Let

$$\mathcal{Q}_\mu = \bigcup_{t\in\mathbb{R}} (t, Q_\mu(t)), \tag{1.3}$$

$$\mathcal{E}_\mu = \bigcup_{t\in\mathbb{R}} (t, E_\mu(t)). \tag{1.4}$$

These results are summarized in the following lemma.

Lemma 1.4. *For System* (1.1), *the sets* $E_\mu(t)$ *and* $E_\mu(s)$ *are diffeomorphic and the sets* $Q_\mu(t)$ *and* $Q_\mu(s)$ *are homeomorphic for all* $t, s \in \mathbb{R}$. *The set* $\mathcal{E}_\mu$ *in Expression* (1.4) *is diffeomorphic to* $\mathbb{R} \times E_\mu(0)$ *and the set* $\mathcal{Q}_\mu$ *is homeomorphic to* $\mathbb{R} \times Q_\mu(0)$. *If* $\partial L(t,\phi)/\partial t$ *is also continuous, then* $\mathcal{Q}_\mu$ *is diffeomorphic to* $\mathbb{R} \times Q_\mu(0)$. *These diffeomorphisms are defined by* $(\alpha, \phi) \mapsto (\alpha, g(\alpha)\phi)$, $\alpha \in \mathbb{R}$, $\phi \in E_\mu(\alpha)$, $(\alpha, \psi) \mapsto (\alpha, h(\alpha)\psi)$, $\alpha \in \mathbb{R}$, $\psi \in Q_\mu(\alpha)$, *where* g *and* h *are continuously differentiable.*

The next result relates the Floquet representation in each eigenspace to the solution $x_t(\phi, s)$ for an arbitrary $\phi \in C$. Suppose $\sigma(U) = \{0\} \cup \{\mu_m\}$ where $\{\mu_m\}$ is either finite or countable and each $\mu_m \neq 0$. For μ_n in $\sigma(U)$, let $P_n(s) : C \to E_{\mu_n}(s)$, $I - P_n(s) : C \to Q_{\mu_n}(s)$, be projections induced by $E_{\mu_n}(s)$ and $Q_{\mu_n}(s)$ defined earlier and satisfying Properties (i)–(iv).

Theorem 1.1. *Suppose* ϕ *is a given element of* C *and the notation is as described earlier. If* α *is an arbitrary real number, then there are constants* $\beta = \beta(\alpha) > 0$, $M = M(\alpha) > 0$ *such that*

$$\Big|x_t(s,\phi) - \sum_{|\mu_n| \geq \exp \alpha\omega} x_t(s, P_n(s)\phi)\Big| \leq M e^{(\alpha-\beta)(t-s)}|\phi|, \qquad t \geq s.$$

Proof. For any integer k, one has

$$C = E_{\mu_1}(s) \oplus \cdots \oplus E_{\mu_k}(s) \oplus F_k(s)$$

for any set $\mu_1, \ldots, \mu_k$ of nonzero elements of $\sigma(U(s))$, where each $E_{\mu_j}(s)$ and $F_k(s)$ are invariant under $U(s)$ and

$$\sigma(U(s)|F_k(s)) = \sigma(U(s)) \setminus \{\mu_1, \ldots, \mu_k\}.$$

Order the μ_n so that $|\mu_1| \geq |\mu_2| \geq \cdots > 0$ and, for any α, let $k = k(\alpha)$ be the integer satisfying $|\mu_k| \geq \exp \alpha\omega$ and $|\mu_{k+1}| < \exp \alpha\omega$. If

$$R_k(s)\phi = \phi - \sum_{j=1}^{k} P_{\mu_j}(s)\phi,$$

then $R_k(s)\phi \in F_k(s)$, $x_{s+m\omega}(s, R_k(s)\phi) \in F_k(s)$ for all $m = 0, 1, \ldots$ and $\phi \in C$. If $e^{\gamma\omega} = |\mu_{k+1}|$, then the spectral radius of

$$U_1 \stackrel{\text{def}}{=} U(s)|F_k(s)$$

is $e^{\gamma\omega}$. Therefore, $\lim_{n\to\infty} |U_1^n|^{1/n} = e^{\gamma\omega}$. The proof is completed exactly as we did for the case when Equation (1.1) was independent of t (see Section 7.6). □

Corollary 1.1. *The solution* $x = 0$ *of Equation* (1.1) *is uniformly asymptotically stable if and only if all characteristic multipliers of Equation* (1.1) *have moduli less than* 1.

Proof. The "if" part follows from Theorem 1.1 with $\alpha = 0$. The "only if" part is a consequence of the Floquet representation associated with any characteristic multiplier. □

In a similar manner, one obtains

Corollary 1.2. *The solution $x = 0$ of Equation* (1.1) *is uniformly stable if and only if all characteristic multipliers of Equation* (1.1) *have moduli ≤ 1 and further, if μ is a multiplier with $|\mu| = 1$, then all solutions of Equation* (1.1) *with initial value in E_μ are bounded.*

Corollary 1.3. *Suppose that $\sum_m P_{\mu_m}(s)\phi$ converges, where the sum is taken over all projections associated with the nonzero eigenvalues. If $R\phi = \phi - \sum_m P_{\mu_m}(s)\phi$, then, for any real number k,*

$$\lim_{t\to\infty} e^{kt}|x_t(s, R\phi)| = 0;$$

that is, the solution corresponding to the initial condition $R\phi$ is a small solution of Equation (1.1).

In Section 8.3 we shall determine the characteristic multipliers and analyze the existence of small solutions in particular cases. In contrast to autonomous linear equations, small solutions need not be identically zero after some finite time as we shall see in Section 8.3.

First we present the decomposition theory.

8.2 Decomposition

In this section, we consider the linear periodic system (1.1) and the problem of the decomposition of the space C using the generalized eigenspaces of the characteristic multipliers of System (1.1). The results are easy consequences of the adjoint theory of Section 6.3.

For a given σ, if $U(\sigma) = T(\sigma+\omega, \sigma)$, then $U^*(\sigma) = T^*(\sigma, \sigma+\omega)$ is the period map corresponding to the periodic Volterra integral equation

$$y(s) + \int_s^{\sigma+\omega} y(\tau)\eta(\tau, s-\tau)\, d\tau = g^{\sigma+\tau}(s), \qquad s \leq \sigma+\omega \tag{2.1}$$

with $g \in B_0$. Here $U^*(\sigma)g$ is the forcing function corresponding to the solution y restricted to $(-\infty, \sigma]$.

We have already seen in Section 8.1 that, for some integer m, $E_\mu(\sigma) = \mathcal{N}(\,(\mu I - U(\sigma))^m\,)$, $Q_\mu(\sigma) = \mathcal{R}(\,(\mu I - U(\sigma))^m\,)$ and

$$C = \mathcal{N}(\,(\mu I - U(\sigma))^m\,) \oplus \mathcal{R}(\,(\mu I - U(\sigma))^m\,). \tag{2.2}$$

Furthermore,

$$B_0 = \mathcal{N}\big((\mu I - U^*(\sigma))^m\big) \oplus \mathcal{R}\big((\mu I - U^*(\sigma))^m\big) \tag{2.3}$$

with $d = \dim \mathcal{N}\big((\mu I - U(\sigma))^m\big) = \dim \mathcal{N}\big((\mu I - U^*(\sigma))^m\big) < \infty$.

Let $\Psi_\mu^*(\sigma) = \mathrm{col}(\psi_1^\mu, \ldots, \psi_d^\mu)$ and $\Phi_\mu(\sigma) = (\phi_1^\mu, \ldots, \phi_d^*)$ be bases for $\mathcal{N}\big((\mu I - U^*(\sigma))^m\big)$ and $\mathcal{N}\big((\mu I - U(\sigma))^m\big)$, respectively. Then

$$\mathcal{R}\big((\mu I - U(\sigma))^m\big) = \{\phi \in C : \langle \Psi_\mu^*(\sigma), \phi \rangle = 0\} \tag{2.4}$$

$$\mathcal{R}\big((\mu I - U^*(\sigma))^m\big) = \{\psi \in B_0 : \langle \psi, \Phi_\mu(\sigma) \rangle = 0\}. \tag{2.5}$$

Relation (2.4) implies that the $d \times d$-matrix $\langle \Psi_\mu^*(\sigma), \Phi_\mu(\sigma) \rangle$ is nonsingular and without loss of generality can be chosen as the identity. Therefore, decompositions (2.2) and (2.3) can be written as

$$C = \{\phi \in C : \phi = \Phi_\mu \langle \Psi_\mu^*(\sigma), \phi \rangle\} \oplus \{\phi \in C : \langle \Psi_\mu^*(\sigma), \phi \rangle = 0\} \tag{2.6}$$

$$B_0 = \{\phi \in C : \psi = \Psi_\mu^* \langle \psi, \Phi_\mu(\sigma) \rangle\} \oplus \{\psi \in B_0 : \langle \psi, \Phi_\mu(\sigma) \rangle = 0\}. \tag{2.7}$$

Relations (2.6) and (2.7) are sufficient for the applications, but some remarks are in order to clarify the relationship between this decomposition and the one given in Section 7.5 for autonomous equations. Define the transpose of System (1.1) to be

$$\begin{aligned} \dot{y}(s) &= \int_{-r}^{0} y(s-\theta)\, d[\eta(s,\theta)], \qquad s \le t, \\ y(t+\xi) &= \psi(\xi), \qquad 0 \le \xi \le r, \quad \psi \in C' \end{aligned} \tag{2.8}$$

In the autonomous case, the adjoint and the transposed equation are related through the mapping $F^T : C' \to B_0$ (see Section 7.1). If one defines the time-dependent analog of F^T to be

$$(F^T(t)\psi)(s) = \psi(0) - \int_s^t \int_{-r}^{\alpha - t} \psi(t - \alpha + \theta)\, d_\theta[\eta(\alpha,\theta)]\, d\alpha,$$

then $F^T(t) : C' \to B_0$, but we no longer have that

$$F^T(t) T^T(t,s) = T^*(s,t) F^T(t).$$

Therefore, we cannot introduce the generalization of the bilinear form defined in (5.9) of Chapter 7. But, there is another way to relate the adjoint equation and the transposed equation when we restrict the forcing function in Equation (2.1) to be continuous (which is no restriction when we compute the eigenfunctions of $U^*(\sigma)$).

Define $S\psi$ to be $(S\psi)(\theta) = \psi(r+\theta)$ for $-r \le \theta \le 0$ and define $I - \Omega(t) : B_0 \to B_0$,

$$g(s) \mapsto g^t(s) - \int_s^t g^t(\alpha) R(\alpha, s)\, d\alpha, \qquad s \le t,$$

where $R(t,s)$ satisfies the formal resolvent equation (3.8) of Chapter 6. From Chapter 6, Theorem 3.1, it follows that $(I - \Omega(t))g = y(\cdot\,; t, g)$ with $y(\cdot\,; t, g)$ the solution of the adjoint equation (2.1). Therefore, $I - \Omega(t)$ is invertible and it is easy to verify that

$$T^*(s,t)g = \Omega(s)^{-1}S^{-1}T^T(s,t)\Omega(t)g \tag{2.9}$$

for continuous g belonging to B_0. Relation (2.9) clarifies the connection between $T^*(s,t)$ and $T^T(s,t)$, when restricted to continuous functions, and can be used to find a representation for the eigenfunctions and generalized eigenfunctions of $U^*(\sigma)$ from $U^T(\sigma)$. This is sufficient for the application and there is no need to introduce a special bilinear form.

To this end, we wish to obtain the same type of decomposition as earlier in the variation-of-constants formula for

$$\dot{x}(t) = L(t)x_t + f(t), \tag{2.10}$$

where $L(t)$ is the same function as in Section 8.1 and f is locally integrable on $(-\infty, \infty)$. The variation-of-constants formula for Equation (2.10) is

$$x_t = T(t,\sigma)x_\sigma + \int_\sigma^t d[K(t,s)]f(s), \qquad t \geq \sigma, \tag{2.11}$$

where the kernel $K(t,\,\cdot\,) : [\sigma, t] \to C$ is given by

$$K(t,s)(\theta) = \int_\sigma^s X(t+\theta, \alpha)\, d\alpha$$

and $X(t,s)$ is the fundamental matrix function to the homogeneous equation.

For any characteristic multiplier $\mu \neq 0$ of Equation (1.1), let

$$x_t = x_t^{E_\mu(t)} + x_t^{Q_\mu(t)} \qquad \text{where} \quad x_t^{E_\mu(t)} \in E_\mu(t),\ x_t^{Q_\mu(t)} \in Q_\mu(t)$$

for any $t \geq \sigma$. To find the integral equation for the components of x_t, let

$$K(t,s) = K(t,s)^{E_\mu(t)} + K(t,s)^{Q_\mu(t)}. \tag{2.12}$$

The first object $K(t,s)^{E_\mu(t)}$ has a simple meaning. Recall that each column of $X(t,s)$ belongs to C for $t \geq s + r$ and

$$X(t,s) = T(t, s+r)X_{s+r}(\,\cdot\,, s), \qquad t \geq s + r.$$

Therefore, each column of $X_{s+r}(\,\cdot\,, s)$ can be decomposed into its components according to the decomposition $E_\mu(s+r) \oplus Q_\mu(s+r)$. Since $T(s+r, s)$ is a homeomorphism on $E_\mu(s+r)$, this allows us to define in a unique manner an $n \times n$ matrix $X_0^{E_\mu(t)}$ whose columns are in $E_\mu(t)$ so that

$$[T(s+r,s)X_0]^{E_\mu(t)} = T(s+r,s)X_0^{E_\mu(t)}.$$

This justifies

$$K(t,s)^{E_\mu(t)} = \int_\sigma^s T(t,\alpha) X_0^{E_\mu(t)}\, d\alpha.$$

Theorem 2.1. *If x is a solution of Equation (2.10) for $t \geq \sigma$ and $\mu \neq 0$ is a characteristic multiplier of Equation (1.1) that decomposes C for any $s \in (-\infty, \infty)$ as $C = E_\mu(s) \oplus Q_\mu(s)$ with $E_\mu(s)$ and $Q_\mu(s)$ as in Section 8.1, then x_t satisfies the integral equations*

$$(2.13) \qquad \begin{aligned} x_t^{E_\mu(t)} &= T(t,\sigma) x_\sigma^{E_\mu(t)} + \int_\sigma^t T(t,s) X_0^{E_\mu(t)} f(s)\, ds, \\ x_t^{Q_\mu(t)} &= T(t,\sigma) x_\sigma^{Q_\mu(t)} + \int_\sigma^t d_s[K(t,s)^{Q_\mu(t)}] f(s), \quad t \geq \sigma, \end{aligned}$$

where $K(t,s)^{Q_\mu(t)} = \Phi(s)\langle \Psi^*(s), K(t,s)\rangle$.

Proof. Suppose $\Phi(0) = \Phi_\mu(0)$ and $\Psi_\mu^*(0)$ are bases chosen as stated before Equations (2.6) and (2.7) with $\langle \Psi_\mu^*(0), \Phi(0)\rangle = I$, the identity. Let $\Phi(t) = T(t,0)\Phi(0)$ and $\Psi^*(t) = T^*(t,0)\Psi^*(0)$ with $\Psi^*(t)$ the matrix solution of the adjoint equation on $(-\infty, t]$. Lemma 1.3 and the corresponding generalization for the formal adjoint equation imply that $\Phi(t)$ is a basis for the solutions of Equation (1.1) on $[t-r, t]$ corresponding to the multiplier μ and $\Psi^*(t)$ is a basis for forcing functions of Equation (2.1) on $(-\infty, t]$ corresponding to the multiplier μ. From Chapter 6, Relation (3.2),

$$\langle T^*(t,0)\Psi^*(0), T(t,0)\Phi(0)\rangle = \text{constant}.$$

Thus $\langle \Psi^*(t), \Phi(t)\rangle = I$ for all $t \in (-\infty, \infty)$. Furthermore,

$$\begin{aligned} x_t^{E_\mu(t)} &= \Phi(t)\langle \Psi^*(t), x_t\rangle \\ &= \Phi(t)\langle \Psi^*(\sigma), x_\sigma\rangle + \Phi(t)\int_\sigma^t \Psi(s)^*(0) f(s)\, ds \\ &= T(t,\sigma)\Phi(\sigma)\langle \Psi^*(\sigma), x_\sigma\rangle + \int_\sigma^t T(t,s)\Phi(s)\Psi(s)^*(0) f(s)\, ds \\ &= T(t,\sigma) x_\sigma^{E_\mu(\sigma)} + \int_\sigma^t T(t,s) X_0^{E_\mu(t)} f(s)\, ds \end{aligned}$$

since $X_0^{E_\mu(s)} = \Phi(s)\langle \Psi^*(s), X_0\rangle = \Phi(s)\Psi^*(s)(0)$. Using the fact that $x_t^{Q_\mu(t)} = x_t - x_t^{E_\mu(t)}$, one completes the proof of the theorem. □

The first of equations (2.13) is equivalent to an ordinary differential equation. In fact, if $\Phi(t) = T(t)\Phi(0) = P(t)e^{Bt}$, $U\Phi = \Phi e^{B\omega}$, and

$$(2.14) \qquad x_t = x_t^{E_\mu(t)} + x_t^{Q_\mu(t)} = P(t)y(t) + x_t^{Q_\mu(t)},$$

then Equations (2.13) are equivalent to the system

$$
\begin{aligned}
&\dot{y}(t) = By + \Psi^*(t)(0)f(t) \\
(2.15)\qquad &x_t^{Q_\mu(t)} = T(t,\sigma)x_\sigma^{Q_\mu(t)} + \int_\sigma^t d_s[K(t,s)^{Q_\mu(t)}]f(s), \quad t \geq \sigma.
\end{aligned}
$$

System (2.15) is now in a form to permit the discussion of problems concerning the perturbation of Equation (1.1) in a manner very similar to that when Equation (1.1) was autonomous. We do not devote any time to a detailed discussion of these questions since they proceed in a manner that is very analogous to ordinary differential equations.

Theorem 2.1 can obviously be employed to make further decompositions of the space C. In fact, for any $\delta > 0$, let $\Lambda = \Lambda(\delta)$ be the characteristic multipliers of Equation (1.1) satisfying $|\mu| > \delta$. As before, one can obtain bases $\Psi_\Lambda^*(\sigma)$ and $\Phi_\Lambda(\sigma)$ for the formal adjoint equation (2.1) and Equation (1.1) corresponding to all multipliers in Λ such that $\langle \Psi_\Lambda^*(\sigma), \Phi_\Lambda(\sigma) \rangle = I$ and

$$
(2.16)\qquad
\begin{aligned}
C &= E_\Lambda(\sigma) \oplus Q_\Lambda(\sigma) \\
E_\Lambda(\sigma) &= \{\phi \in C : \phi = \Phi_\mu \langle \Psi_\mu^*(\sigma), \phi \rangle\} \\
Q_\Lambda(\sigma) &= \{\phi \in C : \langle \Psi_\mu^*(\sigma), \phi \rangle = 0\}.
\end{aligned}
$$

The corresponding Formulas (2.13) become

$$
(2.17)\qquad
\begin{aligned}
x_t^{E_\Lambda(t)} &= T(t,\sigma)x_\sigma^{E_\Lambda(t)} + \int_\sigma^t T(t,s)X_0^{E_\Lambda(t)}f(s)\,ds, \\
x_t^{Q_\Lambda(t)} &= T(t,\sigma)x_\sigma^{Q_\Lambda(t)} + \int_\sigma^t d_s[K(t,s)^{Q_\Lambda(t)}]f(s), \quad t \geq \sigma.
\end{aligned}
$$

It is clear that the spectral radius of $T(t,\sigma)|Q_\Lambda(\sigma)$ is less than $\delta = \exp \gamma\omega$ and the spectrum of $T(t,\sigma)|E_\Lambda(\sigma)$ is equal to Λ. Therefore, for any $\beta > 0$, there is an $M > 0$ such that

$$
(2.18)\qquad
\begin{aligned}
|T(t,\sigma)\phi| &\leq Me^{(\gamma+\beta)(t-\sigma)}|\phi|, \qquad t \geq \sigma,\ \sigma \in \mathbb{R},\ \phi \in Q_\Lambda(\sigma), \\
|K(t,s)^{Q_\Lambda(s)}| &\leq Me^{(\gamma+\beta)(t-s)}, \qquad t \geq s,\ s \in \mathbb{R}, \\
\mathrm{Var}_{[\sigma,t]}\, K(t,\cdot)^{Q_\Lambda(s)} &\leq Me^{(\gamma+\beta)t}, \qquad t \geq s,\ s \in \mathbb{R}.
\end{aligned}
$$

The results in this and the previous section have natural generalizations to NFDE. Consider the linear periodic NFDE

$$
(2.19)\qquad \frac{d}{dt}D(t)x_t = L(t)x_t,
$$

where $D(t)\phi$, $L(t)\phi$ are continuous in t, ϕ, linear in ϕ, $D(t+\omega)\phi = D(t)\phi$, $L(t+\omega)\phi = L(t)\phi$ for all t, ϕ and some fixed $\omega > 0$, and $D(t)\phi$ is atomic at zero. Let $T_{D,L}(t,\sigma)$, $t \geq \sigma$, be the solution operator of Equation (2.19).

A complex number ρ is called a *characteristic multiplier* of the NFDE (2.19) if ρ is an eigenvalue of finite type of $T_{D,L}(\sigma+\omega,\sigma)$, that is, an isolated point of the spectrum of $T_{D,L}(\sigma+\omega,\sigma)$ with finite-dimensional generalized eigenspace. As in Section 1, one shows that the characteristic multipliers are independent of σ. By definition, for each characteristic multiplier ρ and each $\sigma \in \mathbb{R}$, there is a decomposition of C as $C = P_\sigma \oplus Q_s$, where $P_\sigma = \mathcal{N}(T_{D,L}(\sigma+\omega,\sigma) - \rho I)^k$, $Q_\sigma = \mathcal{R}(T_{D,L}(\sigma+\omega,\sigma) - \rho I)^k$, and P_σ has finite dimension, say d. If Φ_σ is a basis for P_σ, then there is a $d \times d$ constant matrix B_σ with the spectrum of $e^{B_\sigma \omega}$ equal to ρ and an $n \times d$ matrix $C_\sigma(t), C_\sigma(t+\omega) = C_\sigma(t)$, such that if $\phi \in P_\sigma$, $\phi = \Phi_\sigma b$, then

$$T_{D,L}(t+\sigma,\sigma) = C_\sigma(t)e^{B_\sigma t)}b. \tag{2.20}$$

In this sense, there is a Floquet representation on the generalized eigenspace P_σ of the characteristic multiplier ρ.

The representation in Theorem 7.3 of Chapter 3 is valid for Equation (2.19). In fact, one can determine a constant k such that for any $\sigma \in \mathbb{R}$, there is a $\Phi^\sigma = (\phi_1^\sigma, \dots, \phi_n^\sigma)$ such that $D(\sigma)\Phi^\sigma = I$, and $|\phi_j^\sigma| \le k$, $j = 1, 2, \dots, n$. If $\Psi^\sigma = I - \Phi^\sigma D(\sigma)$, then

$$T_{D,L}(t,\sigma) = T_D(t,\sigma)\Psi^\sigma + U(t,\sigma), \tag{2.21}$$

where $U(t,\sigma)$ is completely continuous for $t \ge \sigma$ and $T_D(t,\sigma)$ is the solution operator generated by the equation $D(t)y_t = 0$.

Using the same type of reasoning as for the autonomous case, if $e^{a_D \omega}$ is the spectral radius of $T_D(\sigma+\omega,\sigma)$, one proves that any ρ in the spectrum of $T_D(\sigma+\omega,\sigma)$ with $|\rho| > e^{a_d \omega}$ must be a normal eigenvalue and, thus, a characteristic multiplier of Equation (2.19). Furthermore, there are only a finite number of characteristic multipliers ρ satisfying $|\rho| \ge e^{a\omega}$ for any constant $a > a_D$. The space C therefore can be decomposed as $C = P_{\sigma,a} \oplus Q_{\sigma,a}$, where $P_{\sigma,a}$ and $Q_{\sigma,a}$ are invariant under $T_D(\sigma+\omega,\sigma)$ and the spectrum of $T_D(\sigma+\omega,\sigma)|P_{\sigma,a}$ consists only of the multipliers $|\rho|$ with $|\rho| \ge e^{a\omega}$, $a > a_D$, and the spectrum of $T_D(\sigma+\omega,\sigma)|Q_{\sigma,a}$ lies inside the disk with center zero and radius $< e^{(a-\epsilon)\omega}$ for some $\epsilon > 0$. Therefore, there is a constant K such that

$$|T_{D,L}(t,\sigma)|Q_{\sigma,a}| \le Ke^{(a-\epsilon)(t-\sigma)}, \quad t \ge \sigma. \tag{2.22}$$

Also, there is a constant matrix $B_{\sigma,a}$ and an ω-periodic matrix $C_{\sigma,a}(t)$ such that

$$T_{D,L}(t+\sigma,\sigma)\Phi_{\sigma,a} = C_{\sigma,a}(t)e^{B_{\sigma,a}t}, \quad t \in \mathbb{R}, \tag{2.23}$$

where $\Phi_{\sigma,a}$ is a basis for $P_{\sigma,a}$ and the only eigenvalues of $e^{B_{\sigma,a}\omega}$ are those characteristic multipliers ρ of Equation (2.19) with $|\rho| \ge e^{a\omega}$.

As for RFDE, similar decompositions hold in the variation-of-constants formula for the nonhomogeneous linear equation. The reader may supply the details.

8.3 An example: Integer delays

In this section, we analyze the following system of linear periodic delay equations

$$\dot{x}(t) = \sum_{j=0}^{m} B_j(t)x(t - j\omega), \qquad x_s = \phi, \quad t \geq s, \tag{3.1}$$

where B_j are continuous matrix-valued functions such that $B_j(t + \omega) = B_j(t)$.

Define $\Omega_s^t(z)$ to be the fundamental matrix solution of the periodic ordinary differential equation

$$\dot{y}(t) = \sum_{j=0}^{m} z^j B_j(t)y(t) \tag{3.2}$$

with $\Omega_s^s(z) = I$. Let $U : C \to C$ be defined by $U\phi = T(\omega, 0)\phi$. Then U is given by

$$(U\phi)(\theta) = \Omega_{-\omega}^{\theta}(0)\phi(0) + \sum_{j=1}^{m} \int_{-\omega}^{\theta} \Omega_{\tau}^{\theta}(0)B_j(\tau)\phi(\tau - (j-1)\omega)\, d\tau \tag{3.3}$$

for $-\omega \leq \theta \leq 0$ and $(U\phi)(\theta) = \phi(\theta + \omega)$ for $-m\omega \leq \theta \leq -\omega$.

Theorem 3.1. *If* $\det B_m$ *has only isolated zeros, then the spectrum of the operator* $U : C[-m\omega, 0] \to C[-m\omega, 0]$ *defined by* (3.3) *consists of eigenvalues of finite type only,*

$$\sigma(U) = \{\mu : \det \Delta(\mu) = 0\}, \tag{3.4}$$

where $\Delta(z) = z - \Omega_0^{\omega}(z^{-1})$. *For* $\mu \in \sigma(U)$, *the algebraic multiplicity of the eigenvalue* μ *equals the order of* μ *as a zero of* $\det \Delta$.

For a proof of the theorem, we are going to apply the abstract results about characteristic matrices from Section 7.3. Before we can do so, we have to make some preparations. Since U^m is compact and one-to-one, the inverse of U is a well-defined unbounded closed operator on C. To avoid technical complications we shall only prove the theorem for $m = 1$.

For $m = 1$, the space C equals $C([-\omega, 0], \mathbb{R}^n)$ and the mapping $U : C \to C$ is given by

$$(U\phi)(\theta) = \Omega_{-\omega}^{\theta}(0)\phi(0) + \int_{-\omega}^{\theta} \Omega_{\tau}^{\theta} B_1(\tau)\phi(\tau)\, d\tau. \tag{3.5}$$

We shall show that $A = U^{-1}$ is the first operator associated with the operators D, L, and M, which we now define. Let $M : \mathcal{D}(M) \to C$ be the operator

$$(M\psi)(\theta) = B_1(\theta)^{-1}\Big(\frac{d\psi}{d\theta}(\theta) - B_0(\theta)\psi(\theta)\Big) \tag{3.6}$$

with maximal domain $\mathcal{D}(M) = \{\psi \in C : M\psi \in C\}$. The kernel of M is given by

$$\mathcal{N}(M) = \{\psi \in C : \psi(\theta) = \Omega^{\theta}_{-\omega}(0)\psi(-\omega)\}.$$

So (H1) of Section 7.3 is satisfied.

Next we define the restriction $M_0 : \mathcal{D}(M_0) \to C$

$$\mathcal{D}(M_0) = \{\psi \in C : \psi \in \mathcal{D}(M),\ \psi(-\omega) = 0\}, \qquad M_0\psi = M\psi. \tag{3.7}$$

To find the resolvent of M_0, put $(z - M_0)^{-1}\phi = \psi$. Then ψ satisfies the equation

$$\dot{\psi} = (B_0 + zB_1)\psi - B_1\phi.$$

The initial condition is $\psi(-\omega) = 0$ and we can solve for ψ. Consequently,

$$(z - M_0)^{-1}\phi(\theta) = -\int_{-\omega}^{\theta} \Omega^{\theta}_{\tau}(z)B_1(\tau)\phi(\tau)\, d\tau. \tag{3.8}$$

So $\Omega = \rho(M_0) = \mathbb{C}$ and (H2) of Section 7.3 is satisfied. Finally, set $\widehat{C} = \mathbb{C}^n \times C$ and

$$L : C \to \mathbb{C}^n, \quad \phi \mapsto \phi(-\omega), \qquad D : C \to \mathbb{C}^n, \quad \phi \mapsto \phi(0).$$

Then the operator $A = U^{-1}$ can be represented by

$$\mathcal{D}(A) = \{\psi \in C : \psi \in \mathcal{D}(M),\quad DM\psi = L\psi\}, \qquad A\psi = M\psi. \tag{3.9}$$

So the operator A defined by (3.9) is the first operator associated with D, L, and M. The second operator $\widehat{A} : \mathcal{D}(\widehat{A}) \to \widehat{C}$ associated with D, L, and M is given by

$$\begin{aligned} &\mathcal{D}(\widehat{A}) = \{(c,\psi) \in \widehat{C} : \psi \in \mathcal{D}(M),\ c = D\psi\}, \\ &\widehat{A}(c,\psi) = (L\psi, M\psi). \end{aligned} \tag{3.10}$$

Theorem 3.2. *The matrix function $\Delta : \mathbb{C} \to \mathcal{L}(\mathbb{C}^n)$ defined by*

$$\Delta(z) = z\Omega^{\omega}_{0}(z) - I \tag{3.11}$$

is a characteristic matrix for $\widehat{A}$. The equivalence relation is given by

$$\begin{pmatrix} \Delta(z) & 0 \\ 0 & I \end{pmatrix} = F(z)(z - \widehat{A})E(z), \tag{3.12}$$

where $E : \mathbb{C} \to \mathcal{L}(\widehat{C}, \widehat{C}(\widehat{A}))$ is given by

$$E(z)(c,\psi) = \Big(\Omega^{0}_{-\omega}(z)c - \int_{-\omega}^{0} \Omega^{0}_{\tau}(z)B_1(\tau)\phi(\tau)\, d\tau, \chi\Big),$$

$$\chi(\theta) = \Omega_{-\omega}^{\theta}(z)c - \int_{-\omega}^{\theta} \Omega_{\tau}^{\theta}(z)B_1(\tau)\phi(\tau)\,d\tau,$$

and $F : \mathbb{C} \to \mathcal{L}(\widehat{C})$ *is given by*

$$F(z)(c,\psi) = \bigl(c + z\int_{-\omega}^{0} \Omega_{\tau}^{0}(z)B_1(\tau)\phi(\tau)\,d\tau, \psi\bigr).$$

Proof. From Theorem 3.1 of Chapter 7, it follows that the characteristic matrix for $\widehat{A}$ is given by

$$\Delta(z) = (zM - L)(j - z(z - M_0)^{-1}j)$$

where $j : \mathbb{C}^n \to \mathcal{N}(D)$, $c \mapsto \Omega_{-\omega}^{\theta}(0)c$. Using Representation (3.8) for the resolvent of M_0, we conclude

$$(zI - M_0)^{-1}j = \frac{1}{z}\bigl(\Omega_{-\omega}^{\theta}(0) - \Omega_{-\omega}^{\theta}(z)\bigr).$$

So $j - z(z - M_0)^{-1}j = \Omega_{-\omega}^{\theta}(z)$ and

$$\Delta(z) = z\Omega_{-\omega}^{0}(z) - I.$$

The concrete representations for E and F are verified in a similar way. □

Proof of Theorem 3.1 Since $U^{-1} = A$, an eigenvalue μ of U corresponds to an eigenvalue μ^{-1} of A and the spectral data of U and A are the same. This implies that the spectral data of U and $\widehat{A}$ are the same. Hence the theorem follows from Corollary 3.1 of Section 7.3. □

In the special case that the matrices B_j, $j = 0, \dots, m$, are diagonal, we find a characteristic equation for the exponents:

$$\det\bigl(e^{\omega\lambda} - e^{\omega\sum_{j=0}^{m}\bar{B}_j e^{-j\omega\lambda}}\bigr) = 0$$

whcrc

$$\bar{B}_j = \frac{1}{\omega}\int_0^{\omega} B_j(s)\,ds, \quad j = 0, \dots, m$$

denotes the average of B_j. So the characteristic matrix for the exponents (since they are determined only up to multiples of $2\pi i/\omega$) maybe taken to be

$$\lambda = \sum_{j=0}^{m} \bar{B}_j e^{-j\omega\lambda},$$

which is just the characteristic matrix for the autonomous equation

$$\dot{y}(t) = \sum_{j=0}^{m} \bar{B}_j x(t - j\omega).$$

Theorem 3.2 leads to a situation where we can apply the results from Section 7.8. To simplify the arguments, we restrict our attention to the scalar case.

Consider a scalar linear periodic delay equation

$$\dot{x}(t) = \sum_{j=0}^{m} b_j(t)x(t - j\omega), \qquad t \geq s, \quad x_s = \phi, \tag{3.13}$$

where b_j, $j = 0, \ldots, m$, are continuous periodic functions with period ω. As before, we will assume that the solution map $T(t,s) : C \to C$ for (3.13) is one-to-one, i.e., the zeros of b_m are isolated. We have the following results.

Theorem 3.3. *Suppose that the zeros of b_m are isolated. Then the system of Floquet solutions is complete if and only if b_m has no sign change.*

As in the autonomous case there exists a relation between the completeness of the system of Floquet solutions and the existence of small solutions.

Theorem 3.4. *Suppose that the zeros of b_m are isolated. Then* (3.18) *has small solutions if and only if b_m has a sign change.*

Define the following subspaces of C:

$$\mathcal{S} = \{\phi \in C : T(t,s)\phi \quad \text{is a small solution of (3.18) } \}$$
$$\mathcal{M} = \bigoplus_{\mu \in P\sigma(U)} \mathcal{M}_\mu(U).$$

Theorem 3.5. *The system of Floquet solutions is complete if and only if there are no small solutions. Furthermore,*

$$C = \overline{\mathcal{M} \oplus \mathcal{S}}.$$

Example 3.1. The equation

$$\dot{x}(t) = \big(\frac{1}{2} + \sin(2\pi t)\big)x(t-1)$$

has small solutions, although there are infinitely many independent Floquet solutions

$$x(t) = e^{\lambda t}, \qquad \text{with} \quad \lambda = \frac{1}{2}e^{-\lambda}.$$

In the remaining part of this section we shall sketch the proof of these results for $m = 1$.

The proof consists of several steps.

(i) Recall that in the scalar case

$$\Omega_{-\omega}^{\theta}(z) = e^{\int_{-\omega}^{\theta} b_0(s)\,ds + z\int_{-\omega}^{\theta} b_1(s)\,ds}.$$

Furthermore

$$\Delta(z) = \det \Delta(z) = z e^{-\omega \bar{b}_0} e^{-\omega \bar{b}_1 z} - 1, \quad \bar{b}_j = \frac{1}{\omega}\int_{-\omega}^{0} b_j(s)\,ds. \tag{3.14}$$

From (3.13), we deduce that the resolvent of A has the following representation

$$R(z, A)\phi = \frac{P(z,\phi)}{\det \Delta(z)}$$

where

$$\begin{aligned} P(z,\phi)(\theta) &= \Omega_{-\omega}^{\theta}(z)\big[c + z\int_{-\omega}^{0} \Omega_{\tau}^{0}(z) b_1(\tau)\phi(\tau)\,d\tau\big] \\ &\quad - \det \Delta(z)\int_{-\omega}^{\theta} \Omega_{\tau}^{\theta}(z) b_1(\tau)\phi(\tau)\,d\tau. \end{aligned} \tag{3.15}$$

(ii) Similar arguments as used in Section 8 of Chapter 7 show that there exists a sequence of simple closed smooth curves Γ_N such that

- There is a complex function $\alpha_N : [0, 2\pi] \to \mathbb{C}$ that is differentiable on $[0, 2\pi]$ such that $\Gamma_N = \{z = \alpha_N(t), t \in [0, 2\pi]\}$, for $t \in [0, 2\pi]$ $|\alpha_N(t)| \to \infty$ as $N \to \infty$ and $|\alpha_N'(t)| \le |\alpha_N(t)|$;
- the contour Γ_N encloses Λ_N but no other eigenvalues of $\sigma(A)$;
- there exists an $\epsilon > 0$ such that dist $(\Gamma_N, \sigma(A)) > \epsilon$ for all $N = 0, 1, \ldots$.

(iii) Define

$$\mathcal{E} = \{\phi \in C : \mathrm{E}(P(z,\phi)) \le |\omega \bar{b}_1|\}. \tag{3.16}$$

Then for every $\phi \in \mathcal{E} \cap \mathcal{D}(A)$ there exists a positive constant M such that

$$|R(z, A)\phi| \le M \qquad \text{for} \quad z \in \Gamma_N,\ N = 0, 1, \ldots. \tag{3.17}$$

The same argument as used in the proof of Theorem 8.2 of Chapter 7 yields that for $\phi \in \mathcal{E} \cap \mathcal{D}(A^3)$

$$\phi = \lim_{N\to\infty} P_{\Lambda_N}\phi \tag{3.18}$$

where Λ_N denotes the set of eigenvalues enclosed by the contour Γ_N.

(iv) Next we show $\overline{\mathcal{M}} = \mathcal{E}$. Theorem 2.3 of Verduyn Lunel [4] yields that $\mathcal{E}$ is a closed subset of C. Since for $\phi \in \mathcal{M}$, the function $z \mapsto R(z, A)\phi$ is rational, we conclude that $\overline{\mathcal{M}} \subset \mathcal{E}$. On the other hand, from (3.18) we conclude that

$$R(\lambda, A)^3 \mathcal{E} \subset \overline{\mathcal{M}} \cap \mathcal{D}(A^3).$$

Thus

$$\mathcal{E} \subset (\lambda - A)^3(\overline{\mathcal{M}} \cap \mathcal{D}(A^3)) \subset \overline{\mathcal{M}}.$$

(v) Define

$$\mathcal{S} = \bigcap_{\lambda \in \sigma(\hat{A})} \mathcal{N}(P_\lambda) = \{\phi : z \mapsto R(z, \widehat{A})\phi \quad \text{is entire }\}.$$

Suppose $\phi \in \overline{\mathcal{M}} \cap \mathcal{S}$ and $\phi \neq 0$. Using the invariance of $\overline{\mathcal{M}}$ and $\mathcal{S}$

$$R(\lambda, A)^2\phi \in \overline{\mathcal{M}} \cap \mathcal{S}.$$

Therefore, it follows from (3.18) that

$$R(\lambda, A)^3\phi = \sum_{\mu \in \sigma(\hat{A})} P_\mu R(\lambda, A)^3\phi = R(\lambda, A)^3 \sum_{\mu \in \sigma(\hat{A})} P_\mu\phi,$$

since P_μ and $R(\lambda, A)$ commute. However $\phi \in \mathcal{S}$ and hence $P_\mu\phi = 0$ for every $\lambda \in \sigma(A)$. This proves $\overline{\mathcal{M}} \cap \mathcal{S} = \{0\}$. To prove the density of the direct sum, we use duality. From the Neumann series for the resolvent

$$\mathcal{S} = \bigcap_{\lambda \in \sigma(A)} \mathcal{N}(P_\lambda).$$

So the density of $\overline{\mathcal{M}} \oplus \mathcal{S}$ is equivalent to $\overline{\mathcal{M}}^* \cap \mathcal{S}^* = \{0\}$, where $\mathcal{M}^*$ is the linear subspace generated by $\mathcal{R}(P_\lambda^*)$, $\lambda \in \sigma(A^*)$, and

$$\mathcal{S}^* = \bigcap_{\lambda \in \sigma(A^*)} \mathcal{N}(P_\lambda^*).$$

Since the characteristic matrix for A^* is $\Delta(z)$ and the hypotheses are invariant under duality, a similar argument yields $\overline{\mathcal{M}}^* \cap \mathcal{S}^* = \{0\}$. This shows

$$C = \overline{\overline{\mathcal{M}} \oplus \mathcal{S}}.$$

(vi) It suffices to compute the exponential type of $z \mapsto P(z, \phi)$. Since

$$\begin{aligned} &z\Omega^\theta_{-\omega}(z) \int_{-\omega}^0 \Omega^0_\tau(z) b_1(\tau)\phi(\tau)\, d\tau - z\Omega^0_{-\omega}(z) \int_{-\omega}^\theta \Omega^\theta_\tau(z) b_1(\tau)\phi(\tau)\, d\tau \\ &= z\Omega^\theta_{-\omega}(z)\Big[\int_{-\omega}^0 \Omega^0_\tau(z) b_1(\tau)\phi(\tau)\, d\tau - z\Omega^0_\theta(z) \int_{-\omega}^\theta \Omega^\theta_\tau(z) b_1(\tau)\phi(\tau)\, d\tau\Big] \\ &= z\Omega^\theta_{-\omega}(z) \int_\theta^0 \Omega^0_\tau(z) b_1(\tau)\phi(\tau)\, d\tau, \end{aligned}$$

we find

$$
\text{(3.19)} \qquad \begin{aligned} P(z,\phi) &= \Omega^{\theta}_{-\omega}(z)c + z\Omega^{\theta}_{-\omega}(z)\int_{\theta}^{0}\Omega^{0}_{\tau}(z)b_1(\tau)\phi(\tau)\,d\tau \\ &\quad + \int_{-\omega}^{\theta}\Omega^{\theta}_{\tau}(z)b_1(\tau)\phi(\tau)\,d\tau. \end{aligned}
$$

First suppose that b_1 does not change sign, say $b_1(s) \geq 0$. The entire function

$$
z \mapsto \int_{-\omega}^{\theta} e^{\int_{\tau}^{\theta} b_0(s)ds + z\int_{\tau}^{\theta} b_1(s)ds} b_1(\tau)\phi(\tau)\,d\tau
$$

is of order 1, has maximal type in the left half plane, and is bounded in the right half plane. Using the Paley-Wiener theorem we find

$$
\begin{aligned} \mathrm{E}\Big(\int_{-\omega}^{\theta} e^{\int_{\tau}^{\theta} b_0(s)ds + z\int_{\tau}^{\theta} b_1(s)ds} b_1(\tau)\phi(\tau)\,d\tau\Big) &= \max_{-\omega\leq\tau\leq\theta}\int_{\tau}^{\theta} b_1(s)ds \\ &\leq \omega\bar{b}_1. \end{aligned}
$$

Furthermore,

$$
\begin{aligned} \mathrm{E}\big(z\Omega^{\theta}_{-\omega}(z)\int_{\theta}^{0}\Omega^{0}_{\tau}(z)b_1(\tau)\phi(\tau)\,d\tau\big) &= \int_{-\omega}^{\theta} b_1(s)\,ds + \max_{\theta\leq\tau\leq 0}\int_{\tau}^{0} b_1(s)\,ds \\ &= \omega\bar{b}_1. \end{aligned}
$$

Similarly, for $b_1(s) \leq 0$, the functions have exponential growth in the right half plane and are bounded in the left half plane, and the same exponential estimates (with b_1 replaced by $-b_1$) hold. In both cases, it follows that $\overline{\mathcal{M}} = C$. On the other hand, if b_1 does change sign, the maximum of the function

$$
\tau \mapsto \int_{-\omega}^{\theta} b_1(s)\,ds + \int_{\tau}^{0} b_1(s)\,ds
$$

on $[\theta, 0]$ is larger than $\omega\bar{b}_1$. Since $z \mapsto z\Omega^{\theta}_{-\omega}(z)\int_{\theta}^{0}\Omega^{0}_{\tau}(z)b_1(\tau)\phi(\tau)\,d\tau$ can not be canceled by any other term in (3.19) the type of $z \mapsto P(z,\phi)(\theta)$ is larger than $\omega\bar{b}_1$. Hence $C \neq \overline{\mathcal{M}}$. Since the eigenfunctions and generalized eigenfunctions of $\widehat{A}$ correspond to the Floquet solutions of (3.1), we obtain Theorem 3.3.

(vii) To prove the remaining results, it suffices to prove that the subspace

$$
\mathcal{S} = \big\{\phi : z \mapsto R(z,A)\phi \quad \text{is entire}\,\big\}
$$

corresponds to the space of initial conditions $\mathcal{S}$ that yield small solutions of (3.1). It is clear from the exponential estimates in Section 1 that $x(t) = T(t,s)\phi$ is a small solution of (3.1) if and only if

$$
P_{\mu}\phi = 0 \qquad \text{for all} \quad \mu \in \sigma(U)
$$

or, equivalently, $z \mapsto z(I - zU)^{-1}\phi$ is an entire function. But $U = A^{-1}$ and hence $z \mapsto -zA(z-A)^{-1}$ must be entire. Hence $x(t) = T(t,s)\phi$ is a small solution of (3.1) if and only if

$$P_\lambda \phi = 0 \qquad \text{for all} \quad \lambda \in \sigma(A).$$

This completes the proof of Theorems 3.4 and 3.5.

8.4 Supplementary remarks

The paper of Stokes [2] contained the first general discussion of the Floquet theory for periodic functional differential equations. Theorem 1.1 is due to Stokes [2]. Shimanov [4] was the first to state the decomposition theorem for periodic systems for the special case when the function $\eta(t, \cdot)$ has no singular part. The extension to the general case was given by Henry [3].

The case of periodic differential difference equations with integer lags has received considerable attention, with the main effort devoted to the expansion of solutions in terms of a series of Floquet solutions. The presentation in Section 8.3 follows Verduyn Lunel [5] and extends results obtained by Hahn [1], Zverkin [3,5,6], and Lillo [3,4,5]. Recently general scalar differential delay equations with one delay were studied by Huang and Mallet-Paret [1,2] under small divisor conditions.

9
Equations of neutral type

In this chapter we discuss a particular class of neutral equations for which a qualitative theory is available and for which one can reproduce a theory similar to the one for RFDE.

9.1 General linear systems

For $(\sigma, \phi) \in \mathbb{R} \times C$, consider the linear system

$$
\begin{aligned}
&\frac{d}{dt}\big[D(t)x_t\big] = L(t)x_t + h(t), \qquad t \geq \sigma, \\
&x_\sigma = \phi
\end{aligned}
\tag{1.1}
$$

where $h \in \mathcal{L}_1^{loc}$ and for all $t \in \mathbb{R}$, $D(t) : C \to \mathbb{R}^n$ and $L(t) : C \to \mathbb{R}^n$ are given by

$$
D(t)\phi = \phi(0) - \int_{-r}^{0} d[\mu(t,\theta)]\phi(\theta), \quad L(t)\phi = \int_{-r}^{0} d[\eta(t,\theta)]\phi(\theta). \tag{1.2}
$$

The following assumptions on the kernels μ and η will be maintained throughout this chapter.

The kernel $\eta : \mathbb{R} \times \mathbb{R} \to \mathbb{R}^{n\times n}$, $(t,\theta) \mapsto \eta(t,\theta)$, is measurable in (t,θ), normalized so that

$$
\eta(t,\theta) = 0 \quad \text{for} \quad \theta \geq 0, \qquad \eta(t,\theta) = \eta(t,-r) \quad \text{for} \quad \theta \leq -r,
$$

$\eta(t,\theta)$ is continuous from the left in θ on $(-r,0)$ and has bounded variation in θ on $[-r,0]$ for each t. Further, there is an $m \in \mathcal{L}_1^{\text{loc}}\big((-\infty,\infty), \mathbb{R}\big)$ such that

$$
\operatorname{Var}_{[-r,0]} \eta(t,\,\cdot\,) \leq m(t).
$$

The kernel $\mu : \mathbb{R} \times \mathbb{R} \to \mathbb{R}^{n\times n}$, $(t,\theta) \mapsto \mu(t,\theta)$, is measurable in (t,θ), normalized so that

$$
\mu(t,\theta) = 0 \quad \text{for} \quad \theta \geq 0, \qquad \mu(t,\theta) = \eta(t,-r) \quad \text{for} \quad \theta \leq -r,
$$

$\mu(t,\theta)$ is continuous from the left in θ on $(-r,0)$ and has bounded variation in θ on $[-r,0]$ uniformly in t, and such that $t \mapsto D(t)\phi$ is continuous for each ϕ. Further, μ is *uniformly nonatomic* at zero, i.e., for every $\epsilon > 0$, there exists a $\delta > 0$ such that

$$\mathrm{Var}_{[-\delta,0]}\ \mu(t,\,\cdot\,) < \epsilon, \qquad \text{for all} \quad t \in \mathbb{R}. \tag{1.3}$$

Following the proofs of the corresponding results in Sections 6.1 and 1.7 one shows that there exists a unique global solution.

Theorem 1.1. *Suppose the conditions on η and μ are satisfied. For any given $\sigma \in \mathbb{R}$, $\phi \in C([-r,0],\mathbb{R}^n)$, and $h \in \mathcal{L}_1^{loc}([\sigma,\infty),\mathbb{R}^n)$, there exists a unique function $x(\,\cdot\,;\sigma,\phi)$ defined and continuous on $[\sigma-r,\infty)$ that satisfies System (1.1) on $[\sigma,\infty)$.*

To derive a representation for the solution, it turns out to be useful to integrate System (1.1). If we split off the part that explicitly depends on the initial data, we find

$$\begin{aligned} x(t) &= \int_{\sigma-t}^{0} d[\mu(t,\theta)]x(t+\theta) + \int_{\sigma}^{t}\int_{\sigma-\tau}^{0} d[\eta(\tau,\theta)]x(\tau+\theta)\,d\tau \\ &\quad + f(t) \\ &= \int_{\sigma}^{t} d\big[\mu(t,s-t) + \int_{s}^{t} \eta(\tau,s-\tau)\,d\tau\big]x(s) + f(t), \end{aligned} \tag{1.4}$$

where we used the Fubini theorem to reverse the order of integration, and

$$\begin{aligned} f(t) &= D(\sigma)\phi + \int_{-r}^{\sigma-t} d[\mu(t,\theta)]\phi(\theta+t-\sigma) \\ &\quad + \int_{\sigma}^{t}\int_{-r}^{\sigma-\tau} d[\eta(\tau,\theta]\phi(\theta+\tau-\sigma)\,d\tau + \int_{\sigma}^{t} h(s)\,ds. \end{aligned} \tag{1.5}$$

Equation (1.4) is a Stieltjes-Volterra equation

$$x(t) - \int_{\sigma}^{t} d[k(t,s)]x(s) = f(t) \tag{1.6}$$

with

$$k(t,s) = \mu(t,s-t) + \int_{s}^{t} \eta(\tau,s-\tau)\,d\tau. \tag{1.7}$$

A function $k : \mathbb{R} \times \mathbb{R}_+ \to \mathbb{R}^{n\times n}$, $(t,s) \mapsto k(t,s)$, is called a *Stieltjes-Volterra* kernel of type B^∞, if k is a bounded function that is measurable in t for each fixed s, vanishes for $s > t$, and is of bounded variation, continuous from the right in s on $(0,t)$ for each fixed t. In addition, the total variation of k in its second argument is bounded uniformly in t.

From the assumptions on η and μ, it is clear that k defined by (1.7) is a Stieltjes-Volterra kernel of type B^∞.

Every function ζ of bounded variation on $[0,T]$, normalized so that $\zeta(0) = 0$ and ζ is left continuous on $(0,T)$ represents a Borel measure on $\mathbb{R}$ with no mass outside $[0,T]$. This measure will be denoted by $d\zeta$. For any $d\zeta$-measurable function $f : [0,T] \to \mathbb{R}$

$$\int_0^t d[\zeta(s)]f(t-s) \tag{1.8}$$

denotes the convolution of f with respect to the measure $d\zeta$. If f is continuous, then (1.8) is just a Riemann-Stieltjes integral. More generally, (1.8) is defined for any $f \in \mathcal{L}_p[0,T]$ and is called a *Lebesgue-Stieltjes* integral.

If k is a kernel of type B^∞, then

$$g(t) = \big(Cf\big)(t) = \int_0^t d[k(t,s)]f(s), \qquad 0 \le t \le T$$

maps the functions of bounded variation into itself and

$$\mathrm{Var}_{[0,T]}\, g \le \|k\|_T \,\mathrm{Var}_{[0,T]}\, f,$$

where

$$\|k\|_T = \sup_{t\in[0,T]} \mathrm{Var}_{[0,T]}\, k(t,\,\cdot\,).$$

A kernel ρ is called a Stieltjes-Volterra resolvent of type B^∞ corresponding to a function k if

$$\begin{aligned}\rho(t,s) &= -k(t,s) + \int_s^t d[\rho(t,\alpha)]k(\alpha,s)\\ &= -k(t,s) + \int_s^t d[k(t,\alpha)]\rho(\alpha,s), \qquad t \ge s.\end{aligned}$$

A simple contraction argument shows that if k is a kernel of type B^∞ with $\|k\|_T < 1$, then k has a Stieltjes-Volterra resolvent of type B^∞, and if f is a function of bounded variation on $[0,T]$, Equation (1.6) has a unique solution of bounded variation given by the variation-of-constants formula

$$x(t) = f(t) - \int_\sigma^t d[\rho(t,\alpha)]f(\alpha). \tag{1.9}$$

Here the integrals are understood as Lebesgue-Stieltjes integrals. To apply this result to Equation (1.4), we use a scaling argument similar to the one used in Section 6.1. If we define

$$\widetilde{\rho}(t,s) = \rho(t,s)e^{-\gamma(t-s)}, \qquad \tilde{k}(t,s) = k(t,s)e^{-\gamma(t-s)},$$

then $\widetilde{\rho}(t,s)$ satisfies the equation

$$\widetilde{\rho}(t,s) = -\tilde{k}(t,s) + \int_s^t d[\widetilde{\rho}(t,\alpha)]\tilde{k}(\alpha,s)$$

and it suffices to prove that we can choose $\gamma > 0$ such that

$$\|\tilde{k}\|_T = \sup_{[0,T]} \operatorname{Var}_{[0,T]} \tilde{k}(t,\,\cdot\,) < 1.$$

From Definition (1.7) for k and the assumptions on η and μ, we find

$$\begin{aligned}\|\tilde{k}\|_T \leq & \sup_{t\in[0,T]} \operatorname{Var}_{-r\leq s-t<\delta}\, \mu(t,s-t)e^{-\gamma(t-s)} \\ & + \sup_{t\in[0,T]} \operatorname{Var}_{-\delta\leq s-t\leq 0}\, \mu(t,s-t)e^{-\gamma(t-s)} \\ & + M_1(t-s)e^{-\gamma(t-s)} \\ \leq & M_0e^{-\gamma\delta} + \frac{M_1}{\gamma\delta} + M_2\epsilon,\end{aligned}$$

where ϵ can be made arbitrarily small according to the uniformly nonatomic condition on μ. So if we choose $\epsilon > 0$ such that $M_2\epsilon < 1/3$ and γ so large that

$$M_0e^{-\gamma\delta} + \frac{M_1}{\gamma\delta} < 2/3,$$

then we have $\|\tilde{k}\|_T < 1$. This proves the following theorem:

Theorem 1.2. *Suppose that the assumptions on η, μ, and h are satisfied. If $x = x(\sigma,\phi,h)$ is the solution of System (1.1) on $[s,\infty)$ such that $x_\sigma = \phi$, then for $t \geq \sigma$,*

$$x(t) = X(t,\sigma)f(\sigma) + \int_\sigma^t X(t,\alpha)\,df(\alpha), \tag{1.10}$$

where

$$\begin{aligned} f(t) = {} & D(\sigma)\phi + \int_{-r}^0 d[\mu(t,\theta+\sigma-t)]\phi(\theta) \\ & + \int_\sigma^t \int_{-r}^0 d_\theta[\eta(\tau,\theta+\sigma-\tau)]\phi(\theta)\,d\tau + \int_\sigma^t h(s)\,ds \end{aligned} \tag{1.11}$$

and $X(t,s) = I + \rho(t,s)$, $t \geq s$, is the matrix solution to the integral equation

$$X(t,s) = I + \int_s^t d[X(t,\alpha)]\mu(\alpha,s-\alpha) - \int_s^t X(t,\alpha)\eta(\alpha,s-\alpha)\,d\alpha. \tag{1.12}$$

Proof. Since Representation (1.9) is well defined for continuous f and System (1.1) has a unique solution, it follows that any solution of System (1.1) can be represented by (1.9). Therefore, we have

$$\begin{aligned}x(t) &= f(t) - \int_\sigma^t d[\rho(t,\alpha)]f(\alpha)\\ &= f(t) + \rho(t,\sigma)f(\sigma) + \int_\sigma^t \rho(t,\alpha)\,df(\alpha)\\ &= X(t,\sigma)f(\sigma) + \int_\sigma^t X(t,\alpha)\,df(\alpha),\end{aligned}$$

where we define $X(t,s) = I + \rho(t,s)$, $t \geq s$. To complete the proof of the theorem, it remains to show that $X(t,s)$ satisfies Equation (1.12). From the resolvent equation, it follows that

$$I + \rho(t,s) = I + \int_\sigma^t d[k(t,\alpha)](I + \rho(\alpha,s)),$$

integration by parts, using Definition (1.7) for k, proves the theorem. □

Corollary 1.1. *Suppose that the assumptions on η, μ, and h are satisfied. If $x(\sigma,\phi,h)$ is the solution of System (1.1) on $[s,\infty)$ such that $x_\sigma = \phi$, then there are positive constants C and γ such that*

$$|x(\sigma,\phi,h)| \leq Ce^{\gamma t}\big(|\phi| + \int_0^t |f(s)|\,ds\big), \qquad t \geq \sigma. \tag{1.13}$$

Proof. From the assumptions on η and μ, it follows that we can choose γ such that

$$\tilde{k}(t,s) = k(t,s)e^{-\gamma(t-s)}$$

satisfies $\|\tilde{k}\| = \sup_{t\geq 0} \operatorname{Var}_{[0,\infty)}\tilde{k}(t,\,\cdot\,) < 1$. Therefore $\tilde{k}$ has a Stieltjes-Volterra resolvent of type B^∞ on $[0,\infty)$. This proves that

$$\big|X(t,s)\big| = \big|I + \rho(t,s)\big| \leq Ce^{\gamma(t-s)} \tag{1.14}$$

Estimate (1.13) now follows from Representation (1.10). □

As we have seen in the retarded case, the matrix solution $X(t,s)$, $t \geq s$, has a natural interpretation for System (1.1). In fact, it is easy to see that $X(t,s)$, $t \geq s$, is the solution to System (1.1) corresponding to the discontinuous initial data $X(\theta,s) = X_0(\theta)$, $-r \leq \theta \leq 0$, where

$$X_0(\theta) = \begin{cases} I, & \text{for } \theta = 0,\\ 0, & \text{for } -r \leq \theta < 0.\end{cases}$$

Note that the fundamental solution $X(t,s)$ is of bounded variation on compact intervals, but not necessarily absolutely continuous for $t > s$, in contrast to the retarded case, where the fundamental solution $X(t,s)$ is absolutely continuous for $t > s$.

From the existence and uniqueness for solutions of Equation (1.1) and the continuity assumptions, it follows that translation along the solution defines a forward evolutionary system on C:

$$T(t,\sigma)\phi = x_t(.;\sigma,\phi), \qquad t \ge \sigma. \tag{1.15}$$

From the variation-of-constants formula, given by Equation (1.10), we find that the solution of System (1.1) is given by

$$x_t(\theta;\sigma,\phi,h) = T(t,\sigma)\phi(\theta) + \int_\sigma^{t+\theta} X(t+\theta,\alpha)h(\alpha)\,d\alpha \tag{1.16}$$

where it is understood that the integral is considered as a family of Euclidean space integrals parameterized by θ, $-r \le \theta \le 0$.

From the results of Section 6.2, it follows that we can write Equation (1.16) as an abstract integral equation in the Banach space C

$$x_t(\,\cdot\,;\sigma,\phi,h) = T(t-\sigma)\phi + \int_\sigma^t d[K(t,\alpha)]h(\alpha), \tag{1.17}$$

where

$$K(t,s)(\theta) = \int_\sigma^s X(t+\theta,\alpha)\,d\alpha \tag{1.18}$$

and $X(t,s)$ is the solution to Equation (1.12).

One can discuss the formal adjoint in the same manner as in Section 6.3. Following similar arguments, one can show that the adjoint evolutionary system $T^*(s,t)$ can be expressed in terms of a backward evolutionary system on forcing functions for the Stieltjes-Volterra equation

$$y(s) + \int_s^t y(\tau)\,d[k(\tau,s)] = g^t(s), \qquad s \le t, \tag{1.19}$$

where $g^t(s) = g(s-t)$ belongs to B_0 and k is given by (1.7).

From the abstract theory, it follows that Equation (1.19) has a unique solution of bounded variation on compact intervals, given by the variation-of-constants formula

$$\begin{aligned} y(s;t,g) &= g^t(s) + \int_s^t d[\rho(t,\alpha)]g^t(\alpha) \\ &= -\int_s^t d[g(\alpha-t)]X(\alpha,s), \qquad s \le t. \end{aligned} \tag{1.20}$$

Translation along a solution of Equation (1.19) induces a two-parameter family of bounded operators $V(s,t)$, $s \le t$ on B_0. In fact, the solution y restricted to $(-\infty,\sigma]$ satisfies a Volterra integral equation given by

$$y(s) + \int_s^t y(\tau)\,d[k(\tau,s)] = \tilde{g}^\sigma, \qquad s \le \sigma,$$

where $\widetilde{g}$ belongs to B_0 and is given by

$$\widetilde{g}(\theta) = \begin{cases} 0 & \text{for } \theta = 0; \\ y_\sigma(\theta) + \int_\theta^0 y_\sigma(\tau) dk(\tau + \sigma, \theta - \tau), & \text{for } -r \le \theta < 0; \\ \widetilde{g}(-r), & \text{for } \theta \le -r. \end{cases}$$

Define $V(\sigma, t)g = \tilde{g}$, so that $V(\sigma, t)$ maps the forcing function for the solution on $(-\infty, t]$ onto the forcing function for the solution on $(-\infty, \sigma]$. From the uniqueness property for the Volterra integral equation, it is easy to see that $V(s, t)$, $s \le t$, defines a backward evolutionary system on B_0.

Next we derive a representation for $V(s, t)$, $s \le t$, from its definition. The integral equation yields

$$\begin{aligned} g(\sigma + \theta - t) &= y(\sigma + \theta) + \int_{\sigma+\theta}^t y(\tau) dk(\tau, \sigma + \theta - \tau) \\ &= y_\sigma(\theta) + \int_\theta^{t-\sigma} y_\sigma(\tau) dk(\tau + \sigma, \theta - \tau). \end{aligned} \tag{1.21}$$

Therefore

$$\widetilde{g}(\theta) = g(\theta - (t - \sigma)) - \int_0^{t-\sigma} y_\sigma(\tau) dk(\tau + \sigma, \theta - \tau)$$

and, using Representation (1.20), $V(\sigma, t) : B_0 \to B_0$ is given by

$$\begin{aligned} V(\sigma, t)g(\theta) &= g(\theta - (t - \sigma)) + \int_\sigma^t y(\tau) dk(\tau, \theta - \tau)\, d\tau \\ &= g(\theta - (t - \sigma)) \\ &\quad + \int_\sigma^t \int_\tau^t d[g(-t + \alpha)] X(\alpha, \tau) dk(\tau, \theta - \tau). \end{aligned} \tag{1.22}$$

By definition $V(\sigma, t)g(0) = 0$ and, using Equation (1.21),

$$V(\sigma, t)g(0-) = y(\sigma).$$

Similar to the retarded case, one can prove the following result.

Theorem 1.3. *Let $T(t, s)$, $t \ge s$ be the evolutionary system associated with System (1.1) on C. If $V(s, t)$, $s \le t$ denotes the backward evolutionary system for the adjoint equation defined by (1.21), then $V(s, t)$, $s \le t$ is the adjoint system of $T(t, s)$, $t \ge s$, that is,*

$$T(t, s)^* = V(s, t).$$

To end this section, we illustrate how the theory can be applied to discuss two-point boundary-value problems for the nonhomogeneous equation (1.1). The theory parallels the theory for RFDE given in Section 6.4.

Suppose V is a Banach space, $\sigma < \tau$ are given real numbers, $M, N : C \to V$ are linear operators with domain dense in C, and $\gamma \in V$ is fixed. The problem is to find a solution x of

$$\frac{d}{dt}\big[D(t)x_t\big] = L(t)x_t + h(t) \tag{1.23}$$

subject to the boundary condition

$$Mx_\sigma + Nx_\tau = \gamma. \tag{1.24}$$

Let V^* be the dual space of V and M^* and N^* the adjoint operators of M and N respectively.

Theorem 1.4. *In order that Equations* (1.23) *and* (1.24) *have a solution, it is necessary that*

$$\int_\sigma^\tau y(\alpha)h(\alpha)\,d\alpha = -\langle\, \delta, \gamma \,\rangle_V \tag{1.25}$$

for all $\delta \in V^*$ and solutions y of the system of adjoint equations

$$\begin{cases} y(s) + \displaystyle\int_s^\tau y(\alpha)d[k(\alpha, s)] = N^*\delta, & s \le \tau, \\ y(s) + \displaystyle\int_s^\sigma y(\alpha)d[k(\alpha, s)] = -M^*\delta, & s \le \sigma, \end{cases} \tag{1.26}$$

where k is given by (1.7). *If $\mathcal{R}\big(M + NT(\tau, \sigma)\big)$ is closed, then the condition is sufficient.*

9.2 Linear autonomous equations

In this section, we consider the theory of a linear autonomous NFDE(D, L) where D and L are continuous linear functions from C into $\mathbb{R}^n$. The theory completely parallels the theory for RFDE given in Chapter 7 and, in fact, the proofs in Chapter 7 were given in such a way that many would carry over almost verbatim for this situation.

Consider the homogeneous linear autonomous NFDE(D, L)

$$\frac{d}{dt}Dx_t = Lx_t \tag{2.1}$$

where $D, L : C \to \mathbb{R}^n$ are continuous and linear and D is atomic at zero. Following Representation (1.2), we write

$$D\phi = \phi(0) - \int_{-r}^0 d[\mu(\theta)]\phi(\theta), \qquad L\phi = \int_{-r}^0 d[\eta(\theta)]\phi(\theta),$$

where $\mathrm{Var}_{[s,0]}\,\mu \to 0$ as $s \to 0$. Let $x(\phi)$ denote the solution of Equation (2.1) with $x_0(\phi) = \phi$. By making slight modifications to the proofs of Lemma 1.2 of Section 7.1 and Lemma 2.1 of Section 7.2 and of the proof of Theorem 2.1 of Section 7.2, one obtains the following result.

Lemma 2.1. *The solution operator $T(t)$, $t \geq 0$, defined by*

$$T(t)\phi \overset{\text{def}}{=} x_t(\phi),$$

is a C_0-semigroup with infinitesimal generator

$$\mathcal{D}(A) = \{\phi \in C : \frac{d\phi}{d\theta} \in C,\ D\frac{d\phi}{d\theta} = L\phi\}, \quad A\phi = \frac{d\phi}{d\theta}. \tag{2.2}$$

If A is defined by Equation (2.2), then $\sigma(A) = \mathrm{P}\sigma(A)$ and λ is in $\sigma(A)$ if and only if λ satisfies the characteristic equation

$$\det \Delta(\lambda) = 0, \qquad \Delta(\lambda) = \lambda D(e^{\lambda\cdot} I) - L(e^{\lambda\cdot} I). \tag{2.3}$$

For any λ in $\sigma(A)$, the generalized eigenspace $\mathcal{M}_\lambda(A)$ is finite dimensional and there is an integer k such that $\mathcal{M}_\lambda(A) = \mathcal{N}\big((\lambda I - A)^k\big)$ and we have the direct sum decomposition:

$$C = \mathcal{N}\big((\lambda I - A)^k\big) \oplus \mathcal{R}\big((\lambda I - A)^k\big). \tag{2.4}$$

Suppose Λ is a finite set $\{\lambda_1, \dots, \lambda_p\}$ of eigenvalues of Equation (2.1) and let $\Phi_\Lambda = \{\Phi_{\lambda_1}, \dots, \Phi_{\lambda_p}\}$, $B_\Lambda = diag(B_{\lambda_1}, \dots, B_{\lambda_p})$, where Φ_{λ_j} is a basis for the generalized eigenspace of λ_j and B_{λ_j} is the matrix defined by $A\Phi_{\lambda_j} = \Phi_{\lambda_j} B_{\lambda_j}$, $j = 1, 2, \dots, p$. Then the only eigenvalue of B_{λ_j} is λ_j and, for any vector a of the same dimension as Φ_Λ, the solution $T(t)\Phi_\Lambda a$ with initial value $\Phi_\Lambda a$ at $t = 0$ may be defined on $(-\infty, \infty)$ by the relation

$$\begin{aligned} T(t)\Phi_\Lambda a &= \Phi_\Lambda e^{B_\Lambda t} a, \\ \Phi_\Lambda(\theta) &= \Phi_\Lambda(0) e^{B_\Lambda \theta}, \qquad -r \leq \theta \leq 0. \end{aligned} \tag{2.5}$$

Furthermore, there exists a subspace Q_Λ of C such that $T(t)Q_\Lambda \subseteq Q_\Lambda$ for all $t \geq 0$ and

$$C = P_\Lambda \oplus Q_\Lambda,$$

where $P_\Lambda = \{\phi \in C \mid \phi = \Phi_\Lambda a,\ for\ some\ vector\ a\}$.

The spectral analysis of unbounded operators developed in Section 7.3 extends to NFDE. Set $\widehat{C} = \mathbb{R}^n \times C$ endowed with the product norm topology

$$|(c, \phi)| = \big(|c|^2 + |\phi|^2\big)^{\frac{1}{2}} \tag{2.6}$$

and define the operator $\widehat{A} : \mathcal{D}(\widehat{A}) \to \widehat{C}$

$$\mathcal{D}(\widehat{A}) = \{(c, \phi) \in \widehat{C} : \frac{d\phi}{d\theta} \in C,\ c = D\phi\},$$
$$\widehat{A}(c, \phi) = \big(L\phi, \frac{d\phi}{d\theta}\big). \tag{2.7}$$

This is a closed unbounded operator, but the domain of $\widehat{A}$ is not dense in $\widehat{C}$, and hence, $\widehat{A}$ is not the generator of a semigroup. The closure of the domain is precisely given by jC and the part of $\widehat{A}$ in jC is given by Relation (2.2).

Since $\mathcal{D}(\widehat{A}) \subset jC$ and A is the part of $\widehat{A}$ in jC, it follows that the spectrum of A and $\widehat{A}$ are the same and $j\mathcal{M}_\lambda(A) = \mathcal{M}_\lambda(\widehat{A})$.

The structure of $\widehat{A}$ is similar to the one studied in Section 7.3 and an application of Theorem 3.1 of Section 7.3 yields the following result.

Theorem 2.1. *The matrix function $\Delta : \mathbb{C} \to \mathbb{C}^{n\times n}$*

$$\Delta(z) = z\big(I - \int_{-r}^{0} e^{zt}\, d[\mu(t)]\big) - \int_{-r}^{0} e^{zt}\, d[\eta(t)], \quad z \in \mathbb{C} \tag{2.8}$$

is a characteristic matrix for $\widehat{A}$ and the equivalence is given by

$$\begin{pmatrix} \Delta(z) & 0 \\ 0 & I \end{pmatrix} = F(z)(zI - \widehat{A})E(z), \quad z \in \mathbb{C}, \tag{2.9}$$

where the explicit formulas for $E : \mathbb{C} \to \mathcal{L}(\widehat{C}, \widehat{C}(\widehat{A}))$ and $F : \mathbb{C} \to \mathcal{L}(\widehat{C}, \widehat{C})$ are given in Theorem 3.1 of Section 7.3.

Corollary 2.1. *The spectrum of the operator $A : \mathcal{D}(A) \to C$ defined by (2.2) consists of eigenvalues of finite type only,*

$$\sigma(A) = \{\lambda : \det \Delta(\lambda) = 0\}. \tag{2.10}$$

For $\lambda \in \sigma(A)$, the algebraic multiplicity of the eigenvalue λ equals the order of λ as a zero of $\det \Delta$, the partial multiplicities of the eigenvalue λ are equal to the zero-multiplicities of λ as a characteristic value of Δ, and the largest partial multiplicity (ascent) of λ equals the order of λ as a pole of Δ^{-1}. Furthermore, a canonical basis of eigenvectors and generalized eigenvectors for A at λ may be obtained in the following way: If

$$\{(\gamma_{i,0}, \ldots, \gamma_{i,k_i-1}) \mid i = 1, \ldots, p\}$$

is a canonical system of Jordan chains for Δ at $\lambda \in \Omega$, then

$$\chi_{i,0}, \ldots, \chi_{i,k_i-1}, \quad i = 1, \ldots, p,$$

where

$$\chi_{i,\nu}(\theta) = e^{\lambda\theta} \sum_{l=0}^{\nu} \gamma_{i,\nu-l} \frac{\theta^l}{l!},$$

yields a canonical basis for A at λ.

From the equivalence relation (2.9), we have the following representation for the resolvent of A.

Corollary 2.2. *Let $A : \mathcal{D}(A) \to C$ be the generator defined by Relation (2.2). The resolvent of A has the following representation*

$$(R(z,A)\phi)(\theta) = e^{z\theta}\big[\Delta^{-1}(z)K(z)\phi + \int_\theta^0 e^{-z\tau}\phi(\tau)d\tau\big] \tag{2.11}$$

where

$$K(z)\phi = M\phi + \int_{-r}^0 [zd\mu(\tau) + d\eta(\tau)]e^{z\tau}\int_\tau^0 e^{-zs}\phi(s)\,ds. \tag{2.12}$$

From the adjoint theory developed in Section 9.1, we know that the adjoint semigroup $T^*(s) = T(s)^*$ corresponds to the Volterra equation

$$y(s) + \int_s^0 y(\tau)\,d[k(\tau - s)] = g(s), \qquad s \leq 0, \tag{2.13}$$

with $g \in B_0$ and k given by

$$k(t) = k(t,0) = \mu(-t) + \int_0^t \eta(-\tau)\,d\tau. \tag{2.14}$$

Recall that B_0 denotes the Banach space of row-valued functions ψ : $(-\infty, 0] \to \mathbb{R}^{n*}$ that are constant on $(-\infty, -r]$, of bounded variation on $[-r, 0]$, continuous from the left on $(0, r)$, and vanishing at zero with norm $\mathrm{Var}_{[-r,0]}\,\psi$. The generator of $T^*(t)$ equals A^* and can be easily computed.

Lemma 2.2. *The adjoint operator $A^* : \mathcal{D}(A^*) \to B_0$ is given by*

$$\mathcal{D}(A^*) = \{f \in B_0 : \frac{d}{d\theta}[f(0-)\mu + f] \in B_0\} \tag{2.15a}$$

$$A^*f(\theta) = f(0-)\eta(\theta) - \frac{d}{d\theta}[f(0-)\mu(\theta) + f(\theta)]. \tag{2.15b}$$

Proof. It is obvious that given the action of A^*, the domain of definition of A^* cannot be larger than the subspace defined in the right-hand side of (2.15a). So assume that $g \in B_0$ with $g(-r) = 0$, and

$$f(0-)\mu(\theta) + f(\theta) = f(0-) - \int_\theta^0 g(s)\,ds, \qquad \text{for} \quad \theta < 0.$$

From Relation (2.2)

$$\begin{aligned}\langle f, A\phi \rangle &= \int_{-r}^{0} A\phi(\theta)\, df(\theta) \\ &= \int_{-r}^{0} d\eta(\theta)\phi(\theta) f(0-) + \int_{-r}^{0} \frac{d\phi}{d\theta}(\theta)\, d[f(0-)\mu(\theta) + f(\theta)] \\ &= \int_{-r}^{0} \phi(\theta)\, d[f(0-)\eta(\theta)] - \int_{-r}^{0} \phi(\theta)\, dg(\theta) = \langle A^* f, \phi \rangle.\end{aligned}$$

This proves the lemma. □

As in the retarded case, the adjoint equation (2.13) is closely related to the transposed equation. Define $C' = C([0,r], \mathbb{R}^{n*})$. For each $s \in [0,\infty)$ let y^s designate the element in C' defined by $y^s(\xi) = y(-s+\xi)$, $0 \le \xi \le r$. The *transpose* of System (2.1) is defined to be

$$\frac{d}{ds}\Big[y(s) - \int_{-r}^{0} y(s-\theta)\, d[\mu(\theta)]\Big] = \int_{-r}^{0} y(s-\theta)\, d[\eta(\theta)], \qquad s \le 0, \qquad y^0 = \psi, \qquad \psi \in C'. \tag{2.16}$$

Let y be a solution of Equation (2.16) on an interval $(-\infty, r]$. We associate a C_0-semigroup $T^T(s)$ with Equation (2.16), defined by translation along the solution,

$$T^T(s)\psi = y^s(\,\cdot\,;\psi), \tag{2.17}$$

where y is the solution of Equation (2.16). By making slight modifications to the proof of Lemma 1.2 of Section 7.1, one can prove the following result.

Lemma 2.3. *The solution operator $T^T(s)$, $s \ge 0$, defined by Relation* (2.17) *is a C_0-semigroup with infinitesimal generator*

$$\begin{aligned}\mathcal{D}(A^T) &= \{\psi \in C' : \frac{d\psi}{d\xi} \in C',\ D\frac{d\psi}{d\xi} = -L\psi\} \\ A^T\psi &= -\frac{d\psi}{d\xi}.\end{aligned} \tag{2.18}$$

Integration of Equation (2.16) yields a Volterra equation

$$y(s) + \int_{s}^{0} y(\tau)\, d[k(\tau - s)] = F^T\psi(s), \qquad s \le 0,$$

where $F^T : C' \to B_0$ is given by

$$\begin{aligned}(F^T\psi)(s) = \psi(0) &+ \int_{s}^{0} \psi(s-\theta)\, d\mu(\theta) \\ &- \int_{s}^{0}\int_{-r}^{\alpha} \psi(\alpha-\theta) d[\eta(\theta)]\, d\alpha\end{aligned} \tag{2.19}$$

and k is given by (2.14). From the definitions, it is not difficult to verify that

$$F^T T^T(s)\psi = T^*(s)F^T\psi. \tag{2.20}$$

One can give an explicit characterization of Decomposition (2.4) via the formal adjoint A^*. Together with the properties of the adjoint operation, the equivalence (2.9) for A implies the following equivalence relation for $\widehat{A}^*$:

$$(\alpha\ f)\begin{pmatrix} \Delta(z)^T & 0 \\ 0 & I \end{pmatrix} = E(z)^*(zI - \widehat{A}^*)F(z)^*(\alpha f), \tag{2.21}$$

where $E(z)^* : \widehat{C}(\widehat{A})^* \to \widehat{B}_0$ and $F(z)^* : \widehat{B}_0 \to \widehat{B}_0$ are bijective mappings that depend analytically on z, $z \in \mathbb{C}$.

In particular, we find that the Jordan chains of length k of $\Delta(z)^T$ are in one-to-one correspondence with the solutions of

$$(\lambda I - A^*)^k\psi = 0.$$

The proof of the following result follows immediately from the corresponding results in the RFDE case; see Lemma 5.1 and Theorem 5.1 of Section 7.5.

Theorem 2.2. *The spectrum of the infinitesimal generator A^* defined by (1.7) consists of eigenvalues of finite type only,*

$$\sigma(A^*) = \{\lambda : \det\Delta(\lambda) = 0\}. \tag{2.22}$$

For $\lambda \in \sigma(A^*)$, the algebraic multiplicity of the eigenvalue λ equals the order of λ as a zero of $\det\Delta$, *the ascent of λ equals the order of λ as a pole of Δ^{-1}. Furthermore, a canonical basis of eigenvectors and generalized eigenvectors for A^* at λ may be obtained in the following way: If*

$$\{(\beta_{i,0},\ldots,\beta_{i,k_i-1}) : i = 1,\ldots,p\}$$

is a canonical system of Jordan chains for Δ^T at $\lambda \in \mathbb{C}$, then

$$\{\chi_{i,0}\ldots,\chi_{i,k_i-1} : i = 1,\ldots,p\}$$

where $\chi_{i,\nu} = F^T\psi_{i,\nu}$ with

$$\psi_{i,\nu}(\xi) = e^{-\lambda\xi}\sum_{l=0}^{\nu}\beta_{i,\nu-l}\frac{(-\xi)^l}{l!},$$

yields a canonical basis for A^* at λ. Here $F^T : C[0,r] \to B_0$ is the mapping defined by (2.19).

From the theorem, we conclude that the (generalized) eigenfunctions of A^* are precisely given by the images under the operator F^T of the (generalized) eigenfunctions of the transposed generator A^T. Furthermore, from (2.20)

$$A^* F^T \psi = F^T A^T \psi, \qquad \psi \in \mathcal{D}(A^T).$$

In general, the mapping F^T is not one-to-one (see Section 3.3), but on the generalized eigenspace of A^T, it is. Hence, if ψ is a (generalized) eigenfunction of A^T, then $F^T\psi$ is a (generalized) eigenfunction of A^*. This motivates the introduction of the following bilinear form. For $\psi \in C'$ and $\phi \in C$ define

$$\begin{aligned}(\psi, \phi) \stackrel{\text{def}}{=} \langle F^T\psi, \phi\rangle &= \int_{-r}^{0} d[F^T\psi(\theta)]\,\phi(\theta) \\ &= \psi(0)\phi(0) + \int_{-r}^{0} d\Big[\int_{\theta}^{0} \psi(\theta-\alpha)\,d[\mu(\alpha)]\Big]\,\phi(\theta) \\ &\quad + \int_{-r}^{0}\int_{r}^{\theta} \psi(\theta-\tau)\,d[\eta(\tau)]\phi(\theta)\,d\theta \\ &= \psi(0)\phi(0) - \int_{-r}^{0} d\Big[\int_{0}^{\theta} \psi(\theta-\alpha)\,d[\mu(\alpha)]\Big]\,\phi(\theta) \\ &\quad - \int_{-r}^{0}\int_{0}^{\alpha} \psi(\theta-\tau)\,d[\eta(\tau)]\phi(\theta)\,d\theta \end{aligned} \tag{2.23}$$

between C and C'. With respect to this bilinear form, the transposed operator A^T satisfies

$$(\psi, A\phi) = (A^T\psi, \phi).$$

Lemma 2.3. *For λ in $\sigma(A)$, let $\Psi_\lambda = \operatorname{col}(\psi_1, \dots, \psi_p)$ and $\Phi_\lambda = (\phi_1, \dots, \phi_p)$ be bases for $\mathcal{M}_\lambda(A^T)$ and $\mathcal{M}_\lambda(A)$, respectively, and let $(\Psi_\lambda, \Phi_\lambda) = (\psi_i, \phi_j)$, $i, j = 1, 2, \dots, p$. Then $(\Psi_\lambda, \Phi_\lambda)$ is nonsingular and thus may be taken as the identity. The decomposition of C given by Lemma 2.1 may be written explicitly as*

$$\begin{aligned} &\phi = \phi^{P_\lambda} + \phi^{Q_\lambda}, \qquad \phi^{P_\lambda} \text{ in } P_\lambda, \phi^{Q_\lambda} \text{ in } Q_\lambda, \\ &P_\lambda = \mathcal{M}_\lambda(A) = \{\phi \in C : \phi = \Phi_\lambda b \text{ for some } p\text{-vector } b\} \\ &Q_\lambda = \{\phi \in C : (\Psi_\lambda, \phi) = 0\}, \\ &\phi^{P_\lambda} = \Phi_\lambda b, \qquad b = (\Psi_\lambda, \phi), \qquad \phi^{Q_\lambda} = \phi - \phi^{P_\lambda}. \end{aligned}$$

If Λ is a finite set of eigenvalues and C is decomposed as in Lemma 2.2, we say C is decomposed by Λ. In Chapter 1, we have seen that the set $\Lambda = \{\lambda \in \sigma(A) : \operatorname{Re}\lambda \geq \beta\}$ is not necessarily finite. So it is not clear that we can decompose C according to this infinite set of eigenvalues. We will return to this question in Section 9.4.

Suppose C can be decomposed by Λ as $C = P_\Lambda \oplus Q_\Lambda$. Can we estimate $T(t)$ on the complementary space Q_Λ? In the next section, we shall address this question.

9.3 Exponential estimates

In Corollary 1.1, we proved that the solutions to System (2.1) are exponentially bounded. It is our aim to prove precise exponential estimates for the solution.

In Section 9.1, we have seen that we can rewrite the System (2.1) with initial condition $x_0 = \phi$ as a Volterra-Stieltjes equation

$$x(t) - \int_0^t d[k(t-s)]x(s) = f(t) \tag{3.1}$$

where

$$k(t) = \mu(-t) + \int_0^t \eta(-\tau)\, d\tau, \tag{3.2}$$

$$f(t) = \phi(0) + \int_0^t d[\mu(\theta)]\phi(t+\theta) + \int_0^t \int_{-r}^{-\alpha} d[\eta(\theta)]\phi(\alpha+\theta)\, d\alpha. \tag{3.3}$$

From Theorem 1.2, we find that the solution of Equation (3.1) has a simple representation

$$x(t) = X(t)f(0) + \int_0^t X(t-\alpha)\, df(\alpha) \tag{3.4}$$

where $X(t)$ is the matrix solution to

$$X(t) = I + \int_0^t d[k(t-\alpha)]X(\alpha). \tag{3.5}$$

To estimate the solution of Equation (3.1) we are going to apply the Laplace-Stieltjes transform to $X(t)$.

First we present some results from the theory of vector-valued Laplace transforms needed in the sequel. Let $\mathcal{B}$ be a complex Banach space, Φ a function on $[0,\infty)$ to $\mathcal{B}$, and let Φ be of strong bounded variation over every compact interval. If there exist constants C and α such that

$$\Phi_*(t) = \mathrm{Var}_{[0,t)}\, \Phi \le Ce^{\alpha t}, \qquad t \ge 0,$$

then the Laplace-Stieltjes transform of a is defined by

$$f(z) = \int_0^\infty e^{-zt}\, d\Phi(t). \tag{3.6}$$

For $\operatorname{Re} z > \alpha$, the integral is absolutely convergent and exists as a Bochner integral. We let $\sigma_a(\Phi)$ denote the abscissa of absolute convergence of (3.6); that is,

$$\sigma_a(\Phi) = \inf\{\sigma \in \mathbb{R} \mid \int_0^\infty e^{-\sigma t} d\Phi_*(t) \quad \text{converges}\}.$$

If $g : \mathbb{R}_+ \to \mathcal{B}$ is bounded on bounded intervals and $\Phi(t) = \int_0^t g(s)\, ds$, then Φ is of strong bounded variation and

$$\Phi_*(t) = \int_0^t |g(s)|\, ds.$$

The Laplace-Stieltjes transform of Φ is given by

$$\mathcal{L}(g)(z) = \int_0^\infty e^{-zt} g(t)\, dt \tag{3.7}$$

and is called the Laplace transform of g.

Since $\operatorname{Var}_{[0,t)} X(\,\cdot\,)$ satisfies the exponential bound (1.14), there exists an $\alpha \in \mathbb{R}$ such that $\mathcal{L}(X)$ exists and is analytic for $\operatorname{Re} z > \alpha$. To compute the Laplace-Stieltjes transform of X, we use the following lemma.

Lemma 3.1. (*Convolution of Laplace-Stieltjes transform.*) *Let η be an $n \times n$ matrix-valued function of bounded variation and let α be an n-vector function of bounded variation. If the Laplace-Stieltjes integral*

$$f(z) = \int_0^\infty e^{-zt}\, d\alpha(t), \qquad g(z) = \int_0^\infty e^{-zt}\, d\eta(t),$$

are absolutely convergent for $\operatorname{Re} z > \sigma_a$, *then*

$$g(z)f(z) = \int_0^\infty e^{-zt}\, d\zeta(t) \qquad \textit{with} \quad \zeta(t) = \int_0^t d[\eta(\theta)]\alpha(t-\theta). \tag{3.8}$$

So from Equation (3.5), we find for $\operatorname{Re} z > \alpha$

$$\begin{aligned}\int_0^\infty e^{-zt} d[X(t)] &= \int_0^\infty e^{-zt} d[k(t)] \int_0^\infty e^{-zt} d[X(t)] \\ &= \frac{1}{z}\Delta(z) \int_0^\infty e^{-zt} d[X(t)]\end{aligned}$$

and a simple computation yields

$$\mathcal{L}(X)(z) = \Delta(z)^{-1} \tag{3.9}$$

where $\Delta(z)$ is given by Formula (2.8). Set $\Delta(z) = z\Delta_0(z) - \int_{-r}^0 e^{zt}\, d[\eta(t)]$ where

$$\Delta_0(z) = I - \int_{-r}^0 e^{zt}\, d[\mu(t)].$$

In order to control the behavior of $|\Delta(z)|$ as $|z| \to \infty$, we make the following assumption on the kernel μ:

(J) The entries μ_{ij} of μ have an atom before they become constant, i.e., there is a t_{ij} with $\mu_{ij}(t) = \mu_{ij}(t_{ij}+)$ for $t \geq t_{ij}$ and $\mu_{ij}(t_{ij}-) \neq \mu_{ij}(t_{ij}+)$.

For example, μ can be a step function. In that case

$$\Delta_0(z) = I - \sum_{j=1}^{\infty} e^{-z r_j} A_j$$

and $\det \Delta_0$ is an almost-periodic function. The jump condition (J) is much more general and implies that $\det \Delta_0$ is asymptotically almost-periodic. Define

$$\mathbb{C}_{\gamma_1,\gamma_2} = \{z \in \mathbb{C} \mid \gamma_1 < \operatorname{Re} z < \gamma_2\} \qquad \text{and} \quad \mathbb{C}_\gamma = \{z \in \mathbb{C} \mid \operatorname{Re} z > \gamma\}.$$

Lemma 3.2. *If μ satisfies* (J)*, then the zeros of* $\det \Delta_0$ *are located in a finite strip $\mathbb{C}_{\gamma_0,\gamma_1}$ and there exist positive constants ϵ, m and M, such that*

$$m|e^{-\tau z}| \leq |\det \Delta_0(z)| \leq M|e^{\tau z}|$$

outside circles of radius ϵ centered around the zeros of $\det \Delta_0$*. Here τ is the exponential type of* $\det \Delta_0$*.*

Proof. From Lemma 3.1, it follows that there is a function μ^* such that

$$\det \Delta_0(z) = \int_{-\tau}^{0} e^{zs} d\mu^*(s). \tag{3.10}$$

If μ satisfies (J), then μ^* has jumps at both endpoints 0 and τ. An application of Verduyn Lunel [2], Theorem 4.6, yields the lemma. □

Theorem 3.2. *If $a_{D,L} = \sup\{\operatorname{Re} z : \det \Delta(z) = 0\}$, then, for any $\alpha > a_{D,L}$, there is a constant $C = C(\alpha)$ such that the fundamental solution X of System* (2.1) *satisfies the exponential estimates*

$$|X(t)| \leq Ce^{\alpha t}, \quad \operatorname{Var}_{[0,t)} X \leq Ce^{\alpha t}, \qquad t \geq 0. \tag{3.11}$$

Before we prove the theorem we need some preparation. Let $\mathcal{S}$ denote the Schwartz space, that is, the space of C^∞-functions such that

$$\sup_{x \in \mathbb{R}} |x^k \Phi^{(q)}(x)| \leq m_{kq} \qquad k, q = 0, 1, \ldots.$$

These pseudonorms can be used to make $\mathcal{S}$ a complete linear topological vector space. Tempered distributions are representations of continuous linear functionals on $\mathcal{S}$.

If $f : \mathbb{R} \to \mathbb{R}$ is a bounded function, then $\Phi \mapsto (f, \Phi)$ denotes the tempered distribution defined by f on $\mathcal{S}$. For $\Phi \in \mathcal{S}$, the Fourier transform

$$\mathcal{F}\Phi(s) = \frac{1}{2\pi} \int_{-\infty}^{\infty} e^{ist} \Phi(t)\, dt$$

exists, and is an injective continuous mapping from $\mathcal{S}$ onto $\mathcal{S}$. Its inverse is also continuous, and is given by

$$\mathcal{F}^{-1}\Phi(t) = \int_{-\infty}^{\infty} e^{-ist} \Phi(s)\, ds.$$

By duality, one defines the Fourier transform $\mathcal{F}f$ of a tempered distribution f on $\mathcal{S}$

$$(\mathcal{F}f, \Phi) = (f, \mathcal{F}\Phi), \qquad \Phi \in \mathcal{S}. \tag{3.12}$$

Note that

$$|\mathcal{F}\Phi|_1 = \frac{1}{2\pi} |\mathcal{F}^{-1}\Phi|_1$$

where $|\cdot|_1$ is the $\mathcal{L}_1$-norm. The following lemma will be useful in the estimations.

Lemma 3.3. *Let $f : \mathbb{R} \to \mathbb{R}$ be a bounded function. The (generalized) Fourier transform of f belongs to $\mathcal{L}_\infty$ if and only if*

$$|(f, \Phi)| \le K |\mathcal{F}\Phi|_1, \qquad \Phi \in \mathcal{S}. \tag{3.13}$$

Proof. From (3.13) one finds

$$|(\mathcal{F}f, \mathcal{F}^{-1}\Phi)| = |(r, \Phi)| \le K |\mathcal{F}\Phi|_1 \le \frac{K}{2\pi} \|\mathcal{F}^{-1}\Phi\|_1, \qquad \Phi \in \mathcal{S}.$$

Since $\mathcal{F}[\mathcal{S}] = \mathcal{S}$ and $\mathcal{S}$ is dense in $\mathcal{L}_1$, we conclude that $\mathcal{F}f \in \mathcal{L}_\infty$. On the other hand, if $\mathcal{F}f$ belongs to $\mathcal{L}_\infty$, then

$$|(f, \Phi)| = |(\mathcal{F}f, \mathcal{F}^{-1}\Phi| \le K \|\mathcal{F}^{-1}\Phi\|_1 = \frac{K}{2\pi} \|\mathcal{F}\Phi\|_1.$$

□

Proof of Theorem 3.2. In any right half plane $\operatorname{Re} z > \gamma$

$$\det \Delta(z) = z^n \det \Delta_0(z) + O(z^{n-1}) \qquad \text{as} \quad |z| \to \infty \quad \text{in } \operatorname{Re} z > \gamma. \tag{3.14}$$

Let $\alpha > a_{D,L}$. From the Rouché theorem, it follows that $\det \Delta_0(z) \ne 0$ for $\operatorname{Re} z > a_{D,L}$. Therefore, there exists a strip $\alpha - \epsilon < \operatorname{Re} z < \alpha + \epsilon$ such that $\det \Delta_0(z) \ne 0$ on this strip. From Lemma 3.2, we derive that $|\det \Delta_0(z)| \ge C$ for $z \in \alpha - \epsilon < \operatorname{Re} z < \alpha + \epsilon$ From Representation (3.11) for $\det \Delta_0(z)$ and the fact that there exists a constant $C > 0$ such that

$|\det \Delta_0(\alpha + i\nu)| \geq C$ for $\nu \in \mathbb{R}$, it follows that there exists a function ζ such that $\zeta(t) = 0$ for $t < 0$, $e^{\alpha \cdot}\zeta$ is of bounded variation on $\mathbb{R}$, and

$$\left(\det \Delta_0(\alpha + i\nu)\right)^{-1} = \int_0^\infty e^{(\alpha+i\nu)t}\, d\zeta(t)$$

(see for example Hille and Phillips [1], pp. 144–150). The cofactors of $\Delta(z)$ are polynomials of degree $n-1$ with entire coefficients that are bounded in $\mathbb{C}_{-\omega,\omega}$. Therefore, we can expand

$$\Delta(\alpha + i\nu)^{-1} = \int_0^\infty e^{(\alpha+i\nu)t}\, d[\zeta(t)]\Big(\frac{A_1}{\alpha + i\nu} + O\big(\frac{1}{(\alpha + i\nu)^2}\big)\Big) \tag{3.15}$$

as $|\nu| \to \infty$. Let $x \in \mathbb{R}^n$ and $y \in \mathbb{R}^{n*}$ and set $f(\nu; x, y)) = y\Delta(\alpha + i\nu)^{-1}x$. From the Laplace inversion formula and a simple contour integration, we find

$$e^{-\alpha t}X(t) = \lim_{N\to\infty} \frac{1}{2\pi} \int_{-N}^{N} e^{i\nu t} \Delta(\alpha + i\nu)^{-1}\, d\nu.$$

So

$$\left|e^{-\alpha t}X(t)x\right| = \sup_{|y|\leq 1} \left|e^{\alpha t}yX(t)x\right| = \sup_{|y|\leq 1} |\mathcal{F}f|_\infty$$

and it remains to prove that $\mathcal{F}f$ belongs to $\mathcal{L}_\infty$. From (3.15) it follows that it suffices to analyze the Fourier transform of

$$\int_0^\infty e^{-(\alpha+i\nu)t}\, d\zeta(t) y \frac{A_1}{\alpha + i\nu} x. \tag{3.16}$$

In order to use Lemma 3.3, we have to estimate

$$\begin{aligned}
&\Big|\int_{-\infty}^\infty \int_0^\infty d\zeta(t) y \frac{A_1 e^{-(\alpha+i\nu)t}}{\alpha + i\nu} x\Phi(\nu)\, d\nu\Big| \\
&= \Big|\int_0^\infty d\zeta(t) \int_{+\infty}^t e^{-\alpha\sigma} \mathcal{F}\Phi(\sigma) y A_1 x\, d\sigma\Big| \\
&\leq \Big|\int_0^\infty e^{-\alpha t}\, d\zeta(t) |\mathcal{F}\Phi|_1\, |y|\, |A_1|\, |x|\Big| \\
&\leq |A| \|x\|\, \|y\|\, \|\mathcal{F}\Phi\|_1,
\end{aligned}$$

where we have used, in order to reverse the order of integration, that $\Phi \in \mathcal{S}$ and $e^{-\alpha \cdot}\zeta$ is of bounded variation on $\mathbb{R}$. So it follows from Lemma 3.3 that the Fourier transform of (3.16) is bounded. This shows that $X(t)$ satisfies the estimate in Formula (3.13).

In order to estimate the variation of $X(t)$ we write Equation (3.5) as follows

$$X(t) = I + \int_0^t d[X(t - \alpha)]\mu(-\alpha) - \int_0^t X(t - \alpha)\eta(-\alpha)\, d\alpha. \tag{3.17}$$

If $X_*(t) = \mathrm{Var}_{[0,t)} X = \int_0^t |dX(\tau)|$ denotes the variation of X, then it follows from Equation (3.17) that X_* satisfies the inequality

$$X_*(t) \le \int_0^t X_*(t-\theta)\, d|\mu(\theta)| + Ce^{\alpha t}.$$

The assertion now follows from Lemma 3.1 of Chapter 1. □

Corollary 3.1. *If $a_{D,L} = \sup\{\mathrm{Re}\, z : \det \Delta(z) = 0\}$, then, for any $\alpha > a_{D,L}$, there is a constant $c = c(\alpha)$ such that the solution $x(\,\cdot\,;\phi)$ of System* (2.1) *with $x_0 = \phi$ satisfies the exponential estimates*

$$|x(t;\phi)| \le ce^{\alpha t}|\phi|, \qquad t \ge 0.$$

In particular, if $\alpha_{D,L} < 0$, then all solutions of System (2.1) *approach zero exponentially.*

Proof. This is an immediate consequence of Representation (3.4). □

We end this section with an application to homogeneous and inhomogeneous difference equations

$$Dy_t = 0, \qquad t \ge 0 \tag{3.18}$$

and

$$Dy_t = h(t), \qquad t \ge 0 \tag{3.19}$$

where $h \in C\big([0,\infty), \mathbb{R}^n\big)$. If

$$C_D = \{\phi \in C : D\phi = 0\} \tag{3.20}$$

then the theory of the previous section implies that Equation (3.18) defines a $\mathcal{C}_0$-semigroup $T_D(t) : C_D \to C_D$.

As before, we can rewrite Equation (3.19) as a Volterra-Stieltjes equation

$$x(t) - \int_0^t d[\mu(t-s)]x(s) = \phi(0) + \int_0^t d[\mu(\theta)]\phi(t+\theta) + h(t). \tag{3.21}$$

So the theory developed earlier can be applied, and we find the following result.

Theorem 3.4. *Let $y(\psi,h)$ denote the solution of Equation* (3.19) *satisfying $y_0(\psi,h) = \psi$. If $a_D = \sup\{\mathrm{Re}\, z : \det \Delta_0(z) = 0\}$, then for any $\alpha > a_D$, there is a constant $C = C(\alpha)$ such that*

$$|y(\psi,h)(t)| \le C(\alpha)\big[|\psi|e^{\alpha t} + \sup_{0\le s\le t} |h(s)|\big]$$

for $t \ge 0$.

Definition 3.1. Suppose $D : C \to \mathbb{R}^n$ is linear, continuous, and atomic at 0. The operator D is said to be *stable* if the zero solution of the homogeneous difference equation (3.18) with $y_0 = \psi \in C_D$ is uniformly asymptotically stable.

The following theorem is a simple consequence of these results.

Theorem 3.5. *The following statements are equivalent:*

(i) D *is stable.*

(ii) $a_D < 0$.

(iii) *There are constants* $C > 0$ *and* $\alpha > 0$ *such that for any* $h \in C\big([0,\infty), \mathbb{R}^n\big)$, *any solution* y *of the nonhomogeneous equation* (3.19) *satisfies*

$$|y(\psi, h)(t)| \leq C(\alpha)\big[|\psi|e^{-\alpha t} + \sup_{0\leq s\leq t} |h(s)|\big]. \tag{3.22}$$

(iv) *If* $D\phi = \phi(0) - \int_{-r}^0 d[\mu(\theta)]\phi(\theta)$, $\mathrm{Var}_{[-s,0]} \to 0$ *as* $s \to 0$, *and* μ *satisfies* (J), *then there is a* $\delta > 0$ *such that all solutions of the characteristic equation*

$$\det \Delta_0(\lambda) = \det\big[I - \int_{-r}^0 e^{\lambda\theta} d[\mu(\theta)]\big] = 0$$

satisfy $\mathrm{Re}\,\lambda \leq -\delta$.

9.4 Hyperbolic semigroups

In Corollary 3.1, we have proved exponential estimates for the solution of System (2.1). These estimates, in particular, imply that for the semigroup associated with System (2.1), we have

$$\|T(t)\| \leq Me^{\alpha t}$$

where $\alpha > a_{D,L} = \sup\{\mathrm{Re}\, z : \det \Delta(z) = 0\}$. In this section, we shall further extend the results in Section 7.6 for neutral equations and give necessary and sufficient conditions such that the space C can be decomposed according to $\Lambda_\beta = \{z : \mathrm{Re}\, z \geq \beta\}$.

Let $\mathcal{B}$ be a complex Banach space and $A : \mathcal{D}(A) \to \mathcal{B}$ be the generator of a C_0-semigroup $(T(t))_{t\geq 0}$ on $\mathcal{B}$. We say that a C_0-semigroup $(T(t))_{t\geq 0}$ on a Banach space $\mathcal{B}$ is *hyperbolic* when the space $\mathcal{B}$ decomposes into $\mathcal{B} = \mathcal{B}_- \oplus \mathcal{B}_+$ such that $T(t)\mathcal{B}_\pm \subset \mathcal{B}_\pm$,

$$T_-(t) : \mathcal{B}_- \to \mathcal{B}_-, \qquad T_-(t)x = T(t)x$$

extends to a $\mathcal{C}_0$-group on $\mathcal{B}_-$ over $-\infty < t < \infty$, and there are positive constants K, α, β such that

$$\|T_-(t)\| \le Ke^{\beta t}\|\Pi x\|, \qquad t \le 0, \tag{4.1}$$

$$\|T_+(t)x\| \le Ke^{-\alpha t}\|(I - \Pi)x\|, \qquad t \ge 0. \tag{4.2}$$

It is clear that a semigroup is hyperbolic if and only if there is an open annulus containing the circle $\{z \in \mathbb{C} : |z| = 1\}$ in the resolvent set of $T(t)$. In many applications, however, only the generator is explicitly known. Therefore, one would like to have necessary and sufficient conditions on the generator such that the semigroup is hyperbolic.

First, we need a more general version of the Laplace inversion formula. Let Φ be a locally (Bochner) integrable function on $[0, \infty)$ with values in $\mathcal{B}$. Denote by

$$f(z) = \int_0^\infty e^{-zt}\Phi(t)\,dt \tag{4.3}$$

the Laplace transform of Φ. If we define $\sigma_a(\Phi)$ to be the abscissa of absolute convergence of the Laplace transform of Φ, that is, $\sigma_a(\Phi)$ is the infinimum of the real numbers σ such that

$$\int_0^\infty \|e^{-zt}\Phi(t)\|\,dt$$

converges for $\operatorname{Re} z > \sigma$, then for $c > \sigma_a(\Phi)$

$$\frac{1}{2\pi i}(C,1)\text{-}\int_{c-i\infty}^{c+i\infty} e^{zt} f(z)\,dz = \begin{cases} 0, & \text{for } t < 0; \\ \frac{1}{2}\Phi(0+), & \text{for } t = 0; \\ \frac{1}{2}[\Phi(t+) + \Phi(t-)], & \text{for } t > 0 \end{cases} \tag{4.4}$$

whenever the expressions in the right-hand side of (4.4) have a meaning. Here, the *Cesàro mean* of integrals is given by

$$(C,1)\text{-}\int_{c-i\infty}^{c+i\infty} e^{zt} f(z)\,dz = \lim_{N\to\infty}\int_{-N}^{N} e^{(c+i\nu)t} f(c+i\nu)(1 - \frac{|\nu|}{N})\,d\nu.$$

The Cesàro mean is a weaker notion than the principal value of the integrals, but if the principal value exists, it equals the Cesàro mean.

For example, (4.4) holds if Φ is locally of bounded variation. To prove (4.4), apply a continuous linear functional to both sides of (4.4), use the scalar version of (4.4) from Widder [1], theorem II.9.2, and apply the Hahn-Banach theorem.

In particular, if $A : \mathcal{D}(A) \to \mathcal{B}$ is the generator of a $\mathcal{C}_0$-semigroup on $\mathcal{B}$ and $\phi(t) = T(t)x$, $\Phi(t) = T(t)x$, then it is known that the order of $T(t)$ is the supremum of $\sigma_a(T(\cdot)x)$ over $x \in \mathcal{B}$. Since the resolvent of A equals the Laplace transform of the semigroup, we conclude that for $c > \omega(A)$,

$$T(t)x = \frac{1}{2\pi i}(C,1)\text{-}\int_{c-i\infty}^{c+i\infty} e^{zt}R(z,A)x\,dz, \qquad t>0 \tag{4.5a}$$

and

$$\frac{1}{2}x = \frac{1}{2\pi i}(C,1)\text{-}\int_{c-i\infty}^{c+i\infty} R(z,A)x\,dz \tag{4.5b}$$

The idea is to use the inversion formula (4.5a) to analyze the semigroup $T(t)$ from properties of the resolvent only.

One has the following abstract characterization result (see the Supplementary remarks for references).

Theorem 4.1. *Let $A : \mathcal{D}(A) \to \mathcal{B}$ be the generator of a $\mathcal{C}_0$-semigroup $(T(t))_{t\geq 0}$. The semigroup $T(t)$ is hyperbolic if and only if there is an open strip containing the imaginary axis on which the resolvent is uniformly bounded and there exists an $\omega > 0$, such that*

(i) *for each $\phi \in \mathcal{B}$ and $0 < |h| < \omega$, the integral*

$$(C,1)\text{-}\int_{h-i\infty}^{h+i\infty} R(z,A)\phi\,dz \qquad \textit{exists;}$$

(ii) *for each $\phi \in \mathcal{B}$, $\alpha \in \mathcal{B}^*$ and $0 < |h| < \omega$, the function $r(\,\cdot\,,h;\phi,\alpha) : \mathbb{R} \to \mathbb{C}$ defined by*

$$r = r(\nu,h;\phi,\alpha) = \langle\, \alpha, R(h+i\nu,A)\phi\,\rangle \tag{4.6}$$

satisfies

$$|(r,\Phi)| \leq K\|\phi\|\,\|\alpha\|\,\|\mathcal{F}\Phi\|_1, \quad \textit{for all} \quad \Phi \in \mathcal{S}$$

where $\mathcal{S}$ denotes the Schwartz space.

The first condition implies that one has the following integral representation for the semigroup

$$T(t)x = \frac{1}{2\pi i}(C,1)\text{-}\int_{\rho-i\infty}^{\rho+i\infty} e^{zt}R(z,A)x\,dz, \quad t>0. \tag{4.7}$$

Therefore, it is easy to see that

$$\begin{aligned} e^{-\rho t}T(t)x &= \frac{1}{2\pi}(C,1)\text{-}\int_{-\infty}^{\infty} e^{i\nu t}R(\rho+i\nu,A)\phi\,d\nu \\ &= \mathcal{F}r(t). \end{aligned}$$

So $e^{-\rho t}T(t)x$ equals the generalized Fourier transform of r where r is given by Equation (4.6). From the second condition and Lemma 3.4, one derives that $\mathcal{F}r \in L^\infty$.

It is not difficult to apply the theorem to the $\mathcal{C}_0$-semigroup associated with System (2.1).

Theorem 4.2. *If μ satisfies* (J), *and* $\det \Delta_0$ *has no zeros in a strip* $-\delta < \operatorname{Re} z < \delta$, *then the* C_0*-semigroup* $T_{D,L}(t)$ *associated with System* (2.1) *is hyperbolic if and only if* $\det \Delta$ *has no zeros on the imaginary axis.*

The proof of the theorem will be an application of Theorem 4.1. The arguments are similar to those given in the proof of Theorem 3.2 and are left to the reader.

We end this section with some simple corollaries of the exponential estimates in Section 9.3.

Theorem 4.3. *If* $a_{D,L}$ *denotes the order of* $T_{D,L}(t)$, *the semigroup associated with System* (2.1), *then*

$$a_{D,L} = \sup\{\operatorname{Re} z : \det \Delta(z) = 0\}.$$

Theorem 4.3 gives a way to determine the asymptotic behavior of the semigroup $T_{D,L}(t)$. In particular, we have the following important result.

Corollary 4.1. *If there is a* $\delta > 0$ *such that the zeros of* $\det \Delta$ *are in the left half plane* $\operatorname{Re} z \leq \delta < 0$, *then there are positive constants* K *and* α *such that*

$$\|T_{D,L}(t)\| \leq Ke^{-\alpha t}, \qquad t \geq 0,$$

that is the zero solution of System (2.1) *is uniformly asymptotically stable.*

In a similar manner, we can discuss the behavior of C_0-semigroup $T_D(t)$ associated with the homogeneous difference equation.

Theorem 4.4. *If* a_D *denotes the order of* $T_D(t)$, *the semigroup associated with System* (2.1), *then*

$$a_D = \sup\{\operatorname{Re} z : \det \Delta_0(z) = 0\}. \tag{4.8}$$

The semigroup $T_D(t)$ yields important information about the perturbed semigroup $T_{D,L}$. Note that from the Rouché theorem, it follows that in the half plane $\operatorname{Re} z > a_D$, the characteristic equation $\det \Delta(z) = 0$ can only have finitely many roots. So if we decompose C by $\Lambda = \{\operatorname{Re} z > a\}$, $a > a_D$, then Λ is finite and C can be decomposed as in Lemma 2.3.

An eigenvalue μ of a bounded linear operator T on a Banach space is called a *normal eigenvalue* if it is an isolated point of the spectrum of $\sigma(T)$ of T and the corresponding generalized eigenspace is finite dimensional. A normal point of T is either a normal eigenvalue or a point of the resolvent set $\rho(T)$ of T. Let $\widetilde{\rho}(A)$ denote the set of normal points of T. The *essential spectrum* of T is the set $\mathbb{C} \setminus \widetilde{\rho}(T)$ in the complex plane.

Theorem 4.5. *The radius $r_e(D,L,t)$ of the essential spectrum of $T_{D,L}(t)$ satisfies*

$$r_e(D,L,t) \le e^{a_D t}, \qquad t \ge 0 \tag{4.9}$$

where a_D is given in (4.8). *If, in addition, D satisfies*

$$\begin{aligned} D\phi &= D_0\phi + \int_{-r}^{0} A(\theta)\phi(\theta)\,d\theta \\ D_0\phi &= \phi(0) - \sum_{k=1}^{\infty} A_k\phi(-r_k), \quad 0 < r_k \le r \\ &\sum_{k=1}^{\infty} |A_k| + \int_0^r |A(\theta)|\,d\theta < \infty, \end{aligned} \tag{4.10}$$

then

$$r_e(D,L,t) \le e^{a_{D_0} t}, \qquad t \ge 0. \tag{4.11}$$

The estimates can easily be proved using the Rouché theorem, see the remarks made before the theorem. A different proof for Estimate (4.9) follows from the representation for $T_{D,L}(t)$ in Theorem 7.3 of Section 3.7. If D satisfies (4.10), then the representation for $T_D(t)$ in Theorem 7.4 of Section 3.7 yields estimate (4.11).

9.5 Variation-of-constants formula

The purpose of this section is to discuss the variation-of-constants formula for linear inhomogeneous systems.

If $G \in C\big([0,\infty),\mathbb{R}^n\big)$, $F \in \mathcal{L}_1^{loc}\big([0,\infty),\mathbb{R}^n\big)$, then the nonhomogeneous linear NFDE$(D-G, L+F)$ is

$$\frac{d}{dt}\big[Dx_t - G(t)\big] = Lx_t + F(t), \qquad t \ge 0. \tag{5.1}$$

Following the proofs of the corresponding results in Section 6.1 and 1.7, one shows there are constants a and b such that the solution $x(\,\cdot\,;\phi,G,F)$ of Equation (4.2) with $x_0(\,\cdot\,;\phi,G,F=\phi)$ satisfies the estimate

$$|x_t(\phi,G,F)| \le be^{at}\big[|\phi| + \sup_{0\le u\le t}|G(u)-G(0)| + \int_0^t |F(s)|\,ds\big] \tag{5.2}$$

for $t \ge 0$.

Following representation (1.2), we write

$$D\phi = \phi(0) - \int_{-r}^{0} d[\mu(\theta)]\phi(\theta), \qquad L\phi = \int_{-r}^{0} d[\eta(\theta)]\phi(\theta), \tag{5.3}$$

where $\mathrm{Var}_{[-s,0]}\, \mu \to 0$ as $s \to 0$. From the results in Section 9.1, it follows that Equation (5.1) has a fundamental solution $X(t)$, $t \geq -r$, i.e., X is of bounded variation on compact sets, $X(t) = 0$ for $-r \leq t < 0$ and for $t \geq 0$ satisfies the equation

$$X(t) = I + \int_0^t d[X(t-\alpha)]\mu(-\alpha) - \int_0^t X(t-\alpha)\eta(-\alpha)\, d\alpha. \tag{5.4}$$

Theorem 5.1. *Let $T(t)$, $t \geq 0$, denote the solution operator defined by the NFDE(D,L). If X is defined by Equation (4.4), then the solution $x = x(\phi, G, F)$ of Equation (5.1) with $x_0(\phi, G, F) = \phi$ satisfies the relation*

$$\begin{aligned} x(t) - X(0)G(t) = T(t)\phi(0) - X(t)G(0) + \int_0^t X(t-s)F(s)\, ds \\ - \int_0^t d_s[X(t-s)]G(s) \end{aligned} \tag{5.5}$$

for $t \geq 0$.

Proof. From Theorem 1.2, it follows that

$$x(t) = X(t)\phi(0) + \int_0^t X(t-\alpha)\, df(\alpha),$$

where

$$\begin{aligned} f(t) = \phi(0) - G(0) + G(t) + \int_0^t d[\mu(\theta)]\phi(t+\theta) \\ + \int_0^t \int_{-r}^{-\alpha} d[\eta(\theta)]\phi(\alpha+\theta)\, d\alpha. \end{aligned}$$

This yields

$$x(t) = T(t)\phi(0) + \int_0^t X(t-s)F(s)\, ds + \int_0^t X(t,\alpha)\, d[G(\alpha) - G(0)].$$

Together with

$$\begin{aligned} \int_0^t X(t-\alpha)\, d[G(\alpha) - G(0)] = X(0)G(t) - X(t)G(0) \\ - \int_0^t d[X(t-\alpha)]G(\alpha). \end{aligned}$$

This proves the theorem. □

If $G \equiv 0$, this formula is the same as the one for RFDE. So with $G \equiv 0$, results on perturbed linear systems will follow in a manner similar to RFDE. If $G \not\equiv 0$, then new ideas must be employed.

First we use Lemma 2.1 of Chapter 6 to write Equation (5.5) as an integral in the Banach space C

$$\begin{aligned} x_t - X_0 G(t) = T(t)\phi - X_t G(0) + \int_0^t d[K(t,s)]F(s) \\ - \frac{d}{d\theta}\int_0^t d[K(t,s)]G(s) \end{aligned} \tag{5.6}$$

where $K(t,s)(\theta) = \int_0^s X(t+\theta-\alpha)\,d\alpha$. Note that if $G \not\equiv 0$, then the variation-of-constants formula is no longer an integral equation.

Note that $x_t - X_0G(t)$ and $T(t)\phi - X_tG(0)$ do not belong to C. In order to estimate the solution, we would like to make a transformation of variables. In order to do so, we have to introduce the product space $\widehat{C} = \mathbb{R}^n \times C$ with the embedding into $\mathcal{L}_\infty$ given by $(c,\phi) = X_0 c + \phi$. On $\widehat{C}$, one defines the following solution operator $\widehat{T}(t) : \widehat{C} \to \mathcal{L}_\infty$

$$\widehat{T}(t)(c,\phi) = X_t c + T(t)\phi.$$

Define the unbounded operator

$$\mathcal{D}(M) = \{\psi \in \mathcal{L}_\infty : \frac{d}{d\theta}\psi \in \mathcal{L}_\infty\}, \qquad M\psi = \frac{d}{d\theta}\psi.$$

If we set $z_t = x_t - X_0G(t)$, then the variation-of-constants formula becomes

$$z_t = \widehat{T}(t)z_0 + \int_0^t d[K(t,s)]F(s) - M\int_0^t d[K(t,s)]G(s). \tag{5.7}$$

Definition 5.1. If $G : C \to \mathbb{R}^n$ is continuous, we say that $G(\phi)$ is independent of $\phi(0)$ if there is an $\epsilon \in [-r, 0)$ such that $G(\phi)$ depends only on the values $\phi(\theta)$ of the function ϕ for $\theta \in [-r, \epsilon]$.

We are now in a position to prove the following result.

Theorem 5.2. *Suppose D, $L : C \to \mathbb{R}^n$ are linear and continuous and the zero solution of the* NFDE(D,L) *is uniformly asymptotically stable. If F, $G : C \to \mathbb{R}^n$ are continuous together with their first derivatives F_ϕ, G_ϕ and $F(0) = G(0) = 0$, $F_\phi(0) = G_\phi(0) = 0$ and $G(\phi)$ is independent of $\phi(0)$, then the zero solution of the equation*

$$\frac{d}{dt}[Dx_t - G(x_t)] = Lx_t + F(x_t), \qquad t \ge \sigma$$

is exponentially asymptotically stable.

Proof. Let $j : \widehat{C} \to \mathcal{L}_\infty$ be the embedding $j(c, \phi) = X_0 c + \phi$. Consider the map

$$l : C \to \widehat{C}, \qquad l\phi = (-G(\phi), \phi).$$

Since $G(\phi)$ does not depend on $\phi(0)$, the mapping l has a continuous inverse, and since h is a homeomorphism in a fixed neighborhood of $\phi = 0 \in C$, $(c, \psi) = (0, 0) \in \widehat{C}$. Furthermore, there are constants $k_1 > 0$ and $k_2 > 0$ such that in this neighborhood, $(c, \psi) = l(\phi)$ implies $|\psi| \leq k_1|\phi|$ and $|\phi| \leq k_2|\psi|$.

For Equation (5.1), the variation-of-constants formula is given by (5.5). Set $h : C \to \mathcal{L}_\infty$, $h = j \circ l$. If $z_t = h(x_t)$ and ψ is $h(\phi)$, then z_t satisfies the equation

$$(5.8) \quad z_t = T(t)\psi + \int_0^t d[K(t, s)]F(h^{-1}(z_s))\, ds - M \int_0^t d[K(t, s)]G(s).$$

Choose δ so that $\alpha - Kk_2\delta > 0$. Applying the estimates from Theorem 3.2 and the hypotheses on F and G, there is an $\epsilon(\delta) > 0$ such that z in Equation (5.8) satisfies

$$|z_t| < Ke^{-\alpha t}|\psi| + \int_0^t Ke^{-\alpha(t-s)}\delta k_2 |z_s|\, ds$$

as long as $|z_s| < \epsilon(\delta)$. Applying the inequality in Lemma 3.1 of Section 1.3 to $|z_t|e^{\alpha t}$, we obtain

$$\frac{1}{k_2}|x_t| \leq |z_t| \leq Ke^{-(\alpha - Kk_2\delta)t}|\psi| \leq Kk_1 e^{-(\alpha - Kk_2\delta)t}|\phi|$$

as long as $|z_s| < \epsilon(\delta)$. Since this clearly can be assured for all $t \geq 0$ if $|\phi|$ is sufficiently small, we obtain the result stated in the theorem. □

As one sees from this proof, the transformation h reduces the discussion to an argument very similar to the one for ordinary differential equations. One could easily generalize the results to more general perturbations $G(t, \phi)$ and $F(t, \phi)$ of the NFDE(D, L) where D and L can even depend on t.

Next we discuss the decomposition in the variation-of-constants formula for the inhomogeneous linear equation (5.1). Observe that we can also write Equation (5.6) as follows

$$(5.9) \quad x_t = T(t)\phi + \int_0^t d[K(t, s)]F(s) - \frac{d}{d\theta} \int_0^t d[K(t, s)]\big(G(s) - G(0)\big).$$

Suppose $\Lambda = \{\lambda_1, \ldots, \lambda_p\}$, $\lambda_j \in \sigma(A)$, and suppose C is decomposed by Λ as $C = P \oplus Q$. Let $\phi \in C$ be written as $\phi = \phi^P + \phi^Q$, $\phi^P \in P$, $\phi^Q \in Q$. Let $X_0^P = \Phi\Psi(0)$, where Φ is a basis for P and Ψ is a basis for P^T, $(\Psi, \Phi) = I$. Then $K(t, s)^P = \int_0^s T(t - \alpha)X_0^P\, d\alpha$ and $K(t, s)^Q = K(t, s) - \Phi(\Psi, K(t, s))$. Exactly as in Section 7.9, one obtains the following result on the decomposition in the variation-of-constants formula.

Theorem 5.3. *If Λ is a finite set of elements of $\sigma(A)$ and C is decomposed by Λ as $C = P \oplus Q$, and $x_t = x_t^P + x_t^Q$, $x_t^P \in P$, $x_t^Q \in Q$, then x satisfies the equations*

(5.10)

$$\begin{aligned} x_t^P &= T(t)\phi^P + \int_0^t T(t-s)X_0^P F(s) \\ &\quad - \frac{d}{d\theta}\int_0^t T(t-s)X_0^P \big(G(s) - G(0)\big) \\ x_t^Q &= T(t)\phi + \int_0^t d[K(t,s)^Q]F(s) - \frac{d}{d\theta}\int_0^t d[K(t,s)^Q]\big(G(s) - G(0)\big) \end{aligned}$$

for $t \geq 0$.

It is now tempting to let $x_t^P = \Phi y(t)$, where Φ is a basis of P and try to determine an ordinary differential equation for y as in Section 7.9. However, the function y need not be differentiable in t. In fact, if $r = 0$; that is, our equation is an ordinary differential equation,

$$\frac{d}{dt}[x(t) - G(t)] = Bx(t),$$

where B is an $n \times n$ matrix, then $x(t) - G(t)$ is differentiable but $x(t)$ is not differentiable unless $G(t)$ is differentiable. For an ordinary equation, one could let $z = x - G$ to obtain

$$\frac{d}{dt}[z(t)] = Bz(t) + BG(t)$$

and $z(t)$ is differentiable in t. This is the same type of transformation that was made earlier in this section. Using this transformation again, we find from the variation-of-constants formula (5.7), that

$$\begin{aligned} z_t^P &= \widehat{T}(t)z_0^P + \int_0^t T(t-s)X_0^P F(s) - M\int_0^t T(t-s)X_0^P G(s) \\ z_t^Q &= \widehat{T}(t)z_0^Q + \int_0^t d[K(t,s)^Q]F(s) + M\int_0^t d[K(t,s)^Q]G(s) \end{aligned} \tag{5.11}$$

for $t \geq 0$. If $z_t = \Phi y(t)$, where Φ is a basis for P and $T(t)\Phi = \Phi \exp Bt$, $t \in (-\infty, \infty)$ then y satisfies the equation

$$\begin{aligned} \Phi y(t) &= \Phi e^{Bt}y(0) + \int_0^t \Phi e^{B(t-s)}F(s)\,ds - \int_0^t d[\Phi e^{B(t-s)}\Psi(0)]G(s) \\ &= \Phi e^{Bt}y(0) + \int_0^t \Phi e^{B(t-s)}F(s)\,ds + \int_0^t \Phi e^{B(t-s)}B\Psi(0)G(s) \end{aligned}$$

for $t \geq 0$. Therefore, $y(t)$ is continuously differentiable and satisfies the ordinary differential equation

$$\dot{y}(t) = By(t) + B\Psi(0)G(t) + \Psi(0)F(t). \tag{5.12}$$

For $G \equiv 0$, this is the same equation as obtained in Section 7.9.

These relations can be generalized to the case of linear periodic equations, but the details are not given.

In the sequel, we must also have an estimate of the exponential growth of X_t^P and $K(t,s)^Q$. In particular, we need an exponential estimate for $X - X^P$, but this follows immediately from the hyperbolicity property of the semigroup $T_{D,L}(t)$ and the proof of Theorem 3.2. We state the fundamental inequalities.

Lemma 5.1. *Let $T_{D,L}(t)$ be the semigroup generated by System (2.1) with infinitesimal generator $A_{D,L}$. If $\Lambda = \{\lambda \in \sigma(A_{D,L}) : \operatorname{Re}\lambda \geq \alpha\}$ and C is decomposed by Λ as $C = P \oplus Q$, then there are positive constants M and ϵ such that for $\phi \in C$*

$$\begin{aligned} &|T_{D,L}(t)\phi^P| \leq Me^{(\alpha-\epsilon)t}|\phi^P|, \quad \operatorname{Var}_{(t,0]}|T(t)X_0^P| \leq Me^{(\alpha-\epsilon)t}, \quad t \leq 0 \\ &|K(t,\cdot)^Q| \leq Me^{(\alpha-\epsilon)t}, \quad \operatorname{Var}_{[0,t)} K(t,\cdot)^Q \leq e^{(\alpha-\epsilon)t}, \quad t \geq 0. \end{aligned}$$

Similar results can be obtained for the periodic case if we use the decomposition theory from Chapter 8.

9.6 Strongly stable D operators

In Definition 3.1 of Section 9.3, we introduced the concept of a stable D operator. In this section, we discuss how variations in parameters in D affect the property of being stable. We shall only be concerned with difference operators $D(r,A)$ of the form

$$D(r,A)\phi = \phi(0) - \sum_{k=1}^{N} A_k\phi(-r_k) \tag{6.1}$$

where each A_k is an $n \times n$ constant matrix, each r_k is a constant, $A = (A_1, \dots, A_k)$, and $r = (r_1, \dots, r_k)$.

In a problem, both the A_k and the r_k are generally parameters that are not known exactly. It is important, therefore, to know the effect that small changes in the parameters will have on the stability of the zero solution of the equation

$$Dy_t = 0. \tag{6.2}$$

From Theorem 3.5, we know that D is stable if and only if there is a $\delta > 0$ such that

$$\text{(6.3)} \qquad \det \Delta(r, A)(\lambda) = 0, \qquad \Delta(r, A)(\lambda) = I - \sum_{k=1}^{N} A_k e^{-\lambda r_k},$$

implies $\operatorname{Re} \lambda \leq -\delta$. It is not too difficult to show the following fact: for any $0 < \delta_1 < \delta$, there is an $\epsilon > 0$ such that all zeros of $\det \Delta(r, B)(\lambda) = 0$ satisfy $\operatorname{Re} \lambda \leq -\delta_1$ if $|A_k - B_k| < \epsilon$, $k = 1, 2, \ldots, N$.

The situation for variations in the r_k is much more complicated. For example, consider the equation

$$y(t) = ay(t-1) + by(t-2), \qquad t \geq 0$$

whose characteristic equation is

$$1 - ae^{-\lambda} - be^{-2\lambda} = 0.$$

The roots of this equation are given by

$$\rho = \frac{a \pm \sqrt{a^2 + 4b}}{2}, \qquad \rho = e^{\lambda}.$$

For $a = b = -1/2$, $|\rho|^2 = 1/2$. Therefore, if $2\delta = -\ln(\frac{1}{2})$, then $\operatorname{Re} \lambda = -\delta < 0$. For a given integer $n > 0$, consider the equation

$$y(t) = -\frac{1}{2} y\big(t - 1 + \frac{1}{2n+3}\big) - \frac{1}{2} y(t-2).$$

It is easy to check that $y(t) = \sin(n + (3/2))\pi t$ is a solution of this equation and it does not approach zero as $t \to \infty$. Therefore, by taking n large, the perturbation $1/(2n+3)$ in the first delay can be made as small as desired and, for each such perturbation, the equation has a solution that does not approach zero. This happens even though the perturbed equation had all roots of the unperturbed equation with real parts bounded away from zero.

In terms of the notation of the semigroups $T_D(t)$, this example shows that a_D is not continuous when one makes changes in the delays. It becomes important, therefore, to characterize those D operators of the form (6.1) for which one can preserve stability when small perturbations are made in the delays. This section is devoted to this characterization.

Definition 6.1. Let $(\mathbb{R}^+)^N$ be the cross product of $\mathbb{R}^+$ by itself N times. The operator $D(r, A)$ is said to be *stable locally in the delays* if there is an open neighborhood $I(r) \subseteq (\mathbb{R}^+)^N$ of r such that $D(s, A)$ is stable for each $s \in I(r)$.

Definition 6.2. The operator $D(r, A)$ is *stable globally in the delays* if it is stable for each $r \in (\mathbb{R}^+)^N$. In this case, we also say $D(\cdot, A)$ is *strongly stable.*

The main result of this section is the following theorem.

Theorem 6.1. *The following statements are equivalent:*
(i) *For some fixed* $r \in (\mathbb{R}^+)^N$, $r = (r_1, \dots, r_N)$ *with* $r_k > 0$ *rationally independent,* $D(r, A)$ *is stable.*
(ii) *If* $\gamma(B)$ *is the spectral radius of a matrix* B, *then* $\gamma_0(A) < 1$ *where*

$$(6.4) \qquad \gamma_0(A) \stackrel{\text{def}}{=} \sup\{\gamma(\sum_{k=1}^{N} A_k e^{i\theta_k}) : \theta_k \in [0, 2\pi], k = 1, 2, \dots, N\}.$$

(iii) $D(r, A)$ *is stable locally in the delays.*
(iv) $D(r, A)$ *is stable globally in the delays.*

Proof. The scheme for the proof is (i) ⇒ (ii) ⇒ (iii) ⇒ (i) and (iii) ⇔ (iv).

(i) ⇒ (ii). Suppose r satisfies the hypotheses stated in (i), $D(r, A)$ is stable, and Statement (ii) does not hold. Since the θ_k vary over a compact set, there are $\gamma_0 \geq 1$ and $\mu_k \in [0, 2\pi]$, $k = 1, 2, \dots, N$, such that $\gamma_0 = \gamma(\sum_k A_k \exp(i\mu_k))$. Let $f(\sigma) = \sum_k A_k \exp(i\mu_k - \sigma r_k)$ for $\sigma \in \mathbb{R}$. Since $r_k > 0$, $\gamma(f(\sigma)) \to 0$ as $\sigma \to \infty$. Since $\gamma(f(0)) = \gamma_0 \geq 1$, there are $\sigma_0 \geq 0$ and $\theta_0 \in [0, 2\pi]$, such that $\gamma(f(\sigma_0)) = 1$ and

$$(6.5) \qquad 0 = \det [I - e^{-i\theta_0} f(\sigma)] = \det [I - \sum_k A_k e^{-\sigma_0 r_k} e^{i(\mu_k - \theta_0)}].$$

Since the r_k are rationally independent, it follows from Kronecker's theorem that there is a sequence $\{t_n\}$ of real numbers such that

$$\lim_{n \to \infty} (t_n r_1, \dots, t_n r_N) = (\theta_0 - \mu_1, \dots, \theta_0 - \mu_N) \pmod{2\pi}.$$

We may also assume that $\{t_n\}$ is such that the sequence $\{\exp(\sigma_0 + it_n)\}$ converges to some $\zeta_0 = e^{\lambda_0}$, $\operatorname{Re} \lambda_0 = \sigma_0$, as $n \to \infty$. Therefore, using Equation (6.5), we have

$$0 = \det [I - \sum_k A_k e^{-(\sigma_0 + it_n) rk} e^{i(\mu_k - \theta_0 + t_n r_k)}]$$

and so

$$0 = \det [I - \sum_k A_k e^{-\lambda_0 r_k}].$$

This contradicts the fact that $D(r, A)$ is stable since $\operatorname{Re} \lambda_0 = \sigma_0 \geq 0$. This completes the proof that (i) implies (ii).

(ii) ⇒ (iii). Assume Equation (6.4) is satisfied and $D(r, A)$ is not stable locally in the delays. Then there exists a sequence $\{s^j\} \subseteq (\mathbb{R}^+)^N$ and a sequence $\{\lambda_j\}$ of complex numbers such that $|r - s^j| < 1/j$, $\operatorname{Re} \lambda_j \geq -1/j$ and $\det \Delta(s^j, A)(\lambda_j) = 0$ for $j = 1, 2, \dots$ Suppose there is a subsequence of the $\{s^j\}, \{\lambda_j\}$, which we label the same way, such that $\operatorname{Re} \lambda_j \to 0$ as $j \to \infty$. If $\lambda_j = \alpha_j + i\beta_j$, then the β_j satisfy the equation

$$\det [I - \sum_{k=1}^{N} A_k^j e^{i\beta_j r_k}] = 0$$

where $A_k^j = A_k \exp(-\lambda_j s_k^j + i\beta_j r_k)$ for all k and $A^j = (A_1^j, \ldots, A_N^j) \to A$ as $j \to \infty$. Since the spectral radius $\gamma(B)$ of a matrix B is continuous in B, this contradicts Equation (6.4).

Therefore, we may assume each element of the sequence $\{\lambda_j\}$ satisfies $\operatorname{Re}\lambda_j \geq \delta > 0$ for some constant δ. Also, we may assume the s^j are rational since this can be accomplished by a small change in A. To simplify notation, let the generic element of the sequences $\{\lambda_j\}$, $\{s^j\}$ be λ, $s = (s_1, \ldots, s_N)$. For any real x, consider the solutions $z = z(x)$ of the equation

$$\det\left[xI - \sum_{k=1}^{N} A_k \exp(-s_k z)\right] = 0.$$

Since the s_k are rational, this is a polynomial in $\mu = \exp(-z/q)$ for some integer q and the solutions $\mu(x)$ of this polynomial equation can be chosen as a continuous function of x. Also, $\operatorname{Re}\mu(x) \to -\infty$ as $x \to \infty$. For $x = 1$, we have a root λ with $\operatorname{Re}\lambda > 0$. Therefore, there is an $x^* \in \mathbb{R}$, $|x^*| > \lambda_0(A)$ in Equation (6.4) and a $\mu^* = \exp i\theta^*$, $\theta^* \in \mathbb{R}$ such that

$$\det\left[x^*I - \sum_{k=1}^{N} A_k \exp i\theta^*\right] = 0.$$

This contradicts Statement (ii) and completes the proof of the assertion that (ii) implies (iii).

The fact that (iii) implies (i) is obvious. Since (iv) implies (i) and (i) is equivalent to (iii), (iv) implies (iii). Since (ii) and (iii) are equivalent and condition (ii) is independent of the delays, it follows that (iii) implies (iv). The proof of the theorem is complete. □

An immediate consequence of Theorem 6.1 is the following result for the scalar case.

Corollary 6.1. *If each A_k is a scalar, then $D(r, A)$ is stable locally in the delays if and only if $\sum_{k=1}^{N} |a_k| < 1$.*

As another example, consider the real scalar difference equation

$$x(t) = ax(t-r) + bx(t-s) + cx(t-r-s) \tag{6.6}$$

where a, b, and c are constants and $r > 0$, $s > 0$. The characteristic equation is

$$1 \quad ac^{-\lambda r} \quad bc^{-\lambda s} \quad cc^{-\lambda(r+s)} = 0. \tag{6.7}$$

Corollary 6.1 does not apply to this equation since the three delays r, s, and $r + s$, cannot be varied independently.

To apply the theory of this section, we transform Equation (6.6) to an equivalent matrix equation (equivalent meaning that the characteristic equation is the same as Equation (6.7)) involving only the two delays r and s. The particular matrix equation is not important since Theorem 6.1 is a statement only about the solutions of the characteristic equation (6.7).

If we let

$$y(t) = \begin{bmatrix} x(t) \\ x(t-r) \end{bmatrix}, \qquad A = \begin{bmatrix} a & 0 \\ 1 & 0 \end{bmatrix}, \qquad B = \begin{bmatrix} b & c \\ 0 & 0 \end{bmatrix}$$

then Equation (6.6) is equivalent to the system

$$y(t) = Ay(t-r) + By(t-s). \tag{6.8}$$

Theorem 6.1 implies that System (6.8) will be stable locally in the delays if and only if

$$\sup\{\gamma(Ae^{i\theta_1} + Be^{i\theta_2}) : \theta_1, \theta_2 \in [0, 2\pi]\} < 1.$$

It is clear that we can take $\theta_1 = 0$ and obtain this supremum. Therefore, it is necessary to discuss the second-order equation

$$0 = \det(zI - A - Be^{i\theta}) = z^2 - (a + be^{i\theta})z - ce^{i\theta} \tag{6.9}$$

for all $\theta \in [0, 2\pi]$.

If we let $z = (1+\lambda)/(1-\lambda)$, then z inside the unit circle is equivalent to λ in the left half-plane. If we make this transformation and multiply the resulting expression by $(1-\lambda)^2$, then λ must satisfy the equation

$$\begin{aligned} f(\lambda) \stackrel{\text{def}}{=} {} & [1 + a + (b-c)e^{i\theta}]\lambda^2 \\ & + 2(1 + ce^{i\theta})\lambda + 1 - a - (b+c)e^{i\theta} = 0. \end{aligned} \tag{6.10}$$

One may apply the Routh-Hurwitz criteria to the polynomial in Equation (6.10). After some rather lengthy but straightforward calculations, one obtains the following necessary and sufficient conditions for the solutions of Equation (6.10) to have real parts negative and bounded away from zero uniformly in θ:

$$\begin{aligned} 1 + a &> |b + c| \\ 1 - a &> |b - c|. \end{aligned} \tag{6.11}$$

This region in the a, b, c space is larger than the region $|a| + |b| + |c| < 1$ obtained in Corollary 6.1 for the equation

$$x(t) = ax(t - r_1) + bx(t - r_2) + cx(t - r_3)$$

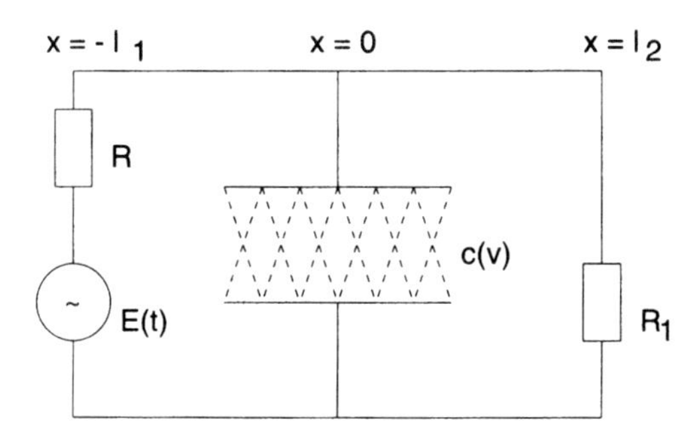

Fig. 9.1.

where the parameters r_1, r_2, and r_3 could vary independently and not through the relation $r_3 = r_1 + r_2$.

As an illustration of where this latter example occurs in the applications, consider the lossless transmission line shown in Figure 9.1 with a nonlinear capacitor connected at $x = 0$; that is, the shunted transmission line. Using the derivative outlined in the introduction, letting L and C be the mutual inductance and capacitance in the line $z = (L/C)^{1/2}$, $a = (LC)^{-1/2}$, $r_1 = 2l_1/a$, $r_2 = 2l_2/a$, $\mu = (R_1 - z)/(R_1 + z)$, and $q = (R - z)/(R + z)$, and supposing $c(v) > 0$, h is the inverse of the function $\int_0^v c(s)ds$, one can show that the function $x(t) = \int_0^{v(t)} c(s)ds$, where v is the voltage at $x = l_2$ satisfies the equation

$$\text{(6.12)} \qquad \frac{d}{dt} Dx_t = -\frac{2}{z}\, g(x(t)) + \frac{2qu}{z}\, g(x(t - r_1 - r_2)) + e(t)$$

where $e(t)$ has the same properties as $E(t)$ and

$$\text{(6.13)} \qquad D\phi = \phi(0) - q\phi(-r_1) - \mu\phi(-r_2) + q\mu\phi(-r_1 - r_2).$$

If $\mu \neq 0$, then the operator contains three delays that are not independent. The parameters r_1 and r_2 are physical parameters and we should have stability of D subject to small changes in these parameters. For $a = q$, $b = \mu$, and $c = -ab$, Inequalities (6.11) are equivalent to $|q| < 1$, $|\mu| < 1$ and these conditions are always satisfied if $R \neq 0$ and $R_1 \neq 0$. Thus, the operator D is stable locally in the delays. For this particular example, it is also easy to observe this fact directly from the characteristic equation since it factors into two simple factors.

9.7 Properties of equations with stable D operators

In this section, we consider the special class of NFDE(D, f) for which D is stable. Also, we assume $f : \Omega \to \mathbb{R}^n$, $\Omega \subseteq C$ is open, that is, f is independent of t. This latter hypothesis simplifies the notation, but some of the results hold when f depends on t. The reader may supply the details for this case or consult the references in Section 9.9.

For an autonomous NFDE(D, f), the concepts of positive orbit $\gamma^+(\phi)$, ω-limit set $\omega(\phi)$, α-limit set $\alpha(\phi)$, and invariant set, are defined as for RFDE.

Theorem 7.1. *If $\Omega \subseteq C$ is open, D is stable and $f : \Omega \to \mathbb{R}^n$ is completely continuous, then a positive orbit $\gamma^+(\phi)$ of the NFDE(D, f) is relatively compact if and only if $\gamma^+(\phi)$ is bounded. If $\gamma^+(\phi)$ is bounded, then $\omega(\phi)$ is a nonempty, compact, connected, invariant set.*

Proof. The latter part of this theorem is proved in essentially the same manner as for the RFDE and is thus omitted.

It remains to prove the first part. Obviously, $\gamma^+(\phi)$ relatively compact implies $\gamma^+(\phi)$ is bounded. Conversely, if $\gamma^+(\phi)$ is bounded, there is a constant M such that $|f(\gamma^+(\phi))| \le M$. Also, for any $\tau \ge 0$, $t \ge 0$, the solution x_t through ϕ satisfies

$$D(x_{t+\tau} - x_t) = \int_t^{t+\tau} f(x_s)ds.$$

From Part (iii) of Theorem 3.4 and, in particular, Relation (3.22), D stable implies

$$|x_{t+\tau} - x_t| \le b|x_\tau - \phi| + bM\tau.$$

Since x_τ is continuous in τ, this implies $x(\phi)(t)$ is uniformly continuous on $[-r, \infty)$. Since $x(\phi)(t)$ is bounded, this implies that $\gamma^+(\phi)$ is precompact and the theorem is proved. □

For an RFDE(f), any solution defined on an interval of length $(k+1)r$, $k \ge 1$, say $[\sigma - r, \sigma + kr]$ must have k derivatives on $[\sigma + (k-1)r, \sigma + kr]$ if $f \in C^{k-1}$. The solution operator for an RFDE(f) continues to smooth the data as time increases. If a solution of an RFDE(f) is defined on $(-\infty, \infty)$, then obviously this solution is C^{k+1} if f is C^k. For an NFDE(D, f), the solution does not smooth on finite intervals. However, if the operator D is stable, the solution operator does tend to smooth on infinite intervals in the following sense.

Theorem 7.2. *Suppose $\Omega \subseteq C$ is open. For any* NFDE(D, f) *with D stable and f having continuous, bounded derivatives on Ω through order $k \ge 0$, any bounded solution on $(-\infty, a]$ must be C^{k+1}. Also, the ω-limit* *set of a bounded orbit consists of uniformly bounded, equicontinuous C^{k+1} functions.*

If f is analytic, we can also obtain the following result (see the references in Section 9.9 for a proof).

Theorem 7.3. *Suppose $\Omega \subseteq C$ is open. For any* NFDE(D, f) *with D stable and f analytic in Ω, any bounded solution on $(-\infty, a]$ must be analytic. Also, the ω-limit of any bounded orbit is analytic.*

As a consequence of Theorems 7.2 and 7.3, we have the following important remark.

Corollary 7.1. *If $\Omega \subseteq C$ is open, D is stable, and $f : \Omega \to \mathbb{R}^n$ has continuous, bounded derivatives of order $k \geq 0$ on Ω (resp. analytic in Ω), then any periodic solution of the* NFDE(D, f) *has continuous, bounded derivatives of order $k + 1$ (resp. analytic).*

Corollary 7.1 is very important in the applications. For example, consider the autonomous equation

$$\frac{d}{dt} Dx_t = f(x_t) \tag{7.1}$$

where D is stable and suppose there exists a nonconstant ω-periodic solution p of this equation. Then p is continuously differentiable. If f also has a continuous derivative then $d^2 Dp_t/dt^2$ exists and is equal to $dD\dot{p}_t/dt$. Therefore,

$$\frac{d}{dt} Dy_t = f_\phi(p_t) y_t \tag{7.2}$$

has the nontrivial ω-periodic solution $\dot{p}$. This remark is fundamental to the discussion of the behavior of solutions of Equation (7.1) near a periodic orbit (see Chapter 10).

To prove this theorem, we make use of the following lemma.

Lemma 7.1. *If D is stable, $h : (-\infty, a] \to \mathbb{R}^n$ is uniformly continuous and bounded, and if x is a bounded solution of*

$$Dx_t = h(t)$$

on $(-\infty, a]$, then x is bounded and uniformly continuous on $(-\infty, a]$. If $\dot{h}$ is also bounded and uniformly continuous on $(-\infty, a]$, then $\dot{x}(t)$ exists and is uniformly continuous and bounded on $(-\infty, a]$.

Proof. Suppose $\tau \geq 0$. Since $D(x_{t+\tau} - x_t) = h(t + \tau) - h(t)$ for all $t \in (-\infty, a - \tau]$, Relation (3.22) implies

$$|x_{t+\tau} - x_t| \le be^{-a(t-\sigma)}|x_{\sigma+\tau} - x_\sigma| + b \sup_{\sigma \le u \le t} |h(u+\tau) - h(u)|$$

Therefore, letting $\sigma \to -\infty$, we have

$$|x_{t+\tau} - x_t| \le b \sup_{-\infty < u \le t} |h(u+\tau) - h(u)|.$$

Thus, x is uniformly continuous on $(-\infty, a]$. Also, using the same argument, for any $\tau > 0$, $s > 0$,

$$\begin{aligned}\left|\frac{x_{t+\tau} - x_t}{\tau} - \frac{x_{t+s} - x_t}{s}\right| &\le b \sup_{u \in (-\infty,t]} \left|\frac{h(u+\tau) - h(u)}{\tau} - \frac{h(u+s) - h(s)}{s}\right| \\ &= b \sup_{u \in (-\infty,t]} \left|\int_0^1 [\dot{h}(u+v\tau) - \dot{h}(u+vs)]dv\right|.\end{aligned}$$

Since $\dot{h}$ is uniformly continuous, this implies $[x(t+\tau) - x(t)]/\tau$ approaches a continuous limit $\dot{x}(t)$ as $\tau \to 0$. Obviously, $D(\dot{x}_t) = \dot{h}(t)$ and, as before, $\dot{x}$ is bounded and uniformly continuous on $(-\infty, a]$. This proves the lemma. □

Proof of Theorem 7.2. Suppose $k = 0$. If $h(t) = Dx_t$, then x bounded and continuous on $(-\infty, a]$ implies h is bounded and continuous on $(-\infty, a]$. Also, $dDx_t/dt = f(x_t)$ implies $\dot{h}$ is bounded and continuous on $(-\infty, a]$. Thus, h is uniformly continuous and Lemma 6.1 implies x is bounded and uniformly continuous. Thus $f(x_t) = \dot{h}(t)$ is bounded and uniformly continuous on $(-\infty, a]$. Lemma 6.1 implies $\dot{x}$ exists, is bounded and uniformly continuous on $(-\infty, a]$. This proves the theorem for $k = 0$. The proof for the case $k > 0$ is left for the reader. □

For the NFDE(D, f) with D stable, we know that $T_{D,f}(t)$ is an α-contracting semigroup (see Theorem 7.3 of Section 3.7 and Theorem 4.5). As a consequence of the results in Chapter 4, if $\{T_{D,f}(t)B,\ t \ge 0\}$ is bounded for B bounded and $T_{D,f}(t)$ is point dissipative, then there exists a compact global attractor $\mathcal{A}_{D,f}$. The functions in $\mathcal{A}_{D,f}$ must satisfy the regularity properties stated in Theorem 7.2 and 7.3. In particular, if f is analytic, then $T_{D,f}(t)$ is one-to-one on the attractor $\mathcal{A}_{D,f}$. Since $\mathcal{A}_{D,f}$ is compact, this implies that $T_{D,f}(t)$ is a group on $\mathcal{A}_{D,f}$.

9.8 Stability theory

In this section, we indicate the changes that are sufficient to adapt the methods of Liapunov and Razumikhin for RFDE to an NFDE(D, f) with a stable D operator. The functional $D\phi = \phi(0)$ played a very important role in all of the results of Chapter 5 on RFDE. For an arbitrary stable D

operator, it turns out that one can obtain similar results by letting $D\phi$ play the role of $\phi(0)$.

If $V : \mathbb{R} \times C \to \mathbb{R}^n$ is continuous and $x(\sigma, \phi)$ is the solution of an NFDE(D, f) through (σ, ϕ) we define

$$\dot{V}(\sigma, \phi) = \limsup_{h \to 0^+} \frac{1}{h} \, [V(t+h, x_{t+h}(\sigma, \phi)) - V(\sigma, \phi)].$$

Theorem 8.1. *Suppose D is stable, $f : \mathbb{R} \times C \to \mathbb{R}^n$, $f : \mathbb{R} \times$ (bounded sets of C) into bounded sets of $\mathbb{R}^n$ and suppose $u(s)$, $v(s)$, and $w(s)$ are continuous, nonnegative, and nondecreasing with $u(s), v(s) > 0$ for $s \neq 0$, and $u(0) = v(0) = 0$. If there is a continuous function $V : \mathbb{R} \times C \to \mathbb{R}^n$ such that*

$$u(|D\phi|) \leq V(t, \phi) \leq v(|\phi|)$$
$$\dot{V}(t, \phi) \leq -w(|D\phi|)$$

then the solution $x = 0$ of the NFDE(D, f) *is uniformly stable. If $u(s) \to \infty$ as $s \to \infty$, the solutions of the* NFDE(D, f) *are uniformly bounded. If $w(s) > 0$ for $s > 0$, then the solution $x = 0$ of the* NFDE(D, f) *is uniformly asymptotically stable. The same conclusion holds if the upper bound on $\dot{V}(t, \phi)$ is given by $-w(|\phi(0)|)$.*

Proof. The proof is omitted since the basic ideas are contained in the proof of Theorem 2.1 of Section 5.2 and that proof can be appropriately modified if one uses Property (3.22) of stable D operators. □

For autonomous equations, one can also generalize the results of Section 5.3.

Definition 8.1. We say $V : C \to \mathbb{R}^n$ is a *Liapunov functional* on a set G in C for an autonomous NFDE(D, f) if V is continuous on $\overline{G}$, the closure of G, and $\dot{V} \leq 0$ on G. Let

$$S = \{\phi \in \overline{G} : \dot{V}(\phi) = 0\}$$

M = largest set in S that is invariant with respect to the NFDE(D, f).

Theorem 8.2. *If D is stable and V is a Liapunov functional on G for the autonomous* NFDE(D, f) *and $\gamma^+(\phi) \subseteq G$ and either $x_t(\phi)$ or $Dx_t(\phi)$ is bounded for $t \geq 0$, then $x_t(\phi) \to M$ as $t \to \infty$. If*

$$G = U_l \stackrel{\text{def}}{=} \{\phi \in C : V(\phi) < l\}$$

and there is a constant $K = K(l)$ such that ϕ in U_l implies either $|\phi(0)| < K$ or $|D\phi| < K$, then any solution $x_t(\phi)$ of the NFDE(D, f) with $\phi \in U_l$ approaches M as $t \to \infty$.

Proof. The proof is essentially the same as the proofs of Theorems 3.1 and 3.2 of Section 5.3, taking into account Inequality (3.22) and Theorem 7.1. □

As a first example, consider the scalar equation

$$\frac{d}{dt}\,[x(t) - cx(t-r)] + ax(t) = 0 \tag{8.1}$$

where $|c| < 1$ and $a > 0$. The operator $D\phi = \phi(0) - c\phi(-r)$ is stable. If

$$V(\phi) = (D\phi)^2 + ac^2 \int_{-r}^{0} \phi^2(\theta)\,d\theta,$$

then

$$\dot V(\phi) = -a(D\phi)^2 - a(1-c^2)\phi^2(0) \le -a(D\phi)^2.$$

Theorem 7.1 implies uniform asymptotic stability.

Since the solution operator for this equation is an α-contraction and therefore has the radius of the essential spectrum less than one for $t \ge 0$, the approach to zero is exponential. Therefore, this proof, using Liapunov functions, implies there is a δ such that all solutions of the equation $\lambda(1 - ce^{-r\lambda}) + a = 0$ have $\operatorname{Re}\lambda \le -\delta < 0$.

As remarked in the introduction, the following equation arises in the theory of transmission lines:

$$\frac{d}{dt}\,Dx_t = -\alpha x(t) - q\alpha x(t-r) - g(Dx_t) \tag{8.2}$$

where $D\phi = \phi(0) - q\phi(-r)$, $|q| < 1$, $\alpha > 0$. The operator D is stable. Let us use Theorem 7.1 to prove the following result.

Theorem 8.3. *The zero solution of Equation* (7.2) *is uniformly asymptotically stable and every solution approaches zero if* $|q| < 1$, $g(0) = 0$, *and*

$$m \stackrel{\text{def}}{=} \inf \frac{g(x)}{x} > -\alpha\frac{1-|q|}{1+|q|}.$$

Proof. Let γ be such that $m > \gamma > -\alpha(1-|q|)/(1+|q|)$. If $\beta = |q|\alpha$,

$$V(\phi) = \frac{1}{2}(D\phi)^2 + \beta \int_{-r}^{0} \phi^2(\theta)\,d\theta.$$

Then

$$\begin{aligned}
\dot V(\phi) &= (D\phi)[-\alpha\phi(0) - q\alpha\phi(-r) - g(D\phi)] + \beta[\phi^2(0) - \phi^2(-r)] \\
&= -\gamma(D\phi)^2 - (D\phi)^2\left[\frac{g(D\phi)}{D\phi} - \gamma\right] + (D\phi)[-\alpha\phi(0) - q\alpha\phi(-r)] \\
&\quad + \beta\phi^2(0) - \beta\phi^2(-r) \\
&\le -\gamma(D\phi)^2 + (D\phi)[-\alpha\phi(0) - q\alpha\phi(-r)] + \beta\phi^2(0) - \beta\phi^2(-r).
\end{aligned}$$

Expanding the expression on the right-hand side of the last inequality, we have

$$\dot{V}(\phi) \le -(\gamma + \alpha - \beta)\phi^2(0) + 2\gamma q\phi(0)\phi(-r) - (\gamma q^2 - \alpha q^2 + \beta)\phi^2(-r).$$

It is easy to check that the right-hand side of this expression is a negative definite quadratic form in $\phi(0), \phi(-r)$. Therefore, there is a positive constant k such that $\dot{V}(\phi) \le -k\phi^2(0)$. Theorem 7.1 implies the stated result on stability. □

In some problems, it is difficult to construct Liapunov functions and one needs an analogue of the Razumikhin-type theorems given in Section 5.4. Before stating such a result for neutral equations, some additional notation is needed.

Suppose D is a stable operator, $\|D\| = K$. Let $0 \le u(s) \le v(s)$, $s \ge 0$, be continuous, nondecreasing functions, $u(s) \to \infty$ as $s \to \infty$, and suppose there is a continuous function $\alpha(\eta)$, $\eta \ge 0$, satisfying $v(K\eta) \le u(\alpha(\eta))$. Let $\beta(\eta) > b(\eta + \alpha(\eta))$ be a continuous function where $b > 0$ is defined in Inequality (3.22). Finally, let $F : [0, \infty) \to \mathbb{R}^+$ be a continuous nondecreasing function such that $F(v(K\eta)) > v(\beta(\eta))$ for $\eta > 0$. Under these conditions, we can state the following result.

Theorem 8.4. *Suppose the preceding notation, D is stable, $f : \mathbb{R} \times C \to \mathbb{R}^n$ is continuous, and takes $\mathbb{R} \times$ (bounded sets of C) into bounded sets of $\mathbb{R}^n$ and consider the* NFDE(D, f). *If there is a continuous, positive function $w(s)$, $s \ge 0$ and a continuous function $V : \mathbb{R}^n \to \mathbb{R}$ such that $u(|x|) \le V(x) \le v(|x|)$ for all $x \in \mathbb{R}^n$ and*

$$\dot{V}(D\phi) \le -w(|D\phi|)$$

for all functions ϕ satisfying $F(V(D\phi)) \ge V(\phi(\theta))$, $-r \le \theta \le 0$, then the solution $x = 0$ of the NFDE(D, f) *is uniformly asymptotically stable and all solutions approach zero at $t \to \infty$.*

Proof. The proof follows along the lines of the proof of Theorem 4.2 of Section 5.4 using properties of stable D operators given in Inequality (3.22). The reader may consult the references in Section 9.9 for the complete proof. □

Let us apply Theorem 8.4 to the shunted transmission line, Equation (6.12), with $\mu = 0$ and $E = 0$; that is, the equation

$$\frac{d}{dt} Dx_t = -g(x(t)) \tag{8.3}$$
$$D\phi = \phi(0) - q\phi(-r), \qquad |q| < 1,$$

and g continuous. Choose $V(x) = x^2$, $u(s) = v(s) = s^2$, and $F(s) = N^2 s$, where

$$N > (1-|q|)^{-1}, \qquad \alpha(\eta) = (1+|q|)\eta, \qquad \beta(\eta) = \frac{1+|q|}{1-|q|}\,\eta.$$

One can show that these functions satisfy the requirements of Theorem 8.4. Also,

$$\dot{V}(D\phi) = -2(D\phi)g(\phi(0)). \tag{8.4}$$

We need to impose conditions that will imply $\dot{V}(D\phi) \le -w(|D\phi|)$ for all ϕ satisfying $F(V(D\phi)) \le V(\phi(\theta))$, $-r \le \theta \le 0$. Suppose $|q| < 1/2$. If ϕ satisfies this latter inequality, then, in particular, $|\phi(-r)| \le N|D\phi|$ and, thus,

$$\frac{\phi(0)}{D\phi} = 1 + q\frac{\phi(-r)}{D\phi} \ge 1 - |q|\frac{|\phi(-r)|}{|D(\phi)|} \ge 1 - |q|N > \frac{1-2|q|}{1-|q|} > 0.$$

If $|g(x)| \to \infty$ as $|x| \to \infty$ and $xg(x) > 0$ for $x \ne 0$, define $\epsilon = 1 - |q|N$ and

$$w_1(s) = \begin{cases} \min g(u), & \epsilon s \le u \le s(1-|q|)^{-1},\ s \ge 0, \\ -\max g(u), & s(1-|q|)^{-1} \le u \le \epsilon s,\ s < 0. \end{cases}$$

If $\phi(0) \ge \epsilon D\phi$ and $D\phi \ge 0$, then $g(\phi(0)) \ge w_1(D\phi)$ and $\dot{V}(D\phi) \le -D\phi w_1(D\phi)$. If $\phi(0) \ge \epsilon D\phi$ and $D\phi \le 0$, then $g(\phi(0)) \le -w_1(D\phi)$ and $\dot{V}(D\phi) \le D\phi w_1(D\phi)$. If $w(s) = \min_{s\ge 0}(sw_1(s), sw_1(-s))$, then $w(s) > 0$ for $s > 0$ and $\dot{V}(D\phi) \le -w(|D\phi|)$ for all ϕ satisfying $\phi(0) \ge \epsilon D\phi$ and, therefore, for all ϕ satisfying $F(V(D\phi)) \ge V(\phi(\theta))$, $-r \ge \theta \le 0$. These computations have proved the following result:

Theorem 8.5. *If $|q| < 1/2$, $xg(x) > 0$ for $x \ne 0$, $|g(x)| \to \infty$ as $|x| \to \infty$, then the solution $x = 0$ of Equation (8.3) is uniformly asymptotically stable and every solution approaches zero as $t \to \infty$.*

9.9 Supplementary remarks

Early work on the properties of NFDE (1.1) are due to Bellman and Cooke [1], Cruz and Hale [3], Hale [9,15], Hale and Meyer [1] and Henry [1,4].

The approach in this chapter is new; it uses the theory of resolvents and makes the approach very similar to the retarded case. The approach generalizes earlier work and allows the kernel μ to have a singular part as well, although we still need a condition on μ. (See also Kappel and Zhang [1].) The Fredholm alternative for periodic systems has been given by Hale [17] and Nosov [2].

The abstract theorem on hyperbolic semigroups is from Kaashoek and Verduyn Lunel [2]. If $\mathcal{B}$ is a Hilbert space, then conditions (i) and (ii) in Theorem 4.1 are automatically satisfied. For details and a proof of Theorem

4.2 see Kaashoek and Verduyn Lunel [2]. Theorem 4.2 is more general than the original result by Henry [1]. See also Greiner-Schwarz [1]. The results in Section 4 can also be formulated as a spectral mapping theorem. If $D\phi = \phi(0) - \sum_{k=1}^{\infty} A_k\phi(r_k)$, then

$$\overline{e^{t\sigma(A_D)}} \subseteq \sigma(T_D(t)) \subseteq e^{t\Lambda + i\mathbb{R}} \bigcup \{0\}$$

where $\Lambda = \overline{\{\mathrm{Re}\,\lambda : \det[I - De^{\cdot\lambda}I]\}}$. (See Greiner-Schwarz [1] and Henry [1].) In Henry [7], this result has been refined and it has been shown that

$$\overline{e^{t\sigma(A_D)}} \setminus \{0\} = \sigma(T_D(t)) \setminus \{0\}$$

for almost all $t \geq 0$.

Neves, Ribeiro and Lopez [1] study mixed initial-value problems for hyperbolic partial differential equations in one space dimension that generalize the NFDE (2.1) studied in this chapter. They obtain similar results and are able to characterize the growth bound (see also Kaashoek and Verduyn Lunel [2]).

Lemma 3.1 and Theorems 3.4 and 4.4 on difference equations are valid for equations $D(t, y_t) = h(t)$. One must, of course, impose conditions that are uniform in t in order to obtain the estimates. For example, the analogue of Theorem 4.4 would assume $a > a_D$, where

$$a_D = \inf\{a : \text{there is a } K \text{ such that } \|T(t,\sigma)\| \leq Ke^{a(t-\sigma)}, t \geq \sigma \geq 0\}$$

where $T(t,\sigma)$ is the solution operator of $D(t, y_t) = 0$. There are more interesting problems in the theory of linear systems. For example, precise conditions for convergence of the spectral projections $P_\lambda\phi$, $\lambda \in \sigma(A)$, to the state ϕ when A is the generator associated with System (2.1). The topology on the state space becomes important, and for precise results we refer to Verduyn Lunel [7].

Some results on perturbed linear systems are also contained in Bellman and Cooke [1] and Nosov [3, 4], but they generally involve more hypotheses than would be required by using ideas similar to the ones in Section 9.5.

Stable D operators in connection with NFDE were introduced in Cruz and Hale [2]. Cruz and Hale proved the equivalence of (i) and (ii) in Theorem 3.5 and gave more information on the asymptotic behavior of solutions of nonhomogeneous difference equations (3.19). This paper also considers the nonautonomous D operators. The equivalence of (i) and (iv) in Theorem 3.5 is due to Henry [2].

The method of proof of Theorem 5.2 is due to Hale and Martinez-Amores [1] (see also Hale and Ize [1]).

Henry [4], Melvin [4], and Moreno [1] were the first to observe that small changes in the delays could drastically change the stability properties of a simple difference equation, and Corollary 6.1 is an immediate consequence of their work. Hale [13] showed that stability locally in the delays implies stability globally in the delays. Theorem 6.1 is contained in the thesis of Silkowski [1]. For the Routh-Hurwitz criteria used in the example after Corollary 6.1, see Coppel [1] or Gantmacher [1]. One also can discuss the analogue of Theorem 6.1 for the preservation with respect to delays of hyperbolicity of the origin. Necessary and sufficient conditions for this property and many examples are given in Avellar and Hale [1].

If the delays in (6.1) are not allowed to vary independently, then it is possible to obtain results with an appropriate modification of Theorem 6.1. In such situations, the region of strong stability in specific examples will be much larger. To formulate the results, let $r = (r_1, \dots, r_M) \in (\mathbb{R}^+)^M$, $\gamma_k = (\gamma_{k1}, \dots, \gamma_{kM})$, γ_{kj} integers, $\gamma_k \neq 0$, $\gamma_k \cdot r = \sum_{j=1}^{M} \gamma_{kj} r_j$, $k = 1, 2, \dots, N$, and consider the difference equation

$$D(r, \gamma, A)y_t \stackrel{\text{def}}{=} y(t) - \sum_{k=1}^{N} A_k y(t - \gamma_k \cdot r) = 0, \tag{9.1}$$

where each A_k is an $n \times n$ constant matrix. Theorem 6.1 remains valid with θ_k replaced by $\gamma_k \cdot \theta$ with $\theta \in \mathbb{R}^M$.

It also is of interest to consider stability globally in the delays for NFDE of the form

$$\frac{d}{dt}[x(t) - \sum_{k=1}^{N} A_k x(t - \gamma_k \cdot r)] = B_0 x(t) + \sum_{k=1}^{N} B_k x(t - \gamma_k \cdot r), \tag{9.2}$$

where the notation is as earlier and each B_k is an $n \times n$ constant matrix. For the retarded case (all matrices $A_k = 0$), such problems have been considered by Koval and Čarkov [1], Repin [2], Cooke and Ferreira [1], Hale, Infante, and Tsen [1]. For NFDE, Zivotovskii [1] has considered the scalar equation with independent delays and Datko [3], Hale, Infante and Tsen have considered the general equation (9.2). For $s_j \in \mathbb{C}$, $j = i, \dots M$, if we define

$$\begin{aligned} &P(\lambda, \gamma, s_1, \dots, s_M, A, B) \\ &\quad = \det[\lambda - (I - \sum_{k=1}^{N} A_k s_1^{\gamma_{k1}} \cdot s_M^{\gamma_{kM}})^{-1} (\sum_{k=1}^{N} B_k s_1^{\gamma_{k1}} \cdot s_M^{\gamma_{kM}})], \end{aligned}$$

then Hale, Infante, and Tsen [1] prove the following result.

Theorem 9.1. *The* NFDE (9.2) *is stable globally in the delays if and only if the following conditions hold:*

(i) *Equation* (9.1) *is stable globally in the delays,*

(ii) $P(iy, \gamma, s_1, \dots, s_M, A, B) \neq 0$, *for all* $y \in \mathbb{R}, y \neq 0$, $|s_j| = 1, j = 1, \dots, M$,

(iii) $\operatorname{Re} \sigma[(I - \sum_{k=1}^{N} A_k)^{-1} \sum_{k=1}^{N} B_k] < 0$, *where* σ *denotes spectrum.*

If we specialize this result to the scalar equation with independent delays

$$\frac{d}{dt}[x(t) - \sum_{k=1}^{N} a_k x(t - r_k)] = b_0 x(t) + \sum_{k=1}^{N} b_k x(t - r_k), \tag{9.3}$$

then Equation (9.3) is stable globally in the delays if and only if

$$\sum_{k=1}^{N} |a_k| < 1, \quad b_0 < 0, \quad \sum_{k=1}^{N} |b_k| \leq |b_0|. \tag{9.4}$$

For an equation with dependent delays,

$$\frac{d}{dt}[x(t) - \sum_{k=1}^{N} a_k x(t - \gamma_k \cdot r)] = b_0 x(t) + \sum_{k=1}^{N} b_k x(t - \gamma_k \cdot r), \tag{9.5}$$

the stability criteria are somewhat more complicated to state, but are determined from the properties of the solutions of the difference equations

$$y(t) - \sum_{k=1}^{N} a_k y(t - \gamma_k \cdot r)] = 0, \quad b_0 z(t) + \sum_{k=1}^{N} b_k z(t - \gamma_k \cdot r) = 0. \tag{9.6}$$

In fact, if we define

$$\alpha(\theta, b) = b_0 + \sum_{k=1}^{N} b_k \cos \gamma_k \cdot \theta, \quad \beta(\theta, b) = \sum_{k=1}^{N} b_k \sin \gamma_k \cdot \theta, \tag{9.7}$$

then Equation (9.6) is stable globally in the delays if and only if the following conditions hold:

$$\begin{gathered} \sum_{k=1}^{N} |a_k| < 1, \quad a_0 < 0, \quad \sum_{k=0}^{N} \neq 0, \\ \text{either } \alpha(\theta, a) \neq 0 \text{ or } \alpha(\theta, a) = 0, \ \beta(\theta, a) = 0. \end{gathered} \tag{9.8}$$

Other results and examples can be found in Hale, Infante, and Tsen [1].

It also is of interest to investigate NFDE when the delays depend on time. Of course, it will be necessary to understand well the corresponding difference equation when the delays depend on time. Very little information of a general nature is known, and the following observations are made to illustrate some of the difficulties.

Suppose that $\gamma_k \cdot r = r_k$, $k = 1, \ldots, M$, and that (9.1) is stable globally in the delays. Will the zero solution of the difference equation,

$$y(t) - \sum_{k=1}^{N} A_k y(t - \gamma_k \cdot r(t)) = 0, \tag{9.9}$$

be asymptotically stable if the function $r(t)$ is continuous and bounded on $\mathbb{R}$? If y is a scalar, this is true as a consequence of the inequalities $\sum_{k=1}^{N} |A_k| < 1$. If y is a vector, an example has been given by Sigueira Marconato and Avellar [1] with $r(t)$ periodic for which a solution of Equation (9.9) is unbounded on $[0, \infty)$. In fact, the system

$$
\begin{aligned}
y_1(t) &= \frac{2}{3}y_1(t - r_1(t)) + \frac{2}{3}y_2(t - r_2(t)) \\
y_2(t) &= -\frac{2}{3}y_1(t - r_2(t)) - \frac{2}{3}y_2(t - r_1(t))
\end{aligned}
$$

is stable globally in the delays if the delays are constant in time and there is an unbounded solution if $r_1(t)$, $r_2(t)$ are continuous periodic functions of period 3 such that $r_1(0) = 1, r_1(1) = 3, r_1(2) = 3$, and $r_2(0) = 2, r_2(1) = 1, r_2(2) = 1$, and $t - r_j(t) \geq -2$ for $t \geq 0$, $j = 1, 2$. For continuous initial data $\psi : [-2, 0] \to \mathbb{R}^2$ with $\psi(-1) = (1, 0)$, $\psi(-2) = (0, 1)$, it is possible to show that the solution $y(t)$ through $(0, \psi)$ is unbounded on $[0, \infty)$.

Theorem 7.1 is due to Cruz and Hale [2]. A special case of Theorem 7.2 for $k = 0$ was proved by Hale [14] and the result as stated is due to Lopes [5]. Theorem 7.3 is due to Hale and Scheurle [1]. For more results on attractors for NFDE with stable D operators and the existence of periodic solutions of systems periodic in time, see Hale [23].

Theorems 8.1 and 8.2 are due to Cruz and Hale [2] (see also Chary [1] and Minsk [1,2]). Example (8.2) is due to Lopes [1]. As remarked in the introduction, one can derive neutral equations from the transmission line problem in different ways. Slemrod [2] used the other form of Problem (8.2) and employed Liapunov functionals to obtain sufficient conditions for stability. Theorem 8.4 is due to Lopes [6]. Example 8.3 is due to Lopes [2]. Lopes [1,2,6] has also generalized these results to obtain uniform ultimate boundedness of solutions of nonautonomous equations. If the equations are also ω-periodic, he has applied the results of Chapter 4 to obtain the existence of ω-periodic solutions.

Infante and Slemrod [1] have used Liapunov functionals and Theorem 8.1 to obtain sufficient conditions on the coefficients in linear autonomous neutral differential difference equations, which will ensure the uniform asymptotic stability of the zero solution. For symmetric systems, Brayton and Willoughby [1] obtained sufficient conditions for the stability of such systems directly from the characteristic equation.

Liapunov functionals also have been used for NFDE where the derivative occurs explicitly and the space of initial data involves the function and its derivative. The reader may consult the volumes of *Trudy Sem. Teorii Diff. Urav Otkl. Argumenton* from the People's Friendship University in Moscow for references.

The relationship between the different types of stability of linear systems of NFDE is not as simple as for RFDE (for a general discussion, see Hale [18]). Even for autonomous equations, many surprising results occur.

It is easy to show that stability implies uniform stability. On the other hand, one can have asymptotic stability and not have uniform asymptotic stability. This follows because uniform asymptotic stability is equivalent to exponential asymptotic stability. On the other hand, one can have all eigenvalues λ_j of the linear system in the left half-plane with $\operatorname{Re}\lambda_j \to 0$ as $j \to \infty$ and all solutions approaching zero as $t \to \infty$. Such a system cannot be uniformly asymptotically stable. One can also have all eigenvalues λ with $\operatorname{Re}\lambda < 0$ and have some solutions unbounded (see Brumley [1]). An even more striking example was given by Gromova and Zverkin [1]. They gave an example of a linear neutral differential difference equation with all eigenvalues simple and on the imaginary axis and yet the equation has unbounded solutions. If the operator D is stable, one obtains the same relationship between the concepts of stability for linear autonomous and periodic equations as for RFDE.

Izé [2] and Izé and de Molfetta [1] have given very general results of the asymptotic behavior of linear equations that are nonautonomous perturbations of autonomous systems.

For the invariance principle using Razumikhin-type Liapunov functions on $\mathbb{R}^n$, see Haddock, Krisztin, Jerj'ecki, and Wu [1].

In certain problems concerning control systems containing gas, steam, or water pipes (see Kobyakov [1] and Solodovnikov [1]), one encounters linear hyperbolic partial differential equations with mixed initial and derivative boundary conditions. The same is true in loss-less transmission lines (see Brayton [2]). Using the process described in the introduction, these problems are equivalent to a system of equations of the following form:

$$\begin{aligned} \dot{x}(t) &= Ax(t) + By(t-r) + f(x(t), y(t), y(t-r)) \\ y(t) - Ex(t) &- Jy(t-r) - g(x(t), y(t), y(t-r)) = 0 \end{aligned}$$

where A, B, E and J are matrices, f and g are given functions and $r > 0$. Brayton [2] discussed the linear version of this equation by means of the Laplace transform. Razvan [1, 2] treated this equation as a special case of a NFDE by letting $y(t) = \dot{z}(t)$. For further remarks, see Zverkin [4]. Hale and Martinez-Amores [1] discussed the equation by writing the equation as

$$\begin{aligned} \dot{x}(t) &= Ax(t) + By(t-r) + f \\ \frac{d}{dt}\,&[y(t) - Ex(t) - Jy(t-r) - g] = 0 \end{aligned}$$

and applying the results of this chapter to the set of initial data $x(0) = a$, $y_0(\theta) = \phi(\theta)$, $-r \le \theta \le 0$, restricted to the set $\phi(0) - Ea - J\phi(-r) - g = 0$. The resulting theory thus follows in a very natural manner from the known results on NFDE. Equations of this type occur also in certain models describing lazer optics (see Chow and Huang [1] for references).

Datko [2] has discussed linear autonomous NFDE in a Banach space setting and has thus obtained some interesting generalizations of the Hille–Yoshida theorem.

10
Near equilibrium and periodic orbits

In this section, we consider autonomous FDE of retarded or neutral type and discuss the behavior of the solutions near equilibrium points and periodic orbits. We concentrate particularly on the existence of stable, unstable, center-stable, and center-unstable manifolds.

10.1 Hyperbolic equilibrium points

Let Ω be a neighborhood of zero in C and let $C_b^p(\Omega, \mathbb{R}^n) \subset C^p(\Omega, \mathbb{R}^n)$ be the subset of functions from Ω into $\mathbb{R}^n$ that have bounded continuous derivatives up through order p with respect to $\phi \in \Omega$. The space $C_b^p(\Omega, \mathbb{R}^n)$ becomes a Banach space if the norm is chosen as the supremum norm over all derivatives up through order p. The norm will be designated by $|\cdot|_p$. Throughout this chapter, we shall assume that $D \in \mathcal{L}(C, \mathbb{R}^n)$ is stable and $F \in C_b^1(\Omega, \mathbb{R}^n)$. Consider the equation

$$\frac{d}{dt}Dx_t = F(x_t). \tag{1.1}$$

If $F(0) = 0$, then 0 is an equilibrium point and the linearization about 0 is

$$\frac{d}{dt}Dx_t = Lx_t, \tag{1.2}$$

where $L \in \mathcal{L}(C, \mathbb{R}^n)$, $L\psi = D_\phi F(0)\psi$. We say that 0 is a *hyperbolic* equilibrium point of (1.1) if the roots of the characteristic equation

$$\det\ \Delta(\lambda) = 0, \quad \Delta(\lambda) = D(e^{\lambda \cdot} I) - L(e^{\lambda \cdot} I), \tag{1.3}$$

have nonzero real parts.

If 0 is a hyperbolic equilibrium point of (1.1) and Λ denotes the set of roots of (1.3) with positive real part, then the space C can be decomposed by Λ as

$$C = U \oplus S.$$

The decomposition of C as $U \oplus S$ defines two projection operators $\pi_U : C \to U$, $\pi_U U = U$, $\pi_S : C \to S$, $\pi_S S = S$, $\pi_S = I - \pi_U$. If Φ is a basis for U and Ψ is a basis for U^T, $(\Psi, \Phi) = 1$, then the projection is given by $\pi_U \phi = \Phi(\Psi, \phi)$. Let $\phi \in C$ be written as $\phi = \phi^U + \phi^S$, $\phi^U \in U$, $\phi^S \in S$. Let $K(t,s)(\theta) = \int_0^s X(t+\theta-\alpha)\, d\alpha$ where $X(\,\cdot\,)$ denotes the fundamental matrix solution to the linear equation (1.2). If $T(t)$ is the semigroup generated by the linear equation (1.2), then U and S are invariant under $T(t)$ and $T(t)$ is defined on U for all $t \in \mathbb{R}$. Define $X_0^U = \Phi\Psi(0)$. Then $K(t,s)^U = \int_0^s T(t-\alpha) X_0^U\, d\alpha$ and $K(t,s)^S = \pi_S K(t,s) = K(t,s) - \Phi(\Psi, K(t,s))$.

From Sections 7.9 and 9.5, it follows that there are positive constants M, α such that for $\phi \in C$

$$
\begin{aligned}
&|T(t)\phi^U| \le M e^{\alpha t} |\phi^U|, \quad \operatorname{Var}_{(t,0]} |T(t) X_0^U| \le M e^{\alpha t}, \quad t \le 0, \\
&|T(t)\phi^S| \le M e^{-\alpha t} |\phi^S|, \quad \operatorname{Var}_{[0,t)} K(t,\,\cdot\,)^S \le M e^{-\alpha t}, \quad t \ge 0.
\end{aligned} \tag{1.4}
$$

Relations (1.4) and the fact that D and L are linear imply that the origin of System (1.2) is a saddle point with the orbits in C behaving as shown in Figure 10.1.

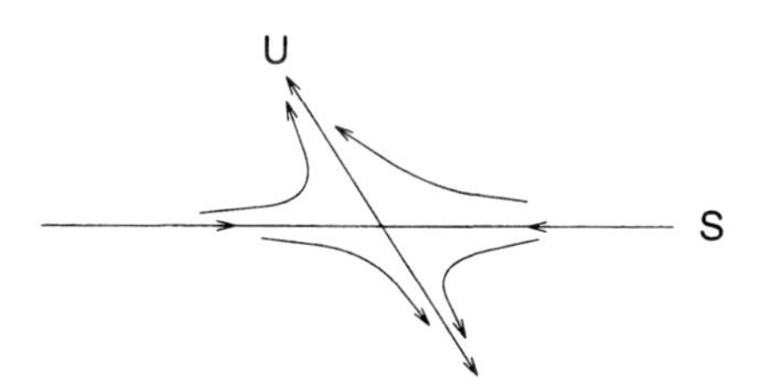

Fig. 10.1.

The set U is uniquely characterized as the set of initial values of those solutions of Equation (1.2) that exist and remain bounded for $t \le 0$. Relations (1.4) imply that these solutions approach zero exponentially as $t \to -\infty$. The set S is characterized as the set of initial values of those solutions of Equation (1.2) that exist and remain bounded for $t \ge 0$. These solutions approach zero exponentially as $t \to \infty$.

It is natural to ask if the solutions of Equation (1.1) have the same qualitative behavior near $x = 0$ as the solutions of Equation (1.2). Of course, the meaning of qualitative behavior must be defined very carefully. It is tempting to say that Equation (1.1) has the same qualitative behavior as Equation (1.2) near $x = 0$ if the orbits of Equation (1.1) can be mapped homeomorphically onto the orbits of (1.2). The following example suggests that such a definition is too strong.

Consider the scalar retarded equation

$$\dot{y}(t) = -y(t-1)[1+y(t)].$$

The constant function $y(t) = -1$, $t \in \mathbb{R}$, is an equilibrium point. If we let $x(t) = y(t) + 1$, then x satisfies the equation

$$\dot{x}(t) = x(t) - x(t)x(t-1),$$

which is a special case of (1.1) with

$$D\phi = \phi(0), \ F(\phi) = \phi(0) - \phi(0)\phi(-1), \ r = 1.$$

The linear equation (1.2) is $\dot{x}(t) = x(t)$, which is a special case of (1.4) with $D\phi = L\phi = \phi(0)$. This ordinary differential equation must be considered as a FDE in C. It is easy to verify that the sets U and S are given by

$$U = \{\,\phi : \phi(\theta) = e^{\theta}\phi(0), \ \theta \in [-1,0]\,\}$$
$$S = \{\,\phi : \phi(\theta) = \psi(\theta) - e^{\theta}\psi(0), \ \theta \in [-1,0], \ \psi \in C\}.$$

If $\phi \in S$, then $T(t)\phi = 0$ for $t \geq 1$. Therefore, each orbit in C must lie on U for $t \geq 1$. The semigroup for the linear equation is not one-to-one. For the particular perturbation $-\phi(0)\phi(-1)$, the semigroup generated by the nonlinear equation also is not one-to-one. In fact, any initial function ϕ with $\phi(0) = 0$ has the property that the solution $x(t,\phi) = 0$ for all $t \geq 0$. On the other hand, if we were to consider an arbitrary higher-order perturbation of the linear equation, it is not unreasonable to expect that the semigroup generated by the nonlinear equation is one-to-one. As a consequence, we cannot expect the orbits of the two systems to be homeomorphic.

On the other hand, we can show that some of the important properties of the trajectories are preserved. More specifically, we show that the set of initial values of those solutions of Equation (1.1) that exist and remain in a δ-neighborhood of $x = 0$ for $t \leq 0$ is diffeomorphic to a neighborhood in U of zero and these solutions approach zero exponentially as $t \to -\infty$. The same result is proved for $t \geq 0$ and S. Any other solution must leave a neighborhood of zero with increasing t and, if it exists for $t \leq -r$, it must also leave a neighborhood of zero with decreasing t.

Let us now be more precise. Let $x(t,\phi)$ be the solution of (1.1) with initial value ϕ at $t = 0$. We define the *stable set* and *unstable set* of the equilibrium point 0 of (1.1) as, respectively,

$$W^s(0) = \{\,\phi \in C : x_t(\,\cdot\,,\phi) \to 0 \text{ as } t \to \infty\,\}$$
$$W^u(0) = \{\,\phi \in C : x_t(\,\cdot\,,\phi) \to 0 \text{ as } t \to -\infty\,\}. \tag{1.5}$$

For a given neighborhood V of 0, we also can define the *local stable* and *local unstable* sets

$$W^s_{\text{loc}}(0) \overset{\text{def}}{=} W^s(0,V) = \{\,\phi \in W^s(0) : x_t(\,\cdot\,,\phi) \in V \text{ for } t \ge 0\,\}$$
$$W^u_{\text{loc}}(0) \overset{\text{def}}{=} W^u(0,V) = \{\,\phi \in W^u(0) : x_t(\,\cdot\,,\phi) \in V \text{ for } t \le 0\,\}. \tag{1.6}$$

We say that $W^u(0,V)$ is a *Lipschitz graph* (resp. C^k*-graph*) over $\pi_U C$ if there is a neighborhood $\widetilde{V}$ of 0 in $\pi_U C$ and a Lipschitz continuous function (resp. C^k-function) g such that

$$W^u(0,V) = \{\,\psi \in C : \psi = g(\phi),\ \phi \in \widetilde{V}\,\}.$$

The set $W^u(0,V)$ is said to be *tangent to* $\pi_U C$ *at* 0 if $|\pi_S\psi|/|\pi_U\psi| \to 0$ as $\psi \to 0$ in $W^u(0,V)$. Similar definitions hold for $W^s(0,V)$.

A basic result on stable and unstable sets of equilibrium points is the following.

Theorem 1.1. *If 0 is a hyperbolic equilibrium point of Equation (1.1), $F \in C^k_b(\Omega, \mathbb{R}^n)$, and D is stable, then there is a neighborhood V of 0 in C such that $W^u(0,V)$ (resp. $W^s(0,V)$) is a C^k-graph over $\pi_U C$ (resp. $\pi_S C$) that is tangent to $\pi_U C$ (resp. $\pi_S C$) at 0.*

To prove this result, we let $f(\phi) = F(\phi) - L\phi$, and rewrite Equation (1.1) as

$$\frac{d}{dt}Dx_t = Lx_t + f(x_t). \tag{1.7}$$

The function f has the property that $f(0) = 0$, $D_\phi f(0) = 0$. We need the following result.

Lemma 1.1. *If $x(t,\phi)$ is a solution of (1.7) that is defined and bounded for $t \le 0$, then $x_t(\,\cdot\,,\phi)$ is a solution of the following integral equation in C:*

$$\begin{aligned} y(t) = T(t)\phi^U &+ \int_0^t T(t-\tau)X_0^U f(y(\tau))\,d\tau \\ &+ \int_{-\infty}^t d[K(t,\tau)^S] f(y(\tau)). \end{aligned} \tag{1.8}$$

If $x(t,\phi)$ is a solution of (1.7) that is defined and bounded for $t \ge 0$, then $x_t(\,\cdot\,,\phi)$ is a solution of the following integral equation in C:

$$\begin{aligned} y(t) = T(t)\phi^S &+ \int_0^t d[K(t,\tau)^S] f(y(\tau)) \\ &- \int_t^\infty T(t-\tau)X_0^U f(y(\tau))\,d\tau. \end{aligned} \tag{1.9}$$

Conversely, if $x(t,\phi)$ is a solution of (1.8) (resp. (1.9)) that is defined and bounded for $t \le 0$ (resp. $t \ge 0$), then $x_t(\,\cdot\,,\phi)$ is a solution of (1.7).

Proof. If $y(t) \stackrel{\text{def}}{=} x_t(\cdot\,,\phi)$, $t \le 0$, is a bounded solution of (1.7), then for any $\bar{t} \in (-\infty, 0]$,

$$\pi_S y(t) = T(t-\bar{t})\pi_S y(\bar{t}) + \int_{\bar{t}}^{t} d[\pi_S K(t,\tau)] f(y(\tau)).$$

If we let $\bar{t} \to -\infty$ and use Relations (1.4), we deduce that

$$\pi_S y(t) = \int_{-\infty}^{t} d[K(t,\tau)^S] f(y(\tau)).$$

Since

$$\pi_U y(t) = T(t)\phi^U + \int_0^t T(t-\tau) X_0^U f(y(\tau))\, d\tau,$$

we see that $y(t)$ must satisfy (1.8). The proof for the case when $\phi \in W^s(0)$ is similar and therefore omitted.

The converse statement is proved by direct computation. □

We now begin the proof of Theorem 1.1, considering first the case where the function is assumed only to be Lipschitz continuous. In this case, we prove that the stable and unstable sets are Lipschitz graphs. More specifically, suppose that there is a continuous function $\eta : [0,\infty) \to [0,\infty)$, $\eta(0) = 0$, and let us consider those functions f in (1.7) that satisfy

$$\begin{aligned} &f(0) = 0 \\ &|f(\phi) - f(\psi)| \le \eta(\sigma)|\phi - \psi| \qquad \text{if} \quad |\phi|,\ |\psi| \le \sigma. \end{aligned} \tag{1.10}$$

With the constants K, K_1, α as defined earlier, choose $\delta > 0$ so that $8KK_1\eta(\delta) < \alpha$, $8K^2K_1^2\eta(\delta) < \alpha$. For $\phi \in \pi_U C$ with $|\phi| \le \delta/2K$, define $\mathcal{S}(\phi,\delta)$ as the set of continuous functions $y : (-\infty, 0] \to C$ such that $|y| = \sup_{-\infty<t\le 0} |y(t)| \le \delta$ and $\pi_U y(0) = \phi$. The set $\mathcal{S}(\phi,\delta)$ is a closed bounded subset of the Banach space of all continuous functions taking $(-\infty, 0]$ into C with the uniform topology. For any $y \in \mathcal{S}(\phi,\delta)$, define

$$\begin{aligned} (\mathcal{T}y)(t) = T(t)\phi &+ \int_0^t T(t-\tau) X_0^U f(y(\tau))\, d\tau \\ &+ \int_{-\infty}^{t} d[K(t,\tau)^S] f(y(\tau)). \end{aligned} \tag{1.11}$$

From the estimates (1.4), we see that

$$\begin{aligned} |\mathcal{T}y(t)| &\le Ke^{\alpha t}|\phi| + KK_1\eta(\delta)\int_t^0 e^{\alpha(t-\tau)}|y(\tau)|\, d\tau \\ &\quad + KK_1\eta(\delta)\int_{-\infty}^{t} e^{-\alpha(t-\tau)}|y(\tau)|\, d\tau \\ &\le \delta\Big(\frac{1}{2} + \frac{2KK_1\eta(\delta)}{\alpha}\Big) \le \frac{3}{4}\delta. \end{aligned}$$

Thus, $\mathcal{T} : \mathcal{S}(\phi, \delta) \to \mathcal{S}(\phi, \delta)$. In the same way, we observe that $\mathcal{T}$ is a contraction mapping with contraction constant $1/4$, independent of ϕ. Therefore, $\mathcal{T}$ has a unique fixed-point $y^*(\,\cdot\,, \phi) \in \mathcal{S}(\phi, \delta)$. This fixed point satisfies (1.8) and thus is a solution of (1.7) from Lemma 1.1. Also, $y^*(\,\cdot\,, \phi)$ is continuous in ϕ.

For $\phi \in \pi_U C$ with $|\phi| \leq \delta/2K$, we now consider the set $\mathcal{S}(\phi, \delta, \alpha)$ of continuous functions $y : (-\infty, 0] \to C$ such that

$$|y|_\alpha = \sup_{-\infty<t\leq 0} e^{-\alpha t/2}|y(t)| \leq \delta$$

and $\pi_U y(0) = \phi$. It is clear that $\mathcal{S}(\phi, \delta, \alpha) \subset \mathcal{S}(\phi, \delta)$ since $\alpha > 0$. From (1.11), we deduce that

$$e^{-\alpha t/2}|\mathcal{T}y(t)| \leq \frac{1}{2}\delta + \frac{KK_1\eta(\delta)}{\alpha}(2 + \frac{2}{3})|y|_\alpha \leq \delta.$$

Therefore, $\mathcal{T}$ maps $\mathcal{S}(\phi, \delta, \alpha)$ into itself. Since $\mathcal{T}$ has a unique fixed point in $\mathcal{S}(\phi, \delta)$ and $\mathcal{S}(\phi, \delta, \alpha) \subset \mathcal{S}(\phi, \delta)$, we infer that $y^*(\,\cdot\,, \phi) \in \mathcal{S}(\phi, \delta, \alpha)$ and $|y^*(t, \phi)| \leq \delta e^{\alpha t/2}$ for $t \leq 0$.

We now obtain better estimates on $y^*(\,\cdot\,, \phi)$. The function $y^*(\,\cdot\,, \phi)$ satisfies the integral inequality

$$\begin{aligned}|y^*(t, \phi)| \leq Ke^{\alpha t}|\phi| &+ KK_1\eta(\delta)\int_t^0 e^{\alpha(t-\tau)}|y^*(\tau, \phi)|\,d\tau \\ &+ KK_1\eta(\delta)\int_{-\infty}^t e^{-\alpha(t-\tau)}|y^*(\tau, \phi)|\,d\tau.\end{aligned}$$

Since $|y^*(\,\cdot\,, \phi)|_\alpha$ is bounded, we have

$$\begin{aligned}e^{-\alpha t/2}|y^*(t, \phi)| &\leq Ke^{\alpha t/2}|\phi| + KK_1\eta(\delta)\int_t^0 e^{\alpha(t-\tau)/2}e^{-\alpha\tau/2}|y(\tau)|\,d\tau \\ &\quad + KK_1\eta(\delta)\int_{-\infty}^t e^{-3\alpha(t-\tau)/2}e^{-\alpha\tau/2}|y(\tau)|\,d\tau \\ &\leq Ke^{\alpha t/2}|\phi| + \frac{KK_1\eta(\delta)}{\alpha}(2 + \frac{2}{3})|y^*(\,\cdot\,, \phi)|_\alpha \\ &\leq K|\phi| + \frac{1}{2}|y^*(\,\cdot\,, \phi)|_\alpha.\end{aligned}$$

From this inequality, we obtain $|y^*(\,\cdot\,, \phi)| \leq 2K|\pi_U\phi|$, which implies that

$$|y^*(t, \phi)| \leq 2Ke^{\alpha t/2}|\phi|, \quad t \leq 0. \tag{1.12}$$

This estimate shows that $y^*(t, 0) = 0$ and that $y^*(0, \phi) \in W^u(0)$.

The same type of computations show that

$$|y^*(t, \phi) - y^*(t, \psi)| \leq 2Ke^{\alpha t/2}|\phi - \psi|, \quad t \leq 0. \tag{1.13}$$

In particular, $y^*(\,\cdot\,, \phi)$ is Lipschitzian in ϕ.

We next observe that

$$
\begin{aligned}
|y^*(0,\phi) - y^*(0,\psi)| &\geq |\phi - \psi| - \int_{-\infty}^{0} KK_1\eta(\delta)e^{\alpha\tau}|y^*(\tau,\phi) - y^*(\tau,\psi)|\,d\tau \\
&\geq |\phi - \psi|[1 - \frac{4K^2K_1^2\eta(\delta)}{3\alpha}] \geq \frac{1}{2}|\phi - \psi|.
\end{aligned}
$$

Thus, the mapping $\phi \mapsto y^*(0,\phi)$ is one-to-one with a continuous inverse.

These estimates, together with the fact that $x(t,\phi)$, $t \leq 0$, $\phi \in W^u(0)$ must satisfy (1.8), imply that there exists a neighborhood V of 0 in C such that the conclusions of Theorem 1.1 are valid for $W^u(0,V)$ if we replace the word C^k-graph by Lipschitz graph.

We can repeat the same type of argument to Equation (1.9) to obtain that $W^s(0,V)$ is a Lipschitz graph over $\pi_S C$.

To show that these local sets are C^1-graphs, we make use of the following result, which is of independent interest.

Lemma 1.2. *Let X, Y be Banach spaces, $Q \subset X$ an open set, and $g : Q \to Y$ be locally Lipschitzian. Then g is continuously differentiable if and only if for each $\phi \in Q$,*

$$
|g(\psi + h) - g(\psi) - g(\phi + h) + g(\phi)| = o(|h|_X) \tag{1.14}
$$

as $(\psi, h) \to (\phi, 0)$.

Proof. It is easy to see that (1.14) holds if g is a C^1-function. Without loss in generality, we take Q to be a ball and g to be Lipschitzian in Q. If the derivative g' of g exists at each point of Q and (1.14) is satisfied, then g' is continuous. Thus, it is enough to prove that g' exists at each point of Q.

Case 1. Let us first suppose that $X = Y = \mathbb{R}$. Since g is absolutely continuous, it is differentiable almost everywhere. For any $\phi \in Q$, $\epsilon > 0$, there is a $\delta > 0$ such that

$$
|g(\psi + h) - g(\psi) - g(\phi + h) + g(\phi)| \leq \epsilon|h| \qquad \text{if} \quad |\psi - \phi| + |h| < \delta.
$$

There is a $\psi^* \in (\phi - \delta, \phi + \delta)$ such that $g'(\psi^*)$ exists. Thus, for $h \neq 0$ sufficiently small,

$$
\left|\frac{g(\phi + h) - g(\phi)}{h} - g'(\psi^*)\right| \leq 2\epsilon,
$$

$$
0 \leq \{\limsup_{h\to 0} - \liminf_{h\to 0}\}\frac{g(\phi + h) - g(\phi)}{h} \leq 4\epsilon.
$$

Since ϵ is arbitrary, this implies that $g'(\phi)$ exists.

Case 2. Now suppose that $X = \mathbb{R}$ and Y is a Banach space with Y^* being the dual space. If $\eta \in Y^*$, then Case 1 implies that ηg is a C^1-function and $|(\eta g)'(\psi)| \leq |\eta| \operatorname{Lip} g$. If $D(\psi) : \eta \mapsto (\eta g)'(\psi)$, then $D(\psi) \in Y^{**}$ is continuous in ψ from (1.14).

For $\phi \in Q, \eta \in Y^*$, $|\eta| \le 1$, we have

$$\eta\Big(\frac{g(\phi+h)-g(\phi)}{h} - D(\phi)\eta\Big) = \frac{1}{h}\int_\phi^{\phi+h} (D(\psi)-D(\phi))\eta \to 0$$

as $h \to 0$ uniformly for $|\eta| \le 1$. Let $\tau : Y \to Y^{**}$ be the canonical inclusion. Then $\tau[g(\phi+h)-g(\phi)]/h \to D(\phi)$ as $h \to 0$. Since τ is an isometry, this implies that $[g(\phi+h)-g(\phi)]/h \to$ a limit in Y as $h \to 0$; that is, $g'(\phi)$ exists.

Case 3. Finally, let X, Y be arbitrary Banach spaces. From Case 2, for any $\psi \in X, h \in X$, the map $t \mapsto g(\psi + th)$ taking $\mathbb{R}$ to Y is C^1 if t is small. Thus, the Gateaux derivative

$$dg(\psi, h) = \frac{dg(\psi+th)}{dt}\Big|_{t=0} \qquad \text{exists.}$$

Condition (1.14) implies that $dg(\psi,h) \to dg(\phi,h)$ in Y as $\psi \to \phi \in Q$, uniformly for $|h| \le 1$. This implies that $h \mapsto dg(\psi,h)$ is linear and continuous. Thus, $dg(\psi,h)$ is the Fréchet derivative at h of g and the proof is complete. □

We now use Lemma 1.2 to prove that $W^u(0,V)$ is a C^1-graph over $\pi_U C$. Let $y^*(\,\cdot\,,\phi)$ be the fixed point in $\mathcal{S}(\phi,\delta)$ of the map $\mathcal{T}$ in (1.11) that defines $W^u(0,V)$. We show that $y^*(\,\cdot\,,\phi)$ is C^1 in ϕ. Define

$$z^*(\psi,\phi,h)(t) = y^*(t,\psi+h) - y^*(t,\psi) - y^*(t,\phi+h) + y^*(t,\phi).$$

From Lemma 1.2, it is sufficient to show that

$$\limsup_{(\psi,h)\to(\phi,0)} \frac{1}{|h|} z^*(\psi,\phi,h)(t) = 0 \tag{1.15}$$

uniformly for $-\infty < t \le 0$.

From the definition of $y^*(t,\phi)$ and the fact that f is a C^1-function, we have, for $t \le 0$,

$$\begin{aligned} z^*(\psi,\phi,h)(t) = &\int_0^t T(t-\tau)X_0^U D_\phi f(y^*(\tau,\phi))z^*(\psi,\phi,h)(\tau)\,d\tau \\ &+ \int_{-\infty}^t d[K(t,\tau)^S] D_\phi f(y^*(\tau,\phi))z^*(\psi,\phi,h)(\tau) \\ &+ \int_0^t T(t-\tau)X_0^U \big[D_\phi f(y^*(\tau,\psi)) \\ &\quad - D_\phi f(y^*(\tau,\phi))\big]\big(y^*(\tau,\psi+h) - y^*(\tau,\psi)\big)\,d\tau \\ &+ \int_{-\infty}^t d[K(t,\tau)^S]\big[D_\phi f(y^*(\tau,\psi)) \\ &\quad - D_\phi f(y^*(\tau,\phi))\big]\big(y^*(\tau,\psi+h) - y^*(\tau,\psi)\big) \\ &+ o(h) \qquad \text{as} \quad |h| \to 0. \end{aligned}$$

Using the estimates (1.4), (1.10), and (1.13), we have

$$
\begin{aligned}
|z^*(\psi,\phi,h)(t)| \le{} & KK_1\eta(\delta)\int_t^0 e^{\alpha(t-\tau)}|z^*(\psi,\phi,h)(\tau)|d\tau \\
& + KK_1\eta(\delta)\int_{-\infty}^t e^{-\alpha(t-\tau)}|z^*(\psi,\phi,h)(\tau)|d\tau \\
& + \frac{8K^2K_1^2}{\alpha}|h|e^{\alpha t/2}\sup_{\tau\le 0}|D_\phi f(y^*(\tau,\psi)) - D_\phi f(y^*(\tau,\phi))| \\
& + o(h) \qquad \text{as} \quad |h|\to 0.
\end{aligned}
$$

As in the proof of the existence of $W^u(0,V)$, one can show that there is a positive constant K_1 such that

$$
\frac{1}{|h|}|z^*(\psi,\phi,h)(t)| \le K_1 \sup_{\tau\le 0}|D_\phi f(y^*(\tau,\psi)) - D_\phi f(y^*(\tau,\phi))|
$$

for $t \le 0$. Since the function $y^*(\,\cdot\,,\psi)$ is continuous in ψ and f is a C^1-function, we have (1.15) and we have proved Theorem 1.1 for $W^u(0,V)$ for $k=1$. The same type of argument applies for $W^s(0,V)$.

The proof that Theorem 1.1 is valid for arbitrary k requires an induction argument going from functions that are C^{k-1} with the $(k-1)$st-derivative Lipschitz continuous to functions that are C^k. This will not be given (see the references in the supplementary remarks).

In Definition (1.5), we defined the global stable and unstable sets for the equilibrium point 0 and Theorem 1.1 asserts that these sets locally near 0 are C^k-manifolds. If we let $\widetilde{T}(t)$ be the semigroup generated by Equation (1.7) (that is, $\widetilde{T}(t)\phi = x_t(\,\cdot\,,\phi)$), then we have the following relations:

$$
\begin{aligned}
W^s(0) &= \bigcup_{t\ge 0}\widetilde{T}(-t)W^s(0,V) \\
W^u(0) &= \bigcup_{t\ge 0}\widetilde{T}(t)W^u(0,V).
\end{aligned}
\tag{1.16}
$$

It is natural to investigate whether or not $W^s(0)$ and $W^u(0)$ are C^k-manifolds. Without further hypotheses on the vector field f, this is not the case. Rather surprisingly, this is not true in general even for the finite-dimensional set $W^u(0)$. In fact, consider the retarded delay equation

$$
\dot{x}(t) = \alpha(x(t))x(t) + \beta(x(t))x(t-1), \tag{1.17}
$$

where $x \in \mathbb{R}$, $\alpha(x)$ and $\beta(x)$ are C^∞-functions defined as

$$
(\alpha(x),\beta(x)) = \begin{cases} (\frac{2e-1}{e-1}, -\frac{e^2}{e-1}), & \text{if } |x|\le 1; \\ (1,0), & \text{if } |x| \ge 2; \end{cases}
$$

and

$$\alpha(x) + \beta(x)e^{-1} = 1 \quad \text{when } 1 \le |x| \le 2.$$

The origin 0 is an equilibrium point of (1.17). Equation (1.17) is linear in a neighborhood of 0 and has $\lambda_1 = 1$ and $\lambda_2 = 2$ as the positive characteristic values. Thus, there is a neighborhood V of 0 such that dim $W^u(0,V) = 2$. Let $x(t) = \epsilon e^t$ be a solution of (1.17) initiating in $W^u(0,V)$. There is a t_0 such that $\inf_{-1\le\theta\le 0} |x(t_0+\theta)| > 2$, and, in a neighborhood of x_{t_0}, the Equation (1.17) becomes $\dot{x}(t) = x(t)$. If ϕ is in a small neighborhood of x_{t_0} and the solution through ψ is defined for negative t, then $\psi(\theta) = \eta e^{t+\theta}$, where η is close to ϵ. Therefore, the unstable set in this neighborhood of x_{t_0} is a smooth manifold but of dimension 1. The local two-dimensional unstable manifold collapses into a one-dimension manifold as we follow the manifold along the solutions.

If it is assumed that the map $\widetilde{T}(t)$ is one-to-one together with $D_\phi \widetilde{T}(t)$, then it is possible to show that both $W^s(0)$ and $W^u(0)$ are embedded submanifolds of C.

We present now an interesting result concerning the manner in which solutions leave the stable manifold. Suppose that the hypotheses of Theorem 1.1 are satisfied, $\Phi = (\phi_1, \phi_2, ..., \phi_d)$ is a basis for U, and let $\pi_U \phi = \Phi b$, $b = b(\phi) \in \mathbb{R}^d$. The mapping $b : C \to \mathbb{R}^d$ is continuous and linear and we take the norm of b to be the Euclidean norm. For any continuous function $V : C \to \mathbb{R}$, we let

$$\dot{V}_-(\phi) = \liminf_{t\to 0^+} \frac{1}{t}[V(x_t(\phi)) - V(\phi)],$$

where $x(\phi)$ is the solution of Equation (1.1) through ϕ.

Lemma 1.3. *Under the assumptions of Theorem* 1.1 *and the preceding notation, there is a positive definite quadratic form* $V(\phi) = b^T E b$ *on* $\mathbb{R}^d$ *with the property that for any constant* $p > 0$, *there is a* $\delta_0 > 0$ *such that for any* δ, $0 < \delta \le \delta_0$, $\dot{V}_-(\phi) > 0$ *if* $V(\phi) \ge p^2\delta^2$, $\phi \in C$, $|\phi| \le \delta$.

Proof. Let $\pi_U x_t = \Phi y(t)$, where $x_t = x_t(\phi)$ is the solution of Equation (1.1) with $x_0 = \phi$. From Theorem 9.1 of Section 7.9, there is a $d \times n$ constant matrix C and a $d \times d$ constant matrix B with the spectrum of B equal to the roots of the characteristic equation (1.3) with positive real parts such that

$$\dot{y}(t) = By(t) + Cf(x_t),$$

where $f(\phi) = F(\phi) - L\phi$. Suppose that E is a $d \times d$ positive definite matrix satisfying $B^T E + EB = I$ and define $V(\phi) = b^T E b$, where $\pi_U \phi = \Phi b$. If $g(\phi) = Cf(\phi)$, then

$$\dot{V}_-(\phi) = b^T b + 2g^T E b.$$

Let $\beta^2 = \min\{\, b^T E b : |b| = 1\,\}$ and $\gamma = \max\{\, b^T E b : |b| = 1\,\}$. Suppose that $\eta : [0, \infty) \to \mathbb{R}$ is a continuous nondecreasing function, $\eta(0) = 0$, such that $|g(\phi)| \le \eta(\delta)|\phi|$ for $|\phi| \le \delta$ and choose $\delta_0 > 0$ so that $4\gamma|E|\eta(\delta_0) < p\delta$. Then as long as $|\phi| \le \delta$, $0 < \delta \le \delta_0$, and $V(\phi) \ge p^2\delta^2$, we have

$$\begin{aligned}\dot{V}_-(\phi) &\ge \frac{1}{\gamma}V(\phi) - 2|E|\eta(\delta)\beta^{-1}V^{1/2}(\phi)\delta \\ &\ge V(\phi)\Big|\frac{1}{\gamma} - \frac{2|E|\eta(\delta)}{p\beta}\Big| \ge \frac{1}{2\gamma}V(\phi) > 0.\end{aligned}$$

This proves the lemma. □

This last lemma holds even if some eigenvalues of Equation (1.3) are on the imaginary axis. It is easily checked that the same proof is valid.

Remark 1.1. It is possible to prove parameterized versions of all of these results; that is, we can consider the equation

$$\frac{d}{dt}D(x_t) = \widetilde{F}(x_t, \lambda)$$

where λ is a parameter in a Banach space, $\tilde{F}(\phi, 0) = F(\phi)$, and prove analogues of Theorem 1.1 and the lemmas for $|\lambda|$ small. We also can consider nonautonomous equations

$$\frac{d}{dt}D(x_t) = \widetilde{F}(x_t, t, \lambda),$$

provided that we suppose that the linear variational equation near the origin has an exponential dichotomy. See the references for the definition of this term.

10.2 Nonhyperbolic equilibrium points

If 0 is a nonhyperbolic equilibrium point of (1.1), then the space C can be decomposed as

$$C = U \oplus N \oplus S$$

where U is finite dimensional and corresponds to the span of the generalized eigenspaces of the roots of (1.3) with positive real parts, N is finite dimensional and corresponds to the span of the generalized eigenspaces of the roots of (1.3) with zero real parts. This decomposition of C defines three projection operators $\pi_U : C \to U$, $\pi_U U = U$, $\pi_N : C \to N$, $\pi_N N = N$, $\pi_S : C \to S$, $\pi_S S = S$, $\pi_U + \pi_N + \pi_S = I$. If Φ is a basis for U and Ψ is a basis for U^T, $(\Psi, \Phi) = 1$, then the projection π_U is given by $\pi_U \phi = \Phi(\Psi, \phi)$. Similarly, the projection π_N can be given by $\pi_N \phi = \Phi_0(\Psi_0, \phi)$, where Φ_0 is a basis for N and Ψ_0 is a basis for N^T, $(\Psi_0, \Phi_0) = 1$. Let $\phi \in C$

be written as $\phi = \phi^U + \phi^N + \phi^S$, $\phi^U \in U$, $\phi^N \in N$, $\phi^S \in S$. Let $K(t,s)(\theta) = \int_0^s X(t+\theta-\alpha)\, d\alpha$ where $X(\cdot)$ denotes the fundamental matrix solution to the linear equation (1.2).

If $T(t)$ is the semigroup generated by the linear equation (1.2), then U, N, and S are invariant under $T(t)$, $T(t)$ is defined on U, N for all $t \in \mathbb{R}$. Define $X_0^U = \Phi\Psi(0)$, $X_0^N = \Phi_0\Psi_0(0)$. Then $K(t,s)^U = \int_0^s T(t-\alpha)X_0^U\, d\alpha$, $K(t,s)^N = \int_0^s T(t-\alpha)X_0^N\, d\alpha$ and $K(t,s)^S = \pi_S K(t,s) = K(t,s) - \Phi(\Psi, K(t,s)) - \Phi_0(\Psi_0, K(t,s))$.

From Sections 7.9 and 9.5, it follows that there are positive constants M, α, and, for any $\epsilon > 0$, a positive constant M_ϵ, such that for $\phi \in C$

$$\begin{aligned}
&|T(t)\phi^U| \le Me^{\alpha t}|\phi^U|, \quad \mathrm{Var}_{(t,0]}\, T(t)X_0^U \le Me^{\alpha t}, \quad t \le 0,\\
&|T(t)\phi^N| \le M_\epsilon e^{\epsilon|t|}|\phi^N|, \quad \mathrm{Var}_{(-t,t)}\, T(t)X_0^N \le Me^{\epsilon|t|}, \quad t \in \mathbb{R},\\
&|T(t)\phi^S| \le Me^{-\alpha t}|\phi^S|, \quad \mathrm{Var}_{[0,t)}\, K(t,\cdot)^S \le Me^{-\alpha t}, \quad t \ge 0.
\end{aligned} \tag{2.1}$$

As in the previous section, the set U is uniquely characterized as the set of initial values of those solutions of Equation (1.2) that exist, remain bounded for $t \le 0$, and approach zero exponentially as $t \to -\infty$. The set S is characterized as the set of initial values of those solutions of Equation (1.2) that exist, remain bounded for $t \ge 0$, and approach zero exponentially as $t \to \infty$. The set N is characterized as the set of initial values of those solutions of Equation (1.2) that exist for all $t \in \mathbb{R}$ and increase less rapidly than any fixed exponential function.

Definition 2.1. For a given neighborhood V of $0 \in C$, the *local strongly stable set (or manifold)* $W_{\text{loc}}^{ss}(0) \overset{\text{def}}{=} W^{ss}(0,V)$ of the equilibrium point 0 of Equation (1.1) is the collection of points $\phi \in C$ with the property that the solution $x_t(\cdot,\phi) \in V$ for $t \ge 0$ and approaches zero exponentially as $t \to \infty$. In the same way, we define the *local strongly unstable set (or manifold)* $W_{\text{loc}}^{su}(0) \overset{\text{def}}{=} W^{su}(0,V)$.

Definition 2.2. For a given neighborhood V of $0 \in C$, a *local center manifold* $W_{\text{loc}}^c(0) \overset{\text{def}}{=} W^c(0,V)$ of the equilibrium point 0 of Equation (1.1) is a C^1-submanifold that is a graph over $V \cap N$ in C, tangent to N at 0, and locally invariant under the flow defined by Equation (1.1). In other words,

$$W_{\text{loc}}^c(0) \cap V = \{\psi \in C : \psi = \phi + h(\phi),\ \phi \in N \cap V\}$$

where $h : N \to U \oplus S$ is a C^1-mapping with $h(0) = 0$, $D_\phi h(0) = 0$. Moreover, every orbit that begins on $W_{\text{loc}}^c(0)$ remains in this set as long as it stays in V.

Definition 2.3. For a given neighborhood V of $0 \in C$, a *local center-stable manifold* $W_{\text{loc}}^{cs}(0) \overset{\text{def}}{=} W^{cs}(0,V)$ of the equilibrium point 0 of (1.1) is a set in C such that $W_{\text{loc}}^{cs}(0) \cap V$ is a C^1-submanifold that is a graph over $(N \oplus S) \cap V$, tangent to $N \oplus S$ at the origin and is locally invariant under the flow. In other words,

$$W_{\text{loc}}^{cs}(0) \cap V = \{\psi \in C : \psi = \phi + h(\phi),\, \phi \in (N \oplus S) \cap V\}$$

where $h : N \oplus S \to U$ is a C^1-mapping with $h(0) = 0$, $D_\phi h(0) = 0$. Moreover, every orbit that begins on $W_{\text{loc}}^{cs}(0)$ remains in this set as long as it stays in V. Furthermore, any orbit that stays in V for all $t \geq 0$ must belong to $W_{\text{loc}}^{cs}(0)$. In the same way, we define the *local center-unstable manifold* $W_{\text{loc}}^{cu}(0) \overset{\text{def}}{=} W^{cu}(0, V)$ of the equilibrium point 0 of (1.1) by replacing $t \geq 0$ by $t \leq 0$, the set $N \oplus S$ by $N \oplus U$, and the set U by S.

The basic result on the existence of the invariant manifolds is the following.

Theorem 2.1. *If F in (1.1) is a C^k-function, then there is a neighborhood V of $0 \in C$ such that each of the sets $W_{\text{loc}}^{ss}(0)$, $W_{\text{loc}}^{su}(0)$, $W_{\text{loc}}^{c}(0)$, $W_{\text{loc}}^{cu}(0)$, and $W_{\text{loc}}^{cs}(0)$ exists and is a C^k-submanifold of C. The manifolds $W_{\text{loc}}^{ss}(0)$ and $W_{\text{loc}}^{su}(0)$ are uniquely defined, whereas the manifolds $W_{\text{loc}}^{c}(0)$, $W_{\text{loc}}^{cu}(0)$, and $W_{\text{loc}}^{cs}(0)$ are not. Furthermore, every invariant set of (1.1) that remains in V must belong to $W_{\text{loc}}^{c}(0)$.*

We do not give all of the details of the proof of Theorem 2.1, but simply outline the major steps. To be specific, we concentrate on the local center manifold. With the decomposition $C = U \oplus N \oplus S$, we can write

$$\phi = \phi^N + \phi^{U\oplus S}, \qquad \phi^N \in N, \quad \phi^{U\oplus S} \in U \oplus S. \tag{2.2}$$

Let Φ_c be a basis for N and Ψ_c be a corresponding basis for the solutions of the adjoint equation with $(\Phi_c, \Psi_c) = I$, where $(\,\cdot\,,\,\cdot\,)$ is the usual bilinear form defined by Relation (2.23) of Chapter 9. We know that $T(t)\Phi_c = \Phi_c e^{B_c t}$, where the eigenvalues of B_c have zero real parts and correspond to the solutions of (1.3) that lie on the imaginary axis. If we let $x_t^N(\,\cdot\,, \phi) = \Phi_c y(t)$, then the solution $x_t = \Phi_c y(t) + x_t^{U\oplus S}$ with $x_0 = \phi$ is a solution of the system

$$\begin{aligned} \dot{y} &= B_c y + \Psi_c(0) f(\Phi_c y + x_t^{U\oplus S}) \\ x_t^{U\oplus S} &= T(t)\phi^{U\oplus S} + \int_0^t d[K(t,\tau)^{U\oplus S}] f(\Phi y(\tau) + x_\tau^{U\oplus S}), \end{aligned} \tag{2.3}$$

and conversely.

It is convenient to modify the function f in the direction of N so that we can consider the arbitrary elements of N rather than those elements of N that are in a small neighborhood of 0. Let $\chi : \mathbb{R}^{d_N} \to [0,1]$ be a C^∞-function with $\chi(y) = 1$ for $|y| \leq 1$, $\chi(y) = 0$ for $|y| \geq 2$. For any $\eta > 0$, define

(2.4) $$\tilde f(y,\phi^{U\oplus S}) = f(\Phi_c y\chi(\frac{y}{\eta}) + \phi^{U\oplus S}).$$

Let $C_\eta = \{\,\phi \in C : \phi = \Phi_c y\chi(y\eta^{-1}) + \phi^{U\oplus S},\ y \in \mathbb{R}^{d_N},\ |\phi^{U\oplus S}| \le \eta\,\}$. For any $\gamma > 0$, there exists a positive number η such that

(2.5) $$\mathrm{Lip}_{C_\eta}\tilde f < \gamma, \quad \sup_{\phi\in C_\eta} |\tilde f(y,\phi^{U\oplus S})| < \eta\gamma.$$

Let

$$\tilde{\mathcal S}(\delta,\Delta) = \{\,h : \mathbb{R}^{d_N} \to U\oplus S : |h| \le \delta,\ |h(y) - h(\bar y)| \le \Delta|y-\bar y|\,\}.$$

For any $h \in \tilde{\mathcal S}(\delta,\Delta)$, define $y(t,y_0,h)$, $y(0,y_0,h) = y_0$, to be the unique solution of the equation

(2.6) $$\dot y = B_c y + \Psi_c(0)\tilde f(y,h(y))$$

and define the operator $\mathcal T$ on $\tilde{\mathcal S}(\delta,\Delta)$ by the formula

(2.7) $$\begin{aligned}(\mathcal T h) = &-\int_0^\infty d[K(-\tau)^U]\tilde f(y(\tau,y_0,h),h(y(\tau,y_0,h)))\,d\tau\\ &+\int_{-\infty}^0 d[K(-\tau)^S]\tilde f(y(\tau,y_0,h),h(y(\tau,y_0,h)))\,d\tau.\end{aligned}$$

After several estimates, it is now possible to show that for appropriate constants δ,Δ (that is, for a sufficiently small η), the operator $\mathcal T$ is a uniform contraction on $\tilde{\mathcal S}(\delta,\Delta)$. Therefore, there is a unique fixed point $h_c \in \tilde{\mathcal S}(\delta,\Delta)$.

By applying the same type of reasoning as in the proof of Theorem 1.1, we deduce that the function h_c is a C^k-function.

The local center manifold $W^c_{\mathrm{loc}}(0)$ is then given by

$$W^c_{\mathrm{loc}}(0) = \{\phi : \phi = \Phi y + h_c(y),\ |y| \le \eta\,\}.$$

The flow on $W^c_{\mathrm{loc}}(0)$ is obtained from the solutions of the ordinary differential equation

(2.8) $$\dot y = B_c y + \Psi_c(0)f\big(\Phi_c y + h_c(\Phi_c y)\big) \overset{\mathrm{def}}{=} B_c y + Y_c(y)$$

with the initial data $y_0 = (\Psi_c,\phi)$. In fact, the solution $x_t(\,\cdot\,,\phi)$ of (1.7) with $\phi \in W^c_{\mathrm{loc}}(0)$ is given by $x_t(\,\cdot\,,\phi) = \Phi_c y(t,y_0) + h_c(\Phi_c y(t,y_0))$, where $y(t,y_0)$ is the solution of (2.8) with $y_0 = (\Psi_c,\phi)$.

In the same way, we obtain the other invariant manifolds in the statement of Theorem 2.1. Also, the solutions of (1.1) on $W^{cu}_{\mathrm{loc}}(0)$ are described by the solutions of an ordinary differential equation. In fact, let $\dim U\oplus N = d_{U\oplus N}$, Φ_{cu} be a basis for $N\oplus U$ and Ψ_{cu} be a corresponding basis for the solutions of the adjoint equation with $(\Phi_{cu},\Psi_{cu}) = I$, where $(\,\cdot\,,\,\cdot\,)$ is the usual bilinear form. We know that $T(t)\Phi_{cu} = \Phi_{cu}e^{B_{cu}t}$, where the eigenvalues of B_{cu} have nonnegative real parts. Then there exists a function $h_{cu} : \mathbb{R}^{d_{U\oplus N}} \to S$ such that

$$W^{cu}_{\text{loc}}(0) = \{\phi \in C : \phi = \Phi_{cu} z + h_{cu}(z),\ |z| \le \eta\}. \tag{2.9}$$

If $\phi \in W^{cu}_{\text{loc}}(0)$, then $x_t(\cdot\,, \phi) = \Phi_{cu} z(t, z_0) + h_{cu}(z(t, z_0))$, where $z(t, z_0)$ is the solution of the ordinary differential equation

$$\dot{z} = B_{cu} z + \Psi_{cu}(0) f(\Phi_{cu} z + h_{cu}(z)) \overset{\text{def}}{=} B_{cu} z + Z_{cu}(z) \tag{2.10}$$

with the initial data $z_0 = (\Psi_{cu}, \phi)$.

The local center-unstable manifold has a certain type of stability property that is sometimes referred to as asymptotic phase. Any solution off the center-unstable manifold decays exponentially toward a solution on the center-unstable manifold as long as it remains in a neighborhood of the origin. The precise description of this property is the following theorem.

Theorem 2.2. *Suppose that the hypotheses of Theorem* 1.1 *are satisfied. Then there exists a neighborhood V of $0 \in C$, positive constants K_1, α_1, and a C^k-function $H : V \to U \oplus N$ such that if $\bar{\phi} = H(\phi)$, then the solution $x_t(\cdot\,, \bar{\phi} + \phi^S)$ of* (1.7) *satisfies the property that*

$$x_t^{U\oplus N}(\cdot\,, \bar{\phi} + \psi) = H\big(\Phi_{cu} z(t, z_\phi) + x_t^S(\cdot\,, \bar{\phi} + \psi)\big), \tag{2.11}$$

where $z(t, z_\phi)$ is the solution of (2.10) *with $z_\phi = (\Psi_{cu}, \phi)$. In addition,*

$$\begin{aligned} |x_t^{U\oplus N}(\cdot\,, \bar{\phi} + \psi) - \Phi_{cu} z(t, z_\phi)| &\le K_1 e^{-\alpha_1 t}, \quad t \ge 0 \\ |x_t^S(\cdot\,, \bar{\phi} + \psi) - h_{cu}(z(t, z_\phi))| &\le K_1 e^{-\alpha_1 t}, \quad t \ge 0 \end{aligned} \tag{2.12}$$

as long as the solution remains in V.

We now outline the proof of Theorem 2.2. The first step is to introduce a coordinate system in which the center-unstable manifold replaces $U \oplus N$ as coordinate. If x_t is a solution of (1.7), let

$$x_t = \Phi_{cu} z(t) + h_{cu}(z(t)) + y_t^S. \tag{2.13}$$

A few computations will show that

$$\begin{aligned} \dot{z} &= B_{cu} z + \Psi_{cu}(0) f(\Phi_{cu} z + h_{cu}(z) + y_t^S) \\ &\overset{\text{def}}{=} B_{cu} z + Z(z, y_t^S), \\ y_t^S &= T(t)(\phi^S - h_{cu}(z_\phi)) + \int_0^t d[K(t,\tau)^S] F(z(\tau), y_\tau^S) \end{aligned} \tag{2.14}$$

where

$$F(z, \psi) = f(\Phi_{cu} z + h_{cu}(z) + \psi) - f(\Phi_{cu} z + h_{cu}(z)). \tag{2.15}$$

We now consider a class of functions $L(z, \psi) = z + M(z, \psi) \in \mathbb{R}^{d_{U\oplus N}}$, $z \in \mathbb{R}^{d_{U\oplus N}}$, $\psi \in S$, with $M(z, 0) = 0$. Our objective is to determine the function $M(z, \psi)$ so that if $\phi \in C$ is given, $z_\phi = (\Psi_{cu}, \phi)$ and

$$\bar{\phi} = \Phi_{cu}(z_\phi + M(z_\phi, \psi)),$$

then the solution $(z(t, z_{\bar{\phi}+\psi}), y_t^S(\bar{\phi}+\psi))$ of (2.14) satisfies the relation

$$\begin{aligned} z(t, z_{\bar{\phi}+\psi}) &= L(z(t,z_\phi), y_t^S(\bar{\phi}+\psi)) \\ &\stackrel{\text{def}}{=} z(t,z_\phi) + M(z(t,z_\phi), y_t^S(\bar{\phi}+\psi)), \\ y_t^S(\bar{\phi}+\psi) &= x_t^S(\,\cdot\,, \bar{\phi}+\psi) - h_{cu}(z(t,z_\phi)) \to 0 \text{ as } t \to \infty. \end{aligned} \tag{2.16}$$

Recall that $z(t, z_\phi)$ is the solution on the center manifold in our coordinate system. This implies that the function $w(t) = M(z(t,z_\phi), y_t(\bar{\phi}+\psi))$ with $w(0) = M(z_\phi, \psi)$ is a solution of the equation

$$\dot{w} = B_{cu} w + Z\big(L(z(t,z_\phi), y_t^S(\bar{\phi}+\psi)), y_t^S(\bar{\phi}+\psi)\big) - Z\big(z(t,z_\phi), 0\big), \tag{2.17}$$

where $y_t^S(\bar{\phi}+\psi)$ is the solution of the equation

$$y_t^S = T(t)\psi + \int_0^t d[K(t,\tau)^S] F\big(L(z(\tau,z_\phi), y_\tau^S), y_\tau^S\big)\, d\tau. \tag{2.18}$$

This suggests that we attempt to obtain the function $M(z,\psi)$ as a fixed point of the operator

$$\begin{aligned} (\mathcal{T}M)(z_\phi, \psi) = \int_0^\infty e^{B_{cu}\tau} \big[Z\big(L(z(\tau,z_\phi), y_\tau^S(\bar{\phi}+\psi)), y_\tau^S(\bar{\phi}+\psi)\big) \\ - Z(z(\tau,z_\phi), 0)\big]\, d\tau, \end{aligned} \tag{2.19}$$

where $y_t^S(\bar{\phi}+\psi)$ is the solution of (2.18).

As in the proof of the center manifold theorem, let $\mathcal{S}(\delta, \Delta)$ be the class of Lipschitz continuous functions $M : V \times \tilde{V} \to \mathbb{R}^{U\oplus N}$, where V (resp. $\tilde{V}$) is a neighborhood of $0 \in \mathbb{R}^{U\oplus N}$ (resp. S), such that $|M(\phi,\psi)| \leq \delta$, $M(\phi, 0) = 0$ and M has Lipschitz constant $< \Delta$. It is possible to show that the neighborhoods V, $\tilde{V}$ and constants δ, Δ can be chosen so that $\mathcal{T}$ is a uniform contraction on $\mathcal{S}(\delta,\Delta)$ and thus has a unique fixed point M^* in $\mathcal{S}(\delta,\Delta)$. In the proof, we use the fact that $|y_t^S(\phi+\psi)| \leq K_1 e^{-\alpha_1 t}$, $t \geq 0$, for some positive constants K_1, α_1. This completes the proof of the theorem.

As in Remark 1.1, we can have parameterized versions of Theorem 2.1 and dependence of the vector field on t.

10.3 Hyperbolic periodic orbits

In this section, we study the neighborhood of a period orbit γ of an FDE of either retarded or neutral type. Since a periodic orbit is a C^1-manifold, we can define a transversal Σ at a point $p \in \gamma$ and, therefore, obtain a Poincaré map on Σ. Since the solutions $T(t)\phi$ of RFDE become smoother in t as t increases, we can be sure that the Poincaré map is C^k if the vector field is C^k. We merely take the period of the periodic orbit large enough. We may then develop the manifold theory near a fixed point for the Poincaré map in a manner analogous to the theory near equilibrium points in Sections 10.1 and 10.2. This artifice will not work for NFDE since the solution operator does not smooth with time. We present an approach that gives partial information in this more general situation.

We will need a special case of the following result that is of independent interest and stated without proof (see the supplementary remarks for references).

Theorem 3.1. *If F is C^k (resp. analytic) and $x : (-\infty, 0] \to \mathbb{R}^n$ is a solution of Equation* (1.1) *on $(-\infty, 0]$ that is bounded, then $x \in C^{k+1}$ (resp. analytic).*

We say that γ is an *ω-periodic orbit* of (1.1) of minimal period $\omega > 0$ if $\gamma = \{p(t), t \in \mathbb{R}\}$, where $p(t)$ is a periodic solution of (1.1) of minimal period $\omega > 0$.

Corollary 3.1. *If F is C^k (resp. analytic) and $\gamma = \{p(t), t \in \mathbb{R}\}$ is an ω-periodic orbit, then $p \in C^{k+1}$ (resp. analytic) and γ is a C^{k+1}-manifold (resp. analytic manifold).*

If $\gamma = \{p(t), t \in \mathbb{R}\}$ is an ω-periodic orbit, then the linear variational equation about $p(t)$ is

$$\frac{d}{dt}Dx_t = L(t)x_t, \qquad L(t) = D_\phi F(p(t)). \tag{3.1}$$

Equation (3.1) is a linear equation with coefficients periodic in t of period ω. Therefore, we can define the Floquet multipliers as in Chapter 8. From Corollary 3.1, if $F \in C^k$, $k \geq 1$, then $p \in C^{k+1}$ and thus, $\dot{p}(t)$ is a nontrivial periodic solution of (3.1). As a consequence, 1 is always a characteristic multiplier of (3.1).

Definition 3.1. The *Floquet multiplier of a periodic orbit* γ are the Floquet multipliers of the linear variational equation (3.1) except that 1 is not a multiplier of γ if 1 is a simple multiplier of (3.1).

Definition 3.2. A periodic orbit γ is *hyperbolic* if each Floquet multiplier of γ has modulus different from 1. The *index* $i(\gamma)$ of a hyperbolic orbit γ is the number (counting multiplicity) of Floquet multipliers with moduli > 1.

Let $x(t, \phi)$ be the solution of (1.1) with initial value ϕ at $t = 0$. We define the *stable set* and *unstable set* of the periodic orbit γ of (1.1) as, respectively,

$$
(3.2)\qquad \begin{aligned} W^s(\gamma) &= \{\,\phi \in C : x_t(\,\cdot\,,\phi) \to \gamma \text{ as } t \to \infty\,\}, \\ W^u(\gamma) &= \{\,\phi \in C : x_t(\,\cdot\,,\phi) \to \gamma \text{ as } t \to -\infty\,\}. \end{aligned}
$$

For a given neighborhood V of γ, we also can define the *local stable* and *local unstable* sets

$$
(3.3)\qquad \begin{aligned} W^s_{\text{loc}}(\gamma) &\stackrel{\text{def}}{=} W^s(\gamma, V) = \{\,\phi \in W^s(\gamma) : x_t(\,\cdot\,,\phi) \in V \text{ for } t \geq 0\,\} \\ W^u_{\text{loc}}(\gamma) &\stackrel{\text{def}}{=} W^u(\gamma, V) = \{\,\phi \in W^u(\gamma) : x_t(\,\cdot\,,\phi) \in V \text{ for } t \leq 0\,\}. \end{aligned}
$$

Theorem 3.2. *If F is a C^k-function, $k \geq 1$, $D\phi = \phi(0)$ (that is, (1.1) is an* RFDE), *and γ is a periodic orbit of the* RFDE (1.1), *then there is a neighborhood V of γ such that $W^s(\gamma, V)$ and $W^u(\gamma, V)$ are C^k-submanifolds of C with dim $W^u(\gamma, V) = i(\gamma) + 1$ and codim $W^s(\gamma, V) = i(\gamma)$.*

Proof. Let $\gamma = \{\,p(t),\ t \in \mathbb{R}\,\}$ have minimal period ω and let $T(t)\phi$ be the solution of (1.1) with initial data ϕ at $t = 0$. Fix $\alpha \in [0, \omega)$ and let Σ_α be a codimension one transversal to γ at p_α. We can choose an ϵ-neighborhood $\mathcal{N}(\gamma, \epsilon)$ of γ and the transversals Σ_α in such a way that, for any $\phi \in \mathcal{N}(\gamma, \epsilon)$, there are a unique $\alpha \in [0, \omega)$ and $\phi_\alpha \in \Sigma_\alpha \cap \mathcal{N}(\gamma, \epsilon)$ such that $\phi = p_\alpha + \phi_\alpha$. The orbit γ and the set of transversals form a coordinate system near γ.

Since the solutions of RFDE smooth in t as t increases, it follows that $T(t)\phi$ is a C^k-function of t, ϕ if $t > kr$. Choose an integer j so that $j\omega > kr$. Let $D(\pi_\alpha) \subset \Sigma_\alpha$ be a neighborhood of p_α in Σ_α with the property that, for each $\phi \in D(\pi_\alpha)$, there is a unique $t = t(\phi)$ near $j\omega$ such that $T(t(\phi))\phi \in \Sigma_\alpha$. If we define $\pi_\alpha : D(\pi_\alpha) \to \Sigma_\alpha$ by $\pi_\alpha(\phi) = T(t(\phi))\phi$, then π_α is a C^k-function and $\pi_\alpha(p_\alpha) = p_\alpha$.

We introduce the stable and unstable sets of the fixed point p_α of π_α:

$$
(3.4)\qquad \begin{aligned} W^s_{\text{loc}}(p_\alpha) &\stackrel{\text{def}}{=} W^s(p_\alpha, V) = \{\,\phi \in C : \pi^m_\alpha \phi \in V,\ m \geq 0,\ \pi^m_\alpha \phi \to p_\alpha \\ &\qquad\qquad \text{as } m \to \infty\,\} \\ W^u_{\text{loc}}(p_\alpha) &\stackrel{\text{def}}{=} W^u(p_\alpha, V) = \{\,\phi \in C : \pi^m_\alpha \phi \in V,\ m \leq 0,\ \pi^m_\alpha \phi \to p_\alpha \\ &\qquad\qquad \text{as } m \to \infty\,\}. \end{aligned}
$$

It is possible to show that the elements of the point spectrum of the linear mapping $D_\phi \pi_\alpha(p_\alpha)$ are the Floquet multipliers of γ. Since γ is hyperbolic, no element of the spectrum of $D_\phi \pi_\alpha(p_\alpha)$ has modulus one. Now, we can use the same type of arguments as in Section 1 for stable and unstable manifolds near equilibrium points to conclude that there is a neighborhood V of p_α such that $W^s(p_\alpha, V)$, $W^u(p_\alpha, V)$ are C^k-submanifolds of C. In the previous proof, every infinite integral is replaced by an infinite sum. Since γ is compact, we also can choose V independent of α. Also, we extend the definition of $W^s(p_\alpha, V)$ to all $\alpha \in \mathbb{R}$ by setting $W^s(p_\alpha, V) = W^s(p_{\alpha+\omega}, V)$, $W^u(p_\alpha, V) = W^u(p_{\alpha+\omega}, V)$.

For any $\beta \in [0, \omega)$ and any neighborhood $\widetilde{V}$ of p_β, define

$$W^s_{\beta,\mathrm{loc}}(\gamma) = W^s(\gamma) \cap (\Sigma_\beta \cap \widetilde{V}),$$
$$W^u_{\beta,\mathrm{loc}}(\gamma) = W^u(\gamma) \cap (\Sigma_\beta \cap \widetilde{V}).$$

We claim that $W^u_{\beta,\mathrm{loc}}(\gamma)$ is a C^k-submanifold of dimension $i(\gamma)$. In fact, for any integer j such that $(j-1)\omega > kr$, it is obvious that there is a neighborhood V of p_0 such that

$$T(j\omega + \beta)W^u(p_0, V) \subset W^u_{\beta,\mathrm{loc}}(\gamma).$$

Also, since $W^u(p_0, V) = W^u(p_{j\omega}, V)$ for all j, there is a neighborhood $\widetilde{V}$ of p_β such that

$$T(j\omega)^{-1}T(j\omega - \beta)W^u_{\beta,\mathrm{loc}}(\gamma) \subset W^u(p_0, V);$$

that is, the inverse map of $T(j\omega + \beta) : W^u(p_0, V) \to W^u_{\beta,\mathrm{loc}}(\gamma)$ is $T(j\omega)^{-1}T(j\omega - \beta)$, which is C^k.

If we extend the definition of $W^u_{\beta,\mathrm{loc}}(\gamma)$ so that it is periodic in β, then the argument shows that $W^u_{\omega,\mathrm{loc}}(\gamma) = T(j\omega - \beta)W^u_{\beta,\mathrm{loc}}(\gamma)$. If we denote the tangent space of a submanifold M of C by $\mathcal{T}M$, then

$$TW^u_{\omega,\mathrm{loc}}(\gamma) = D_\phi T(j\omega - \beta)TW^u_{\beta,\mathrm{loc}}(\gamma). \tag{3.5}$$

Using the fact that Σ_β is transversal to γ at p_β (that is, $\Sigma_\beta \oplus [\dot{p}_\beta] = C$) and the implicit function theorem, we can find an $\epsilon > 0$ such that the set $\bigcup_{\delta \in (\beta-\epsilon, \beta+\epsilon)} W^u_{\delta,\mathrm{loc}}(\gamma)$ is a C^k-submanifold modeled on $[\dot{p}_\beta] \oplus TW^u_{\beta,\mathrm{loc}}(\gamma)$. Since γ is compact, this shows that $W^u(\gamma)$ satisfies the properties stated in Theorem 3.2.

We need to show the same properties for the stable manifold of γ. For any integer $j \geq 1$, we have $W^s_{\beta,\mathrm{loc}}(\gamma) = T(j\omega - \beta)W^s_{\omega,\mathrm{loc}}(\gamma)$. Choose j so that $(j-1)\omega > kr$. Let $Y \subset C$ be the set of $\phi \in C$ such that $D_\phi T(j\omega - \beta)\phi \in TW^s_{\omega,\mathrm{loc}}(\gamma)$. Since $TW^s_{\omega,\mathrm{loc}}(\gamma)$ is a closed subspace of C, it follows that Y is a closed subspace of C. Also, $[\dot{p}_\beta] \oplus TW^s_{\beta,\mathrm{loc}}(\gamma) \oplus Y = C$ from (3.5). We now use the implicit function theorem to obtain that $W^s_{\beta,\mathrm{loc}}(\gamma)$ is a C^k-graph over Y and is thus a C^k-submanifold of C.

The stable set of γ in a neighborhood of β is defined by

$$W^{s,\epsilon}_{\beta,\mathrm{loc}}(\gamma) = \{\, p_\delta + \phi : T(j\omega - \delta)\phi \in W^s_{\omega,\mathrm{loc}}(\gamma),\ \delta \in (\beta - \epsilon, \beta + \epsilon) \,\}$$

for some $\epsilon > 0$. Using the fact that we have $[\dot{p}_\beta] \oplus TW^s_{\beta,\mathrm{loc}}(\gamma) \oplus Y = C$, we can use the implicit function theorem to show that $W^{s,\epsilon}_{\beta,\mathrm{loc}}(\gamma)$ is a C^k-submanifold modeled on $[\dot{p}_\alpha] \oplus Y = [\dot{p}_\alpha] \oplus TW^s_{\beta,\mathrm{loc}}(\gamma)$. Since γ is compact, we can obtain the conclusion in Theorem 3.2 for the stable manifold of γ. This completes the proof of the theorem. □

Let us now turn to the analogue of Theorem 3.2 for NFDE. In this case, we have been unable to find a way to use the Poincaré map

$$\phi \mapsto \pi_\alpha(\phi) = T(t(\phi))\phi.$$

If $t(\phi)$ is not a constant, then $\pi_\alpha(\phi)$ is not differentiable. As a consequence, we proceed in a different manner. We introduce the *synchronized stable* and *synchronized unstable* sets of a point p_β on the periodic orbit γ:

$$\begin{aligned} \mathcal{W}^s_\beta(\gamma) &= \{\phi \in C : x_t(\cdot, \phi) - p_{t+\beta} \to 0 \text{ as } t \to \infty\}, \\ \mathcal{W}^u_\beta(\gamma) &= \{\phi \in C : x_t(\cdot, \phi) - p_{t+\beta} \to 0 \text{ as } t \to -\infty\}. \end{aligned} \tag{3.6}$$

For a given neighborhood V of γ, we also can define the *local synchronized stable* and *local synchronized unstable* sets of a point p_β on the orbit γ:

$$\begin{aligned} \mathcal{W}^s_{\beta,\mathrm{loc}}(\gamma) &\stackrel{\mathrm{def}}{=} \mathcal{W}^s_\beta(\gamma, V) = \{\phi \in \mathcal{W}^s_\beta(\gamma) : x_t(\cdot, \phi) \in V \text{ for } t \geq 0\}, \\ \mathcal{W}^u_{\beta,\mathrm{loc}}(\gamma) &\stackrel{\mathrm{def}}{=} \mathcal{W}^u_\beta(\gamma, V) = \{\phi \in \mathcal{W}^u_\beta(\gamma) : x_t(\cdot, \phi) \in V \text{ for } t \leq 0\}. \end{aligned} \tag{3.7}$$

We define the *local synchronized stable* and *local synchronized unstable* sets of γ as

$$\begin{aligned} \mathcal{W}^s_{\mathrm{loc}}(\gamma) &= \mathcal{W}^s(\gamma, V) = \bigcup_{\beta \in \mathbb{R}} \mathcal{W}^s_{\beta,\mathrm{loc}}(\gamma), \\ \mathcal{W}^u_{\mathrm{loc}}(\gamma) &= \mathcal{W}^u(\gamma, V) = \bigcup_{\beta \in \mathbb{R}} \mathcal{W}^u_{\beta,\mathrm{loc}}(\gamma). \end{aligned} \tag{3.8}$$

Theorem 3.3. *If F is a C^k-function, $k \geq 2$, and γ is a periodic orbit of (1.1), then there is a neighborhood V of γ such that the synchronized stable manifold $\mathcal{W}^s(\gamma, V)$ is a C^{k-1}-submanifold of C and the synchronized unstable manifold $\mathcal{W}^u(\gamma, V)$ is a C^k-submanifold of C with dim $\mathcal{W}^u(\gamma, V) = i(\gamma) + 1$ and codim $\mathcal{W}^s(\gamma, V) = i(\gamma)$.*

The proof of Theorem 3.3 involves several results of independent interest.

Lemma 3.1. *If F is a C^k-function, $k \geq 1$, and γ is a periodic orbit of (1.1), then there is a neighborhood V of γ such that $\mathcal{W}^s_\beta(\gamma, V)$ and $\mathcal{W}^u_\beta(\gamma, V)$ are C^k-submanifolds of C with dim $\mathcal{W}^u_\beta(\gamma, V) = i(\gamma)$ and codim $\mathcal{W}^s_\beta(\gamma, V) = i(\gamma) + 1$.*

Proof. To simplify the notation, we first take $\beta = 0$. If $x(t) = p(t) + z(t)$ in (1.1), then z satisfies the equation

$$\frac{d}{dt} Dz_t = L(t, z_t) + G(t, z_t), \tag{3.9}$$

$$L(t, \psi) = D_\phi F(p_t)\psi, \quad G(t, \phi) = F(p_t + \phi) - F(p_t) - L(t, \phi). \tag{3.10}$$

The variation-of-constants formula for Equation (3.9) is

$$z_t = T(t,\sigma)\phi + \int_0^t d[K(t,\tau)]G(\tau, z_\tau). \tag{3.11}$$

For each fixed $\tau \in \mathbb{R}$, the results in Chapter 8 imply that the space C can be decomposed as

$$C = U(\tau) \oplus N(\tau) \oplus S(\tau)$$

where $U(\tau)$ is finite dimensional and corresponds to the span of the generalized eigenspaces of the Floquet multipliers of (3.1) with moduli > 1, $N(\tau)$ is one dimensional and is the span $[\dot{p}_\tau]$ of $\dot{p}_\tau$. This decomposition of C defines three projection operators

$$\begin{aligned}
&\pi_{U(\tau)} : C \to U(\tau), \quad \pi_{U(\tau)}U(\tau) = U(\tau),\\
&\pi_{N(\tau)} : C \to N(\tau), \quad \pi_{N(\tau)}N(\tau) = N(\tau),\\
&\pi_{S(\tau)} : C \to S(\tau), \quad \pi_{S(\tau)}S(\tau) = S(\tau)
\end{aligned}$$

where $\pi_{U(\tau)} + \pi_{N(\tau)} + \pi_{S(\tau)} = I$. If $T(t,\sigma)$ is the solution operator of the linear equation (3.9), then

$$T(t,\sigma)U(\sigma) = U(t), \quad T(t,\sigma)N(\sigma) = N(t), \quad T(t,\sigma)S(\sigma) = S(t).$$

This relation holds on $S(\,\cdot\,)$ for all $t \geq \sigma$ and it holds on $U(\,\cdot\,)$, $N(\,\cdot\,)$ for all $t \in \mathbb{R}$.

Also, there are positive constants M, α such that for $\phi \in C$,

$$\begin{aligned}
|T(t,\sigma)\pi_{U(\sigma)}\phi| &\leq Me^{\alpha(t-\sigma)}|\phi|, \quad t \leq \sigma,\\
\mathrm{Var}_{(t,0]}\ \pi_{U(\sigma)}K(t,\sigma) &\leq Me^{\alpha(t-\sigma)}, \quad t \leq \sigma,\\
|T(t,\sigma)\pi_{S(\sigma)}\phi| &\leq Me^{-\alpha(t-\sigma)}|\phi|, \quad t \geq \sigma,\\
\mathrm{Var}_{[0,t)}\ \pi_{S(\sigma)}K(t,\sigma) &\leq Me^{-\alpha(t-\sigma)}, \quad t \geq \sigma.
\end{aligned} \tag{3.12}$$

With this notation, consider the integral equation

$$\begin{aligned}
z_t = T(t,0)\phi^S &+ \int_0^t d[\pi_{S(\tau)}K(t,\tau)]G(\tau, z_\tau)\\
&- \int_t^\infty d[(\pi_{U(\tau)K(t,\tau)} + \pi_{N(\tau)})K(t,\tau)]G(\tau, z_\tau)
\end{aligned} \tag{3.13}$$

for $t \geq 0$, $\phi^S \in \pi_{S(0)}C$ arbitrary, and the integral equation

$$\begin{aligned}
z_t = T(t,0)\phi^U &+ \int_0^t d[\pi_{U(\tau)}K(t,\tau)]G(\tau, z_\tau)\, d\tau\\
&+ \int_{-\infty}^t d[(\pi_{N(\tau)} + \pi_{S(\tau)})K(t,\tau)]G(\tau, z_\tau)
\end{aligned} \tag{3.14}$$

for $t \leq 0$, $\phi^U \in \pi_{U(0)}C$ arbitrary. Any solution of these integral equations will be a solution of Equation (3.9).

The proof of Lemma 3.1 will be complete if we make use of the following result, which is verified using the same type of arguments as in the proof of Theorem 1.1.

Lemma 3.2. *If $F \in C^k, k \geq 1$, and γ is a hyperbolic periodic orbit of (1.1), then there are positive constants δ, ν, such that for any $\phi^S \in \pi_{S(0)}C$, $|\phi^S| < \nu$, there is a unique solution $z_t^*(\phi^S)$ of Equation (3.13) with $|z_t^*(\phi^S)| \leq \delta e^{-\alpha t/2}$ for $t \geq 0$. Furthermore, if $H^s(\phi^S) = z_0^*(\phi^S)$, then the set $\{ H^s(\phi^S), |\phi^S| < \nu \}$ is a C^k-graph over the set $\{ \phi^S : |\phi^S| < \nu \}$ through the projection operator $\pi_{S(0)}$. Also, $D_\phi H^s(0) = I$.*

For any $\phi^U \in \pi_{U(0)}C$, $|\phi^U| < \nu$, there is a unique solution $z_t^{**}(\phi^U)$ of Equation (3.14) with $|z_t^{**}(\phi^U)| \leq \delta e^{\alpha t/2}$ for $t \leq 0$. Furthermore, if $H^u(\phi^U) = z_0^{**}(\phi^U)$, then the set $\{ H^u(\phi^U), |\phi^U| < \nu \}$ is a C^k-graph over the set $\{ \phi^U : |\phi^U| < \nu \}$ through the projection operator $\pi_{U(0)}$. Also, $D_\phi H^u(0) = I$.

Proof of Theorem 3.3. If we fix $\beta \in \mathbb{R}$ and let $x(t) = p(t+\beta) + z(t)$ in (1.1), then $z(t)$ satisfies the equation

$$\frac{d}{dt} Dz_t = L(t+\beta, z_t) + G(t+\beta, z_t), \tag{3.15}$$

where L and G are given in (3.10). Proceeding exactly as in the proof of Lemma 3.1, we obtain C^k-functions

$$H^s(\phi_\beta^S, \beta),\ z^*(\phi_\beta^S, \beta), \quad \text{and } H^u(\phi_\beta^U, \beta),\ z^{**}(\phi_\beta^U, \beta),$$

defined for $\phi_\beta^S \in \pi_{S(\beta)}C$, $|\phi_\beta^S| < \nu$, $\phi_\beta^U \in \pi_{U(\beta)}C$, $|\phi_\beta^U| < \nu$. Also, $D_\phi H^s(0,\beta) = I$, $D_\phi H^u(0,\beta) = I$. Since γ is compact, the constants δ, ν can be chosen to be independent of β. It is clear from the preceding construction that

$$\begin{aligned} \mathcal{W}_{\beta,\text{loc}}^s(\gamma) &= \{ p_\beta + H^s(\phi_\beta^S, \beta),\ |\phi_\beta^S| < \nu \}, \\ \mathcal{W}_{\beta,\text{loc}}^u(\gamma) &= \{ p_\beta + H^u(\phi_\beta^U, \beta),\ |\phi_\beta^U| < \nu \}. \end{aligned}$$

It also is clear that

$$\mathcal{W}_{\text{loc}}^s(\gamma) = \bigcup_{\beta \in \mathbb{R}} \mathcal{W}_{\beta,\text{loc}}^s(\gamma), \qquad \mathcal{W}_{\text{loc}}^u(\gamma) = \bigcup_{\beta \in \mathbb{R}} \mathcal{W}_{\beta,\text{loc}}^u(\gamma). \tag{3.16}$$

It remains to show that the sets $\mathcal{W}_{\text{loc}}^s(\gamma)$ and $\mathcal{W}_{\text{loc}}^u(\gamma)$ are submanifolds satisfying the properties stated in Theorem 3.3. We discuss only the stable set since the unstable set is treated in a similar way.

We claim first that, if $\nu > 0$ is sufficiently small and $\eta \in \mathcal{W}_{\text{loc}}^s(\gamma)$, then there are unique $\beta \in [0, \omega)$, $\phi_\beta^S \in \pi_{S(\beta)}C$, $|\phi_\beta^S| < \nu$, such that $\eta = p_\beta + H^s(\phi_\beta^S)$. Define the function

$$f(\eta, \beta, \phi_\beta^S) \stackrel{\text{def}}{=} \eta - p_\beta - H^s(\phi_\beta^S, \beta)$$

for $(\eta, \beta, \phi_\beta^S) \in C \times \mathbb{R} \times \pi_{S(\beta)}C$. It is clear that $f(p_\beta, 0, 0) = 0$ and the derivative $D_\beta f$ with respect to β, ϕ_β^S evaluated at $(h, \psi) \in \mathbb{R} \times \pi_{S(\beta)}C$ satisfies the relation $D_\beta f(p_\beta, 0, 0)(h, \psi) = -\dot{p}_\beta h - \psi$. Therefore, the map $D_\beta f(p_\beta, 0, 0)$ is an isomorphism on $\mathbb{R} \times \pi_{S(\beta)}C$. The implicit function theorem implies that there are a positive number ν and C^k-functions $\beta(\eta), \phi_\beta^S(\eta)$ with $|\eta| < \nu$, $|\beta(\eta)| < \nu$, $|\phi_\beta^S(\eta)| < \nu$, such that $f(\eta, \beta(\eta), \phi_\beta^S(\eta)) = 0$. Since γ is compact, the claim is proved.

Our next assertion is that $\mathcal{W}_{\text{loc}}^s(\gamma)$ is a C^{k-1}-submanifold of C. Since γ is compact, it is sufficient to show that, for any $\beta_0 < \omega$, the set

$$\mathcal{S}(\beta_0) = \bigcup_{\beta \in [0, \beta_0]} \mathcal{W}_{\beta,\text{loc}}^s(\gamma)$$

is C^{k-1}-diffeomorphic to $[0, \beta_0] \times B_0(\nu)$, where $B_0(\nu) = \{\phi_0^S \in \pi_{S(0)}C : |\phi_0^S| < \nu\}$. From Lemma 1.4 in Section 8.1, we know that there is a C^{k-1}-isomorphism $\mathcal{I}_\beta : \pi_{S(0)}C \to \pi_{S(\beta)}C$. For any $\eta \in \mathcal{W}_{\text{loc}}^s(\gamma)$, there are unique $\beta(\eta) \in [0, \omega)$, $\phi_\eta^S \in \pi_{S(\beta(\eta))}$ such that $\eta = p_{\beta(\eta)} + \phi_{\beta(\eta)}^S$. The mapping $\eta \mapsto (\beta(\eta), \mathcal{I}_{\beta(\eta)}^{-1}\phi_{\beta(\eta)}^S)$ is a C^{k-1}-diffeomorphism.

In the same way, we conclude that $\mathcal{W}_{\text{loc}}^u(\gamma)$ is a C^k-submanifold of C. We obtain C^k rather that C^{k-1} because the isomorphism from Lemma 1.4 of Section 8.1 is C^k. This completes the proof of Theorem 3.3. □

At this time, we have been unable to show the following natural conjecture.

Conjecture. *$W_{\text{loc}}^s(\gamma) = \mathcal{W}_{\text{loc}}^s(\gamma)$ and $W_{\text{loc}}^u(\gamma) = \mathcal{W}_{\text{loc}}^u(\gamma)$.*

It is clear that $W_{\text{loc}}^s(\gamma) \supset \mathcal{W}_{\text{loc}}^s(\gamma)$. However, the proof that $W_{\text{loc}}^s(\gamma) \subset \mathcal{W}_{\text{loc}}^s(\gamma)$ is not so easy. It is not obvious that a solution of (1.1) that approaches γ must satisfy the integral equation (3.13) or, more precisely, the one corresponding to (3.15).

10.4 Nondegenerate periodic orbits of RFDE

Suppose that (1.1) has a periodic orbit γ of period ω. We say that γ is *nondegenerate* if 1 is not a Floquet multiplier of γ. We remark that this is equivalent to saying that the linear variational equation (3.1) has 1 as a simple characteristic multiplier. The concept of nondegeneracy depends on ω. More precisely, if ω_m is the minimal period of γ, then it is possible for γ to be nondegenerate and have other Floquet multipliers that are roots of unity. If this were the case, then the orbit would not be nondegenerate if we choose $\omega = j\omega_m$ for j an appropriate integer.

If Λ is a Banach space and $\widetilde{F} : C \times \Lambda \to \mathbb{R}^n$, $(\phi, \lambda) \mapsto \widetilde{F}(\phi, \lambda)$, is continuous in (ϕ, λ) and continuously differentiable in ϕ, $\widetilde{F}(\phi, 0) = F(\phi)$, we consider the RFDE

$$\dot{x}(t) = \widetilde{F}(x_t, \lambda) \tag{4.1}$$

as a perturbation of the RFDE

$$\dot{x}(t) = F(x_t). \tag{4.2}$$

Our objective is to prove the following result.

Theorem 4.1. *If γ is a nondegenerate periodic of (4.2) of period ω and the function $\widetilde{F}$ satisfies the hypotheses, then there exist a positive constant ν and a neighborhood V of γ such that for each $\lambda \in \Lambda$, $0 \leq |\lambda| < \nu$, Equation (4.1) has a nondegenerate periodic orbit $\gamma_\lambda \in V$ of period ω_λ, γ_λ and ω_λ depend continuously on λ, $\gamma_0 = \gamma$, $\omega_0 = \omega$, and γ_λ is the only periodic orbit in V whose period approaches ω as $\lambda \to 0$.*

Remark 4.1. Under further smoothness properties on the function $\widetilde{F}$, it is possible to obtain that the orbit and period in Theorem 4.1 are smooth in λ. More precisely, if $\widetilde{F}$ is C^k, then these functions are C^k. One can even take $k = \infty$, but it is not known if we can extend the result to analyticity.

Remark 4.2. If $\omega > r$, then we can use the Poincaré map introduced in the proof of Theorem 3.2 together with the implicit function theorem to obtain the conclusion in Theorem 4.1. On the other hand, if $\omega \leq r$, we cannot proceed in this way since this map is not differentiable. If we take some multiple $j\omega$ so that $j\omega > r$, then, as remarked earlier, the orbit γ may not be nondegenerate with respect to the period $j\omega$.

Remark 4.2 suggests that the standard implicit function theorem may not be appropriate to prove Theorem 4.1. We will make use of the following version of the parametric implicit function theorem, which involves only the differentiability with respect to the parameter along the fixed-point set. The proof may be supplied by the reader as an application of the contraction mapping principle.

Lemma 4.1. *Let E be an open subset of a Banach space Y, F be a closed subset of a Banach space X, $\operatorname{int} F \neq \emptyset$, where $\operatorname{int} F$ denotes the interior of F. Assume that $T : F \times E \to F$ satisfies the following hypotheses:*

(i) *$T(x, \cdot) : E \to F$ is continuous.*

(ii) *$T(\cdot, y) : F \to F$ is continuous and, for each $y \in E$, has a unique fixed point $x(y)$ that depends continuously on y.*

(iii) *If $x(E) = F_1 \subset F$, then $T(x, y)$ is continuously differentiable in y for $(x, y) \in F_1 \times E$.*

(iv) *There is an open set* $F_2 \subset X$ *such that* $F \subset F_2$ *and the derivative* $D_xT(x,y)$ *of* $T(x,y)$ *with respect to* x *is continuous and satisfies* $|D_xT(x,y)| \le \delta < 1$ *for all* $(x,y) \in F_2 \times E$.

Then the fixed point $x(y) \in F$, $y \in E$, *of* $T(x,y)$ *is continuously differentiable in* y.

Proof of Theorem 4.1. For any real number $\beta > -1$, if $t = (1+\beta)\tau$ and $x(t) = y(\tau)$, then $x(t+\theta) = y(\tau + \theta/(1+\beta))$, $-r \le \theta \le 0$. If we define $y_{\tau,\beta}(\theta) = y(\tau + \theta/(1+\beta))$, $-r \le \theta \le 0$, then Equation (4.1) becomes

$$\dot{y}(\tau) = (1+\beta)\widetilde{F}(y_{\tau,\beta}, \lambda). \tag{4.3}$$

If there is a periodic solution of Equation (4.1) of period ω, then there is a periodic solution of Equation (4.3) of period $(1+\beta)\omega$, and conversely. Let $\gamma = \{\, p(t) : 0 \le t < \omega \,\}$, where p is a periodic solution of (4.3) of period ω. If $y(\tau) = p(\tau) + z(\tau)$ in (4.3), then

$$\dot{z}(\tau) = L(\tau)z_{\tau,0} + H(\tau, z, \lambda, \beta), \tag{4.4}$$

where

$$L(\tau)\psi = D_\phi F(p_t)\psi, \tag{4.5}$$

and

$$H(\tau, z, \lambda, \beta) = -F(p_{\tau,0}) - L(\tau)z_{\tau,0} + (1+\beta)\widetilde{F}(p_{\tau,\beta} + z_{\tau,\beta}, \tau). \tag{4.6}$$

We remark that $H(\tau,0,0,0) = 0$ and the derivative $D_zH(\tau,z,\lambda,\beta)$ exists and satisfies $D_zH(\tau,0,0,0) = 0$.

Let $C(2r) = C([-2r,0], \mathbb{R}^n)$. For $|\beta| \le 1/2$, we can interpret Equation (4.4) as an RFDE on $C(2r)$. The linear equation

$$\dot{z}(\tau) = L(\tau)z_{\tau,0} \tag{4.7}$$

has one as a simple characteristic multiplier corresponding to the periodic solution $\dot{p}$. Therefore, we may decompose $C(2r)$ relative to the multiplier 1 as $C(2r) = E \oplus K$, where $E = E_1(0)$ and $K = K_1(0)$ (see Chapter 8). Suppose that $P : C(2r) \to C(2r)$ is the projection induced by this decomposition and P takes $C(2r)$ onto K.

Let Ω_0 be the set of continuous ω-periodic functions in $\mathbb{R}^n$ with $|z|_0 = \sup_t |z(t)|$ for $z \in \Omega_0$. Corollary 4.1 of Section 6.4 implies that the nonhomogeneous linear equation

$$\dot{z}(\tau) = L(\tau)z_{\tau,0} + h(\tau), \quad h \in \Omega_0,$$

has a solution in Ω_0 if and only if

$$\int_0^\omega q(\tau)h(\tau)\,d\tau = 0$$

where $q(\tau)$ is a nonzero ω-periodic solution of the equation adjoint to (4.7). We may assume that $\int_0^\omega q(\tau)q^T(\tau)\,d\tau = 1$. For any $h \in \Omega_0$, let $\Gamma(h) = \int_0^\omega q(\tau)h(\tau)\,d\tau$. Then $\Gamma : \Omega_0 \to \mathbb{R}$ is a continuous linear mapping.

For any $h \in \Omega_0$, the equation

$$\dot{z}(\tau) = L(\tau)z_{\tau,0} + h(\tau) - \Gamma(h)q'(\tau)$$

has a solution in Ω_0 and it has a unique solution whose $(I-P)$-projection is zero. If we designate this solution by $\mathcal{K}h$, the $\mathcal{K}$ is a continuous linear operator taking Ω_0 into Ω_0.

For any λ, β, and $u \in \Omega_0$, consider the equation

$$R(u,\lambda,\beta) \overset{\text{def}}{=} u - \mathcal{K}[H(\,\cdot\,,u,\lambda,\beta) - \Gamma(H(\,\cdot\,,u,\lambda,\beta))q^T] = 0. \tag{4.8}$$

Using the fact that $H(\tau,0,0,0) = 0$, $D_zH(\tau,0,0,0) = 0$, we deduce from the implicit function theorem that there are positive constants ν_0, β_0, δ_0, such that Equation (4.8) has a unique solution $u^*(\lambda,\beta) \in \Omega_0$, $|u^*(\lambda,\beta)| \le \delta_0$, $|\lambda| \le \nu_0$, $|\beta| \le \beta_0$, $u^*(\lambda,\beta)$ is continuous in λ, β, $u^*(0,0) = 0$ and $u^*(\lambda,\beta)$ satisfies the equation

$$\dot{z}(\tau) = L(\tau)z_{\tau,0} + H(\tau,z,\lambda,\beta) - B(\lambda,\beta)q^T(\tau) \tag{4.9}$$

where we have put

$$B(\lambda,\beta) = \int_0^\omega q(\tau)H(\tau,u^*(\lambda,\beta),\lambda,\beta)\,d\tau. \tag{4.10}$$

In particular, $u^*(\lambda,\beta)(\tau)$ is continuously differentiable in τ.

If u is a C^1-function in Ω_0, then the function $H(\,\cdot\,,u,\lambda,\beta)$ is continuously differentiable in β. Thus, Lemma 4.2 implies that $u^*(\lambda,\beta)$ has a continuous first derivative with respect to β. This implies that the function $B(\lambda,\beta)$ is continuously differentiable in β. To compute this derivative, we observe that the function $\partial u^*(\lambda,\beta)/\partial\beta$ is a solution of the equation

$$\dot{v}(\tau) = L(\tau)v_{\tau,0} + L_1(\tau,v,\lambda,\beta) - \frac{\partial B(\lambda,\beta)}{\partial\beta}q^T(\tau) \tag{4.11}$$

where

$$\begin{aligned} L_1(\tau,v,\lambda,\beta) = {} & (1+\beta)D_\phi\widetilde{F}(p_{\tau,\beta} + u^*_{\tau,\beta},\lambda)v_{\tau,\beta} - D_\phi F(p_{\tau,0})v_{\tau,0} \\ & + \widetilde{F}(p_{\tau,\beta} + u^*_{\tau,\beta},\lambda) \\ & - \frac{1}{1+\beta}D_\phi\widetilde{F}(p_{\tau,\beta} + u^*_{\tau,\beta},\lambda)((\,\cdot\,)\dot{p}_{\tau,\beta} + (\,\cdot\,)\dot{u}^*_{\tau,\beta}) \end{aligned}$$

where $((\,\cdot\,)\dot{p}_{\tau,\beta})(\theta) = \theta\dot{p}(\tau+\theta/(1+\beta))$ and $((\,\cdot\,)\dot{u}^*_{\tau,\beta})(\theta) = \theta\dot{u}^*(\tau+\theta/(1+\beta))$. From (4.6) and (4.9), we see that $\dot{u}^*_{\tau,\beta} \to 0$ as $\lambda,\beta \to 0$. From this fact and the relation $\dot{p}(t) = F(p_t)$, it follows that $L_1(t,v,0,0) = J(t,\dot{p})$, where

$$J(t,\dot{p}) = \dot{p}(t) - L(t)(\,\cdot\,)\dot{p}_t. \tag{4.12}$$

Since Equation (4.11) has the ω-periodic solution $\partial u^*(\lambda,\beta)/\partial\beta$, we must have

$$\frac{\partial B(\lambda,\beta)}{\partial\beta} = \int_0^\omega q(\tau)L_1(\tau,v,\lambda,\beta)\,d\tau$$

and so

$$\frac{\partial B(0,0)}{\partial\beta} = \int_0^\omega q(\tau)J(\tau,\dot p)\,d\tau \tag{4.13}$$

where J is given in (4.12).

We show now that Expression (4.13) is not zero. If Expression (4.13) is zero, then there is a nontrivial ω-periodic solution z of the equation

$$\dot z(t) = L(t)z_t + J(t,\dot p).$$

If $x(t) = z(t) - t\dot p(t)$, then a few elementary computations imply that x is a solution of (4.7) with z and $\dot p$ being ω-periodic functions. Since $\dot p$ also is a solution of (4.7), we deduce that the characteristic multiplier one is not simple. This is a contradiction to the assumption that the orbit γ is elementary and proves that the Expression (4.13) is not zero.

Since $B(0,0) = 0$ and $\partial B(0,0)/\partial\beta \neq 0$, the implicit function theorem implies the existence of positive constants $\beta_1 \le \beta_0$, $\nu_1 \le \nu_0$, and a continuous function $\beta(\lambda)$, $|\lambda| \le \nu_1$, $|\beta(\lambda)| \le \beta_1$, so that $\beta(0) = 0$ and $B(\lambda,\beta(\lambda)) = 0$. Since $u^*(\lambda,\beta)$ is a solution of Equation (4.9), it follows that $u^*(\lambda,\beta(\lambda))$ is an ω-periodic solution of Equation (4.4). This proves the existence of a periodic solution $y^*(\lambda)$ of (4.3) and thus a solution $x^*(\lambda)$ of Equation (4.1) of period $\omega(\lambda) = (1+\beta(\lambda))^{-1}\omega$, which is continuous in λ for $0 \le |\lambda| \le \nu_1$, $x^*(0) = p$. The linear variational equation associated with this periodic solution $x^*(\lambda)$ is a continuous function of λ and, therefore, the multiplier one will have a generalized eigenspace of dimension one for $0 \le |\lambda| \le \nu_2 \le \nu_1$.

Since each periodic orbit of (4.1) near γ with a period close to ω can be obtained in this manner, we have completed the proof of Theorem 4.1. □

Corollary 4.1. *If γ is a nondegenerate periodic orbit of the* RFDE (4.2), *then there are a neighborhood V of γ and a positive number $\delta > 0$ such that* (4.2) *contains no periodic orbits in $V \setminus \{\gamma\}$ of period $\tilde\omega$ satisfying $|\tilde\omega - \omega| < \delta$. In particular, $V \setminus \{\gamma\}$ contains no ω-periodic orbits.*

Remark 4.3. It should be possible to extend this result to NFDE with a stable D operator. It will be necessary to use the method of Liapunov-Schmidt to obtain the analogue of (4.9) and then to show that the periodic functions that satisfy that relation are C^1-functions of t. The method of Hale and Scheurle [1] could perhaps be used to prove this fact.

10.5 Supplementary remarks

For RFDE, the original formulation of Theorem 1.1 for Lipschitz perturbations of a linear vector field is due to Hale and Perelló [1]. Lemma 1.2 is due to Henry [6]. In ordinary differential equations, we have the classical result of Hartman and Grobman that the flow near a hyperbolic equilibrium point is topologically equivalent to its linearization. Sternberg [2,3] has given the appropriate extension of this theorem for the situation where the RFDE has a global attractor. The topological equivalence is relative to the solutions on the attractor.

Chafee [1, 2] discussed the existence of center manifolds of an equilibrium point of RFDE in connection with the bifurcation of periodic orbits from an equilibrium point for equations containing a small parameter. Kurzweil [1, 2] and Fodčuk [1, 2], have also considered this general problem. Diekmann and van Gils [1] also have proved the existence of center manifolds for RFDE by using the sun-star theory of dual semigroups and the variation-of-constants formula. Ruiz-Claeyssen [1] has discussed the case in which the initial data are chosen from the space W_1^∞ of absolutely continuous functions with essentially bounded derivatives in order to study the effects of delays on the behavior near equilibrium. Lima [1] and Hale [24] have considered the same problem in the space C and more general fading memory spaces by making use of the fixed-point theorem in Lemma 4.1.

The center manifold theorem allows one to prove the following result:

For any FDE of the form (1.1) *with* p *eigenvalues of Equation* (1.2) *on the imaginary axis, there is an ordinary differential equation* $\dot{u} = h(u)$, $u \in \mathbb{R}^p$, *defined in a neighborhood of zero in* $\mathbb{R}^p$ *such that* $h(0) = 0$ *and the stability properties of the solution* $u = 0$ *of this equation are the same as the stability properties of the solution* $x = 0$ *of Equation* (1.1)

(see Hale [8]). Special cases of this general result were used earlier by Prokopev and Shimanov [1], Hale [9] and Hausrath [1] to obtain sufficient conditions for the stability of the solution of Equation (1.1) when no roots of (1.2) have positive real parts. In these latter papers, it was necessary to extend some of the classical transformation theory of Liapunov to equations in infinite-dimensional spaces. The remarks of Hausrath [1] were later used by Henry [5] and Carr [1], for parabolic equations as well as ODE. Further extension of this transformation theory was used by Chow and Mallet-Paret [1] to understand the the Hopf bifurcation for RFDE. See also the remarks in Section 12.10.

Theorem 3.1 is due to Nussbaum [1] for RFDE and to Hale and Scheurle [1] for NFDE. Hale [10] was the first to formulate a version of Theorem 3.2. The proof in the text is based on Hale and Lin [2]. The proof of Theorem 3.3 is based on ideas from Hale [10].

There are several approaches that could possibly be used to verify the conjecture in Section 10.3 that the local stable and unstable sets of a periodic orbit coincide with the synchronized local and unstable sets of the orbit. The functions H^s and H^u can be used as part of a coordinate system in a neighborhood of γ. In fact, using the implicit function theorem and the compactness of γ, we can deduce that there are positive constants μ, $\nu > 0$ such that for any $\eta \in C$ with $\operatorname{dist}(\eta, C) < \mu$, there are unique $\beta \in [0, \omega)$, $\phi_\beta^S \in \pi_{S(\beta)}C$, $\phi_\beta^U \in \pi_{U(\beta)}C$, such that $|\phi_\beta^S| < \nu$, $|\phi_\beta^U| < \nu$, and

$$\eta = p_\beta + H^s(\phi_\beta^S, \beta) + H^u(\phi_\beta^U, \beta).$$

It is reasonable to expect that one could prove the conjecture with this coordinate system and appropriate modifications of the arguments used by Hale and Raugel [2] in their study of convergence to equilibrium of gradient-like systems.

Another approach would be to develop a more extensive theory of invariant manifolds (center stable, center unstable, and the theory of foliations near the periodic orbit) or even more generally near an invariant set. There seem to be some difficulties in using the approach in Hirsch, Pugh, and Shub [1] for maps adapted for flows. If we attempt to use something similar to the approach in the text by considering the variation of solutions from a particular solution on the invariant manifold, then we encounter nonautonomous equations for which we can construct the synchronized manifolds. The same difficulty as in the conjecture occurs. On the other hand, if the nonautonomous equation can be considered as a skew product flow, then it is feasible that invariant manifolds in the skew product flow can be used to determine the invariant manifolds in the original space. If this is the case, then the conjecture could be proved. For finite-dimensional problems, Chow and Yi are investigating this problem by this approach. Success in the finite dimensional case will certainly lead to some results in infinite dimensions.

Theorem 4.1 was given by Hale [10] and generalizes an earlier result of Halanay [1]. Stokes [3] has given a different proof of this result based on a general theory of when approximate periodic solutions of an equation imply the existence of an exact periodic solution. His procedure also is effective numerically.

Stokes [1,4,5] has studied the orbital stability of periodic orbits in a general setting. Stokes [6] has introduced a local coordinate system around a periodic orbit γ of a RFDE and has done this in such a way as to be able to determine the many properties of the solutions in a neighborhood of γ. This result has implications for nonautonomous perturbations of the vector field. The coordinate system is not a natural generalization of the finite dimensional case in the sense that the "angle" coordinate appears with retarded terms. At the present time, we are investigating the possibility of obtaining a better coordinate system that will permit an easier transcription of results from the finite-dimensional setting to RFDE and NFDE.

11
Periodic solutions of autonomous equations

The purpose of this chapter is to give a procedure for determining periodic solutions of some classes of autonomous RFDE.

For equations that are close to linear, the analysis in Chapter 10 can be effectively applied using the period and amplitude as undetermined parameters chosen in such a way as to satisfy the bifurcation equations. In some applications, it is desirable to determine other parameters so that these equations are satisfied. Such a situation arises from the case of a Hopf bifurcation from a constant solution to a nonconstant periodic solution as some parameter varies. This is discussed in Section 11.1

Sections 11.2 and 11.3 are devoted to general fixed-point theorems that apply to equations that are not necessarily close to linear. Three types of equations are used as illustrations of these general results.

11.1 Hopf bifurcation

In this section, we discuss one of the simplest ways in which nonconstant periodic solutions of autonomous equations can arise—the so-called Hopf bifurcation. More specifically, we consider a one-parameter family of RFDE of the form

$$\dot{x}(t) = F(\alpha, x_t) \tag{1.1}$$

where $F(\alpha, \phi)$ has continuous first and second derivatives in α, ϕ for $\alpha \in \mathbb{R}$, $\phi \in C$, and $F(\alpha, 0) = 0$ for all α. Define $L : \mathbb{R} \times C \to \mathbb{R}^n$ by

$$L(\alpha)\psi = D_\phi F(\alpha, 0)\psi \tag{1.2}$$

where $D_\phi F(\alpha, 0)$ is the derivative of $F(\alpha, \psi)$ with respect to ϕ at $\phi = 0$ and define

$$f(\alpha, \phi) = F(\alpha, \phi) - L(\alpha)\phi. \tag{1.3}$$

Additional hypotheses also will be imposed.

(H1) The linear RFDE($L(0)$) has a simple purely imaginary characteristic root $\lambda_0 = i\nu_0 \neq 0$ and all characteristic roots $\lambda_j \neq \lambda_0, \bar{\lambda}_0$, satisfy $\lambda_j \neq m\lambda_0$ for any integer m.

Since $L(\alpha)$ is continuously differentiable in α, Lemma 10.1 of Section 7.10 implies that there is an $\alpha_0 > 0$ and a simple characteristic root $\lambda(\alpha)$ of the linear RFDE($L(\alpha)$) that has a continuous derivative $\lambda'(\alpha)$ in α for $|\alpha| < \alpha_0$. We suppose

(H2) Re $\lambda'(0) \neq 0$.

We will show that Hypotheses (H1) and (H2) imply there are nonconstant periodic solutions of Equation (1.1) for α small that have period close to $2\pi/v_0$. Before stating the result precisely, we introduce some notation that will be needed in the proof. The additional notation will also make the statement of the result more specific.

By taking α_0 sufficiently small, we may assume Im $\lambda(\alpha) \neq 0$ for $|\alpha| < \alpha_0$ and obtain a function $\phi_\alpha \in C$ that is continuously differentiable in α and is a basis for the solutions of the RFDE($L(\alpha)$) corresponding to $\lambda(\alpha)$. The functions

$$(\mathrm{Re}\,\phi_\alpha, \mathrm{Im}\,\phi_a) \stackrel{\mathrm{def}}{=} \Phi_a$$

form a corresponding basis for the characteristic roots $\lambda(\alpha), \bar{\lambda}(\alpha)$. Similarly, we obtain a basis Ψ_α for the adjoint equation with $(\Psi_\alpha, \Phi_\alpha) = I$. If we decompose C by $(\lambda(\alpha), \bar{\lambda}(\alpha))$ as $C = P_\alpha \oplus Q_\alpha$, then Φ_α is a basis for P_α. We know that

$$\Phi_\alpha(\theta) = \Phi_a(0)e^{B(\alpha)\theta}, \quad -r \le \theta \le 0, \tag{1.4}$$

and the eigenvalues of the 2×2 matrix $B(\alpha)$ are $\lambda(\alpha)$ and $\bar{\lambda}(\alpha)$. By a change of coordinates and perhaps redefining the parameter α, we may assume that

$$\begin{gathered} B(\alpha) = v_0 B_0 + \alpha B_1(\alpha) \\ B_0 = \begin{bmatrix} 0 & 1 \\ -1 & 0 \end{bmatrix}, \quad B_1(\alpha) = \begin{bmatrix} 1 & \gamma(\alpha) \\ -\gamma(\alpha) & 1 \end{bmatrix} \end{gathered} \tag{1.5}$$

where $\gamma(\alpha)$ is continuously differentiable on $0 \le |\alpha| < \alpha_0$.

We can now state the Hopf bifurcation theorem; we refer to the conclusions stated in this theorem as a *Hopf bifurcation.*

Theorem 1.1. *Suppose $F(\alpha, \phi)$ has continuous first and second derivatives with respect to $\alpha, \phi, F(\alpha, 0) = 0$ for all α, and Hypotheses (H1) and (H2) are satisfied. Then there are constants $a_0 > 0$, $\alpha_0 > 0$, $\delta_0 > 0$, functions $\alpha(a) \in \mathbb{R}$, $\omega(a) \in \mathbb{R}$, and an $\omega(a)$-periodic function $x^*(a)$, with all functions being continuously differentiable in a for $|a| < a_0$, such that $x^*(a)$ is a solution of Equation* (1.1) *with*

$$x_0^*(a)^{P_\alpha} = \Phi_{\alpha(a)} y^*(a), \qquad x_0^*(a)^{Q_\alpha} = z_0^*(a), \tag{1.6}$$

where $y^*(a) = (a,0)^T + o(|a|)$, $z_0^*(a) = o(|a|)$ *as* $|a| \to 0$. *Furthermore, for* $|\alpha| < \alpha_0$, $|\omega - (2\pi/\nu_0)| < \delta_0$, *every* ω*-periodic solution of Equation* (1.1) *with* $|x_t| < \delta_0$ *must be of this type except for a translation in phase.*

Proof. We prove this result by applying a classical procedure in ordinary differential equations. Let $\beta \in [-1,1]$, $\omega_0 = 2\pi/\nu_0$, $t = (1+\beta)\tau$, $x(t+\theta) = u(\tau + \theta/(1+\beta))$, $-r \le \theta \le 0$, and define $u_{\tau,\beta}$ as an element of space $C([-r,0],\mathbb{R}^n)$ given by $u_{\tau,\beta}(\theta) = u(\tau + \theta/(1+\beta))$, $-r \le \theta \le 0$. Equation (1.1) is then equivalent to the equation

$$\frac{du(\tau)}{d\tau} = (1+\beta)F(\alpha, u_{\tau,\beta}). \tag{1.7}$$

If this equation has an ω_0-periodic solution, then Equation (1.1) has a $(1+\beta)\omega_0$-periodic solution, and conversely.

Let us rewrite Equation (1.7) as

$$\begin{aligned} \frac{du(\tau)}{d\tau} &= L(0)u_\tau + N(\beta,\alpha,u_\tau,u_{\tau,\beta}) \\ N(\beta,\alpha,u_\tau,u_{\tau,\beta}) &= (1+\beta)L(\alpha)u_{\tau,\beta} - L(0)u_\tau + (1+\beta)f(\alpha,u_{\tau,\beta}) \end{aligned} \tag{1.8}$$

This means that we are going to consider Equation (1.8) as a perturbation of the autonomous linear equation

$$\frac{du(\tau)}{d\tau} = L(0)u_\tau. \tag{1.9}$$

We know that the columns of $U(\tau) \stackrel{\text{def}}{=} \Phi_0(0)\exp(B(0)\tau)$, $\tau \in \mathbb{R}$, form a basis for the ω_0-periodic solutions of Equation (1.9) and the rows of $V(\tau) \stackrel{\text{def}}{=} \exp(-B(0)\tau)\Psi_0(0)$, $\tau \in \mathbb{R}$ form a basis for the ω_0-periodic solutions of the formal adjoint equation of Equation (1.9).

We may now apply Corollary 4.1 of Section 6.4 to obtain necessary and sufficient conditions for the existence of ω_0-periodic solutions of Equation (1.8). In fact, a direct application of that corollary shows that every ω_0-periodic solution of Equation (1.8), except a translation in phase, is a solution of the equations

$$u(\tau) = U(\tau)(a,0)^T + \mathcal{K}(I-J)N(\beta,\alpha,u_{\cdot},u_{\cdot,\beta}) \tag{1.10a}$$

$$JN(\beta,\alpha,u_{\cdot},u_{\cdot,\beta}) = 0, \tag{1.10b}$$

and conversely. The operators $\mathcal{K}, J$ are defined in Corollary 4.1 of Section 6.4.

One can now apply the implicit function theorem to solve Equation (1.10a) for $u = u^*(a,\beta,a)$ for a,β,α in a sufficiently small neighborhood of zero, $u^*(a,0,0) - U(\cdot)(a,0)^T = o(|a|)$ as $|a| \to 0$. The function $u^*(a,\beta,\alpha)$ is continuously differentiable in a,α from the implicit function theorem. Since $u^*(a,\beta,\alpha)(t)$ satisfies Equation (1.10a), it also satisfies a differential integral equation and is, therefore, continuously differentiable in t. From Lemma 4.1 of Section 10.4, it follows that $u^*(a,\beta,\alpha)$ is also continuously differentiable in β.

Therefore, all ω_0-periodic solutions of Equation (1.8) are obtained by finding the solutions a,β,α of the bifurcation equations

$$JN(\beta,\alpha,u^*_\cdot(a,\beta,\alpha),u^*_{\cdot,\beta}(a,\beta,\alpha))=0. \tag{1.11}$$

Using the definition of J in Equation (4.12) of Section 6.4, Equation (1.11) is equivalent to the equation

$$G(a,\beta,\alpha)=0$$

where

$$G(a,\beta,\alpha)\stackrel{\text{def}}{=}\int_0^{\omega_0} e^{-B(0)s}\Psi_0(0)N(\beta,\alpha,u^*_s(a,\beta,\alpha),u^*_{s,\beta}(a,\alpha,\beta))\,ds. \tag{1.12}$$

From the preceding discussion, it follows that it remains to solve the equation $G(a,\beta,\alpha)=0$. Since $G(0,\beta,\alpha)=0$ for all β and α, let $H(a,\beta,\alpha)=G(a,\beta,\alpha)/a$. Noting properties of $u^*(a,\beta,\alpha)$ and the definition of $G(a,\beta,\alpha)$ in Expression (1.12), one easily observes that

$$H(0,0,\alpha)=\int_0^{\omega_0} e^{-B(0)s}\Psi_0(0)\{L(\alpha)U_se_1-L(0)U_se_1\}ds$$

where $e_1=(1,0)^T$. One may now apply Lemma 10.3 of Section 7.10 directly to obtain

$$\frac{\partial H(0,0,\alpha)}{\partial\alpha}=\omega_0\begin{bmatrix}1\\-\gamma(0)\end{bmatrix}.$$

Furthermore,

$$H(0,\beta,0)=\int_0^{\omega_0} e^{-B(0)s}\Psi_0(0)\{(1+\beta)L(0)U_{s,\beta}e_1-L(0)U_se_1\}\,ds.$$

Writing this as two separate integrals, changing s into $s/(1+\beta)$ in the first integral, and noting that

$$\frac{dU(s/(1+\beta))}{d(s/(1+\beta))}=L(0)U_{s/(1+\beta),\beta},$$

one sees that

$$H(0,\beta,0)=\beta\int_0^{\omega_0} e^{-B(0)s}\Psi_0(0)\Phi_0(0)e^{B(0)s}B(0)e_1ds.$$

If x is a solution of Equation (1.9) and y is a solution of the adjoint equation, then $(y^t,x_t)=$ constant for all t. Therefore,

$$
\begin{aligned}
I &= \left(e^{-B(0)(s+\cdot)}\Psi_0(0), \Phi_0(0)e^{B(0)(s+\cdot)}\right) \\
&= e^{-B(0)s}\Psi_0(0)\Phi_0(0)e^{B(0)s} \\
&\quad - \int_{-r}^{0}\int_{0}^{\theta} e^{-B(0)(s+\xi-\theta)}\Psi_0(0)\, d[\eta(\theta)]\Phi_0(0)e^{B(0)(s+\xi)}\, d\xi
\end{aligned}
$$

for all $s \in \mathbb{R}$. Integrating this from 0 to ω_0 and using the fact that the second integral is zero, one obtains

$$H(0, \beta, 0) = \beta\omega_0 B(0)e_1.$$

Combining all of this information, one has

$$H(0, 0, 0) = 0,$$

$$\frac{\partial H}{\partial(\beta, \alpha)}(0, 0, 0) = \omega_0 \begin{bmatrix} 0 & 1 \\ -\nu_0 & -\gamma(0) \end{bmatrix}.$$

Therefore, the implicit function theorem implies the existence of $\beta(a)$ and $\alpha(a)$ such that $\beta(0) = 0$, $\alpha(0) = 0$, and $H(a, \beta(a), \alpha(0)) = 0$ and the solution is unique in a neighborhood of zero. The fact that the corresponding ω-periodic function $x((1 + \beta)t) = u(\tau)$ satisfies Equation (1.1) and the properties stated in the theorem are obvious. □

11.2 A periodicity theorem

Our objective in this section is to give a general fixed-point theorem that has been very useful in obtaining periodic solutions of autonomous functional differential equations that are not necessarily perturbations of linear systems as Section 11.1. Proofs are given only for those properties that relate directly to the functional differential equations.

Definition 2.1. Suppose X is a Banach space, U is a subset of X, and x is a given point in U. Given a map $A : U\backslash\{x\} \to X$, the point $x \in U$ is said to be an *ejective point of* A if there is an open neighborhood $G \subseteq X$ of x such that for every $y \in G \cap U$, $y \neq x$, there is an integer $m = m(y)$ such that $A^m y \notin G \cap U$.

For any $M > 0$, we let $S_M = \{x \in X : |x| = M\}$, and $B_M = \{x \in X : |x| < M\}$. Then $S_M = \partial B_M$.

For references to the following two theorems, which are stated without proof, see Section 11.7.

Theorem 2.1. *If K is a closed, bounded, convex, infinite-dimensional set in $X, A : K\backslash\{x_0\} \to K$ is completely continuous, and $x_0 \in K$ is an ejective point of A, then there is a fixed point of A in $K\backslash\{x_0\}$. If K is finite dimensional and x_0 is an extreme point of K, then the same conclusion holds.*

Theorem 2.2. *If K is a closed convex set in $X, A : K\backslash\{0\} \to K$ is completely continuous, $0 \in K$ is an ejective point of A, and there is an $M > 0$ such that $Ax = \lambda x$, $x \in K \cap S_M$ implies $\lambda < 1$, then A has a fixed point in $K \cap B_M\backslash\{0\}$ if either K is infinite dimensional or 0 is an extreme point of K.*

Remark 2.1. Theorems 2.1 and 2.2 remain valid for mappings A that are α-contractions. This generalization should play a role in studying periodic solutions of equations with a period smaller than twice the delay, equations with infinite delays, and certain neutral functional differential equations. This area has not been exploited very much at the present time.

In the application of Theorems 2.1 and 2.2 to retarded functional differential equations, the mapping A is usually similar to the "transversal" map of Poincaré in ordinary differential equations. In fact, one obtains a set $K \subseteq C$ such that every solution $x(\phi)$, $\phi \in K$, of the RFDE(f), $f : C \to \mathbb{R}^n$, returns to K in some time $\tau(\phi) > 0$; that is, $x_{\tau(\phi)}(\phi) \in K$ if $\phi \in K$. The mapping $A : K \to K$ is then defined by $A\phi = x_{\tau(\phi)}(\phi)$. If A were completely continuous and K were closed, convex, and bounded, then there would be a $\phi \in K$ such that $A\phi = \phi$, and thus, a periodic solution of the RFDE(f). It looks as if the problem is over, but it is not. We wish to obtain nonconstant periodic solutions of the RFDE(f), and, if there is a constant $a \in \mathbb{R}^n$ such that the constant function $\tilde{a} \in C$ defined by $\tilde{a}(\theta) = a$, $-r \leq \theta \leq 0$, satisfies $\tilde{a} \in K$, $f(\tilde{a}) = 0$, then the only fixed point of A in K could be $\tilde{a}$. If K contains no such constant functions, then the problem of the existence of nonconstant periodic solutions is solved. Unfortunately, in the applications, the construction of such a K is very difficult and, often, the set K contains only one constant solution x_0, and it is an extreme point. The theorems assert that if x_0 is ejective, then there is a nonconstant periodic solution of the RFDE(f).

From the preceding discussion, it is clear that an efficient method is needed for determining when a constant solution of an RFDE(f) is ejective relative to some set K and the mapping A defined earlier. Such a result will now be given.

Suppose $L : C \to \mathbb{R}^n$ is linear and continuous, $f : C \to \mathbb{R}^n$ is completely continuous together with a continuous derivative f' and $f(0) = 0$, $f'(0) = 0$. Consider the equations

$$\dot{x}(t) = Lx_t + f(x_t) \tag{2.1}$$

$$\dot{y}(t) = Ly_t. \tag{2.2}$$

For any characteristic root λ of Equation (2.2), there is a decomposition of C as $C = P_\lambda \oplus Q_\lambda$, where P_λ and Q_λ are invariant under the solution operator $T_L(t)$ of Equation (2.2), $T_L(t)\phi = y_t(\phi)$, $\phi \in C$. Let the projection operators defined by the decomposition of C be π_λ, $I - \pi_\lambda$ with the range of π_λ equal to P_λ.

Theorem 2.3. *Suppose the following conditions are fulfilled:*

(i) *There is characteristic root λ of Equation (2.2) satisfying $\operatorname{Re}\lambda > 0$.*

(ii) *There is a closed convex set $K \subseteq C$, $0 \in K$, and $\delta > 0$, such that*

$$v = v(\delta) \overset{\text{def}}{=} \inf\{|\pi_\lambda \phi| : \phi \in K,\ |\phi| = \delta\} > 0.$$

(iii) *There is a completely continuous function $\tau : K\backslash\{0\} \to [\alpha, \infty)$, $0 \leq \alpha$ such that the map defined by*

$$A\phi = x_{\tau(\phi)}(\phi), \qquad \phi \in K\backslash\{0\} \tag{2.3}$$

takes $K\backslash\{0\}$ into K and is completely continuous.

Then 0 is an ejective point of A.

Proof. If $\Phi_\lambda = (\Phi_{1\lambda}, \ldots, \Phi)_{d\lambda})$ is a basis for P_λ and $\pi_\lambda \Phi = \Phi b$, then $b = b(\phi) \in \mathbb{R}^d$ is a continuous d-vector linear functional on $\mathbb{R}^d$ and we take the norm of b to be the Euclidean norm. For any continuous $V : C \to \mathbb{R}$, let

$$\dot{V}(\phi) = \liminf_{t \to 0^+} \frac{1}{t}\,[V(x_t(\phi)) - V(\phi)].$$

Exactly as in the proof of Lemma 1.3 of Section 10.1, one can prove the following

Lemma 2.1. *There is a positive definition quadratic form $V(\phi) = b^T Bb$ with the property that for any $p > 0$, there is a $\delta_0 > 0$ such that for any δ, $0 < \delta < \delta_0$, $\dot{V}(\phi) > 0$ if $V(\phi) \geq p^2\delta^2$, $\phi \in \bar{B}_\delta$.*

With the function $V(\phi)$ in Lemma 2.1, observe that Condition (ii) implies that

$$v^2 \overset{\text{def}}{=} \inf\{V(\phi/|\phi|) : \phi \in K, |\phi| \neq 0\} > 0$$

since there are $\alpha > 0$, $\beta > 0$, such that $|\pi_\lambda \phi|^2 \leq \alpha |b|^2 \leq \alpha\beta^{-2}V(\phi)$.

Fix $p < v$ and choose δ_0 as in Lemma 2.1. Lemma 2.1 implies $V(x_t(\phi))$ is increasing in t as long as $V(x_t(\phi)) \geq p^2\delta^2$ and $|x_t(\phi)| \leq \delta$, $0 < \delta < \delta_0$. If

$$G(\delta) = \{\phi \in C : |\phi| < \delta, V(\phi) < p^2\delta^2\},$$

then $G(\delta)$ is an open set, $0 \in G$.

Since τ is completely continuous, there is a $k > 0$ such that $\tau(\phi) \leq k$ for all $\phi \in G(\delta_0) \cap K$. Choose $0 < \delta_1 < \delta_0$ sufficiently small so that the solution $x(\phi)$ through $\phi \in \mathrm{Cl}\, G(\delta_1)$ satisfies $x_t(\phi) \in G(\delta_0)$ for $0 \leq t \leq k$. For any $\phi \in G(\delta_0) \cap K$, $|\phi| = \epsilon$, then the fact that $p < v$ implies

$$V(\phi) = |\phi|^2 V(\phi/|\phi|) = \epsilon^2 V(\phi/|\phi|) \geq \epsilon^2 v^2 > \epsilon^2 p^2.$$

Consequently, Lemma 2.1 implies $V(\phi)$ is increasing along the solutions of Equation (2.1) for $|\phi| < \delta_0$, $\phi \in K\backslash\{0\}$. Therefore, for any $\phi \in G(\delta_1) \cap K$, $\phi \neq 0$, there are $t_1 > t_2 > 0$ such that $x_t(\phi) \in G(\delta_1)$, $0 \leq t < t_2$, $x_t(\phi) \in G(\delta_0)\backslash G(\delta_1)$, $t_2 \leq t < t_1$ and $x_{t_1}(\phi) \in \partial G(\delta_0)$. Let $n_0(\phi)$ be the least integer such that $A^k(\phi) \in \{x_t(\phi), 0 \leq t \leq t_2\}$, $k = 0, 1, \ldots, n_0(\phi)$. Since $t_1 - t_2 \geq k$ by the choice of δ_1, it follows that $A^{n_0(\phi)+1}(\phi) \notin G(\delta_1) \cap K$. This proves 0 is an ejective point of A. The proof of the theorem is complete. □

We are now in a position to state a useful tool for obtaining periodic solutions of an RFDE by combining Theorems 2.1, 2.2, and 2.3.

Theorem 2.4. *Suppose K is a closed convex set in C, $0 \in K$, such that Conditions* (i)–(iii) *of Theorem* 2.3 *are satisfied. If either of the conditions*

(iv′) *K is bounded and infinite-dimensional,*

(iv″) *K is bounded, finite-dimensional, and 0 is an extreme point of K,*

(iv‴) *there is an $M > 0$ such that $Ax = \lambda x$, $x \in K \cap S_M$ implies $\lambda < 1$;*

hold, then there is a nonzero periodic solution of Equation (2.1) *with initial value in $K\backslash\{0\}$.*

11.3 Range of the period

In this section, we return to the discussion of the one-parameter family of RFDE considered in Section 11.1. More specifically, suppose the hypotheses of Theorem 1.1 are satisfied so that a Hopf bifurcation occurs. We will show how to determine some information about the range of the periods of Equation (1.1) as α is varied by combining the Hopf bifurcation with some results on fixed points of mappings that depend on a parameter α.

Definition 3.1. Suppose X is a Banach space, $x_0 \in X$ is given, $U \subseteq X$ is an open neighborhood of x_0, and $A : U \to X$ is a continuous map. The point x_0 is said to be *a uniform attractive point for A* if there is neighborhood U_0 of x_0, $U_0 \subseteq U$, such that for any neighborhood V of x_0, there is an integer $m(V)$ such that $A^j(U_0) \subseteq U$ for $j \geq 0$, $A^j(U_0) \subseteq V$, $j \geq m(V)$.

For references to the following theorem, which is stated without proof, see Section 11.7.

Theorem 3.1. *Suppose K is a closed convex set of a Banach space X, 0 is an extreme point of K, $\{0\} \neq K$, and $J = (a, \infty)$, $-\infty \leq a < \infty$. Suppose $A : K \times J \to K$ satisfies the following hypotheses:*

(i) $A(0, \alpha) = 0$ *for* $\alpha \in J$.

(ii) A *is completely continuous.*

(iii) *There is an* $\alpha_0 \in J$ *such that for any compact interval* $J_0 \subset J$, $\alpha_0 \notin J_0$, *there is an* $\epsilon = \epsilon(J_0) > 0$ *such that* $A(x, \alpha) \neq x$ *for* $\alpha \in J_0$, $0 < |x| \leq \epsilon$.

(iv) *There is an open interval* I_0 *about* α_0 *such that* 0 *is a uniform attractive point of* $A(\cdot, \alpha)$ *for* $\alpha \in I_0$, $\alpha < \alpha_0$, *and* 0 *is an ejective point of* $A(\cdot, \alpha)$ *for* $\alpha \in I_0$, $\alpha > \alpha_0$, *or conversely.*

(v) *There is an* $\alpha_1 \in J$ *such that* $A(x, \alpha) = x$ *for* $x \neq 0$ *implies* $\alpha \geq \alpha_1$.

If the hypotheses are satisfied and $S \subseteq K \times J$ *is defined by*

$$S = \mathrm{Cl}\{(x, \alpha) \in K \times J : x \neq 0, A(x, \alpha) = x\},$$

then S_0, *the maximal closed connected component of* S *that contains* $(0, \alpha_0)$ *is unbounded.*

We now discuss the way in which one can apply Theorem 3.1 to the RFDE in Section 11.1. For System (1.1), suppose the following conditions are satisfied:

(3.1) There is a closed convex set $K \subseteq C$ and $a < 0$ such that 0 is an extreme point of K, $\{0\} \neq K$, and, for any $\phi \in K$, $\phi \neq 0$ and $\alpha \in (a, \infty)$, there is a completely continuous function $\tau(\phi, \alpha) > 0$ such that the solution $x(\phi, \alpha)$ of Equation (1.1) through ϕ satisfies $x_{\tau(\phi,\alpha)}(\phi, \alpha) \in K$.

(3.2) If $A(\alpha)\phi = x_{\tau(\phi,\alpha)}(\phi, \alpha)$, $\phi \in K$, $\phi \neq 0$, $A(\alpha)0 = 0$, $\alpha \in (a, \infty)$, then $A(\alpha) : K \to C$ is completely continuous.

Hypotheses (3.1) and (3.2) imply Hypotheses (i) and (ii) of Theorem 3.1 are satisfied. For the linear RFDE($L(0)$) with $L(\alpha)$ defined in Equation (1.2), suppose

(3.3) the linear RFDE($L(0)$) has a simple purely imaginary characteristic root $\lambda_0 = i\nu_0 \neq 0$ and all characteristic roots $\lambda_j \neq \lambda_0, \bar{\lambda}_0$ satisfy $\mathrm{Re}\,\lambda_j < 0$.

With Hypothesis (3.3), Lemma 10.1 of Section 7.10 implies there is an $\alpha_2 > 0$ and a simple characteristic root $\lambda(\alpha)$ of the linear RFDE($L(\alpha)$), which has a continuous first derivative $\lambda'(\alpha)$ in α for $|\alpha| < \alpha_2$. Suppose

$$\mathrm{Re}\,\lambda'(0) \neq 0. \tag{3.4}$$

Let $\pi_{\lambda(\alpha)}$ be the projection operator defined by the decomposition of C as $C = P_{\lambda(\alpha)} \oplus Q_{\lambda(\alpha)}$ with the range of $\pi_{\lambda(\alpha)}$ equal to $P_{\lambda(\alpha)}$, the generalized eigenspace of $\lambda(\alpha)$. Suppose

(3.5) for any compact interval $J_0 \subseteq (a, \infty)$, there is a $\delta > 0$ such that

$$v \stackrel{\text{def}}{=} \inf\{|\pi_{\lambda(\alpha)}\phi| : \phi \in K,\ |\phi| = \delta,\ \alpha \in J_0\} > 0.$$

If Hypotheses (3.3), (3.4), and (3.5) are satisfied, then Theorem 1.1 of Section 10.1 and Theorem 2.3 imply that Condition (iv) of Theorem 3.1 is satisfied. Furthermore, the proof of Theorem 2.3 and the uniformity of Hypothesis (3.5) in α imply that Condition (iii) is satisfied.

If Conditions (3.4) and (3.5) are satisfied, then Theorem 1.1 implies there are $b_0 > 0$, $\alpha_2 > 0$, $\delta_0 > 0$ and functions $\alpha(b) \in (a, \infty)$, $\omega(b) \in \mathbb{R}$, $\phi^*(b) \in C$, $|b| < b_0$, $\alpha(0) = 0$, $\omega(0) = 2\pi/\nu_0$, $\phi^*(0) = 0$, such that each function is continuous and continuously differentiable in b, the solution $x^*(\phi^*(b), \alpha(b))$ of Equation (1.1) for $\alpha = \alpha(b)$ through $\phi^*(b)$ is an $\omega(b)$-periodic solution. Furthermore, any periodic solution of Equation (1.1) for $|\alpha| < \alpha_0$, $|x| < \delta_0$, must be of the preceding form. We can now prove the following.

Theorem 3.2. *Suppose System* (1.1) *satisfies Hypotheses* (3.1)–(3.5) *and*

(3.6) *there exists* $\alpha_1 \in (a, \infty)$ *such that* $A(\phi, \alpha) = \phi$ *for* $\phi \neq 0$ *implies* $\alpha \geq \alpha_1$,

then S_0 *is unbounded where* S_0 *is defined in Theorem* 3.1 *with* $\alpha_0 = 0$. *Also, if*

$$(\phi^*(b), \alpha(b)) \in S \qquad \textit{for} \quad |b| < b_0 \tag{3.7}$$

then the period $\omega(\phi, \alpha)$ *of each periodic solution* $x(\phi, \alpha)$ *of Equation* (1.1), *with initial value* ϕ *satisfying* $(\phi, \alpha) \in S_0$, *is a continuous function of* α.

Proof. With Conditions (3.1)–(3.6) satisfied, we have already shown that all conditions of Theorem 3.1 are satisfied. Therefore, S_0 is unbounded. If Condition (3.7) is satisfied, then in a sufficiently small neighborhood U of $(0, 0) \in S_0$, the only periodic solutions of Equation (1.1) with initial value ϕ satisfying $(\phi, \alpha) \in U$ must have $\phi = \phi^*(b)$, $\alpha = \alpha(b)$. Therefore, the period $\omega(\phi, \alpha)$ is continuous at $\phi = 0$, $\alpha = 0$. For other values of (ϕ, α) the continuity follows because we have assumed $\tau(\phi, \alpha)$ is continuous. □

Theorem 3.2 shows that the range of the period ω on S_0 is an interval containing $2\pi/v_0$. If one knows that there exist $(\phi, \alpha) \in S_0$ for each $\alpha \in [0, \infty)$ and $\omega_\infty = \limsup \omega(\phi, \alpha)$ as $\alpha \to \infty$, then $\omega(S_0) \supseteq [2\pi/\nu_0, \omega_\infty)$ if $\omega_\infty > 2\pi/\nu_0$ and $\omega(S_0) \supseteq (\omega_\infty, 2\pi/\nu_0]$ if $\omega_\infty < 2\pi/\nu_0$.

11.4 The equation $\dot{x}(t) = -\alpha x(t-1)[1+x(t)]$

In this section, we apply the results of the previous sections to the scalar equation

$$\dot{x}(t) = -\alpha x(t-1)[1+x(t)] \tag{4.1}$$

where $\alpha > 0$. Let us first show that a Hopf bifurcation occurs at $\alpha = \pi/2$. To do this, we will apply Theorem 1.1 and, therefore, we need information about the behavior of the zeros of the characteristic polynomial

$$\lambda e^{\lambda} = -\alpha \tag{4.2}$$

of the linear part of Equation (4.1),

$$\dot{y}(t) = -\alpha y(t-1). \tag{4.3}$$

This information is contained in the following lemma.

Lemma 4.1. *If $0 < \alpha < \pi/2$, every solution of Equation (4.2) has a negative real part. If $\alpha > e^{-1}$, there is a root $\lambda(\alpha) = \gamma(\alpha) + i\sigma(\alpha)$ of Equation (4.2), which is continuous together with its first derivative in α and satisfies $0 < \sigma(\alpha) < \pi$, $\sigma(\pi/2) = \pi/2$, $\gamma(\pi/2) = 0$, $\gamma'(\pi/2) > 0$, and $\gamma(\alpha) > 0$ for $\alpha > \pi/2$.*

Proof. The assertion concerning the real parts for $0 < \alpha < \pi/2$ is in Theorem A.5 of the Appendix. To prove the remaining part of the lemma, let $\rho(\mu) = -\mu e^{\mu}$. Then $\rho'(\mu) = -(1+\mu)e^{\mu}$ and, therefore, $\rho'(\mu) > 0$, $-\infty < \mu < -1$, $\rho'(-1) = 0$, $\rho'(\mu) < 0$, $-1 < \mu < \infty$. Consequently, $\rho(\mu)$ has a maximum at $\mu = -1$, $\rho(-1) = e^{-1}$. Therefore, Equation (4.2) has no real roots for $\alpha > e^{-1}$. If $\alpha > e^{-1}$, $\lambda = \gamma + i\sigma$, $\mu = -\gamma$, satisfies Equation (4.2), then $\mu - i\sigma = \alpha \exp(\mu - i\sigma)$ and

$$\mu = \alpha e^{\mu} \cos \omega, \qquad \sigma = \alpha e^{\mu} \sin \sigma$$

or

$$\mu = \sigma \cot \sigma, \qquad \alpha = \frac{\sigma e^{-\sigma \cot \sigma}}{\sin \sigma} \stackrel{\text{def}}{=} f(\sigma).$$

Let us consider $f(\sigma)$ for $0 < \sigma < \pi$. It is clear that $f(\sigma) > 0$.

$$\frac{f'(\sigma)}{f(\sigma)} = \frac{1}{\sigma} - 2\cot\sigma + \sigma \operatorname{cosec}^2 \sigma = \frac{(1-\sigma\cot\sigma)^2 + \sigma^2}{\sigma} > 0.$$

Furthermore, $f(\sigma) \to \infty$ as $\sigma \to \pi$ and $f(\sigma) \to e^{-1}$ as $\sigma \to 0$. Therefore, there is exactly one value of σ say $\sigma_0 = \sigma_0(\alpha)$, $0 < \sigma_0(\alpha) < \pi$, for which $f(\sigma_0(\alpha)) = \alpha$ if $\alpha > e^{-1}$. Let $\gamma_0(\alpha) = -\sigma_0(\alpha)\cot\sigma_0(\alpha)$. The functions $\sigma_0(\alpha)$ and $\gamma_0(\alpha)$ are clearly differentiable in α.

Also, $f(\pi/2) = \pi/2$, $\gamma_0(\pi/2) = 0$, and $\gamma_0(\alpha) > 0$ if $\alpha > \pi/2$. From the equation $\lambda(\alpha)\exp\lambda(\alpha) = -\alpha$, one observes that $\gamma'(\pi/2) > 0$. The lemma is proved. □

Using Lemma 4.1 and Theorem 1.1, we may now state

Theorem 4.1. *Equation* (4.1) *has a Hopf bifurcation at* $\alpha = \pi/2$.

Our next objective is to show that Equation (4.1) has a nonzero periodic solution for *every* $\alpha > \pi/2$.

Let $x(\phi, \alpha)$ be the solution of Equation (4.1) through ϕ. It is clear that $x(\phi, \alpha)(t) > -1$, $t \geq 0$, if $\phi(0) > -1$. Also, it is clear that there is no $t_0 > 0$ such that $x(\phi, \alpha)(t) = 0$ for $t \geq t_0$ unless $\phi = 0$. We say the zeros of $x(\phi, \alpha)$ are *bounded* if $x(\phi, \alpha)(t)$ has only a finite number of positive zeros.

Lemma 4.2.

(i) *If* $\phi(0) > -1$ *and the zeros of* $x(\phi, \alpha)$ *are bounded, then* $x(\phi, \alpha)(t) \to 0$ *as* $t \to \infty$.

(ii) *If* $\phi(0) > -1$, *then* $x(\phi, \alpha)(t)$ *is bounded. Furthermore, if the zeros of* $x(\phi, \alpha)$ *are unbounded, then any maximum of* $x(\phi, \alpha)(t)$, $t > 0$, *is less than* $e^{\alpha} - 1$.

(iii) *If* $\phi(0) > -1$ *and* $\alpha > 1$, *then the zeros of* $x(\phi, \alpha)$ *are unbounded.*

(iv) *If* $\phi(\theta) > 0$, $-1 < \theta < 0$ [*or if* $\phi(0) > -1$, $\phi(\theta) < 0$, $-1 < \theta < 0$], *then the zeros (if any) of* $x(\phi, \alpha)(t)$ *are simple and the distance from a zero of* $x(\phi, \alpha)(t)$ *to the next maximum or minimum is* ≥ 1.

Proof. (i) Suppose there is a $t_1 > 0$ such that

$$x(t) \stackrel{\text{def}}{=} x(\phi, \alpha)(t)$$

is of constant sign for $t \geq t_1 - 1$. Since $x(t) > -1$ for all $t \geq 0$, $\dot{x}(t)x(t-1) < 0$, $t \geq t_1$. Therefore, $x(t)$ is bounded and approaches a limit monotonically. This implies $\dot{x}(t)$ is bounded and therefore $\dot{x}(t) \to 0$ as $t \to \infty$. This implies $x(t) \to 0$ or -1, but -1 is obviously excluded.

(ii) x satisfies

$$1 + x(t) = [1 + x(t_0)] \exp\left[-\alpha \int_{t_0 - 1}^{t-1} x(\xi) d\xi\right] \tag{4.4}$$

for any $t \geq t_0 \geq 0$. If the zeros of $x(t)$ are bounded, then Part (i) implies x is bounded. If there is a sequence of nonoverlapping intervals I_k of $[0, \infty)$ such that x is zero at the endpoints of each I_k and has constant sign on I_k, then there is a t_k such that $\dot{x}(t_k) = 0$. Thus, $x(t_k - 1) = 0$. Consequently, Equation (4.4) implies for $t_0 = t_k - 1$, $t = t_k$.

$$\ln(1 + x(t_k)) = -\alpha \int_{t_k - 2}^{t_k - 1} x(\xi)\, d\xi < \alpha$$

since $x(t) > -1$, $t \geq 0$. Finally, $x(t_k) \leq e^{\alpha} - 1$ for all t_k. This proves Part (ii),

(iii) If the zeros of x are bounded, then Part (i) implies $x(t) \to 0$ as $t \to \infty$ and, thus, the existence of a $t_0 > 0$ such that $\alpha(1 + x(t)) > 1$ for $t \geq t_0$ and $x(t)$ has constant sign for $t \geq t_0$. Thus

$$\dot{x}(t)x(t-1) = -\alpha x^2(t-1)(1+x(t)) < -x^2(t-1) < 0, \qquad t \geq t_0 + 1,$$

and $x(t) \to 0$ monotonically as $t \to \infty$. If x is positive on $[t_0, \infty)$, then

$$x(t_0+3) - x(t_0+2) = \int_{t_0+2}^{t_0+3} \dot{x}(t)\, dt < \int_{t_0+1}^{t_0+2} x(t)\, dt < -x(t_0+2)$$

and $x(t_0+3) < 0$. This is a contradiction. If $x(t)$ is negative on $[t_0, \infty)$, then a similar contradiction is obtained.

(iv) Suppose $x(t_0) = 0$ and $x(t) > 0$, $t_0 - 1 < t < t_0$. For $t_0 < t < t_0+1$, $\dot{x}(t) < 0$. Similarly, if $x(t) < 0$ for $t_0 - 1 < t < t_0$ and $x(t_0) = 0$, then $\dot{x}(t) > 0$, $t_0 < t < t_0 + 1$. Thus, the assertions of (iv) are obvious and the lemma is proved. □

Let K be the class of all functions $\phi \in C$ such that $\phi(\theta) \geq 0$, $-1 < \theta \leq 0$, $\phi(-1) = 0$, ϕ nondecreasing. Then K is a *cone*; that is, K is a closed convex set in C with the following properties: if $\phi \in K$, then $\lambda\phi \in K$ for all $\lambda > 0$ and if $\phi \in K$, $\phi \neq 0$, then $-\phi \notin K$. If $\alpha > 1$, $\phi \in K$, $\phi \neq 0$, let

$$z(\phi, \alpha) = \min\{t : x(\phi,\alpha)(t) = 0, \dot{x}(\phi,\alpha)(t) > 0\}.$$

This minimum exists from Lemma 4.2, Parts (iii) and (iv). Also $z(\phi, \alpha) > 2$. Furthermore, Lemma 4.2, Part (iv) implies $x(\phi,\alpha)(t)$ is positive and nondecreasing on $(z(\phi,\alpha), z(\phi,\alpha)+1)$. Consequently, if $\tau(\phi,\alpha) = z(\phi,\alpha)+1$, then the mapping

$$\begin{aligned} A(\alpha)0 &= 0 \\ A(\alpha)\phi &= x_{\tau(\phi,\alpha)}(\phi,\alpha), \qquad \phi \neq 0, \end{aligned}$$

is a mapping of the cone K into itself. Since $\dot{x}(\phi,\alpha)(\tau(\phi,\alpha) - 1) > 0$, continuity of $x(\phi,\alpha)(t)$ in t, ϕ, α implies that $\tau(\phi,\alpha)$ is continuous in $K\backslash\{0\} \times (1,\infty)$.

Lemma 4.3. *The map* $\tau : (K\backslash\{0\}) \times (1,\infty) \to (0,\infty)$ *defined by* $\tau(\phi,\alpha) = z(\phi,\alpha) + 1$ *is completely continuous.*

Proof. First of all, we claim a solution $x = x(\phi,\alpha)$, $\phi \in K$, cannot take a time longer than 2 to become negative because, if $x(1) = \eta > 0$, we have

$$1 + x(2) = (1+\eta)\exp\left[-\alpha \int_0^1 x(s)\, ds\right] \leq (1+\eta)e^{-\alpha\eta}$$

and so $x(2) \leq (1+\eta)e^{-\alpha\eta} - 1$, and this quantity is negative because the function $h(\eta) = (1+\eta)e^{-\alpha\eta} - 1$ satisfies $h(0) = 0$ and $h'(\eta) = (-\alpha - \alpha\eta + 1)e^{-\alpha\eta} < 0$ for $\eta > 0$.

For any bounded set $B \subseteq K$ and any $\psi \in B$, $\alpha \in (1,\infty)$, let $t_0(\phi,\alpha) \leq 3$ denote the point where the solution $x = x(\phi,\alpha)$ has a minimum. Since $t_0(\phi,\alpha) \geq 1$, the set $H(\alpha) = \mathrm{Cl} \bigcup_{\phi\in B} x_{t_0(\phi,\alpha)}(\phi,\alpha)$ is compact and

$$H(\alpha) \subseteq K_1 \stackrel{\text{def}}{=} \{\psi \in C : -1 < \psi(\theta) \leq 0, -1 \leq \theta \leq 0, \psi \text{ nonincreasing}\}.$$

For any $\psi \in K_1$, define the continuous function

$$\tau_1 : [K_1 \backslash \{0\}] \times (1, \infty) \to (0, \infty)$$

by the relation $\tau_1(\psi, \alpha) = \min\{t > 0 : x(\psi, \alpha)(t) = 0\}$. If we prove $\tau_1(H(\alpha) \backslash \{0\}, \alpha)$ is bounded for each $\alpha \in (1, \infty)$, then $\tau(B \backslash \{0\}, \alpha)$ is bounded for each α. Since $H(\alpha)$ is compact, it is therefore only necessary to prove that τ_1 is bounded on a neighborhood of zero in K_1. This is easy to verify in the following manner. If $\psi \in K_1 \backslash \{0\}$ and $\tau_1(\psi, \alpha) > 1$, then $x(\psi, \alpha)(1) = \beta < 0$ and

$$1 + x(\psi, \alpha)(2) = (1 + \beta) \exp\left[-\alpha \int_0^1 x(s, \psi, \alpha)\, ds\right] \geq (1 + \beta) e^{-\alpha\beta}.$$

So

$$x(\psi, \alpha)(2) \geq h(\beta) \stackrel{\text{def}}{=} (1 + \beta) e^{-\alpha\beta} - 1.$$

If β is small and negative, then this function is positive since $h(0) = 0$ and $h'(0) = 1 - \alpha < 0$. This shows that τ_1 is bounded.

Since $\tau(\phi, \alpha)$ is continuous for $(\phi, \alpha) \in [K \backslash \{0\}] \times (1, \infty)$, and $\mathbb{R}$ is locally compact, one obtains the conclusion stated in the lemma. □

From Parts (ii) and (iv) of Lemma 4.2, it follows that $|A(\alpha)\phi| \leq e^\alpha - 1$ for each $\phi \in K$ and $A(\alpha)$ takes any bounded set B in $K \backslash \{0\}$ into the set $\{\phi \in C : |\phi| < e^\alpha - 1\}$. Since $\tau : [K \backslash \{0\}] \times (1, \infty) \to (0, \infty)$ is completely continuous and $\tau(\phi, \alpha) > 1$, it follows that the closure of $A(\alpha)B$ is compact. Also, if $\phi_k \in K \backslash \{0\}$, $\phi_k \to 0$, then we may assume $\tau(\phi_k, \alpha) \to \tau_0(\alpha)$ as $k \to \infty$. Continuity of solutions $x(\phi, \alpha)(t)$ of Equation (4.1) with respect to t, ϕ implies $x_{\tau(\phi_k, \alpha)}(\phi_k, \alpha) \to x_{\tau_0(\alpha)}(0, \alpha) = 0$. Therefore, $A(\alpha)$ is continuous at 0 and $A(\alpha)$ is completely continuous.

Let $\lambda(\alpha)$ be the root of Equation (4.2) given in Lemma 4.1, let C be decomposed as $C = P_{\lambda(\alpha)} \oplus Q_{\lambda(\alpha)}$ in the usual manner, and let $\pi_{\lambda(\alpha)}$ be the usual projection on $P_{\lambda(\alpha)}$.

Lemma 4.4. *If J_0 is a compact set of $(1, \infty)$ then*

$$\mu = \inf\{|\pi_{\lambda(\alpha)}\phi|,\ \phi \in K,\ |\phi| = 1,\ \alpha \in J_0\} > 0.$$

Proof. Let $\lambda = \lambda(\alpha)$ be the solution of Equation (4.2) given by Lemma 4.1, $\phi(\theta) = e^{\lambda\theta}/(1 + \lambda)$, $-1 \leq \theta \leq 0$, $\psi(s) = e^{-\lambda s}$, $0 \leq s \leq 1$, $\Phi = (\phi, \bar{\phi})$, $\Psi = (\psi, \bar{\psi})$. The formal adjoint of Equation (4.3) is

$$\dot{z}(t) = \alpha z(t + 1)$$

and the bilinear form is

$$(\psi, \phi) = \psi(0)\phi(0) + \alpha \int_{-1}^{0} \psi(\xi + 1)\phi(\xi)\, d\xi.$$

It is easily seen that (Ψ, Φ) is the identity. Therefore, for any $\phi \in C$, $\pi_\lambda \phi = \Phi(\Psi, \phi)$. To show the conclusion of the lemma is true, it is therefore sufficient to show that

$$\inf\{|(\Psi, \phi)|,\ \phi \in K,\ |\phi| = 1,\ \alpha \in J_0\} > 0.$$

Since $(\psi, \phi), (\bar{\psi}, \bar{\phi})$ are complex conjugate, it is sufficient to look at (ψ, ϕ). If $\phi \in K$, $|\phi| = 1$, then $\phi(0) = 1$ and

$$\begin{aligned}(\psi, \phi, \alpha) &\stackrel{\text{def}}{=} (\psi, \phi) = R(\phi) + iI(\phi)\\ R(\phi) &= 1 - \alpha \int_{-1}^{0} \phi(\theta) e^{-\gamma(\theta+1)} \cos\sigma(\theta + 1)\, d\theta\\ I(\phi) &= \alpha \int_{-1}^{0} \phi(\theta) e^{-\gamma(\theta+1)} \sin\sigma(\theta + 1) d\theta\end{aligned}$$

Since J_0 is compact, there is an $\epsilon > 0$ such that $\epsilon < \sigma = \sigma(\alpha) < \pi - \epsilon$ for $\alpha \in J_0$.

Now suppose there exist sequences $\phi_n \in K$, $\phi_n(0) = 1$, $\alpha_n \in J_0$ such that $(\psi, \phi_n, \alpha_n) \to 0$ as $n \to \infty$. We may assume $\alpha_n \to \beta$ as $n \to \infty$ since J_0 is compact. Since $I(\phi_n) \to 0$, we have $\phi_n(\theta) \to 0$, $-1 \le \theta \le 0$. Thus, $R(\phi_n) \to 1$ as $n \to \infty$. This is a contradiction to the fact that $R(\phi_n) \to 0$ as $n \to \infty$ and the lemma is proved. □

If we now take $M > e^\alpha - 1$, then the lemmas imply that all of the conditions of Theorems 2.2 and 2.3 are satisfied. Therefore, we have proved the following result.

Theorem 4.2. *If $\alpha > \pi/2$, Equation (4.1) has a nonzero periodic solution.*

The lemmas also contain more information. Let

$$\begin{aligned}S &= \text{Cl}\{(\phi, \alpha) \in K \times (1, \infty) : A(\alpha)\phi = \phi, \phi \neq 0\}\\ S_0 &= \text{maximal closed connected component of } S \text{ that}\\ &\quad\ \text{contains } (0, \pi/2).\end{aligned} \tag{4.5}$$

Since the root $\lambda(\alpha) = \gamma(\alpha) + i\sigma(\alpha)$ of Equation (4.2) given by Lemma 4.1 satisfies $\sigma(\pi/2) = \pi/2$ and the solution x^* obtained by the Hopf bifurcation Theorem 1.1 can be chosen (by a phase shift) to satisfy $x^*(t) = \cos\pi(t/2)$ for $\alpha = \pi/2$, it is clear that the initial value of Hopf bifurcating solution belongs to K for α close to $\pi/2$. With this remark and the lemmas we have shown that all conditions of Theorem 3.2 are satisfied except Hypothesis (3.6). The next lemma shows this condition is also satisfied.

Lemma 4.5. *There is an $\alpha_1 > 1$ such that for any $\alpha \in (1, \alpha_1)$, the only solution of $A(\alpha)\phi = \phi$ in K is $\phi = 0$.*

Proof. Suppose $\phi \in K$, $A(\alpha)\phi = \phi$, and let $z_1(\phi, \alpha) = z_1$ and $z_2(\phi, \alpha) = z_2$ be the first and second zeros of the periodic solution $x(\phi, \alpha)$ of Equation (4.1). From Equation (4.4) and the fact that $|\phi| = |\phi(0)| = |x(z_2 + 1)|$, $x(\xi) \geq x(z_1 + 1)$ for $\xi \in [z_2 - 1, z_2]$, we have

$$-x(z_1 + 1) = 1 - \exp\left[-\alpha \int_{z_1-1}^{z_1} x(\xi)\, d\xi\right] \leq 1 - e^{-\alpha|\phi|} \stackrel{\text{def}}{=} \gamma(|\phi|)$$

$$\begin{aligned} |\phi| = x(z_2 + 1) &= \exp\bigl[-\alpha \int_{z_2-1}^{z_2} x(\xi)\, d\xi\bigr] - 1 \\ &\leq e^{-\alpha x(z_1+1)} - 1 \leq e^{\alpha\gamma(|\phi|)} - 1. \end{aligned}$$

If $f(\beta) = \exp(+\alpha\gamma(\beta)) - 1 - \beta$, then there is a unique solution $\beta(\alpha)$ of $f(\beta) = 0$ for each $\alpha > 1$ and $\beta(\alpha) \to 0$ as $\alpha \to 1$. Furthermore, $f(\beta) < 0$ if $\beta > \beta(\alpha)$. Therefore, if there is a ϕ such that $A(\alpha)\phi = \phi$, $\phi \in K$, then $|\phi| \leq \beta(\phi)$. Since the zero solution of Equation (4.1) is asymptotically stable for $0 < \alpha < \pi/2$, there are $\delta > 0$ and $\epsilon > 0$, such that the only periodic solution of Equation (4.1) with $\alpha \in (1-\epsilon, 1+\epsilon)$, $|\phi| < \delta$, is $\phi = 0$. Therefore, if α_1 is chosen such that $\beta(\alpha_1) < \delta$, the lemma is proved. □

The following result is now an immediate consequence of Theorem 3.2 and the fact that fixed points of $A(\alpha)$ satisfy $|\phi| < e^{\alpha} - 1$.

Theorem 4.3. *If S_0 is defined as in Expression (4.5), then S_0 is unbounded and for any $\alpha_2 > 1$, there is an $\alpha > \alpha_2$ and $\phi \in K$ such that $(\phi, \alpha) \in S_0$.*

As described in Section 11.3, Theorem 3.2 together with Theorem 1.1 can now be used to obtain an estimate of the range of the period $\omega(\phi, \alpha)$ of periodic solutions of Equation (4.1) with initial values ϕ in K. Since we know that S_0 contains an element for each $\alpha \in [\pi/2, \infty)$, and $\omega(0, \pi/2) = 4$ for $\alpha = \pi/2$, we need only estimate $\alpha = \limsup \omega(\phi, \alpha)$ as $\alpha \to \infty$.

The following is stated without proof and is obtained from Theorem 3.2 and estimates showing that $\omega_\infty = \limsup\{\omega(\phi, \alpha) : (\phi, \alpha) \in S_0, \alpha \to \infty\} = \infty$ (see Section 11.7 for references).

Theorem 4.4. *For any $p > 4$, there is a periodic solution of Equation (4.1) of period p.*

11.5 The equation $\dot{x}(t) = -\alpha x(t-1)[1 - x^2(t)]$

In this section, we consider the equation

$$\dot{x}(t) = -\alpha x(t-1)[1 - x^2(t)] \tag{5.1}$$

where $\alpha > 0$. Our purpose is only to show the modifications of the previous section necessary to obtain a nonconstant periodic solution of Equation (5.1).

Since the linear part of this equation is the same as in the previous section, the following result holds.

Theorem 5.1. *Equation* (5.1) *has a Hopf bifurcation at* $\alpha = \pi/2$.

If $x(\phi, \alpha)$ is the solution of Equation (5.1), then $-1 < x(\phi, \alpha)(t) < 1$ for $t \geq 0$ if $-1 < \phi(0) < 1$. Also, there is a $t_0 > 0$ such that $x(\phi, \alpha)(t) = 0$ for $t \geq t_0$ only if $\phi = 0$. The analogue of Lemma 4.2 is

Lemma 5.1.

(i) *If* $-1 < \phi(0) < 1$ *and the zeros of* $x(\phi, \alpha)(t)$ *are bounded, then* $x(\phi, \alpha)(t) \to 0$ *as* $t \to \infty$.

(ii) *If* $-1 < \phi(0) < 1$, *then* $-1 \leq x(\phi, \alpha)(t) \leq 1$, $t \geq 0$, *and if the zeros of* $x(\phi, \alpha)$ *are unbounded, then any maximum [or minimum] of* $x(t)$, $t > 0$, *is less than* $(e^{2\alpha} - 1)/(e^{2\alpha} + 1)$ *[greater than* $-(e^{2\alpha} - 1)/(e^{2\alpha} + 1)$*].*

(iii) *If* $-1 < \phi(0) < 1$ *and* $\alpha > 1$, *then the zeros of* $x(\phi, \alpha)$ *are unbounded.*

(iv) *If* $1 > \phi(\theta) > 0$, $-1 < \theta < 0$ *[or if* $\phi(0) > -1$, $\phi(\theta) < 0$, $-1 < \theta < 0$*], then the zeros (if any) of* $x(\phi, \alpha)(t)$ *are simple and the distance from a zero of* $x(\phi, \alpha)(t)$ *to the next maximum or minimum is* ≥ 1.

Proof. The proof of Parts (ii), (iii) and (iv) are the same as the proof in Lemma 4.2 except for obvious modifications. To prove Part (ii), observe that $x = x(\phi, \alpha)$ satisfies

$$\ln \frac{1 + x(t)}{1 - x(t)} - \ln \frac{1 + x(t_0)}{1 - x(t_0)} = -2\alpha \int_{t_0 - 1}^{t-1} x(\xi)\, d\xi$$

for any $t \geq t_0 \geq 1$. Using the same argument as in Lemma 4.2. Part (iii), for this equation, one proves Part (ii). □

Let

$$K = \{\phi \in C : \phi(-1) = 0,\ 0 \leq \phi(\theta) < 1,\ -1 \leq \theta \leq 0,$$
$$\phi \text{ nondecreasing}\}.$$

Then K is a *truncated cone*; that is, K is the intersection of a cone with a ball with center zero. Define the operator $A : K \to K$ as in Section 11.4.

As before, one shows that τ and A are completely continuous mappings. Lemma 4.4 is true for K.

From Lemma 5.1, Part (ii),

$$|A\phi| \leq \frac{(e^{2\alpha}-1)}{(e^{2\alpha}+1)} \stackrel{\text{def}}{=} \beta < 1$$

for all $\phi \in K$. If we choose $M > \beta$, then the conditions of Theorem 2.2 and 2.3 are satisfied. Therefore, we have proved

Theorem 5.2. *If $\alpha > \pi/2$, then Equation (5.1) has a nonconstant periodic solution.*

One can use the same type of arguments as in Section 11.4 to prove the analogue of Lemma 4.5 and obtain

Theorem 5.3. *S_0 is unbounded and for any $\alpha_2 \in (1, \infty)$, there are $\alpha > \alpha_2$ and $\phi \in K$ such that $(\phi, \alpha) \in S_0$.*

Some comments are made in Section 11.7, concerning the range of the period of the periodic solutions of Equation (5.1).

11.6 The equation $\ddot{x}(t) + f(x(t))\dot{x}(t) + g(x(t-r)) = 0$

In this section, we discuss the existence of nonconstant periodic solutions of the equation

$$\ddot{x}(t) + f(x(t))\dot{x}(t) + g(x(t-r)) = 0 \tag{6.1}$$

where $r \geq 0$, f is continuous, and g is continuous together with its first derivative, $f(0) = -k$, $g'(0) = 1$.

As in the previous sections, we first discuss the Hopf bifurcation. To do this we must consider the characteristic equation for the linear part of Equation (6.1); namely, the equation

$$\lambda^2 - k\lambda + e^{-\lambda r} = 0. \tag{6.2}$$

Lemma 6.1. *Let $\sigma_0(r)$, $0 < \sigma_0(r) < \pi/2r$, be the unique solution of $\sigma^2 = \cos\sigma r$, and let $k_0(r) = [\sigma_0(r)]^{-1}\sin\sigma_0(r)r$. If $k < -k_0(r)$, then all roots of Equation (6.2) have negative real parts. There is an $\epsilon > 0$ and a root $\lambda(k)$ of Equation (6.2) that is continuous together with its first derivative in k for $k \in (-k_0(r)-\epsilon, -k_0(r)+\epsilon)$, $\lambda(-k_0(r)) = i\sigma_0(r)$, $\operatorname{Re}\lambda'(-k_0(r)) > 0$. Finally, for each $k > -k_0(r)$, there are precisely two roots λ of Equation (6.2) with $\operatorname{Re}\lambda > 0$ and $-\pi/r < \operatorname{Im}\lambda < \pi/r$.*

Proof. Suppose $\lambda = \mu + i\sigma$ is a solution of Equation (6.2). Separating the real and imaginary parts of this equation, one obtains the following equations for μ, σ:

$$\begin{aligned} \mu^2 - \sigma^2 - k\mu + e^{-\mu r}\cos\sigma r &= 0 \\ 2\sigma\mu - k\sigma - e^{-\mu r}\sin\sigma r &= 0. \end{aligned} \tag{6.3}$$

It is shown in Theorem A.6 of the Appendix that all roots of Equation (6.3) have negative real parts if $k < -k_0(r)$, where $k_0(r)$ is specified as in the lemma. If $k = -k_0(r)$, then $\mu = 0$ and $\sigma = \sigma_0(r)$ are solutions of Equations (6.3). Also, the implicit function theorem implies there is an $\epsilon > 0$ and a unique solution $\mu(k)$, $\sigma(k)$, $\mu(-k_0(r)) = 0$, $\sigma(-k_0(r)) = \sigma_0(r)$, for $k \in (-k_0(r) - \epsilon, -k_0(r) + \epsilon)$, which is continuous and continuously differentiable in k. Also, it is easy to compute the derivatives of these functions at $k = -k_0(r)$ and observe that $\mu'(-k_0(r)) > 0$. Thus, all of the lemma is proved except the last assertion.

Let $U_{ab} = \{\lambda \in C : a \le \text{Re }\lambda \le b,\ |\text{Im }\lambda| < \pi/r\}$. The application of the implicit function theorem and the fact that all roots of Equations (6.2) have negative real parts for $k < -k_0(r)$ show there is a $k_1 > -k_0(r)$ such that Equation (6.2) with $k \in [-k_0(r), k_1]$ has exactly two roots in the region $U_{0,\infty}$. Now suppose $k \ge k_1$. Our first observation is that there is no solution of Equation (6.2) with either $\lambda = \mu + (i\pi/r)$, $\mu \ge 0$ or $\lambda = i\sigma$, $0 \le \sigma \le \pi/r$. The last assertion is obvious from Equations (6.3) since we would have $\sigma = \sigma_0(r)$ and $k = -k_0(r)$, which is a contradiction to the fact that $k \ge k_1 > -k_0(r)$. If $\lambda = \mu + (i\pi/r)$, then Equations (6.3) imply that $\mu = k/2$ and

$$0 = \mu^2 - \left(\frac{\pi}{r}\right)^2 - k\mu - \exp(-r\mu) = \frac{-k^2}{4} - \left(\frac{\pi}{r}\right)^2 - \exp(-r\mu) < 0$$

which is a contradiction.

For any fixed $k_2 \in \mathbb{R}$, there is a real number $b(k_2) > k_1$ such that there are no solutions of Equation (6.2) with $\text{Re}\,\lambda \ge b(k_2)$ for any $k \in [k_1, k_2]$. Therefore, for any $k \in [k_1, k_2]$, there are no solutions of Equation (6.2) on the boundary of $U_{0,b(k_2)}$. Let $f_t(\lambda) = \lambda^2 - [(1-t)k_1 + tk]\lambda + \exp(-r\lambda)$, $0 \le t \le 1$, $k \in [k_1, k_2]$. Since $(1-t)k_1 + tk \in [k_1, k_2]$ for $t \in [0,1]$, and $k \in [k_1, k_2]$, it follows from Rouche's theorem that the number of zeros of $f_t(\lambda)$ in $U_{0,b(k_2)}$ is constant for $t \in [0,1]$. In particular, the number of zeros of $f_1(\lambda)$ (Equation (6.2) for k_1) in $U_{0,b(k_2)}$ is the same as the number of zeros of $f_0(\lambda)$ (Equation (6.2) for k_1) in $U_{0,b(k_2)}$. Since we have already observed this latter number is precisely two, the proof of the lemma is complete. □

Using Lemma 6.1 and Theorem 1.1, we may now state

Theorem 6.1. *Equation* (6.1) *has a Hopf bifurcations at* $k = -k_0(r)$, *where* $k_0(r)$ *is defined in Lemma* (6.1).

Under some additional hypotheses on f and g in Equation (6.1), we will prove that Equation (6.1) has a nonconstant periodic solution for every $k > -k_0(r)$.

The additional hypotheses on f and g are the following:

(6.4a) $F(x) = \int_0^x f(s)\,ds$ is odd in x.

(6.4b) $F(x) \to \infty$ as $|x| \to \infty$ and there is a $\beta > 0$ such that $F(x) > 0$ and is monotone increasing for $x > \beta$.

(6.4c) $g'(x) > 0$, $xg(x) > 0$, $x \neq 0$, $g(x) = -g(-x)$, $g'(0) = 1$.

(6.4d) $g(F^{-1}(x))/x \to 0$, $F^{-1}(x)/x \to 0$ as $x \to \infty$.

In this notation, Equation (6.1) is equivalent to the system of equations

$$\begin{aligned} \dot{x}(t) &= y(t) - F(x(t)), \\ \dot{y}(t) &= -g(x(t-r)). \end{aligned} \tag{6.5}$$

Let $C_0 = C([-r,0],\mathbb{R}) \times \mathbb{R}$ and designate elements in C_0 by $\psi = (\phi, a)$, $\phi \in C([-r,0],\mathbb{R})$, $a \in \mathbb{R}$. For any $\psi \in C_0$. Equation (6.5) has a unique solution $z(\psi)$, $z = (x,y)$, through ψ at zero. In the following, the symbol $z_t \in C_0$ shall designate a solution of Equation (6.5) and $z_t = (x_t, y(t))$. Also, $z(t)$ will designate $z(t) = (x(t), y(t))$, and $\psi(0) = (\phi(0), a)$ if $\psi \in C_0$.

Let

$$K = \{\psi = (\phi, a) \in C_0 : 0 \le a < \infty, 0 = \phi(-r) \le \phi(\theta),\ -r \le \theta \le 0,$$
$$\phi(\theta) \text{ nondecreasing in } \theta\}.$$

Lemma 6.2. *If Hypotheses* (6.4) *are satisfied, then the following assertions hold:*

(i) *There is a continuous $\tau_1 : K\backslash\{0\} \to (r,\infty)$ such that*

$$z_{t_1(\psi)}(\psi) \in -K \overset{\text{def}}{=} \{-\psi : \psi \in K\}.$$

(ii) *There is a continuous $\tau_2 : K\backslash\{0\} \to (r,\infty)$ such that $z_{\tau_2(\psi)}(\psi) \in K$.*

(iii) *For any $\psi \in K\backslash\{0\}$, the solution $z(\psi)$ of Equations* (6.5) *is oscillatory; that is, both $x(\psi)(t)$ and $y(\psi)(t)$ have infinitely many zeros.*

Proof. If $\psi \in K$ and $z(\psi)$ is a solution of Equations (6.5), then $-z(\psi) = z(-\psi)$ is also a solution. Therefore Statement (i) implies Statement (ii). Since Statements (i) and (ii) imply Statement (iii), it is only necessary to prove Statement (i).

If $\psi \in K$ and $z = z(\psi)$ is the solution of Equations (6.5) through ψ, we analyze the curve in the (x,y)-plane traced out by $z(t)$, $t \ge 0$. Let

$$\Gamma = \{(x,y) \in \mathbb{R}^2 : y = F(x), x \in \mathbb{R}\}.$$

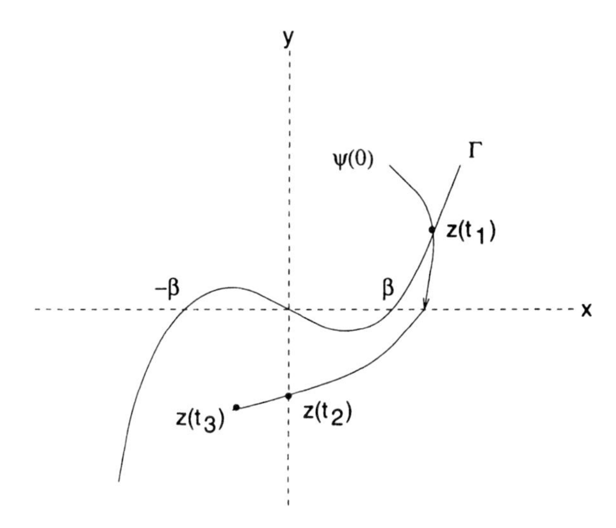

Fig. 11.1.

The accompanying Figure 11.1 will be helpful in understanding the following argument.

Suppose $\psi(0)$ is above Γ. As long as $z(t)$ is above Γ, $\dot{x}(t) = y(t) - F(x(t)) > 0$ and $\dot{y}(t) = -g(x(t-r)) \le 0$. If $z(t)$ does not intersect Γ, then $x(t) \ge x(r) > 0$ for $t \ge r$ and $\dot{y}(t) = -g(x(t-r)) = g(-x(t-r)) \le -x(r) < 0$ from Hypothesis (6.4c). This is clearly impossible and so there is a first time t_1 such that $z(t_1) \in \Gamma$. Also $\dot{x}(t_1) = 0$ and $\dot{y}(t_1) < 0$.

As long as $z(t)$ is below Γ and $x(t) \ge 0$, we have $\dot{x}(t) < 0$ and $\dot{y}(t) \le 0$. Since any crossing of Γ must occur with a vertical slope, it follows that $z(t)$ cannot cross Γ for $t > t_1$ and $x(t) \ge 0$. Therefore, there are $\delta > 0$ and $\epsilon > 0$, such that $\dot{x}(t) < -\delta < 0$ for $t > t_1 + \epsilon$ and $x(t) \ge 0$. Thus, there is a first value $t_2 > t_1$ such that $x(t_2) = 0$ and $\dot{x}(t_2) < -\delta$.

An argument similar to this implies there is a first $t_4 > t_2$ such that $z(t_4) \in \Gamma$ and $x(t) < 0$ for $t_2 < t \le t_4$. But, this clearly implies $\dot{y}(t_4) > 0$ and, thus, $x(t_4 - r) < 0$. If $t_3 = t_2 + r$, then $z_{t_3} \in K$. If $\tau_1(\psi) = t_3$, then $\tau_1 : K\backslash\{0\} \to (r, \infty)$ is continuous and $z_{\tau_1(\psi)} \in K$. If $\psi(0)$ is between Γ and the x-axis, the same argument may be applied to complete the proof of the lemma. □

For any $\psi \in K\backslash\{0\}$, let $\tau_1(\psi)$ be the number given by Lemma 6.3, Part (i), and define $A : K\backslash\{0\} \to K$ by $A\psi = -z_{\tau_1(\psi)}(\psi)$. If $\psi \ne 0$, $A\psi = \psi$, then the symmetry in Equations (6.5) implies that

$$z_{2\tau_1(\psi)}(\psi) = z_{\tau_1(\psi)}(z_{\tau_1(\psi)}(\psi)) = z_{\tau_1(\psi)}(-\psi) = -z_{\tau_1(\psi)}(\psi) = \psi$$

and ψ corresponds to a nontrivial periodic solution of Equations (6.5) of period $2\tau_1(\psi)$.

Lemma 6.3. *The map $A : K\backslash\{0\} \to K$ defined by $A\psi = -z_{\tau_1(\psi)}(\psi)$ is completely continuous.*

Proof. Let $B = \{\psi \in K\backslash\{0\} : |\psi(0)| \leq M_0, a \leq a_0\}$. Since $\tau_1(\psi) > r$ and $A\psi = -z_{\tau_1(\psi)}(\psi)$ for $\psi \in K\backslash\{0\}$, we will have shown that $A(B)$ has compact closure if $A(B)$ is bounded. Let $\xi(a)$ be the largest positive root of $a + F(\xi) = 0$ and let $M > \max(M_0, \xi(a))$. Then $0 \leq x(t) < M$ for

$$0 \leq t \leq t_2(\psi) \stackrel{\text{def}}{=} \tau_{z_1(\psi)}(\psi) - r.$$

Also, $0 \leq -\dot{y}(t) \leq g(M)$ for $0 \leq t \leq t_2(\psi) + r$. Now suppose there is a $\tau \in (0, t_2(\psi))$ such that $F_m - y(\tau) = M/r$ where $F_m = \inf\{F(x) : 0 \leq x \leq M\}$. Then $\dot{x}(t) \leq -M/r$ as long as $\tau \leq t \leq \tau + r$ and $x(t) \geq 0$. Since $x(0) \leq M$, it follows that $\tau \leq t_2(\psi) \leq \tau + r$. Since $-\dot{y}(t) = g(x(t-r)) \leq g(M)$ for $0 \leq t \leq t_2(\psi) + r$, it follows that $-y(t) \leq -y(\tau) + g(M)(t - \tau)$ and so

$$-y(t) \leq 2rg(M) + \frac{M}{r} - F_m \stackrel{\text{def}}{=} y_m$$

for $0 \leq t \leq t_2(\psi) + r$. Of course, if $-y(t) > (-M/r) + F_m$ for $0 \leq t \leq t_2(\psi)$, one obtains the better estimate $-y(t) \leq rg(M) + (M/r) - F_m$.

If $\beta = \inf\{s : F(s) = 0\}$, then $\beta \leq 0$. If $x(t) > \beta$ for $t_2(\psi) \leq t \leq t_2(\psi) + r$, then $|x(t)| \leq \max(M, |\beta|)$ and we are done. If there is a $\tau \in [t_2(\psi), t_2(\psi) + r)$ such that $x(\psi) = \beta$, then $\dot{x}(t) \geq y(t) \geq -y_m$ for $\tau \leq t \leq t_2(\psi) + r$. Therefore $x(t) \geq -y_m r + \beta$ and

$$|x(t)| \leq \max(M, |\beta| + |y_m| r) \stackrel{\text{def}}{=} x_m.$$

Thus, on $[0, t_2(\psi) + 1]$, both $x(t)$ and $y(t)$ are uniformly bounded for $\psi \in B$. Thus, $A(B)$ is bounded and the lemma is proved. □

Lemma 6.4. *There is a constant $M > 0$ such that if $A\psi = \mu\psi$, $\psi \in K\backslash\{0\}$, $|\mu| = M$, then $\mu < 1$.*

Proof. We use the notation in the proof of Lemma 6.2. If $\psi \in K\backslash\{0\}$, $A\psi = \mu\psi$, and $\psi(0)$ is below Γ, then $(A\psi)(0)$ is above Γ since F is an odd function.

Suppose now that $\psi(0)$ is above Γ. For a given $\alpha > 0$, let $\xi = \xi(\alpha)$ be the largest positive solution of $\alpha - F(\xi) = 0$. If $\psi = (\phi, a)$, then the fact that $\psi(0)$ is above Γ implies that $|\psi| = a$ and $\phi(0) \leq a$. In the proof of Lemma 6.3, we obtained the following estimates on $z_{\tau_1(\psi)} = (x_{\tau_1(\psi)}, y(\tau_1(\psi)), \tau_1(\psi) = t_2(\psi) + r$:

$$
\begin{aligned}
|y(\tau_1(\psi))| &\le 2rg(\xi(a)) + \xi(a) + |F_m| \\
&= \Big[\frac{2rg(F^{-1}(a))}{a} + \frac{F^{-1}(a)}{a} + |F_m|\Big]a \\
|x_{\tau_1(\psi)}| &\le \max\Big(\frac{F^{-1}(a)}{a}, \frac{|\beta|}{a} + \frac{|y(\tau_1(\psi))|r}{a}\Big)a.
\end{aligned}
$$

Hypotheses (6.4d) imply there are $a_0 > 0$ and $k < 1$ such that $|y(\tau_1(\psi))| \le ka$, for $a \ge a_0$ and $\psi \in K$, $\psi(0)$ above Γ. Therefore, if $A\psi = \mu\psi$ for $\psi \in K$, $\psi(0)$ above Γ and $a \ge a_0$, then $\mu < 1$. This proves the lemma. □

Our next objective is to show that 0 is an ejective point of A. To do this, we will make use of the decomposition of the space C as $C = U \oplus S$ where U is the subspace of dimension 2 of C spanned by the initial values of the two eigenvalues with positive real parts given in Lemma 6.1. Let π_U be the associated projection onto U.

Lemma 6.5. *If $k_0(r)$ is as in Lemma 6.1, then $\inf_{\phi \in \partial B(1) \cap K} |\pi_U \phi| > 0$ if $f(0) < k_0(r)$.*

Proof. For $k = -f(0) > -k_0(r)$, the linear part of Equations (6.5) is

$$
\begin{aligned}
\dot{x}(t) &= y(t) + kx(t) \\
\dot{y}(t) &= -x(t-r)
\end{aligned}
$$

and the formal adjoint equations are

$$
\begin{aligned}
\dot{u}(t) &= -ku(t) + v(t+r) \\
\dot{v}(t) &= -u(t)
\end{aligned}
$$

and the associated bilinear form is

$$
(\zeta, \psi) = b\phi(0) + \xi(0)a - \int_{-r}^{0} \xi(\theta + r)\phi(\theta)\, d\theta
$$

where $\zeta = (b, \xi)$, $b \in \mathbb{R}$, $\xi \in C([0,r], \mathbb{R})$ and $\psi = (\phi, a)$, $\phi \subset C$, $a \subset \mathbb{R}$.

If the eigenvalues in Lemma 6.1 are distinct, the appropriate basis for the solutions of the adjoint equation defining the projection π is

$$
\zeta_j(s) = b_j e^{-\lambda_j s}, \qquad b_j = (-\lambda_j, -1), \qquad j = 1, 2. \tag{6.6}
$$

Let us consider first the case where $\lambda = \gamma + i\sigma$, $0 < \sigma < \pi/r$. If $b = \alpha - i\beta$ where α and β are real two vectors, then the real and imaginary parts of (ζ, ψ) are given by

$$
\begin{aligned}
\mathrm{Re}\,(\zeta, \psi) &= \alpha\psi(0) + \int_{-r}^{0} e^{-\gamma(\xi - r)} \cos\sigma(\xi + r)\phi(\xi)\, d\xi, \\
-\mathrm{Im}\,(\zeta, \psi) &= \beta\psi(0) + \int_{-r}^{0} e^{-\gamma(\xi - r)} \sin\sigma(\xi + r)\phi(\xi)\, d\xi
\end{aligned}
$$

where $\beta\psi(0) = \sigma|\lambda|\,\phi(0) \geq 0$. If there is a sequence $\psi_n = (\phi_n, a_n) \in \partial B(1) \cap K$ such that $\pi_U \psi_n \to 0$ as $n \to \infty$, then Im $(\zeta, \psi_n) \to 0$, Re$(\zeta, \psi_n) \to 0$ as $n \to \infty$. But this implies $\phi_n \to 0$ as $n \to \infty$ and $a_n \to 0$ as $n \to \infty$. Therefore, $\psi_n \to 0$ as $n \to \infty$, which is a contradiction.

If now $\lambda_1 > \lambda_2 > 0$ are real roots of Equation (6.2) and ζ_j are defined in Equations (6.6), then

$$\begin{aligned}(\zeta_1, \psi) &= b_1\psi(0) + \int_{-r}^{0} e^{-\lambda_1(\xi+r)}\phi(\xi)\,d\xi \\ &= -\lambda_1\phi(0) - a + \int_{-r}^{0} e^{-\lambda_1(\xi+r)}\phi(\xi)\,d\xi, \\ (\zeta_2, \psi) &= b_2\psi(0) + \int_{-r}^{0} e^{-\lambda_2(\xi+r)}\phi(\xi)\,d\xi \\ &= -\lambda_2\phi(0) - a + \int_{-r}^{0} e^{-\lambda_2(\xi+r)}\phi(\xi)\,d\xi.\end{aligned}$$

Suppose there is a sequence $\psi_n = (\phi_n, a_n) \in \partial B(1) \cap K$ such that $(\zeta_1, \psi_n) \to 0$, $(\zeta_2, \psi_n) \to 0$ as $n \to \infty$. Without loss in generality we may assume $\phi_n(0) \to \phi_0$, $a_n \to a_0$ as $n \to \infty$. Since $\lambda_1 > \lambda_2$, the assertion implies

$$a_0 + \lambda_2\phi_0 \geq a_0 + \lambda_1\phi_0.$$

Since $k > \lambda_1 > \lambda_2$, this is a contradiction unless $\phi_0 = 0$. If $\phi_0 = 0$, then the assertion implies

$$\int_{-r}^{0} [e^{-\lambda_2(\xi+r)} - e^{-\lambda_1(\xi+r)}]\phi_n(\xi)\,d\xi \to 0 \qquad \text{as } n \to \infty.$$

Since $\lambda_1 > \lambda_2$, this implies $\phi_n \to 0$ as $n \to \infty$. Then $(\zeta_1, \psi_n) \to 0$ implies $a_n \to 0 = a_0$ and $\psi_n \to 0$, which is a contradiction.

If the two eigenvalues of Lemma 6.1 are complex, then one can show that the map τ_1 in Lemma 6.2 is completely continuous. Therefore Lemma 6.5 and Theorem 2.3 imply 0 is an ejective point of A. If these eigenvalues are real, then τ_1 is unbounded in a neighborhood of $\phi = 0$. However, one can use Lemma 6.2 to show there are $\epsilon > 0$, $\delta > 0$ such that $|A\phi| > \delta$ for $\phi \in K\backslash\{0\}$, $|\phi| < \epsilon$. Thus, 0 is an ejective point of A. The proofs of these facts are left to the reader.

The case where $\zeta_1(t) = b\exp(\lambda_1 t)$, $\zeta_2(t) = (bt + c)\exp(\lambda_1 t)$, $\lambda_1 > 0$, remains to be considered. This case is left for the reader. With this remark, the proof of the lemma is complete. □

Using these results and Theorems 2.2 and 2.3, we have

Theorem 6.2. *If F and g satisfy Conditions* (6.4a)–(6.4d) *and* $f(0) < k_0(r)$, *where $k_0(r) > 0$ is given in Lemma* 6.1, *then Equation* (6.1) *has a nonconstant periodic solution.*

A special case of this theorem is $f(x) = k(x^2 - 1)$, $k > 0$; that is, the van der Pol equation with a retardation. It is interesting to look at the latter equation in more detail. If $x = u/\sqrt{k}$, then u satisfies the equation

$$\ddot{u}(t) + (u^2(t) - k)\dot{u}(t) + u(t-r) = 0.$$

This equation satisfies the conditions of Theorem 6.2 for every $k > 0$. Therefore, there is a periodic solution $u^*(k)$ with $|u^*(k)| \geq c > 0$ for $0 \leq k \leq 1$. Thus, the solution $x^*(k) = u^*(k)/\sqrt{k}$ of

$$\ddot{x}(t) + k(x^2(t) - 1)\dot{x}(t) + x(t-r) = 0$$

approaches ∞ as $k \to 0$.

The conditions of Theorem 6.2 are also satisfied by $f(x) = ax^2 + b$, $a > 0$, $b < k_0(r)$, $g(x) = x$, and, in particular, for $f(x) = x^2$, $g(x) = x$.

11.7 Supplementary remarks

Hopf [1] was the first to state Theorem 1.1 for ordinary differential equations. Many generalizations to infinite-dimensional systems have been given (see Marsden and MacCracken [1] for references). To the author's knowledge, the first statement similar to Theorem 1.1 for RFDE was given by Chow and Mallet-Paret in a course at Brown University in 1974 with a proof different from the one in the text. The proof in the text is easily adapted to certain types of partial differential equations. It is of interest to determine the stability and amplitude of the bifurcating periodic orbit. Efficient procedures for doing this using a method of averaging have been given by Chow and Mallet-Paret [1]. For Equation (4.1), Gumowski [1] has computed the period and amplitude of the solution using the method of undetermined parameters (see, also Strygin [1]). Ginzburg [1] has also considered the bifurcation problem. Other results giving more properties of the orbits are contained in Chafee [2]. The global existence of the Hopf bifurcation as a function of the initial data, λ, and the period has been discussed by Chow and Mallet-Paret [2] and Nussbaum [9].

It is also possible that a bifurcation from a constant occurs as one varies the delays in an equation. In the case of a single delay as in Equation (4.1) of Section 10.4, one can reduce the discussion of Equation (4.2) of Section 10.4 and apply the Hopf bifurcation Theorem 1.1. The case of several delays has been discussed in Hale [24].

As remarked in Section 10.5, much research has been devoted to the behavior of solutions of differential difference equations with a small delay. Perelló [1] has also discussed equations with a small delay and proved the following interesting "continuity" result. Suppose one has shown there is an invariant torus for an ordinary differential equation and a cross section of this torus is mapped into itself by the flow induced by the differential equation. Then the Brouwer fixed-point theorem implies there is a periodic solution. Perelló [1] has shown the differential difference equation obtained from this ordinary equation by the introduction of small delays must also have a periodic solution.

If it were possible to construct a closed bounded convex set K in C that is free of equilibrium points and has the property that for each $\phi \in K$, the trajectory through ϕ of an RFDE returns to K, then the usual fixed-point theorems could be applied to obtain the existence of nonconstant periodic solutions. However, this construction is very difficult in an infinite-dimensional space.

In a fundamental paper, Jones [4] introduced the idea of finding a cone that maps into itself under the flow rather than a cross section of a torus. The cone is easier to construct than a torus but some other complications arise because most problems have zero as an equilibrium point. Therefore, one is interested in finding nonzero fixed points of a mapping of a cone into itself knowing that zero is a fixed point. This problem has been the motivation for a number of interesting fixed-point theorems for cone mappings (see Krasnoselskii [2] and Grafton [1]) as well as mappings of a convex set into itself with an ejective fixed point as an extreme point of the convex set (Browder [2]). Using the ideas of Browder [2], Nussbaum [3] proved Theorem 2.1. Theorem 2.2 is also due to Nussbaum [3] and combines in an interesting way ejectivity as in Theorem 2.1 and the concept of eigenvalue as in the work of Krasnoselskii [2] and Grafton's theorem (see the statement in Hale [11] and Lopes [4]). Theorem 2.3 is essentially due to Chow and Hale [2]. Smith [1] has also developed the ideas of Krasnoselskii to obtain fixed-point theorems and periodic solutions.

Theorem 3.1 is due to Nussbaum [5] and was proved with the intention of making applications to the range of the period. The usefulness of the Hopf bifurcation theorem to assist in this effort was pointed out by Chow and Hale [2].

Lemmas 4.1 and 4.2 are due to Wright [2]. Theorems 4.2 and 5.2 are due to Jones [4, 5] with the proof following the ideas in Grafton [1]. Theorems 4.3, 4.4, and 5.3 are due to Nussbaum [5]. If we let $1 + x(t) = e^{u(t)}$ in Equation (4.1), then u satisfies the equation

$$\dot{u}(t) = -\alpha f(u(t-1)) \tag{7.1}$$

where $f(u) = e^u - 1$. In a series of papers, Nussbaum [3,4,5,6,7] has discussed Equation (7.1) under very general hypotheses on f and has given sophisticated theorems on the existence of periodic solutions and the range of the periods of these solutions as a function of α. For the equation

$$\dot{x}(t) = -\alpha x(t-1)[a - x(t)][b + x(t)], \tag{7.2}$$

the results of Nussbaum imply that the range of the period of the periodic solutions of Equations (7.2) contains the interval $4 < p < 2 + (a/b) + (b/a)$.

If $a = b = 1$, that is, Equation (5.1), then no information is obtained about the range of the period. For Equation (5.1), Jones [6] has shown there is a periodic solution of period 4 for every $\alpha > 0$. Kaplan and Yorke [1] have given more general conditions on f in Equation (7.1) to have periodic solutions of period 4.

A special case of Theorem 6.2 was first given by Grafton [1]. The proof in the text is in the same spirit as Grafton [1] but with many corrections and simplifications taken from Nussbaum [3]. One can also relax the conditions on f and g in Equation (6.1) and prove an analogue of Theorem 6.2 (see Grafton [2] and Nussbaum [3]). Nussbaum [7] has also given some results on the range of the period of periodic solutions of Equation (6.1).

Periodic solutions of many other types of equations are known to exist. Among these are equations of the form

$$\dot{x}(t) = -\alpha\Big[\int_{-r}^{0} x(t+\theta)\, d\beta(\theta)\Big][1 + x(t)] \tag{7.3}$$

$$\dot{x}(t) = -\alpha\left[\int_{-r}^{0} x(t+\theta)\, d\beta(\theta)\right][1 - x(t)]^2 \tag{7.4}$$

where β is a nondecreasing function, $\beta(0) = 0$, and $\beta(1) = 1$. Some remarks and numerical results for β a step function with two jumps is contained in Jones [6]. Chow and Hale [2] discussed the case of Equation (7.3) where $\beta(\theta)$ is absolutely continuous on $[-1, 0]$ and $\beta(\theta) = 1$ on $[-\frac{1}{2}, 0]$ and showed there is a nonconstant periodic solution for $\alpha > \alpha_0 > 0$. Walther [1] has eliminated the hypothesis that $\beta(\theta)$ be absolutely continuous. The proof in Walther [1] could possibly be simplified by using the ideas in Chow and Hale [2].

In the analysis of more general Equations (7.3) and (7.4), it is necessary to have detailed information about the behavior of the eigenvalues of the linear equation as the function β varies in some class of functions. Some interesting results on this general question are contained in Walther [2]. In this same paper, Walther also gives sufficient conditions for all roots of the linear equation to lie in the left half-plane for every α. In these cases, any nonconstant periodic solution that arises by a variation in α cannot be due to a Hopf bifurcation.

Some results on the uniqueness and stability of periodic solutions of special cases of Equation (7.3) also have been obtained (see Jones [6], Kaplan and Yorke [2], and Walther [3]).

Chow [2] considers the equation

$$\dot{x}(t) = -\sigma x(t) + e^{-x(t-r)},$$

$0 < \sigma < e^{-1}$, and shows there is a nonconstant periodic solution for $r \geq r_0$ sufficiently large. The proof could be simplified by using the results in this chapter.

Numerical studies have been made by Grafton [3] on the equations

$$\ddot{x}(t) + a[x^2(t) - 1]\dot{x}(t) + Ax(t-r) + Bx(t-s) = 0$$
$$\ddot{x}(t) + a[x^n(t) - 1]\dot{x}(t) + Ax(t-r) + Bx^3(t-r) = 0$$
$$\ddot{x}(t) + a[x^2(t-r) - 1]\dot{x}(t-r) + x(t) = 0$$

for various values of the constants a, A, B, r, s, and n. A variety of types of oscillations occur that have been mathematically explained. Nussbaum [7] has shown the existence of a periodic solution of the first equation for $a < 0$, $r > 0$, $s = 0$, $A = 1$, and $-1 < B \leq 0$. In [7], Nussbaum has also discussed some other cases of equations of this type.

Considerable attention has been devoted to the following equation, which can be considered as a generalization of Equation (7.1):

$$\dot{x}(t) = -\frac{\alpha}{\delta}\int_{-1}^{-1+\delta} f(x(t+\theta))\,d\theta, \tag{7.5}$$

where α and δ are positive numbers, $\delta \leq \frac{1}{2}$, and the function f is C^1, $f'(0) = 1$, $xf(x) > 0$ for $x \neq 0$ and $f(x) \geq -B > -\infty$ for all x. The following result has been proved by Walther [4] (see also Nussbaum [10]):

Theorem 7.1. *There is a constant $\alpha_0 > 0$ such that for $\alpha > \alpha_0$, there are real numbers $z_1 > 1-\delta$, $z_2 - z_1 > 1-\delta$, and a periodic solution $x_\alpha(t) = x(t)$ of (7.5) of period z_2 such that $x(0) = 0$, $x(t) > 0$ on $(0, z_1)$, $x(t) < 0$ on (z_1, z_2).*

The constant α_0 is the value of α at which the zero solution of the linear variational equation becomes unstable. The method of proof uses cone maps and the ejectivity properties of the origin.

Alt [1] also has considered the existence of periodic solutions for more general equations with one time delay or distributed time delays.

Another interesting class of equations are those for which the delay is determined by a threshold condition. More specifically, Alt [2] has considered periodic solutions of the equation

$$\dot{x}(t) = -f(x(t), x(t - \sigma(x_t))), \tag{7.6}$$

where the function $\sigma(\phi)$ is determined by a threshold condition

$$\int_{-\sigma(\phi)}^{0} k(\phi(s))\,ds = k_0, \tag{7.7}$$

where $k : \mathbb{R} \to \mathbb{R}^+$ is continuous, $k \geq c > 0$ and $k(0) = k_0$. The function f is assumed to satisfy $f(x, y) \geq 0$ (resp. ≤ 0) for $x \geq 0, y \geq 0$ (resp. $x \leq 0, y \leq 0$), $f(0, y) < 0$ for $y < 0$, $f(x, x) \neq 0$ for $x \neq 0$, $f(0, x)) \geq -B > -\infty$ for $x \in (-\infty, 0]$. Assuming also that the linear variational equation has an eigenvalue with positive real part, Alt [2] shows that there is a periodic solution of (7.6) and (7.7). The method of proof is to show first that there is set $K \subset C$ that is homeomorphic to a convex bounded set and such that for any $\phi \in K$, there is a $\tau(\phi)$ such that the solution through ϕ at 0 is in K at $\tau(\phi)$. The set K has 0 on the boundary of K and it must be shown that it is ejective. Since $\sigma(\phi)$ is not differentiable in ϕ, the obvious linearization cannot be used for the ejectivity. Alt [2] transforms the equation to a neutral functional differential equation for which the method using projections onto the unstable manifold can be applied.

Equations of the form (7.6) and (7.7) often serve as models in biological problems. In the manner in which it is presented, the delay depends on the state of the system. Smith [2] has shown how to eliminate the state-dependent delay for certain types of threshold-type problems.

For a functional differential equation $\dot{x}(t) = f(x_t)$ with delay interval of length $r > 0$ and an equilibrium point 0, a solution $x(t)$ is said to be *slowly oscillating* about the point 0 if it has a sequence of zeros approaching ∞ and the distances between the zeros of $x(t)$ are $> r$. Most of the results on the existence of periodic solutions of delay differential equations are for slowly oscillating ones. In many situations, the slowly oscillating periodic solutions are the ones that enjoy stability properties and thus are the ones that will be observed in the system. To substantiate this remark, let us restrict attention to Equation (7.1) and always assume that $f'(0) = 1$, $xf(x) > 0$ for $x \neq 0$, $f(x)$ is bounded below, and $\alpha > \frac{\pi}{2}$. In this situation, we know that (7.1) has a slowly oscillating periodic solution. We denote by S_f the set of initial data for the slowly oscillating solutions of (7.1). Walther [5] has shown that all of the slowly oscillating solutions of (7.1) eventually enter a solid torus; more specifically, there are constants $a > 0$, and $r > 0$ such that for any solution $x(t)$ of (7.1) with $x_t \in S_f$ of (7.1), there is a t_0 such that $x_t \in \{\phi \in S_f : a \leq \|\phi\| \leq r\}$ for $t \geq t_0$. If we assume in addition that $f'(x) > 0$ for all x, then Walther [6] has shown that there is an $a > 0$ such that ϕ is in the closure of S_f if the solution $x(t, \phi)$ through ϕ at 0 satisfies $\limsup_{t \to \infty} |x(t, \phi)| > a$. He shows also that there is an $a' > 0$ such that the set $\{\phi \in C : \limsup_{t \to \infty} |x(t, \phi)| \geq a'\}$ is dense in C. Since the set S_f is open, we obtain openness and denseness of the slowly oscillating solutions.

For the bifurcation of slowly oscillating periodic solutions of (7.1) under some symmetry conditions on f, see Walther [7]. Under the assumption that f is odd and some restrictive monotonicity conditions, Chow and Walther [1] have proved the hyperbolicity of the slowly oscillating periodic solution (of period 4). When the function f is not odd, there may be several slowly oscillating periodic solutions (see Saupe [1,2]). If α is large, then there is such a solution with a very long period. Xie [1] has shown that this solution is linearly stable. He accomplishes this by obtaining a good expression for the eigenvalues of the Poincaré map.

Another very interesting class of equations has the form

$$\dot{x}(t) = -\mu x(t) + \mu f(x(t-1)), \tag{7.8}$$

where $\mu > 0$ is a constant and the function f has *negative feedback*; that is, $f(0) = 0$, $f'(0) < 0$ and $xf(x) < 0$ for $x \neq 0$. Also, suppose that there is an interval I such that $f(I) \subset I$. Under the additional assumption that the origin of (7.8) is linearly unstable, Hadeler and Tomiuk [1] have shown that there is a slowly oscillating periodic solution. Some earlier results were obtained by Pesin [1].

The linearization of (7.8) about the origin has a two-dimensional invariant manifold W_0 consisting of the span of the eigenfunctions of an eigenvalue with positive real part and the solutions corresponding to these eigenfunctions are slowly oscillating. For the complete equation near the origin, there is a two-dimensional local invariant manifold $W_{\mathrm{loc}}(0)$ that is tangent to W_0 at 0. The solutions on this manifold are slowly oscillating. Let $W(0)$ be the global extension of $W_{\mathrm{loc}}(0)$ by following the solutions of (7.8). Under the assumption that f is monotone and bounded either from above or below, Walther [8] has shown that the boundary of $W(0)$ is a periodic orbit that attracts all orbits of (7.8) with initial data in $W(0) \setminus \{0\}$.

As remarked earlier, there may be more than one slowly oscillating periodic solution of (7.8). Let γ be a hyperbolic unstable periodic orbit of a slowly oscillating periodic solution and let W_γ be the unstable manifold of this orbit. Walther [9] has shown that W_γ is a smooth annulus-like graph of dimension two whose boundary consists either of two other slowly oscillating periodic orbits or one slowly oscillating periodic orbit and the equilibrium solution 0.

A further generalization of (7.8) that is important in physiology and laser optics is

$$\left(\epsilon_m \frac{d}{dt} + 1\right) \cdots \left(\epsilon_0 \frac{d}{dt} + 1\right) x(t) = f(x(t-1)), \tag{7.9}$$

where each $\epsilon_j > 0$, $j = 1, \ldots, m$ and f satisfies the same conditions as in (7.8). Assuming that the origin is unstable and $m = 1$ (that is, a second-order equation), an der Heiden [1] has proved the existence of a slowly oscillating periodic solution. For any m, Hale and Ivanov [1] have proved the same result, provided that the constants ϵ_j, $0 \leq j \leq m$, are sufficiently small.

For a recent survey on oscillations in scalar delay differential equations, see Ivanov and Sharkovsky [1].

Except for second-order differential equations, there are few results on the existence of periodic solutions of systems of delay differential equations. Táboas [1] has considered the system

$$\begin{aligned}
\dot{x}_1(t) &= -x_1(t) + \alpha F_1\big(x_1(t-1), x_2(t-1)\big), \\
\dot{x}_2(t) &= -x_2(t) + \alpha F_1\big(x_1(t-1), x_2(t-1)\big)
\end{aligned}$$

where $\alpha > 0$ is a constant, $F = (F_1, F_2)$ is a bounded C^3-map with $\partial F_1/\partial x_2$ and $\partial F_2/\partial x_1$ different from 0 at the origin and F satisfies a *negative feedback condition*:

$$x_2 F_1(x_1, x_2) > 0, \quad x_2 \neq 0, \quad x_1 F_2(x_1, x_2) < 0, \quad x_1 \neq 0.$$

Under these hypotheses, Táboas [1] shows that there is an $\alpha_0 > 0$ such that, for any $\alpha > \alpha_0$, there is a periodic solution of period > 4. The method of proof is to show that the negative feedback condition causes the solutions of the equation to rotate in the (x_1, x_2)-plane and then he uses modifications of the method of cone maps and ejective fixed points for the cone

$$K = \{ (\phi_1, \phi_2) \in C : e^{\theta}\phi_j(\theta) \text{ nondecreasing} - 1 \leq \theta \leq 0, \ j = 1, 2, \\ \phi_1(-1) = 0, \ \phi_2(-1) \geq 0\}$$

after renormalization so that $\partial F_1(0)/\partial x_2 = -\partial F_2(0)/\partial x_1$.

In the proof of ejectivity, he found the following interesting modification of Theorem 2.3.

Theorem 7.2. *Suppose that the following conditions are fulfilled:*

(i) *There is an eigenvalue* λ *of Equation* (2.2) *satisfying* $\operatorname{Re}\lambda > 0$.

(ii) *There exists a closed convex subset* K *of* C, $0 \in K$, *and a continuous function* $\tau : K \setminus \{0\} \to [\alpha, \infty)$, $\alpha > 0$, *such that the map*

$$A\phi = x_{\tau(\phi)}(\phi), \quad \phi \in K \setminus \{0\}, \quad A0 = 0,$$

is completely continuous and $AK \subset K$.

(iii) $\inf\{ |\pi_\lambda x_t| : x_t = x_t(\phi), \phi \in K, |\phi| = \delta, 0 \leq t \leq \tau(\phi) \} > 0$.

(iv) *Given* $G \subset C$ *open,* $0 \in G$, *there is a neighborhood* V *of* 0 *such that* $x_t(\phi) \in G$ *if* $\phi \in V \cap K$, $\phi \neq 0$ *and* $0 \leq t \leq \tau(\phi)$.

Then 0 *is an ejectivity fixed point of* A.

Leung [1] has proved the existence of a periodic solution of period $> 2r$ for the prey predator model

$$\dot{x}(t) = x(t)[a - bx(t) - cy(t)]$$
$$\dot{y}(t) = \alpha y(t)[x(t - r) - \beta]$$

provided that all constants are positive, $a > \alpha^{-1} b\beta(b + \alpha)$ and $r > [2\alpha(a - b\beta)]^{-1}$.

Recently, there have been several problems in ecology and physiology for which the delay depends on the state of the system and it cannot be eliminated by any change of variables. In this case, no general qualitative theory is available. The primary difficulty arises from the fact that the solutions of the equation are not differentiable with respect to the delay function. Therefore, the theory given in the text is not applicable. In spite of this fact, there have appeared recently some interesting results dealing with special types of solutions of equations with state-dependent delays and, in particular, the slowly oscillating periodic solutions. Consider Equation (7.8), with the delay $r = r(x(t))$ depending on the present state of the system. Under the same assumptions (negative feedback and the existence of a positively invariant interval containing the origin), it has been shown that there is a slowly oscillating periodic solution if μ is large (see Mallet-Paret and Nussbaum [5], Kuang and Smith [1,2]). The existence of the periodic solutions is proved by using an appropriate extension of the method of cone maps mentioned before. The papers of Kuang and Smith [1,2] apply also to threshold-type problems. Mallet-Paret and Nussbaum [5,6] have discussed detailed properties of the slowly oscillating periodic solution as $\epsilon \to 0$. In particular, for the equation

$$\epsilon \dot{x}(t) = -\mu x(t) - k\mu x(t - r(x(t)), \quad r(x) = 1 + cx,$$

they have shown that the solution approaches a saw-tooth function of period $k+1$ given by the straight line segment $y = c^{-1}x$ on the interval $-1 \leq x \leq k$. This is in marked contrast with what occurs with a constant delay (see Chapter 12 for a discussion of this case).

For equations with several delays or RFDE with distributed delays with the distribution function having several large peaks, the problem of existence of periodic solutions is much more difficult than the same problem for the equations that have been discussed earlier.

Even the local Hopf bifurcation theorem is difficult. If the bifurcation parameters are taken to be the delay parameters, there is a difficulty due to the fact that the solutions of an RFDE are not differentiable with respect to the delays. However, using the fact that the periodic solutions of an RFDE will be as smooth as the vector field, it is possible to use a variant of the usual implicit function theorem to show that the Hopf bifurcation theorem remains valid with the delays being the parameters (see, for example, Hale [24]). For the types of differentiability properties of solutions of RFDE with respect to delays, see Hale and Ladeira [1,2].

A greater difficulty arises in the understanding of the manner in which the roots of the characteristic equation of a linear system with several delays depend on parameters. For the linear equation,

$$\dot{x}(t) + ax(t)bx(t - r) + cx(t - \tau) = 0, \tag{7.10}$$

where a, b, c, $r \geq 0$, and $\tau \geq 0$, are constants, there have been many attempts to understand the region in the parameter space for which the solutions approach zero (for example, Bellman and Cooke [1], Hale [24], Mahaffey [1], Marriott, Vallée, and Delisle [1], Nussbaum [11], Ragazzo and Perez et al. [1], Ruiz-Claeyssen [1], Stech [1]). The boundary of this region is of primary interest since it represents the points at which the origin can undergo a bifurcation from stability to instability. The complete description of this boundary is not available at this time. However, Hale and Huang [1] have given a good description in the case when a, b, and c are fixed and the delays r, τ are considered as the parameters. To be more precise, the *stable region* is defined as a maximal connected set $D \subset [0, \infty) \times [0, \infty)$ that contains the origin $(0, 0)$ such that for each $(r, \tau) \in D$, the zero solution of (7.10) is asymptotically stable. Hale and Huang [1] show the following:

Theorem 7.3. *If a stable region is unbounded, then its boundary will approach a straight line parallel to the r-axis or τ-axis as $r + \tau \to \infty$.*

As a consequence of this result, we see that, if a half line in the first quadrant of the (r, τ)-plane contains an unstable point, then the intersection of this line and the boundary of the stable region contains at most finitely many points and eventually leaves the stable region.

More detailed properties of the boundary of a stable region are discussed in Hale and Huang [1]. For a point on the boundary, if the characteristic equation has only one pair of imaginary roots, then there is a Hopf bifurcation to a periodic orbit as the delays cross the boundary. The determination of whether it is supercritical or subcritical is very difficult since it requires considerable computation (see Stech [2], Franke and Stech [1] for how to use MACSYMA to do these computations and more complicated ones). It is possible to have multiple eigenvalues on the boundary of a stable region. In this case, very complicated oscillatory phenomena can arise as the boundary is crossed—even chaos. The special situation where $r = 3\tau$ has been analyzed in some detail by Stech [1], where he showed that three periodic solutions could bifurcate from the origin. Nussbaum [11] had previously observed that, in this case for a special type of nonlinear equation, there were two periodic orbits that existed globally with respect to the delays larger than some value. He conjectured that there should be another periodic solution as well, which Stech confirmed near the bifurcation point on the boundary of the stable region.

In modeling of physical systems, the manner in which the equations depend on the past history is taken to be as simple as possible due to the severe difficulties in the analysis of RFDE. In many situations, authors have chosen the memory functions to be of the type that repeated differentiations will reduce the problem to an ordinary differential equation of high order. This leads to a simpler system, but it leaves open an important question. If some property is discovered from the ordinary differential equation, does it remain valid for the original problem if we choose a memory function that is close in some sense to the one that was analyzed, but for which no reduction to an ordinary differential equation is possible? Hines [1] has many results in this direction, including the preservation of stability and the upper and lower semicontinuity of attractors. Farkas and Stépán [1] has similar results for the preservation of stability.

12
Additional topics

In the previous chapters, we have touched only the surface of the theory of functional differential equations. In recent years, the subject has been investigated extensively and now there are several topics that can be classified as a field in itself. In this chapter, we give an introduction to some of these areas, describe the main results, and occasionally give indications of proofs. We cover generic theory, equations with negative feedback and Morse decompositions, slowly oscillating periodic solutions, singularly perturbed delay equations, averaging, and abstract phase spaces associated with equations with infinite delay. In the supplementary remarks, we give references for the detailed proofs and indicate other areas of functional differential equations that are currently being investigated.

12.1 Equations on manifolds–Definitions and examples

In this section, we begin with a few examples that will serve as motivation for the consideration of functional differential equations on finite-dimensional manifolds.

Example 1.1. For any constant c, the scalar equation

$$\dot{\theta}(t) = c\sin(\theta(t-1)) \tag{1.1}$$

can be considered as an RFDE on the circle $S^1 = \{\, y \in \mathbb{R}^2 : y_1^2 + y_2^2 = 1 \,\}$ by considering θ as an angle variable only determined up to a multiple of 2π.

Example 1.2. If b, c are constants, then we write the second-order RFDE

$$\ddot{\theta}(t) + b\dot{\theta}(t) = c\sin(\theta(t-1)) \tag{1.2}$$

as a system of first-order RFDE

$$\begin{aligned} \dot{x}_1(t) &= x_2(t) \\ \dot{x}_2(t) &= c\sin x_1(t-1) - bx_2(t). \end{aligned} \tag{1.3}$$

By considering x_1 as an angle variable only determined up to a multiple of 2π, Equation (1.3) is an RFDE on the cylinder $S^1 \times \mathbb{R}$. We remark that we can take the space of initial data for the solution x of (1.2) as $C([-1,0], S^1) \times \mathbb{R}$.

Example 1.3. Let $S^2 = \{ (x,y,z) \in \mathbb{R}^3 : x^2 + y^2 + z^2 = 1 \}$ and consider the following system of RFDE:

$$\begin{aligned} \dot{x}(t) &= -x(t-1)y(t) - z(t) \\ \dot{y}(t) &= x(t-1)x(t) - z(t) \\ \dot{z}(t) &= x(t) + y(t). \end{aligned} \tag{1.4}$$

If $(x(t), y(t), z(t))$ is a solution of Equation (1.4), it is easy to see that

$$x(t)\dot{x}(t) + y(t)\dot{y}(t) + z(t)\dot{z}(t) = 0$$

for all $t \geq 0$. As a consequence, for $t \geq 0$, $x^2(t) + y^2(t) + z^2(t) = a^2$, a constant. Thus, if an initial condition $\phi = (\phi_1, \phi_2, \phi_3)$ satisfies $\phi(\theta) \in S^2$ for all $\theta \in [-1,0]$, we conclude that the solution $(x,y,z)(t;\phi) \in S^2$ for all $t \geq 0$.

With this remark, we can define an RFDE on S^2 by the map

$$F : (\phi_1, \phi_2, \phi_3) \in C([-1,0], S^2) \mapsto F(\phi)$$

where $F(\phi)$ is the tangent vector to S^2 at the point $\phi(0)$ defined by

$$F(\phi) = \big(-\phi_1(-1)\phi_2(0) - \phi_3(0),\ \phi_1(-1)\phi_1(0) - \phi_3(0),\ \phi_1(0) + \phi_2(0)\big).$$

We now formalize the notions in these examples to obtain an RFDE on an n-dimensional manifold. Roughly speaking, an RFDE on a manifold M is a function F mapping each continuous path ϕ lying on M, $\phi \in C([-r,0], M)$, into a vector $F(\phi)$ tangent to M at the point $\phi(0) \in M$.

Let M be a separable C^∞ finite n-dimensional manifold, I the interval $[-r,0]$, $r \geq 0$, and $C(I,M)$ the totality of continuous maps ϕ of I into M. Let TM be the tangent bundle of M and $\tau_M : TM \to M$ its C^∞-canonical projection. Assume that there is given on M a complete Riemannian structure (it exists because M is separable) with δ_M the associated complete metric. This metric on M induces an admissible metric on $C(I,M)$ by

$$\delta(\phi, \bar{\phi}) = \sup_{\theta \in I} \delta_M(\phi(\theta), \bar{\phi}(\theta)).$$

The space $C(I,M)$ is separable (since M is complete and separable) and is a C^∞-manifold modeled on a separable Banach space. If M is imbedded as a closed submanifold of a Euclidean space V, then $C(I,M)$ is a closed C^∞-submanifold of the Banach space $C(I,V)$.

If $\rho : C(I, M) \to M$ is the evaluation map, $\rho(\phi) = \phi(0)$, then ρ is C^∞, and for each $a \in M$, $\rho^{-1}(a)$ is a closed submanifold of $C(I, M)$ of codimension $n = \dim M$. A *retarded functional differential equation* (RFDE) *on* M is a continuous function $F : C(I, M) \to TM$ such that $\tau_M \circ F = \rho$. If we want to emphasize the function F defining the RFDE, we write RFDE(F).

A solution of RFDE(F) is defined in the obvious way, namely, as a continuous function $x : [-r, \alpha) \to M$, $\alpha > 0$, such that $\dot{x}(t)$ exists and is continuous for $t \in [0, \alpha)$ and $(x(t), \dot{x}(t)) = F(x_t)$ for $t \in [0, \alpha)$. Locally, if $F(\phi) = (\phi(0), f(\phi))$ for an appropriate function f, then this is equivalent to $\dot{x}(t) = f(x_t)$.

We remark that it is just as easy to define a *neutral functional differential* (NFDE) *on* M. We simply choose a map $D : C(I, M) \to M$ such that D is atomic at zero and ask that $\tau_M \circ F = D$.

The basic theory of existence, uniqueness, and continuous dependence on initial data for general RFDE on manifolds is the same as the theory when $M = \mathbb{R}^n$.

Example 1.4. Any C^k-vector field on M defines a C^k-RFDE on M. In fact, if $X : M \to TM$ is a C^k-vector field on M, it is easy to see that $F = X \circ \rho$ is a C^k-RFDE on M.

Example 1.5. To show that the equation considered in Example 1.2 is an RFDE according to the definition, we need the concept of a *second-order* RFDE *on* M. Let $\bar{F} : C(I, TM) \to TM \times TM$ be a continuous function that locally has the representation

$$\bar{F}(\phi, \psi) = \big((\phi(0), \psi(0)), (\psi(0), f(\phi, \psi))\big).$$

The solutions $(x(t), y(t))$ of the RFDE($\bar{F}$) on TM satisfy the equations

$$\dot{x}(t) = y(t), \quad \dot{y}(t) = f(x_t, y_t) \tag{1.5}$$

where $x(t) \in M$. If it is possible to perform the differentiations, then we obtain the second-order equation

$$\ddot{x}(t) = f(x_t, \dot{x}_t).$$

If we now return to Example (1.2), we see that the formulation requires that we consider initial data in the space $C(I, S^1) \times C(I, \mathbb{R})$. However, this does not affect the dynamics since the solution will be in the space $C(I, S^1) \times \mathbb{R}$ after one unit of time.

Example 1.6. (*Delay differential equations on* M). Let $g : M \times M \to TM$ be such that $(\tau_M \circ g)(x, y) = x$ and let $d : C(I, M) \to M \times M$ be defined by $d(\phi) = (\phi(0), \phi(-r))$. The function $F = g \circ d$ is an RFDE on M that can be written locally as

$$\dot{x}(t) = \bar{g}(x(t), x(t - r)),$$

where $g(x,y) = (x, \bar{g}(x,y))$.

For notational purposes, it is convenient to let

$$\mathcal{X}^k = \{ F \in BC^k(C(I,M), TM) : \tau_M \circ F = \rho \},$$

where BC^k denotes that the function and derivatives up through order k are bounded. An $F \in \mathcal{X}^k$ corresponds to a C^k-RFDE. As for an RFDE on $\mathbb{R}^n$, we let $T_F(t)$ denote the solution operator for the RFDE(F) and introduce the concept of dissipativeness, globally defined solutions, attractors, etc. For an RFDE(F), we denote the set of globally defined solutions by $\mathcal{A}(F)$. We know that if the equation is point dissipative, then $\mathcal{A}(F)$ is compact, invariant and is the global attractor. Another important concept is the following. An element $\psi \in \mathcal{A}(F)$ is called a *nonwandering point* of F if for any neighborhood U of ψ in $\mathcal{A}(F)$ and any $T > 0$, there exist $t = t(U,T) > T$ and $\tilde{\psi} \in U$ such that $T_F(t)\tilde{\psi} \in U$. The set of all nonwandering points of F is called the *nonwandering set* and is denoted by $\Omega(F)$. The following result is easy to prove but very important for the generic theory to be discussed in a later section.

Lemma 1.1. *If F is a C^1-RFDE on M that is point dissipative, then $\Omega(F)$ is closed and, moreover, if $T_F(t)$ is one-to-one on $\mathcal{A}(F)$, then $\Omega(F)$ is invariant.*

If M is a compact manifold without boundary, then $\mathcal{A}(F)$ is always the global attractor for a C^1-RFDE on M. Furthermore, it is not difficult to prove the following result.

Theorem 1.1. *If M is a compact manifold and $F \in \mathcal{X}^1$, then $\mathcal{A}(F)$ is upper semicontinuous in F; that is, for any neighborhood U of $\mathcal{A}(F)$ in M, there is a neighborhood V of F in $\mathcal{X}^1$ such that $\mathcal{A}(G) \subset U$ if $G \in V$.*

The following result also is true.

Theorem 1.2. *If M is a compact manifold and $F \in \mathcal{X}^1$, then* $\dim \mathcal{A}(F) \geq \dim M$ *and the restriction of ρ to $\mathcal{A}(F)$ is onto; that is, through any $x \in M$, there is at least one global solution. Furthermore, if M is without boundary and $\mathcal{A}(F)$ is a compact manifold without boundary, then $\mathcal{A}(F)$ is homeomorphic to M.*

If X is a C^1-vector field on M and M is a compact manifold without boundary, then $\mathcal{A}(F) = M$. A stronger version of the last part of Theorem 1.2 is

Theorem 1.3. *If M is a compact manifold without boundary and X is a C^1-vector field on M, $F = X \circ \rho$, then there is a neighborhood U of F in*

$\mathcal{X}^1$ such that for each $G \in U$, the attractor $\mathcal{A}(G)$ is diffeomorphic to M, $\mathcal{A}(G) \to \mathcal{A}(F)$ in the Hausdorff sense as $G \to F$, and the restriction of $T_G(t)$ to $\mathcal{A}(G)$ is a one-parameter family of diffeomorphisms.

Example 1.1. (Revisited). In Equation (1.1), if $c = 0$, we have the ordinary differential equation $\dot{x}(t) = 0$ on S^1. Theorem 1.3 implies that the attractor for (1.1) is diffeomorphic to S^1 if $|c|$ is sufficiently small. The same remark also is true if we consider the equation

$$\dot{x}(t) = \sin x(t) + cg(x(t-1)),$$

where $g(x + 2\pi) = g(x)$ is a given C^1-function and $|c|$ is sufficiently small.

The following result also can be useful.

Theorem 1.4. *Let M be a compact manifold without boundary and let $F \in \mathcal{X}^2$. If there is a constant $k > 0$ such that $\|D_\phi T_F(\phi)\| \leq k$, $\|D^2_\phi T_F(\phi)\| \leq k$ for all $t \geq 0$ and all $\phi \in C(I, M)$, then the attractor $\mathcal{A}(F)$ is a connected, compact C^1-manifold (without boundary) that is homeomorphic to M.*

Furthermore, the restriction of $T_F(t)$ to $\mathcal{A}(F)$ is a one-parameter family of diffeomorphisms.

Finally, there is a neighborhood U of F in $\mathcal{X}^2$ such that for any $G \in U$, the attractor $\mathcal{A}(G)$ is a manifold that is diffeomorphic to $\mathcal{A}(F)$.

Example 1.7. Consider the scalar equation

$$\dot{x}(t) = b(t) \sin[x(t) - x(t-1)], \tag{1.6}$$

where $b \in C^1(\mathbb{R}, \mathbb{R})$ is such that $b(0) = 0$ and $|db/dx| \leq \alpha < 1$. As in Example 1.1, this equation defines an RFDE on S^1. Any constant function is a global solution of (1.6). Conversely, any global solution is a constant function. To see this, one considers the map

$$B : z(t) \mapsto \int_{t-1}^{t} b(u) \sin[z(u)]\, du$$

acting in the Banach space of all bounded continuous functions $z(t)$ with the sup norm. It is easy to see that B is a contraction map and that $z(t) \equiv 0$ are its fixed points. On the other hand, if $x(t)$ is a global solution of (1.6), then $[x(t) - x(t-1)]$ is bounded and

$$x(t) - x(t-1) = \int_{t-1}^{t} b(u) \sin[x(u) - x(u-1)]\, du,$$

which shows that $x(t) - x(t-1) \equiv 0$. The equation gives $\dot{x}(t) \equiv 0$ and $x(t) =$ constant. In this case, the attractor $\mathcal{A}(F)$ is a circle. Let us indicate how to show that for any $\phi \in C$, there is a unique constant function $\alpha(\phi)$

such that $T_f(t) \to \alpha(\phi)$ as $t \to \infty$. For any constant function c, the linear variational equation is

$$\dot{y}(t) = b'(0)[y(t) - y(t-1)].$$

The corresponding characteristic equation is $\lambda = b'(0) - b'(0)e^{-\lambda}$. From the Appendix and the fact that $|b'(0)| < 1$, we know that all solutions of this equation have negative real parts except for $\lambda = 0$. Furthermore, $\lambda = 0$ is simple. This is enough to prove the assertion (for details, as well as a reference to another proof, see the references in the supplementary remarks). The map α is a C^1-retraction, $\alpha^2 = \alpha$, $\alpha \circ T_f(t) = T_f(t) \circ \alpha$. The image $\alpha(C)$ of α is the attractor $\mathcal{A}(F)$ and, for any constant function c, the set $\alpha^{-1}(c)$ is a submanifold of C of codimension 1.

Example 1.8. Consider the scalar equation

$$\dot{\theta}(t) = \frac{\pi}{2}(1 - \cos\theta(t)) + \frac{\pi}{2}(1 - \cos\theta(t-1)) \tag{1.7}$$

on S^1. The only equilibrium point on S^1 is $\theta = 0$. On the other hand, the function $\theta(t) = \pi t$ is a periodic solution of (1.7). The attractor $\mathcal{A}(F)$ must contain the point 0 as well as this periodic orbit. As a consequence, $\mathcal{A}(F)$ cannot be homeomorphic to S^1. If $\mathcal{A}(F)$ is a manifold, it must have dimension > 1.

12.2 Dimension of the global attractor

In this section, we present some results on the size of the global attractor $\mathcal{A}(F)$ of an RFDE(F) in terms of limit capacity and Hausdorff dimension. The principal results are applicable not only to RFDE but to many other types of evolutionary equations. As we will see, the theory applies to the situation where the solution operator $T(t)$ can be written as $T(t) = S(t) + U(t)$, where $U(t)$ is compact and $S(t)$ is a linear operator whose norm approaches zero as $t \to \infty$.

Let K be a topological space. We say that K is *finite-dimensional* if there exists an integer n such that for every open covering A of K, there exists another open covering A' of K refining A such that every point of K belongs to at most $n+1$ sets of A'. In this case, the *dimension* of K, dim K, is defined as the minimum n satisfying this property. Then dim $\mathbb{R}^n = n$ and, if K is a compact finite-dimensional space, it is homeomorphic to a subset of $\mathbb{R}^n$ with $n = 2\dim K + 1$. If K is a metric space, its *Hausdorff dimension*, $\dim_H(K)$ is defined as follows: for any $\alpha > 0$, $\epsilon > 0$, let

$$\mu_\epsilon^\alpha(K) = \inf \sum_i \epsilon_i^\alpha$$

where the inf is taken over all coverings $B_{\epsilon_i}(x_i)$, $i = 1, 2 \ldots$, of K with $\epsilon_i < \epsilon$ for all i, where $B_{\epsilon_i}(x_i) = \{ x : d(x, x_i) < \epsilon_i \}$. Let $\mu^\alpha(K) = \lim_{\epsilon \to 0} \mu^\alpha_\epsilon(K)$. The function μ^a is called the *Hausdorff measure of dimension* α. For $\alpha = n$ and K a subset of $\mathbb{R}^n$ with $|x| = \sup |x_j|$, μ^n is the Lebesgue outer measure. It is not difficult to show that if $\mu^\alpha(K) < \infty$ for some α, then $\mu^\beta(K) = 0$ if $\beta > \alpha$. Thus,

$$\inf\{ \alpha : \mu^\alpha(K) = 0 \} = \sup\{ \alpha : \mu^\alpha(K) = \infty \},$$

and we define the Hausdorff dimension of K as

$$\dim_H(K) = \inf\{ \alpha : \mu^\alpha(K) = 0 \}.$$

It is known that $\dim(K) \le \dim_H(K)$, and these numbers are equal when K is a submanifold of a Banach space. For general K, there is little that can be said relating these numbers.

To define another measure of the size of a metric space K, let $n(\epsilon, K)$ be the minimum number of open balls of radius ϵ needed to cover K. Define the *limit capacity* of K, $c(K)$, by

$$c(K) = \limsup_{\epsilon \to 0} \frac{\log n(\epsilon, K)}{\log(1/\epsilon)}.$$

In other words, $c(K)$ is the minimum real number such that for every $\sigma > 0$, there is a $\delta > 0$ such that

$$n(\epsilon, K) \le \left(\frac{1}{\epsilon}\right)^{c(K)+\sigma} \qquad \text{if} \quad 0 < \epsilon < \delta.$$

It is not difficult to show that $\dim_H(K) \le c(K)$.

Theorem 2.1. *Suppose that X is a Banach space, $T : X \to X$ is an α-contraction, point dissipative and orbits of bounded sets are bounded. If T is a C^1-map and $D_xT = S + U$, where U is compact and the norm of S is less than 1, then the global attractor $\mathcal{A}$ of T has the following properties:*

(i) *$c(\mathcal{A}) < \infty$;*

(ii) *If $d = 2c(\mathcal{A}) + 1$ and S is any linear subspace of X with $\dim S > d$, then there is a residual set Π of the space of all continuous projections P of X onto S (taken with the uniform operator topology) such that $P|\mathcal{A}$ is one-to-one for every $P \in \Pi$.*

The first part of this theorem says that the limit capacity $c(\mathcal{A})$ of $\mathcal{A}$ is finite, which in turn implies that the Hausdorff dimension is finite. The second part of the theorem says that the attractor can be "flattened" in a residual set of directions onto a finite-dimensional subspace of dimension $> 2c(\mathcal{A}) + 1$.

The proof of Theorem 2.1 is not given. However, it is worthwhile to point out that in proving that $c(\mathcal{A})$ is finite, explicit estimates are obtained (better estimates often can be obtained in a specific problem). The estimates for the limit capacity of a compact attractor for a map T can be obtained by an application of general results for the capacity of compact subsets of a Banach space E with the property that $T(K) \supset K$ for some C^1-map $T : U \to E$, $U \supset K$, whose derivative can be decomposed as a sum of compact map and a contraction—a special case of an α-contraction. To describe the nature of these estimates, we need some notation. For $\lambda > 0$, the subspace of $\mathcal{L}(E)$ consisting of all maps $L = L_1 + L_2$ with L_1 compact and $\|L_2\| < \lambda$ is denoted by $\mathcal{L}_\lambda(E)$. Given a map $L \in \mathcal{L}_\lambda(E)$, we define $L_S = L|S$ and

$$\nu_\lambda(L) = \min\{\, \dim S : S \text{ is a linear subspace of } E \text{ and } \|L_S\| < \lambda\,\}.$$

It is easy to prove that $\nu_\lambda(L)$ is finite for $L \in \mathcal{L}_{\lambda/2}(E)$. The basic result for the estimate of $c(K)$ is contained in

Theorem 2.2. *Let E be a Banach space, $U \subset E$ an open set, $T : U \to E$ a C^1-map and $K \subset U$ a compact set such that $T(K) \supset K$. If the Fréchet derivative $D_xT \in \mathcal{L}_{1/4}(E)$ for all $x \in K$, then*

$$c(K) \leq \frac{\log(\delta[2(\lambda(1+\sigma)+k^2)/\lambda\sigma]^\delta)}{\log[1/2\lambda(1+\sigma)]},$$

where $k = \sup_{x\in K} \|D_xT\|$, $0 < \lambda < 1/2$, $0 < \sigma < (1/2\lambda) - 1$, $\delta = \sup_{x\in K} \nu_\lambda(D_xT^2)$. If $D_xT \in \mathcal{L}_1(K)$ for all $x \in K$, then $c(K) < \infty$.

We may apply Theorem 2.1 directly to RFDE. In fact, if $T_F(t)$ is the solution map of RFDE(F), then the map $T \stackrel{\text{def}}{=} T_F(r)$ satisfies the conditions of Theorem 2.1 since $T_F(r)$ is compact. We summarize this remark in

Theorem 2.3. *Suppose $F \in \mathcal{X}^1$ and $\mathcal{A}(F)$ is the set of globally defined and bounded solutions of* RFDE(F). *For any $\beta > 0$, there is a positive number d_β such that $c(\mathcal{A}_\beta(F)) \leq d_\beta$, where $\mathcal{A}_\beta(F) = \mathcal{A}(F) \cap \{\, \phi \in C^0(I, M) : |\phi| \leq \beta\,\}$. If $\mathcal{A}(F)$ is the global attractor (which happens if the* RFDE *is point dissipative), then $c(\mathcal{A}(F)) < \infty$. The same remarks apply to the Hausdorff dimension.*

Of course, there is a transcription of Theorem 2.2 to RFDE and we leave this for the reader to do.

The finite dimensionality of the sets $\mathcal{A}_\beta(F)$ implies the finite dimensionality of the period module of any almost periodic solution of RFDE(F), generalizing the same result for ordinary differential equations. Any almost periodic function $x(t)$ has a Fourier expansion $x(t) \sim \Sigma a_n exp(-i\lambda_n t)$, where $\Sigma|a_n|^2 < \infty$. The *period module* of $x(t)$ is the vector space $\mathcal{M}$

spanned by the set $\{\lambda_n\}$ over the rationals. Saying the period module is finite-dimensional is equivalent to saying that the function x is quasi-periodic. As a consequence of Theorem 2.3, we have the following result.

Theorem 2.4. *For a C^1-RFDE(F) on a manifold M, every almost periodic solution is quasi-periodic. If RFDE(F) has a global attractor, then there is an integer N such that the period module of any almost periodic solution has dimension $\leq N$.*

The conclusion that $\mathcal{A}(F)$ has finite dimension depends upon the fact that the RFDE is at least C^1. In fact, let Q_L be the set of functions $\gamma : \mathbb{R}^n \to \mathbb{R}^n$ with global Lipschitz constant L. For each $\gamma \in Q_L$, each solution of the ordinary differential equation $\dot{x} = \gamma(x)$ is defined for all $t \in \mathbb{R}$.

Theorem 2.5. *For each $L > 0$, there is a continuous RFDE(F), depending only on L, such that for every $\gamma \in Q_L$, every solution of the ODE $\dot{x} = \gamma(x)$ is also a solution of the RFDE(F). In particular, $\mathcal{A}(F)$ has infinite dimension.*

12.3 $\mathcal{A}$-stability and Morse-Smale maps

The primary objective in the qualitative theory of discrete dynamical systems T on a Banach space X is to study the manner in which the flow changes when T changes. Due to the infinite dimensionality of the space and the fact that T may not be one-to-one, a comparison of all orbits of two different maps is very difficult and is likely to lead to severe restrictions on the maps that are to be considered. For this reason, we restrict our discussion to mappings that have a global attractor and then make comparisons of orbits on the attractors.

Let $C^r(X,X)$, $r \geq 1$, be the space of C^r-maps from X to X. Let $KC^r(X,X)$ be the subset of $C^r(X,X)$ with the property that

(i) $T \in KC^r(X,X)$ implies that T has a global attractor $\mathcal{A}(T)$.

(ii) $\mathcal{A}(T)$ is upper semicontinuous on $KC^r(X,X)$.

For $T, S \in KC^r(X,X)$, we say that T is *equivalent* to S, $T \sim S$, if there is a homeomorphism $h : \mathcal{A}(T) \to \mathcal{A}(S)$ such that $hT = Sh$ on $\mathcal{A}(T)$. We say that T is *$\mathcal{A}$-stable* if there is a neighborhood V of T in $C^r(X,X)$ such that $T \sim S$ for every $S \subset V \cap KC^r(X,X)$.

A fixed point x of a map $T \in C^r(X,X)$ is *hyperbolic* if the spectrum of $D_xT(x)$ does not intersect the unit circle in $\mathbb{C}$ with center zero. For any hyperbolic fixed point of T, let

$$\begin{aligned} W^s(x,T) &= \{\, y \in X : T^n y \to x \text{ as } n \to \infty \,\}, \\ W^u(x,T) &= \{\, y \in X : T^{-n} y \text{ is defined for } n \geq 0 \\ &\qquad\qquad \text{and } T^{-n}y \to x \text{ as } n \to \infty \,\}. \end{aligned}$$

The sets $W^s(x,T)$, $W^u(x,T)$ are called, respectively, the stable and unstable sets of T. If x is a hyperbolic fixed point of T, then there is a neighborhood V of x such that

$$\begin{aligned} W^s_{\text{loc}}(x,T) &= W^s_{\text{loc}}(x,T,V) \stackrel{\text{def}}{=} \{\, y \in W^s(x,T) : T^n y \in V, n \geq 0 \,\}, \\ W^u_{\text{loc}}(x,T) &= W^u_{\text{loc}}(x,T,V) \stackrel{\text{def}}{=} \{\, y \in W^u(x,T) : T^{-n} y \in V,\ n \geq 0 \,\} \end{aligned}$$

are C^r-manifolds. These sets are called the local stable and unstable manifolds. If the maps T and DT are one-to-one on X, then $W^s(x,T)$, $W^u(x,T)$ are immersed in X.

A point $x \in X$ is a *periodic point of period* p *of* T if $T^p x = x$, $T^j x \neq x$, $j = 1, 2, \ldots, p-1$. A periodic point x of period p is *hyperbolic* if the spectrum of $DT^p(x)$ does not intersect the unit circle in $\mathbb{C}$ with center zero.

A map $T \in KC^r(X,X)$ is *Morse-Smale* if

(i) T, $D_x T$ are one-to-one on $\mathcal{A}(T)$.

(ii) $\Omega(T)$, the nonwandering set of T, is finite and consists of the periodic points $\text{Per}\,(T)$ of T.

(iii) All periodic points are hyperbolic with finite-dimensional unstable manifolds.

(iv) $W^s(x,T)$ is transversal to $W^u(x,T)$ for all $x, y \in \text{Per}\,(T)$.

The following result is basic in stability theory.

Theorem 3.1. *If $T \in KC^r(X,X)$ is Morse-Smale, then T is $\mathcal{A}$-stable.*

For a Morse-Smale map $T \in KC^r(X,X)$, the global attractor $\mathcal{A}(T)$ can be written as

$$\mathcal{A}(T) = \bigcup_{x \in \text{Per}\,(T)} W^u(x,T).$$

When we apply these results to an RFDE(F), we fix a $t_0 > 0$ (for example, we could take $t_0 = r$) and then define $T \equiv T_F(t_0)$. If T is Morse-Smale, then Theorem 3.1 asserts the topological equivalence on the attractors of the discrete flow defined by the iterates of T to that defined by a map S close to T. It does not say very much about the continuous flows defined by two RFDE. If the continuous flows are gradient (that is, the α- and ω-limit sets of any bounded orbit belong to the set of equilibrium points), then the consideration of the map captures the essential properties of the flow. On the other hand, if there are periodic orbits for the continuous flow, then the discrete map will not capture this behavior. If we assume that there are only a finite number of equilibrium points and periodic orbits for the continuous flow, each of which is hyperbolic, all stable and unstable manifolds intersect transversally, and the nonwandering set consists only of the equilibrium points and periodic orbits, then we would like to call the continuous flow a *Morse-Smale* flow. It is reasonable to expect that a continuous version of Theorem 3.1 is true, but no one has written the proof.

We now present an example of a nontrivial Morse-Smale gradient system for which we can give a considerable amount of information about the flow on the attractor.

Suppose $b \in C^2([-1,0], \mathbb{R}), b(-1) = 0, b(\theta) > 0, b'(\theta) \geq 0,\ b''(\theta) \geq 0$ for $\theta \in (-1, 0]$ and

$$b''(\theta_0) > 0 \qquad \text{for some} \quad \theta_0 \in [-1,0]. \tag{3.1}$$

Let $g \in C^1(\mathbb{R}, \mathbb{R})$ be such that

$$G(x) \stackrel{\text{def}}{=} \int_0^x g(s)\, ds \to \infty \qquad \text{as} \quad |x| \to \infty \tag{3.2}$$

and consider the equation

$$\dot{x}(t) = -\int_{-1}^{0} b(\theta) g(x(t+\theta))\, d\theta. \tag{3.3}$$

Let $T_{b,g}(t)\phi \stackrel{\text{def}}{=} T(t)\phi$ be the solution of (3.3) with initial value ϕ at $t = 0$. We say that $\mathcal{A}_{b,g}$ is the *minimal global attractor* for (3.3) if $\mathcal{A}_{b,g}$ is invariant and attracts any bounded set of C. We have the following result.

Theorem 3.2. *If* (3.1), (3.2) *are satisfied, then there exists a locally compact minimal global attractor for System* (3.3). *Furthermore, System* (3.3) *defines a gradient flow with the ω-limit set of any orbit being a single equilibrium point. If the zero set E of g is bounded, then $\mathcal{A}_{b,g}$ is compact. If, in addition, each element of E is hyperbolic, then* dim $W^u(x_0) = 1$ *for each* $x_0 \in E$ *and*

$$\mathcal{A}_{b,g} = \bigcup_{x_0 \in E} W^u(x_0). \tag{3.4}$$

In the proof of Theorem 3.4 of Chapter 5, we used a Liapunov functional to show that positive orbits of bounded sets are bounded and that the ω-limit set of any orbit belongs to the set of equilibrium points, that is, the zeros of g. Therefore, if we define $\mathcal{A}_{b,g} = \bigcup_{r>0} \omega(\mathcal{B}(0,r))$, where $\mathcal{B}(0,r)$ is the ball in C of center 0 and radius r, then it is clear that $\mathcal{A}_{b,g}$ is the minimal global attractor. The existence of the Liapunov functional V shows also that the α-limit set of any orbit in $\mathcal{A}_{b,g}$ belongs to the set of equilibrium points. If the zero set E of g is bounded, then System (3.3) is point dissipative and there is a compact global attractor. This set is $\mathcal{A}_{b,g}$. If x_0 is an equilibrium point (that is, $g(x_0) = 0$), we consider the linear variational equation about x_0:

$$\dot{y}(t) = -\int_{-1}^{0} b(\theta)g'(x_0)y(t+\theta)\,d\theta. \tag{3.5}$$

The characteristic equation is

$$\lambda + \int_{-1}^{0} b(\theta)g'(x_0)e^{\lambda\theta}\,d\theta = 0. \tag{3.6}$$

It is possible to show that the equilibrium point x_0 is hyperbolic if $g'(x_0) \neq 0$ and x_0 is asymptotically stable if $g'(x_0) > 0$ and unstable if $g'(x_0) < 0$. Furthermore, if $g'(x_0) < 0$, it is possible to show that dim $W^u(x_0) = 1$. Therefore, if all of the equilibrium points are hyperbolic, we have (3.4). Since the ω-limit set of an orbit is connected, the only situation in which this limit can contain more than one equilibrium point x_0 is when $g'(x_0) = 0$. In this case, the only characteristic value is $\lambda = 0$, and it is simple. It follows from general convergence theorems for gradient systems that the ω-limit set is a singleton (see the supplementary remarks for references). This completes the proof of the theorem.

Let us now suppose that the zero set E of g is finite and each zero of g is simple. In this case, the compact global attractor $\mathcal{A}_{b,g}$ is represented as in Equation (3.4) and is one-dimensional. If $g'(x_0) < 0$, then there are two orbits in C leaving the point x_0; that is, there are two distinct solutions $\phi(t)$, $\psi(t)$ of System (3.3), defined for $t \leq 0$ that approach x_0 as $t \to -\infty$. The problem is to determine the limits of $\phi(t)$, $\psi(t)$ as $t \to \infty$. We know that these limits must be equilibrium points of System (3.3). It is natural to suspect that one of these limits will be larger than x_0 and one will be less than x_0 (that is, the ordering of the real numbers is preserved on the attractor). If $E = \{x_1, x_2, x_3\}$ with $x_1 < x_2 < x_3$, this is obviously true since $\mathcal{A}_{b,g}$ is connected.

It is a surprising fact that this is not in general true if the set E consists of five zeros of g. To describe this situation, it is convenient to systematize the notation. Let us use the symbol $j[k,l]$ to mean that the unstable point x_j is connected to x_k, x_l by an orbit. If g has only the five simple zeros, $x_1 < x_2 < x_3 < x_4 < x_5$, then x_2, x_4 are unstable with one-dimensional unstable manifolds and x_1, x_2, x_3 are asymptotically stable. The flow on $\mathcal{A}_{b,g}$ is determined by the manner in which the points x_2, x_4 are connected by orbits to the other equilibrium points. The main result is the following.

Theorem 3.3. *For a given function b, one can realize each of the following flows on the attractor $\mathcal{A}_{b,g}$ by choosing an appropriate function g with five simple zeros:*

(i) $2[1,3], 4[3,5]$,

(ii) $2[1,4], 4[3,5]$,

(iii) $2[1,5], 4[3,5]$,

(iv) $2[1,3], 4[2,5]$,

(v) $2[1,3], 4[1,5]$.

The only situation in which the flow on $\mathcal{A}_{b,g}$ preserves the natural order of the reals is case (i) (it is also Morse-Smale). Case (ii) has a nontransverse intersection between $W^u(x_2)$ and $W^s(x_4)$ and case (iv) has a nontransverse intersection between $W^s(x_2)$ and $W^u(x_4)$. In cases (iii) and (v), there is transversal intersection of the stable and unstable manifolds (thus, Morse-Smale) and the natural order of the reals is not preserved by the flow on the attractor.

12.4 Hyperbolicity is generic

The aim of the generic theory of differential equations is to study qualitative properties of solutions that are typical in the sense that they hold for all equations defined by functions of a residual set of the function space being considered. More precisely, if $\mathcal{X}$ is a complete metric space, then a property $\mathcal{P}$ on the elements $x \in \mathcal{X}$ is said to be *generic* if there is a residual set $\mathcal{Y} \subset \mathcal{X}$ such that each element of $\mathcal{Y}$ has property $\mathcal{P}$. We recall that a residual set is either an open dense set or, more generally, a countable intersection of such sets.

For any Banach space E and any C^∞ n-dimensional manifold M with a complete Riemannian metric, we let $C^k(E, M)$ denote the space of functions from E to M that are continuous together with derivatives up through order k. The metric on $C^k(E, M)$ is the one induced by the Riemannian metric on M, taking into account the differences between functions and their derivatives up through order k.

For ordinary differential equations, the most basic result in the generic theory is the *Kupka-Smale theorem.* To be specific, for the ordinary differential equation

$$\dot{x} = f(x), \tag{4.1}$$

where $f \in C^k(\mathbb{R}^n, \mathbb{R}^n)$, let $\mathcal{P}$ be the property that all critical points (equilibrium points) and periodic orbits are hyperbolic and the stable and unstable manifolds intersect transversally. The Kupka-Smale theorem asserts that this property $\mathcal{P}$ is generic. This theorem shows that the hyperbolicity requirement in a Morse-Smale system is generic.

The complete proof of the Kupka-Smale theorem for retarded FDE is not available at the present time. However, the following result is true.

Theorem 4.1. *The set of $f \in C^k(C, \mathbb{R}^n)$, for which all critical points and all periodic orbits of the* RFDE,

$$\dot{x}(t) = f(x_t), \tag{4.2}$$

are hyperbolic, is residual.

We give an indication of the proof, emphasizing only those parts that are distinctly different from the one for ordinary differential equations. We recall that an equilibrium point is *nondegenerate* if zero is not an eigenvalue of the corresponding linear variational equation. A periodic orbit defined by a periodic function $p(t)$ is called nondegenerate if the linear variational equation for $p(t)$ has 1 as a simple characteristic multiplier. For a fixed compact set $K \subset \mathbb{R}^n$ and a fixed $T > 0$, we define the following subsets of $C^k(C, \mathbb{R}^n)$:

$$\mathcal{G}_0(K) = \{f : \text{all critical points in } K \text{ of (4.2) are nondegenerate}\},$$

$$\mathcal{G}_1(K) = \{f : \text{all critical points in } K \text{ of (4.2) are hyperbolic}\},$$

$$\mathcal{G}_{3/2}(T, K) = \{f \in \mathcal{G}_1(K) : \text{all periodic orbits of (4.2) in } K \text{ and having a period in } (0, T] \text{ are nondegenerate}\}$$

$$\mathcal{G}_2(T, K) = \{f \in \mathcal{G}_1(K) : \text{all periodic orbits of (4.2) in } K \text{ and having a period in } (0, T] \text{ are hyperbolic}\}.$$

For each T, each of the sets $\mathcal{G}_0(K)$, $\mathcal{G}_1(K)$, $\mathcal{G}_{3/2}(T, K)$, $\mathcal{G}_2(T, K)$ is open as a consequence of the results in Chapter 10. If for each T, we show that $\mathcal{G}_2(T, K)$ is dense, then it follows that $\mathcal{G}_2(K) = \cap_{N=1}^{\infty} \mathcal{G}_2(N, K)$ is residual, and this will show that Theorem 4.1 holds at least for those solutions lying in K. By taking a sequence K_m of compact sets whose union is $\mathbb{R}^n$, and intersecting the corresponding residual sets $\mathcal{G}_2(K_m)$, the theorem is proved.

To show that $\mathcal{G}_0(K)$ and $\mathcal{G}_1(K)$ are dense, first take any $f \in \mathcal{X}^1$ and make two perturbations as follows:

(i) By Sard's theorem, there is an f^1 near f for which all critical points in K are isolated and nondegenerate. Thus, $f^1 \in \mathcal{G}_0(K)$.

(ii) By perturbing locally near each critical point, one may obtain f^2 near f^1 such that each critical point is hyperbolic. Thus, $f^2 \in \mathcal{G}_1(K)$.

To show that $\mathcal{G}_{3/2}(T, K)$ and $\mathcal{G}_2(T, K)$ are dense for any T, we begin with some $f \in \mathcal{G}_1(K)$. One may easily find a lower bound $T_0 > 0$ for periods of periodic solutions of Equation (4.2) in K and so, trivially, $f \in \mathcal{G}_2(T, K)$

for $T < T_0$. An induction argument on the period T is then used. This induction also involves two steps.

(a) An argument involving Sard's theorem can be used to prove that $\mathcal{G}_2(T, K)$ is dense in $\mathcal{G}_{3/2}(3T/2, K)$.

(b) A local perturbation around each nondegenerate periodic orbit of some $f \in \mathcal{G}_{3/2}(T, K)$ yields a hyperbolic orbit, thereby showing that $\mathcal{G}_{3/2}(T, K)$ is dense in $\mathcal{G}_2(T, K)$.

The most difficult step is (a), since we must be concerned with all of the periodic solutions with period in $(T, 3T/2]$. These may not be nondegenerate. Let $x(t)$ be a periodic solution with least period $\tau \in (T, 3T/2]$ and consider the map

$$\Phi : (\phi, t, g) \to x_t(\phi, g) - \phi,$$

where $x_t(\phi, g) \in C$ is the solution of

$$\dot{x}(t) = g(x_t), \quad x_0 = \phi. \tag{4.3}$$

Clearly, zeros of Φ correspond to the initial data for periodic solutions of Equation (4.3). Therefore, $\Phi(x_0, \tau, f) = 0$. What is of particular interest are the periodic solutions in a neighborhood of (x_0, τ) for g in a neighborhood of f. To study the zeros of Φ, the Fréchet derivative of Φ must be analyzed and, in particular, the derivative $\Gamma = D_g\Phi(x_0, \tau, f)$ with respect to the variable $g \in \mathcal{X}^1$. This leads to the study of the inhomogeneous variational equation

$$\dot{y}(t) = D_\phi f(x_t)y_t + h(x_t), \quad y_0 = 0, \tag{4.4}$$

about the known periodic solution $x(t)$ since $\Gamma : \mathcal{X}^1 \to C$ is given by $(\Gamma h)(t) = y_t$. For simplicity, suppose that $\tau > r$. In order to use Sard's theorem, we must verify that the range of the linear map Γ is dense in C: that is, for any $\psi \in C$ and any $\epsilon > 0$, there is an $h \in \mathcal{X}^1$ such that the solution of (4.4) satisfies $|y_\tau - \psi| < \epsilon$. We may always choose such an h of the form

$$h(\phi) = H(\phi(0), \phi(\frac{-r}{N}), \phi(\frac{-2r}{N}), \ldots, \phi(-r)) \tag{4.5}$$

for some N, because of the following result.

Lemma 4.1. *Let $x(t)$ be a periodic solution of* (4.2) *of least period $\tau > 0$. Then, for sufficiently large N, the map*

$$\gamma(t) = (x(t), x(t - \frac{r}{N}), x(t - \frac{2r}{N}), \ldots, x(t - r))$$

is a one-to-one regular (that is, $\dot{\gamma}(t) \neq 0$ for all t) mapping of the reals mod τ into $\mathbb{R}^{n(N+1)}$.

We remark that it is possible to prove a generalization of Theorem 4.1 by using the C^k-Whitney topology on the functions from C to $\mathbb{R}^n$.

It is interesting to restrict the class of functions $\mathcal{X}^1$. For example, suppose that the systems under investigation are differential difference equations of the form

$$\dot{x}(t) = F(x(t), x(t-1)). \tag{4.6}$$

To obtain a generic theorem about this restricted class of equations is more difficult since there is less freedom to construct perturbations. For instance, the function h in (4.5) cannot be used. Nevertheless, Theorem 4.1 remains true for these equations. Instead of using Lemma 4.1 to produce h, one first approximates Equation (4.6) with an analytic F and then uses the following

Lemma 4.2. *If $x(t)$ is a periodic solution of Equation (4.6) of least period $\tau > 0$ and F is analytic, then the map*

$$\gamma(t) = (x(t), x(t-1))$$

is one-to-one and regular except at a finite number of t values in the reals mod τ.

One may consider an even more restrictive class of equations of the form

$$\dot{x}(t) = F(x(t-1)).$$

For this class of equations, Theorem 4.1 is not known.

It would be very interesting to obtain the complete Kupka-Smale theorem for any of these situations, that is, assuming that there is a residual set of f such that the conclusions of Theorem 4.1 hold and the stable and unstable manifolds of critical points and periodic orbits intersect transversally. Some ideas distinctly different from the ones used for ordinary differential equations seem to be required.

12.5 One-to-oneness on the attractor

The following result is almost obvious.

Proposition 5.1. *Let $F \in \mathcal{X}^k$, $k \geq 1$. If $\mathcal{A}$ is a compact invariant set of* RFDE(F) *and $T_F(t)$ is one-to-one on $\mathcal{A}$, then $T_F(t)$ is a continuous group of operators on $\mathcal{A}$.*

In particular, if $\mathcal{A}(F)$ is the global attractor for $T_F(t)$ and $T_F(t)$ is one-to-one on $\mathcal{A}(F)$, then $T_F(t)$ is a group on $\mathcal{A}(F)$. This is certainly sufficient reason to study the following question: When is $T_F(t)$ one-to-one? As we have noted in Chapter 3, this need not be true on all of C. Therefore, it is natural to ask if perhaps the property of one-to-oneness is generic. At this time, this question has not been answered completely, but there are some results, which are discussed in this section.

Theorem 5.1. *For an analytic* RFDE(F), *any globally defined bounded solution* $x(t)$, $t \in \mathbb{R}$, *is analytic. In particular,* $T_F(t)$ *is one-to-one on the set* $\mathcal{A}(F)$ *of globally defined bounded solutions.*

For a class of linear nonautonomous RFDE, analyticity is not required. Let $\mathcal{L} = \mathcal{L}(C, \mathbb{R}^n)$ be the Banach space of all continuous linear mappings $L : C \to \mathbb{R}^n$ with the usual norm and let $C^1(\mathbb{R}, \mathcal{L})$ be the space of continuously differentiable mappings from $\mathbb{R}$ to $\mathcal{L}$ with the uniform C^1 topology on compact sets of $\mathbb{R}$. It is possible to prove the following result.

Theorem 5.2. *The set of* $L \in C^1(\mathbb{R}, \mathcal{L})$ *such that the corresponding solution operator of* RFDE(L) *is one-to-one on* $\mathbb{R}$ *is dense. For any compact set* $K \subset \mathbb{R}$, *the set of* $L \in C^1(\mathbb{R}, \mathcal{L})$ *for which the solution operator of* RFDE(L) *is one-to-one on* K *is not open.*

It is not known if the set of L in Theorem 5.2 is residual. For a more restricted class, better results are obtained. Let $\mathcal{D}(\mathbb{R}) \subset C^1(\mathbb{R}, \mathcal{L})$ be the set of L such that there is an integer N such that the corresponding measure $\eta(t, \theta)$ of L has at most N discontinuities in θ for all $t \in \mathbb{R}$ and, in addition, $\eta(t, \theta)$ is a step function in θ. For any compact set $K \subset \mathbb{R}$, the set $\mathcal{D}(K)$ is defined in a similar way.

Theorem 5.3. *The set of* L *in* $\mathcal{D}(\mathbb{R})$ *for which the solution operator of RFDE(L) is one-to-one on* $\mathbb{R}$ *is residual. Also, for any compact set* $K \subset \mathbb{R}$, *the corresponding set is open in* $\mathcal{D}(K)$.

In the case where we restrict the solution map $T_F(t)$ to the global attractor, the one-to-oneness can be related to other interesting concepts. If E, X are Banach spaces and $B(e) \subset X$, $e \in E$, is a family of subsets, then we say that $B(e)$ is *lower semicontinuous* at e_0 if $\text{dist}\,(B(e_0), B(e)) \to 0$ as $e \to e_0$. We recall that the distance between two subsets B, C is defined by $\text{dist}\,(B, C) = \sup_{x \in B} \inf_{y \in C} \|x - y\|$.

Theorem 5.4. *If* M *is a compact manifold and* $F_0 \in \mathcal{X}^1$ *is given, then the following conclusions hold:*

(i) *There is a compact set* $K \supset \mathcal{A}(F)$ *such that* F_0 *can be uniformly approximated on* K *by a* C^1-RFDE(F) *for which* $T_F(t)$ *is one-to-one on* $\mathcal{A}(F)$.

(ii) *If* F_0 *is* $\mathcal{A}$*-stable, then* $T_{F_0}(t)$ *is one-to-one on* $\mathcal{A}(F_0)$.

(iii) *If* F_0 *is given and* $\mathcal{A}(F)$ *is lower semicontinuous at* F_0, *then* $T_{F_0}(t)$ *is one-to-one on* $\mathcal{A}(F_0)$.

For delay differential equations, this result can be modified in the following way.

Theorem 5.5. *If M is a compact manifold and $F_0 \in \mathcal{X}^1$ is a given delay differential equation with a finite number of delays, then the following statements hold:*

(i) *F_0 can be uniformly approximated by a C^1-delay differential equation F with the same number of delays and $T_F(t)$ is one-to-one on $\mathcal{A}(F)$.*

(ii) *If F_0 is $\mathcal{A}$-stable, then $T_{F_0}(t)$ is one-to-one on $\mathcal{A}(F_0)$.*

(iii) *If F_0 is given and $\mathcal{A}(F)$ is lower semicontinuous at F_0, then $T_{F_0}(t)$ is one-to-one on $\mathcal{A}(F_0)$.*

The proof of these theorems use some auxiliary results of independent interest.

Proposition 5.2. *If $F, G \in \mathcal{X}^1$, $F \sim G$ and $T_F(t)$ is one-to-one on $\mathcal{A}(F)$, then $T_G(t)$ is one-to-one on $\mathcal{A}(G)$.*

This proposition yields immediately assertion (b) in the theorems.

Proposition 5.3. *Suppose that F_i, $F \in \mathcal{X}^1$, $i \geq 1$, and that $F_i \to F$ as $i \to \infty$. If $\mathcal{A}(F)$ is lower semicontinuous in F and $T_{F_i}(t)$ is one-to-one on $\mathcal{A}(F_i)$, then $T_F(t)$ is one-to-one on $\mathcal{A}(F)$.*

These propositions, together with the Stone-Weierstrass approximation theorem, are the main ingredients of the proof of Theorems 5.4 and 5.5.

We remark that the conclusions of Theorem 5.4 remain valid for RFDE on noncompact manifolds. In this case, we restrict consideration to the class of RFDE that has a global attractor that is upper semicontinuous.

An RFDE is said to be *gradient-like* if all positive orbits are bounded, the ω-limit set of each positive orbit and the α-limit set of each bounded negative orbit belongs to the set E of equilibrium points. If E is bounded, then there is a global attractor. It is possible to prove the following result.

Theorem 5.6. *If an* RFDE(F_0) *on a compact manifold M is gradient-like and all equilibrium points are hyperbolic, then $\mathcal{A}(F)$ is lower semicontinuous at F_0. Thus, $\mathcal{A}(F)$ is Hausdorff continuous in F from Theorem* 5.5.

Theorem 5.6 is intuitively obvious for the following reason. If the stable and unstable manifolds were transversal, then the RFDE(F) would be $\mathcal{A}$-stable and, in particular, $\mathcal{A}(F)$ is Hausdorff continuous in F. If the stable and unstable manifolds are not transversal along an orbit γ_F whose α-limit set is ϕ_- and whose ω-limit set is ϕ_+, then, under a small perturbation G of

F, and for any neighborhood U of γ_F, there is an orbit $\gamma_G = \bigcup_{t\in\mathbb{R}} T_G(t)\phi$ of the RFDE(G) such that $T_G(t)\phi$ remains in U for as long as we like. This certainly suggests that $\mathcal{A}(F)$ should be lower semicontinuous. Supplying a precise proof requires considerably more effort.

12.6 Morse decompositions

Suppose that $T_0(t)$, $t \geq 0$, is a C^0-semigroup on a Banach space X for which there is a compact global attractor $\mathcal{A}$. A *Morse decomposition* of the attractor $\mathcal{A}$ is a finite ordered collection $\mathcal{A}_1 < \mathcal{A}_2 < \cdots < \mathcal{A}_M$ of disjoint compact invariant subsets of $\mathcal{A}$ (called *Morse sets*) such that for any $\phi \in \mathcal{A}$, there are positive integers N and K, $N \geq K$, such that $\alpha(\phi) \subset \mathcal{A}_N$ and $\omega(\phi) \subset \mathcal{A}_K$ and $N = K$ implies that $\phi \in \mathcal{A}_N$. In the case $N = K$, we have $T(t)\phi \in \mathcal{A}_N$ for $t \in \mathbb{R}$.

The Morse sets, together with the *connecting orbits*

$$C_K^N \stackrel{\text{def}}{=} \{\, \phi \in \mathcal{A} : \alpha(\phi) \subset \mathcal{A}_N,\ \omega(\phi) \subset \mathcal{A}_K \,\}$$

for $N > K$, give the global attractor $\mathcal{A}$.

There are two obvious Morse decompositions, namely, the set $\mathcal{A}$ itself and the empty set. Neither of these decompositions is interesting. A Morse decomposition becomes important when it gives some additional information about the flow defined by the semigroup restricted to the attractor. In this section, we consider in some detail a Morse decomposition for a special class of differential difference equations. We only give some ideas of the proofs; the reader may consult the references in the section on supplementary remarks for details.

Consider the equation

$$\dot{x}(t) = -\beta x(t) - g(x(t-1)), \tag{6.1}$$

where $\beta \geq 0$ and the following hypotheses are satisfied:

(H_1) $g \in C^\infty(\mathbb{R}, \mathbb{R})$ and has *negative feedback*; that is, $xg(x) > 0$ for $x \neq 0$ and $g'(0) > 0$,

(H_2) there is a constant k such that $g(x) \geq -k$ for all x,

(H_3) the zero solution of Equation (6.1) is hyperbolic.

The results to be stated hold in more general situations. The right-hand side of Equation (6.1) can be replaced by $f(x(t), x(t-1))$ with a modified definition of negative feedback. Hypothesis (H_2) can be replaced by the assumption that the semigroup associated with Equation (6.1) is point dissipative. We need only the existence of a compact global attractor. We will prove that (H_1) and (H_2) imply this property. Hypothesis (H_3) is not necessary, but to eliminate it requires more complicated definitions.

In Chapter 11, and especially in the supplementary remarks of that chapter, we mentioned several examples that have the form (6.1) and have negative feedback.

Lemma 6.1. *Under hypotheses* (H_1) *and* (H_2), *the semigroup* $T(t)$ *generated by Equation* (6.1) *is a bounded map and is point dissipative. Thus, there is a compact global attractor* $\mathcal{A}$.

Proof. By integrating Equation (6.1) over $[0,1]$, one deduces that $T(1)$ is a bounded map. Thus, $T(t)$ is a bounded map for any $t > 0$.

To show that $T(t)$ is point dissipative, we observe that

$$\frac{d}{dt}(e^{\beta t}x(t)) \leq e^{\beta t}k$$

for all $t \geq 0$. Therefore,

$$x(t) \leq x(0)e^{-\beta t} + k(1 - e^{-\beta t})/\beta$$

and $\limsup_{t\to\infty} x(t) \leq 2k/\beta$. Since $x(t)$ is bounded above, it follows that $-g(x(t-1))$ is bounded below by a constant K_1. Therefore, arguing as earlier, one obtains that $\liminf_{t\to\infty} x(t) \geq -2K_1/\beta$. This shows that $T(t)$ is point dissipative and Corollary 4.3.2 completes the proof of the lemma. □

We now describe an interesting Morse decomposition of the flow on the attractor $\mathcal{A}$ defined by the semigroup $T(t)$ corresponding to (6.1). To do this, it is convenient to think of the flow on $\mathcal{A}$ in the following way. For any $\phi \in \mathcal{A}$, we know that $T(t)\phi \in \mathcal{A}$ for all $t \in \mathbb{R}$. Since $(T(t)\phi)(\theta) = (T(t+\theta)\phi)(0)$ for $\theta \in [-1,0]$, the orbit $T(t)\phi$, $t \in \mathbb{R}$ can be identified with the function $x(t,\phi) = (T(t)\phi)(0)$, $t \in \mathbb{R}$. With this observation, we make the following definition. For any $\phi \in \mathcal{A}$, $\phi \neq 0$, let $\sigma \geq t$ be the first zero of $x(t,\phi)$ in $[t,\infty)$, if it exists. We define $V(T(t)\phi)$ as the number of zeros of $x(t,\phi)$ (counting multiplicity) in the half-open interval $(\sigma-1,\sigma]$. If σ does not exist, we define $V(T(t)\phi) = 1$. Thus, $V(T(t)\phi)$ is either a positive integer or ∞. We will refer to $V(T(t)\phi)$ either as the *Liapunov function for* (6.1) or as the *zero number of* $x(\cdot,\phi)$.

Theorem 6.1.

(i) $V(T(t)\phi)$ *is nonincreasing in* t *for each* $\phi \in \mathcal{A}$, $\phi \neq 0$.

(ii) $V(T(t)\phi)$ *is an odd integer for each* $\phi \in \mathcal{A}$, $\phi \neq 0$.

(iii) *There is a constant* K *such that* $V(T(t)\phi) \leq K$ *for all* $t \in \mathbb{R}$, $\phi \in \mathcal{A}$, $\phi \neq 0$.

We give only some of the intuitive ideas of why Theorem 6.1 is true. The complete proof is very difficult and the interested reader should consult the references. Let us first suppose that the zeros of $x(t, \phi)$ are simple. Let $\sigma_0 < \sigma_1$ be consecutive zeros with $x(t, \phi) > 0$ in between. Then $\dot{x}(\sigma_0, \phi) > 0$ and $\dot{x}(\sigma_1, \phi) < 0$. From the negative feedback condition (H_1), it follows that $x(\sigma_0 - 1, \phi) < 0$ and $x(\sigma_1 - 1, \phi) > 0$. Thus, $x(t, \phi) = 0$ at some point in $(\sigma_0 - 1, \sigma_1 - 1)$; that is, $x(t, \phi)$ can have no more zeros in $(\sigma_1 - 1, \sigma_1]$ than it does in $(\sigma_0 - 1, \sigma_0]$. This shows that $V(T(t)\phi)$ is nonincreasing in t. Again, if we assume that the zeros of $x(t, \phi)$ are simple, then $x(\sigma, \phi) = 0$, $\dot{x}(\sigma, \phi) > 0$ (resp. $\dot{x}(\sigma, \phi) < 0$) imply that $x(\sigma - 1, \phi) < 0$ (resp. $x(\sigma - 1, \phi) > 0$), which in turn implies that the number of zeros of $x(t, \phi)$ in $(\sigma - 1, \sigma]$ is odd.

If the zeros of $x(t, \phi)$ are not simple, one first proves that $V(x(\cdot, \phi) < \infty$ for $\phi \in \mathcal{A}, \phi \neq 0$. This requires several technical estimates. Also, if $\phi \in \mathcal{A}, \phi \neq 0$ has a zero of order exactly k at $t = \sigma$, then it is easy to see that $t = \sigma - 1$ is a zero of exactly order $k - 1$ and $D^{k-1}x(\sigma - 1, \phi)D^k x(\sigma, \phi) < 0$. The proofs of (i) and (ii) are completed by noting sign changes near the zeros.

The proof of property (iii) in the theorem is technical and difficult. The difficulties arise in the determination of the behavior of the flow on $\mathcal{A}$ near the origin. We can recognize the problem even in the case where the origin is hyperbolic; that is, the solutions of the characteristic equation for the linear variational equation around zero of (6.1),

$$\lambda + \beta + g'(0)e^{-\lambda} = 0, \tag{6.2}$$

have nonzero real parts. If $\phi \in \mathcal{A}$, $\phi \neq 0$ and the solution $x(t, \phi)$ stays in a small neighborhood of the origin for $t \leq -\tau$, then it must lie on the unstable manifold W^u of zero. Since this set is finite-dimensional, the solution will approach the origin as $t \to -\infty$ along an eigenspace of the linear equation and therefore should have the same type of oscillatory properties as the eigenfunctions. Therefore, we should have an integer N^* such that $V(x(t, \phi) \leq N^*$ for all $\phi \in W^u \setminus \{0\}$. If $x(t, \phi)$ remains in a small neighborhood of the origin for $t \geq \tau$, then it must lie on the stable manifold W^s of 0. Again, one would expect that this solution would approach the origin along *one* of the eigenspaces of the linear equation. If this were the case, then the oscillatory properties would be the same as those on the eigenspace. However, since W^u is infinite-dimensional, this is far from obvious and it is conceivable that there is a solution that approaches zero faster than any exponential (the so-called small solutions mentioned in Chapter 3). It is true that no such solutions exist, but the proof is very difficult. Even knowing this does not prove part (iii) of Theorem 6.1. If one takes into account the fact that the orbits of (6.1) that are of interest lie on the compact attractor $\mathcal{A}$, then one can prove the following result.

Theorem 6.2. *If the origin is hyperbolic and* $\dim W^u = N^*$, *then* N^* *is even and there is a neighborhood* $U \subset \mathcal{A}$ *of the origin such that*

$$\phi \in W^s \setminus \{0\} \text{ implies } V(x(t, \phi)) > N^* \text{ for all } t \in \mathbb{R},$$

$$\phi \in W^u \setminus \{0\} \text{ implies } V(x(t, \phi)) < N^* \text{ for all } t \in \mathbb{R}.$$

In particular, $W^s \cap W^u = \{0\}$, *and so there is no orbit homoclinic to the origin.*

Using Theorem 6.2, one can show part (iii) of Theorem 6.1 in the case when the origin is hyperbolic. If the origin is not hyperbolic, more care is needed.

With these results, we are now in a position to define a Morse decomposition of the attractor $\mathcal{A}$. It is tempting to consider, for each odd integer N, the following sets as part of a Morse decomposition:

$$\{\phi \in \mathcal{A},\, \phi \neq 0 : V(x(t, \phi)) = N \text{ for all } t \in \mathbb{R}\}.$$

However, this will not work because the function V is not defined at the origin and these sets in general are not closed. In fact, several of them may contain the point 0 in their closure. The definition must be refined to keep the orbits away from the origin.

We assume that the origin is hyperbolic (the definition can be given without this hypotheses but is more complicated) and let $N^* = \dim W^u(0)$. For any odd integer N, define

$$\mathcal{A}_N = \{\phi \in \mathcal{A},\, \phi \neq 0 : V(x(t, \phi)) = N \text{ for } t \in \mathbb{R} \text{ and } 0 \notin \alpha(\phi) \cup \omega(\phi)\}.$$

Let $\mathcal{A}_{N^*} = \{0\}$. With this definition, the sets $\mathcal{A}_N$ for N odd are compact and do not contain the origin. We remark that $\mathcal{A}_N = \emptyset$ for large N by (iii) of Theorem 6.1. It is now possible to prove

Theorem 6.3. *If the origin is hyperbolic, the sets* $\mathcal{A}_N$, $N \in \{N^*, 1, 3, 5, \ldots\}$, *form a Morse decomposition of* $\mathcal{A}$ *with the ordering* $\mathcal{A}_N < \mathcal{A}_K$ *if and only if* $K < N$.

Further properties also are known about the Morse sets $\mathcal{A}_N$. In particular, for N an odd integer, if $\phi \in \mathcal{A}_N$, then the zeros of $x(t, \phi)$ are simple. This allows one to prove that each $\mathcal{A}_N$ for $N < N^*$ is not empty and contains a periodic orbit $x_N(t)$ with least period τ satisfying $2/N < \tau < 2/(N-1)$ and $x_N(t)$ has exactly two zeros in $[0, \tau)$.

The proof of this last fact uses a special type of Poincaré map. Consider the map $\Theta : \mathcal{A}_N \to S^1$ from the Morse set $\mathcal{A}_N$ to the unit circle S^1 in the plane with center 0, induced by the map

$$\phi \in \mathcal{A}_N \mapsto (x(0, \phi), \dot{x}(0, \phi)) \in \mathbb{R}^2 \setminus \{0\}.$$

From the properties mentioned earlier, the image of the orbit winds around the circle infinitely often as $t \to \pm\infty$. In particular, it has a transversal cross section, namely, the half-line $x = 0$, $\dot{x} > 0$ in $\mathbb{R}^2 \setminus \{0\}$ and has a corresponding Poincaré map. It is this map that is used to prove the existence of the periodic solutions mentioned before.

In order to obtain more information about the structure of the flow on the attractor $\mathcal{A}$, it is first necessary to understand the existence of connecting orbits C_K^N for various N and K. Using the Conley index and the theory of connection matrices for isolated invariant sets, it has been recently shown that $C_K^N \neq \emptyset$ for all $N > K$ (see the supplementary remarks for references).

As we have seen, the existence of the Morse decomposition and of the connecting orbits gives a much better picture of the flow on the attractor $\mathcal{A}$. This does not mean that the flow is simple. In fact, numerical studies suggest that, in many cases, the flow in the set $\mathcal{A}_N$ may have a very complicated structure involving multiple periodic orbits arising from period-doubling bifurcations and even chaotic dynamics (see the supplementary remarks for references).

12.7 Singularly perturbed systems

Consider the equation

$$\epsilon\dot{x}(t) = -x(t) + f(x(t-1)), \tag{7.1}$$

where $\epsilon > 0$ is a parameter and $f \in C^1(\mathbb{R}, \mathbb{R})$.

If we formally take the limit of Equation (7.1) as $\epsilon \to 0$, then we obtain a difference equation

$$x(t) = f(x(t-1)), \tag{7.2}$$

which can be considered as a discrete dynamical system defined by the map

$$x \mapsto f(x). \tag{7.3}$$

It is an interesting problem to determine how the dynamics of Equation (7.1) mirror the dynamics of the difference equation (7.2) or the discrete dynamical system (7.3) when ϵ is small. In this section, we investigate some of the known similarities and dissimilarities.

For any interval $I \subset \mathbb{R}$ (closed or open), let $X_I \stackrel{\text{def}}{=} C([-1,0], I)$. We let $T_\epsilon(t)$ denote the semigroup on C. The following result is very easy to prove.

Proposition 7.1. (Positive invariance). *If I is an interval such that $f(I) \subset I$, then $T_\epsilon(t)X_I \subset X_I$ for $t \geq 0$.*

If x_0 is a fixed point of f, then the constant function $x_0 \in C$ is an equilibrium point of Equation (7.1), and conversely. If x_0 is an attracting fixed point of f, we say that an interval J is the *maximal interval of attraction* of x_0 if $x_0 \in J$, $f(J) \subset J$, $f^n(x) \to x_0$ as $n \to \infty$ for each $x \in J$ and there is no interval $J' \supset J$ with this property. We remark that the maximal interval of attraction is open. It is possible to prove the following property.

Proposition 7.2. (Stability). *If x_0 is an attracting fixed point of f with maximal interval of attraction J, then the equilibrium solution x_0 of Equation (7.1) is asymptotically stable and, for each $\psi \in X_J$ and every $\epsilon > 0$, we have*

$$\lim_{t \to \infty} T_\epsilon(t)\psi = x_0.$$

Under the assumption that f has negative feedback, we have seen in the previous section that there is a Morse decomposition of the attractor. Also, in the supplementary remarks to Chapter 11, we asserted that for each $\epsilon > 0$, there is a slowly oscillating periodic solution if the origin is unstable and if there is an interval I such that $f(I) \subset I$. It is natural to discuss the limit of this solution as $\epsilon \to 0$. To state a precise result, we say that a point (a, b) is a period-two point of f if $a \neq b$ and $f(a) = b$, $f(b) = a$. We say that a function $w(t)$, $t \in \mathbb{R}$ is a *square wave* if there are constants $a \neq b$ such that $w(t) = a$ for $t \in (2n, 2n+1)$, $w(t) = b$ for $t \in (2n+1, 2n)$ for all integers n.

For the statement of the next result, we recall (see Section 11.7) that a solution of (7.1) is slowly oscillating (about zero) if it has a sequence of zeros approaching infinity and the distance between zeros is > 1.

Theorem 7.1. *Suppose that there is an interval I such that $0 \in I$, $f(I) \subset I$, f has negative feedback on I, and $f'(0) < -1$. Then there exists an $\epsilon_0 > 0$ such that for $0 < \epsilon < \epsilon_0$, there is a slowly oscillating periodic solution x^ϵ of Equation (7.1) that is continuous in ϵ.*

Furthermore, if $f'(x) < 0$ for $x \in I$ and (a, b) is a period-two point of f in I that is asymptotically stable, then $x^\epsilon(t)$ has exactly one maximum and one minimum over a period and approaches a square wave uniformly on all compact sets of $\mathbb{R} \setminus \{n = 0, \pm 1, \pm 2, \dots\}$ with the values (a, b) of the square wave corresponding to the period-two point of the map f.

It might be expected that the conclusion of Theorem 7.1 would remain true without the severe restriction that f is monotone on the interval I. However, this is typically the exception. If f is not monotone, it can be shown that the function $x^\epsilon(t)$ approaches a square wave uniformly on all compact sets of $\mathbb{R} \setminus \{n = 0, \pm 1, \pm 2, \dots\}$ with the values (a, b) of the square wave corresponding to a period-two point of the map f. However, at the points of transition near the integers, the function $x^\epsilon(t)$ begins to oscillate with the number of oscillations increasing to infinity as $\epsilon \to 0$. The amplitudes of the oscillations around the point a (resp. b) are bounded but do not approach zero as $\epsilon \to 0$. Thus, the limiting process exhibits a Gibbs' type of phenomenon at the integers. This fact is easily observed numerically but very difficult to prove. However, the underlying reason for the Gibbs phenomenon has a very simple dynamical and geometric interpretation. It is possible to write down some equations that serve to determine the transition curves that allow the solutions to pass from point a to point b. These equations are essentially the same as Equations (7.10). The problem is to determine the constant r so that there is a solution of these equations that has a (resp. b) as its α-limit set (resp. ω-limit set). If the function f is monotone, then the dominant eigenvalue near a (resp. b) is real and the transition curve should be monotone (and, thus, the conclusion of the theorem). On the other hand, if f is not monotone, then the dominant eigenvalue is complex and the transition curve will oscillate. See the bibliography for details.

If the mapping f were to be a function of a parameter λ, the period-two points often arise through a period-doubling bifurcation from a fixed point. Therefore, it is of interest to understand the implications of a period-doubling bifurcation of the map on the dynamics of the flow of Equation (7.1) for ϵ near zero. We now describe this situation more precisely.

For $\epsilon > 0$ and small and $f \in C^k(\mathbb{R} \times \mathbb{R}), k \geq 3$, we consider periodic solutions of the equation

$$\epsilon \dot{x}(t) = -x(t) + f(x(t-1), \lambda) \tag{7.4}$$

under the assumption that the point $\lambda = 0$ corresponds to a generic period-doubling point for the map $x \mapsto f(x, 0)$. More specifically, we assume that

$$f(x, \lambda) = -(1+\lambda)x + ax^2 + bx^3 + o(x^3) \quad \text{as } x \to 0, \tag{7.5}$$

where a, b are constants such that $\beta = a^2 + b \neq 0$. Under this assumption on f, for each small value of λ for which $\lambda\beta > 0$, there are nonzero constants $d_{1\lambda}$, $d_{2\lambda}$, $d_{1\lambda} \neq d_{2\lambda}$, such that $f(d_{1\lambda}, \lambda) = d_{2\lambda}$, $f(d_{2\lambda}, \lambda) = d_{1\lambda}$ and so $f^2(d_{1\lambda}, \lambda) = d_{1\lambda}$. Furthermore, $d_{1\lambda}, d_{2\lambda} \to 0$ as $\lambda \to 0$. The points $d_{1\lambda}, d_{2\lambda}$ are periodic points of period two of the map $f(\cdot, \lambda)$. If $\beta > 0$, we say that the bifurcation is *supercritical* (the fixed point 0 of the map $f(\cdot, 0)$ is stable) and if $\beta < 0$, we say that the bifurcation is *subcritical* (the fixed point 0 of the map $f(\cdot, 0)$ is unstable).

We are interested in how the period-doubling bifurcation of the map is reflected into the bifurcation from the origin of periodic solutions of Equation (7.4) of period approximately 2. The principal result is

Theorem 7.2. *Suppose that $f(x, \lambda)$ satisfies (7.5). Then there is a a neighborhood U of $(0,0)$ in the (λ, ϵ) plane and a sectorial region S in U such that if $(\lambda, \epsilon) \in U$, then there is a periodic solution $\tilde{x}_{\lambda,\epsilon}$ of Equation (7.4)* *with period* $2\tau(\lambda, \epsilon) = 2 + 2\epsilon + O(|\epsilon|(|\lambda| + |\epsilon|))$ *as* $(\lambda, \epsilon) \to (0,0)$ *if and only if* $(\lambda, \epsilon) \in S$. *Furthermore, this solution is unique. If, in addition,* $f(x, \lambda) = -f(-x, \lambda)$, *then* $\tilde{x}_{\lambda,\epsilon}(t + \tau(\lambda, \epsilon)) = -\tilde{x}_{\lambda,\epsilon}(t)$.

Of course, the sector S must belong to the set $\epsilon > 0$ in the (λ, ϵ) plane. If the period-two doubling bifurcation of the map is supercritical, then the sector $S \subset \{ (\lambda, \epsilon) : \epsilon > 0, \lambda > 0 \}$ and, for $\lambda = \lambda_0 > 0$, fixed, the set $\{ \epsilon : (\epsilon, \lambda_0) \in S \}$ is an interval $\lambda_0 \times (0, \epsilon_0(\lambda_0))$. At the point $(\lambda_0, \epsilon_0(\lambda_0))$, there is a Hopf bifurcation and the periodic solution approaches a square wave as $\epsilon \to 0$; that is, the periodic solution $\tilde{x}_{\lambda_0,\epsilon}(t)$ has the property that $\tilde{x}_{\lambda_0,\epsilon}(t) \to d_{1\lambda}$ (respectively, $d_{2\lambda}$) as $\epsilon \to 0$ uniformly on compact sets of $(0, 1)$ (respectively, $(1, 2)$). Part of this result is contained in Theorem 7.1.

In the supercritical case, the sector S is completely different and the periodic orbits have a different structure as $\epsilon \to 0$. The sector S contains points (ϵ, λ) with λ both negative and positive. More precisely, for $\lambda = \lambda_0 > 0$, fixed, the set $\{ \epsilon : (\epsilon, \lambda_0) \in S \}$ is an interval $\lambda_0 \times (\epsilon_0(\lambda_0), \beta_0(\lambda_0))$. At the point $(\lambda_0, \epsilon_0(\lambda_0))$, there is a Hopf bifurcation. For $\lambda = \lambda_0 < 0$, fixed, the set $\{ \epsilon : (\epsilon, \lambda_0) \in S \}$ is an interval $\lambda_0 \times (0, \alpha_0(\lambda_0)$. As $\epsilon \to 0$, the unique periodic solution becomes *pulse-like* in the following sense: the periodic solution $\tilde{x}_{\lambda_0,\epsilon}(t)$ has the property that $\tilde{x}_{\lambda_0,\epsilon}(t) \to 0$ as $\epsilon \to 0$ uniformly on compact sets of $(0, 1) \cup (1, 2)$. The magnitude of the pulse exceeds $\max\{|d_{1\lambda}|, |d_{2\lambda}| \}$. The part of the period doubling in the map that is reflected in the pulse-like solution is that the jumps in the solution occur near the integers and are opposite in direction.

Let us briefly outline the proof since it makes use of so much of the local theory that we have developed in the previous chapters. The linear variational equation around the equilibrium solution 0 of Equation (7.4) is

$$\epsilon \dot{y}(t) = -y(t) - (1 + \lambda)y(t - 1). \tag{7.6}$$

By analyzing the characteristic equation

$$\epsilon\mu + 1 + (1 + \lambda)e^{-\mu} = 0, \tag{7.7}$$

it is possible to see that if $\lambda \leq 0$, then the origin is asymptotically stable for all $\epsilon > 0$. On the other hand, if $\lambda > 0$, then there is an $\epsilon_0(\lambda) > 0$ such that for $\epsilon > \epsilon_0(\lambda)$, the origin is asymptotically stable, and, for $0 < \epsilon < \epsilon_0(\lambda)$, the origin is unstable with a pair of complex solutions of (7.7) with positive real part. For $\epsilon = \epsilon_0(\lambda)$, there are two purely imaginary solutions of (7.7). Furthermore, if the complex roots near $\epsilon = \epsilon_0(\lambda)$ are denoted by $\mu(\lambda, \epsilon)$, $\bar{\mu}(\lambda, \epsilon)$, then $\partial \text{Re}\mu(\lambda, \epsilon_0(\lambda))/\partial\epsilon > 0$. Therefore, there is a Hopf bifurcation in Equation (7.4) at the origin at the point $(\lambda, \epsilon_0(\lambda))$. It can be shown also that there is a unique periodic orbit bifurcating from the origin under the assumption that $\beta \neq 0$ and the period is approximately 2.

The basic problem now is to determine the region near the origin in the parameter space (λ, ϵ) for the existence of this bifurcating periodic orbit and to determine the behavior of this orbit as $\epsilon \to 0$. To accomplish this, we introduce some scalings. We suppose that Equation (7.4) has a periodic solution $x(t)$ with period $2+2r\epsilon$ and let

$$w_1(t) = x(-\epsilon rt), \quad w_2(t) = x(-\epsilon rt + 1 + \epsilon r). \tag{7.8}$$

Since $x(t)$ has period $2+2r\epsilon$, we see that

$$\begin{aligned} w_2(t) &= x(-\epsilon r(t+1) - 1) \\ w_2(t-1) &= x(-\epsilon rt - 1). \end{aligned} \tag{7.9}$$

If we use (7.8) and (7.9) in (7.4), we deduce that

$$\begin{aligned} \dot{w}_1(t) &= rw_1(t) - rf(w_2(t-1), \lambda) \\ \dot{w}_2(t) &= rw_2(t) - rf(w_1(t-1), \lambda). \end{aligned} \tag{7.10}$$

This equation now is independent of ϵ. We now look for periodic solutions of System (7.10) in a neighborhood of the origin regarding it as a two-parameter bifurcation problem with (λ, r) as parameters.

Some caution must be exercised at this point. Every periodic solution $x(t)$ of Equation (7.4) of period $2+2r\epsilon$ leads to a periodic solution of System (7.10) through the transformation (7.9). In addition, the corresponding solution of System (7.10) must encircle the origin. The following converse also is true: any periodic solution of System (7.10) that encircles the origin and has period $\omega > 2$ corresponds to a periodic solution of Equation (7.4) of period $2+2r\epsilon$ if and only if ϵ satisfies the equation $r(\omega-2)\epsilon = 2$.

The next step is to determine the approximate value of the constant r in the period $2+2r\epsilon$. The appropriate approximate value of r is obtained by considering the linear variational equation around the zero solution of System (7.10) for $\lambda = 0$,

$$\begin{aligned} \dot{w}_1(t) &= rw_1(t) + rw_2(t-1) \\ \dot{w}_2(t) &= rw_2(t) + rw_1(t-1). \end{aligned} \tag{7.11}$$

The eigenvalues of System (7.11) are the roots of the characteristic equation,

$$\det\left(\mu I - rL(e^{\mu\cdot} I)\right) = (\mu - r)^2 - r^2 e^{-2\mu} = 0, \tag{7.12}$$

where

$$L\phi = \phi(0) + \begin{bmatrix} 0 & 1 \\ 1 & 0 \end{bmatrix} \phi(-1). \tag{7.13}$$

The left-hand side of Equation (7.12) always has $\mu = 0$ as a zero. It is a simple zero if $r \neq 1$ and a double zero if $r = 1$. Bifurcation from a simple zero can never lead to any periodic orbits. Therefore, we are forced to take $r = 1$ in the first approximation. For $r = 1$, the remaining eigenvalues of (7.11) have negative real parts. If we let $r = 1 + h$, $w = (w_1, w_2)$, where h is a small parameter, then (7.10) can be written as

$$\dot{w}(t) = Lw_t + hLw_t - F_{\lambda,h}(w_t) + o(|w(t-1)|^3), \tag{7.14}$$

where

$$F_{\lambda,h}(\phi) = (1+h)\begin{bmatrix} \phi_2(-1) + f(\phi_2(-1),\lambda) \\ \phi_1(-1) + f(\phi_1(-1),\lambda) \end{bmatrix}, \qquad \phi = \begin{bmatrix} \phi_1 \\ \phi_2 \end{bmatrix}, \tag{7.15}$$

and $w_t(\theta) = w(t+\theta)$ for $-1 \le \theta \le 0$.

We now consider Equation (7.14) as a perturbation of the linear equation

$$\dot{v}(t) = Lv_t. \tag{7.16}$$

Of course, we will consider Equation (7.14) with initial data in the space $C = C([-1,0], \mathbb{R}^2)$. Since the characteristic equation for the linear part of (7.14) for $(\lambda, h) = (0,0)$ has a zero as a root of multiplicity two, we know that the small periodic orbits of Equation (7.14) will lie on a two-dimensional center manifold that is tangent to the subspace generated by generalized eigenvectors ζ_1, ζ_2 associated with the eigenvalue zero of (7.16). Therefore, the first problem is to determine the approximate vector field on the center manifold. If we let $w_t = z_1\zeta_1 + z_2\zeta_2 + \bar{w}_t$, where $\bar{w}_t$ lies in the natural linear space complementary to the span of ζ_1, ζ_2, then we show that the approximate flow on the center manifold is given by the system of ordinary differential equations (there are several nontrivial computations here)

$$\begin{aligned} \dot{z}_1 &= 2hz_1 + 2\lambda(\frac{2}{3}z_1 + z_2) - 2\beta(\frac{2}{3}z_1 + z_2)^3 - a^2(\frac{2}{3}z_1 + z_2)z_1^3 \\ \dot{z}_2 &= -z_1. \end{aligned} \tag{7.17}$$

For $(h, \lambda) = (0,0)$, we can use the theory of normal forms to make a nonlinear change of variables to obtain the equation

$$\begin{aligned} \dot{z}_1 &= (2h + \lambda\frac{4}{3})z_1 + 2\lambda z_2 - 2\beta z_2^3 - 4\beta z_1 z_2^2 \\ \dot{z}_2 &= -z_1 \end{aligned} \tag{7.18}$$

up through terms of order $(h+\lambda)^2|z| + |z|^4$.

To analyze the periodic solutions of System (7.18), it is now convenient to rescale variables

$$\mu = |\lambda|^{1/2}, \ \ h = \mu^2\delta, \ \ z_2 = \mu u_2, \ \ z_1 = \mu^2 u_1, \ \ t \mapsto -\mu^{-1}t, \tag{7.19}$$

to obtain the new equations

$$\begin{aligned} \dot{u}_1 &= -\mu(2\delta + (\mathrm{sgn}\lambda)\frac{4}{3})u_1 - 2(\mathrm{sgn}\lambda)u_2 + 2\beta u_2^3 + 4\beta\mu u_1 u_2^2 \\ \dot{u}_2 &= u_1. \end{aligned} \tag{7.20}$$

Equation (7.20) is equivalent to the second-order scalar equation

$$(7.21)\quad \ddot{W} + \mu(2\delta + (\mathrm{sgn}\lambda)\frac{4}{3})\dot{W} + 2((\mathrm{sgn}\lambda)W - \beta W^3) - 2\mu\beta W^2\dot{W} = 0,$$

where we have put $W = u_2/\sqrt{2}$. For $\mu = 0$, this is a conservative system

$$(7.22)\qquad \ddot{W} + 2((\mathrm{sgn}\lambda)W - \beta W^3) = 0.$$

The bifurcation diagram in (λ, h) space for the periodic orbits of System (7.18) (or the (μ, h) space for (7.20)) are well known. In spite of this fact, there is still more work to do. In fact, we have remarked before that not all periodic orbits of System (7.18) are valid candidates for periodic orbits of Equation (7.4). We must seek those periodic orbits of System (7.18) that encircle the origin and have period > 2.

In the case $\beta > 0$, there are periodic orbits of Equation (7.21) only if $\lambda > 0$. Each periodic orbit of Equation (7.21) encircles the origin and has period > 2 and, therefore, corresponds to a periodic orbit of Equation (7.4). The periodic orbits that approach a square wave correspond to periodic orbits of System (7.20) that approach the heteroclinic cycle of the conservative system Equation (7.21).

If $\beta < 0$, the periodic orbits of System (7.18) that are candidates for periodic orbits of Equation (7.4) encircle the origin and are outside the figure eight (the homoclinic orbits) for the conservative system (7.22). The analysis in this case is very complicated and involves very delicate estimates of Abelian integrals. The pulse-like solution of Equation (7.4) corresponds to the periodic orbits of System (7.20) for $\lambda < 0$ that approach the figure eight for the conservative system (7.22).

It is possible to extend Theorem 7.2 to the matrix case. Let us describe the setup and results. Suppose that $\epsilon > 0$ is a real parameter, A is an $n \times n$ nonsingular real constant matrix, $f \in C^k(\mathbb{R}^m \times \mathbb{R}, \mathbb{R}^m)$, $k \geq 4$, $f(0, \lambda) = 0$ for all λ, and consider the vector equation

$$(7.23)\qquad \epsilon\dot{x}(t) + Ax(t) = Af(x(t-1), \lambda).$$

As for the scalar case, we impose conditions on f so that the m-dimensional map $x \to f(x, \lambda)$ undergoes a generic period-doubling bifurcation at the point $(x, \lambda) = (0, 0)$ and then investigate under what conditions Equation (7.23) possesses a periodic solution of period approximately 2 for ϵ small and discuss the limiting behavior of this solution as $\epsilon \to 0$. If $\sigma(L)$ denotes the spectrum of an $n \times n$ matrix L, then our first hypothesis is

$$(\mathrm{H}_1)\qquad \sigma(f_x(0,0)) \cap S^1 = \{-1\}$$

where S^1 is the unit circle in the complex plane with center at the origin. We also suppose that $\sigma(f_x(0, \lambda))$ contains the point $-(1 + \lambda)$ for λ small. With these hypotheses, we can make a change of coordinates in Equation (7.23) and write, without loss of generality,

$$
\begin{aligned}
f(x,\lambda) &= (f_1(x,\lambda), f_2(x,\lambda))^T \in \mathbb{R}\times\mathbb{R}^{m-1} \\
f_1(x,\lambda) &= -(1+\lambda)x_1 + c_1x_1^2 + x_1C_2x_2 + c_3x_1^3 \\
&\qquad + O(|x_2|^2 + |x|^4 + |\lambda|\,|x|^2) \\
f_2(x,\lambda) &= G_0x_2 + x_1^2G_1 + x_1G_2x_2 \\
&\qquad + O(|\lambda|\,|x_2|\,|x_2|^2 + |x|^3 + |\lambda|\,|x|^2),
\end{aligned}
\tag{7.24}
$$

where $x = (x_1, x_2) \in \mathbb{R}\times\mathbb{R}^{m-1}$, $c_1, c_3 \in \mathbb{R}$, $C_2 \in \mathbb{R}^{1\times(m-1)}$, $G_0, G_2 \in \mathbb{R}^{(m-1)\times(m-1)}$, $G_1 \in \mathbb{R}^{(m-1)\times 1}$, and

$$\sigma(G_0) \cap S^1 = \emptyset.$$

If we now apply the method of Liapunov-Schmidt for the existence of period-two points of the map $f(x,\lambda)$ near $(0,0)$, then the generic condition for existence of such points is that

$$R_0 \stackrel{\text{def}}{=} C_2(I_{m-1} - G_0)^{-1}G_1 + c_1^2 + c_3 \neq 0, \tag{H$_2$}$$

where I_{m-1} is the identity matrix in $\mathbb{R}^{m-1}$.

To be able to say that these period-two points of the map are carried over into periodic orbits of (7.23) of period $2+2r\epsilon$ for (ϵ,λ) small, we need some additional conditions that relate the matrix A^{-1} to the operator G_0. To motivate the hypotheses, we introduce in the matrix case the coordinates and scaling (7.8), (7.9) to obtain

$$
\begin{aligned}
\dot w_1(t) &= rAw_1(t) - rAf(w_2(t-1),\lambda) \\
\dot w_2(t) &= rAw_2(t) - rAf(w_1(t-1),\lambda).
\end{aligned}
\tag{7.25}
$$

As for the scalar case, we regard (7.25) as a two-parameter bifurcation problem with (λ, r) as parameters.

To determine the approximate value of the constant r in the period $2+2r\epsilon$, we consider the linear variational equation around the zero solution of (7.25) for $\lambda = 0$,

$$
\begin{aligned}
\dot w_1(t) &= rAw_1(t) - rADw_2(t-1) \\
\dot w_2(t) &= rAw_2(t) - rADw_1(t-1).
\end{aligned}
\tag{7.26}
$$

The eigenvalues of (7.26) are the roots of the characteristic equation,

$$\Delta(\mu, r) \stackrel{\text{def}}{=} \det(\mu I_{2m} - rLe^{\mu\cdot}I_{2m}) = 0 \tag{7.27}$$

where L is a continuous linear map from $C([-1,0],\mathbb{R}^{2m})$ into $\mathbb{R}^{2m}$,

$$L\phi = \begin{bmatrix} A & 0 \\ 0 & A \end{bmatrix}\phi(0) - \begin{bmatrix} 0 & AD \\ AD & 0 \end{bmatrix}\phi(-1)$$

$$D = f_x(0,0).$$

The left-hand side of Equation (7.26) always has $\mu = 0$ as a zero. We determine r so that Equation (7.27) has $\mu = 0$ as a double zero. We make the following hypotheses:

$$(\mathrm{H}_3) \qquad A^{-1} = \begin{bmatrix} r_0 & A_{12} \\ A_{21} & A_{22} \end{bmatrix}, \qquad 0 \neq r_0 \in \mathbb{R},$$

where $A_{21} \in \mathbb{R}^{(m-1)\times 1}$, $A_{12} \in \mathbb{R}^{1\times(m-1)}$, $A_{22} \in \mathbb{R}^{(m-1)\times(m-1)}$,

$$(\mathrm{H}_4) \qquad R_1 \stackrel{\text{def}}{=} r_0^2 + 2A_{12}(I_{m-1} + G_0)^{-1}A_{21} \neq 0.$$

We suppose also that

$$(\mathrm{H}_5) \qquad \det\left[i\omega I_m - r_0 A(I_m \pm De^{-i\omega})\right] \neq 0 \text{ for } \omega \in \mathbb{R}\setminus\{0\}.$$

The justification of hypotheses (H_3), (H_4), and (H_5) is contained in the following result.

Lemma 7.1. *Let $\Delta(\mu, r)$ be as in Equation (7.27). Under the hypotheses (H_3), (H_4), the point $\mu = 0$ is a double zero of $\Delta(\cdot, r)$ if and only if $r = r_0$. With the additional hypothesis (H_5), no other zeros of $\Delta(\mu, r)$ lie on the imaginary axis.*

The main result for the matrix case that generalizes Theorem 7.2 is

Theorem 7.3. *Suppose that (H_1)–(H_5) are satisfied. Then there is a neighborhood U of $(0,0)$ in the (λ,ϵ)-plane and a sectorial region S in U such that if $(\lambda,\epsilon) \in U$, then there exists a periodic solution $\tilde{x}_{\lambda,\epsilon}$ of Equation (7.23) with period $2\tau(\lambda,\epsilon) = 2 + 2r_0\epsilon + O(|\epsilon|(|\lambda| + |\epsilon|))$ as $(\lambda,\epsilon) \to (0,0)$ if and only if $(\lambda,\epsilon) \in S$. Furthermore, this solution is unique. The set $\partial S \equiv \{(\lambda,\epsilon) \in S : \epsilon > 0, \lambda R_1 > 0\}$ corresponds to a Hopf bifurcation. Let $S_\lambda = \{\epsilon : (\lambda,\epsilon) \in S\}$. If $R_0R_1 > 0$ (respectively, $R_0R_1 < 0$) and $\lambda R_1 > 0$ (respectively, $\lambda R_1 < 0$), then $\tilde{x}_{\lambda,\epsilon}$ approaches a square wave (respectively, a pulse-like wave) as $\epsilon \to 0$.*

The basic idea for the proof of this result is the same as for the scalar case—treat (7.25) as a perturbation of (7.26), obtain the vector field on a center manifold, and relate the periodic solutions on the center manifold to periodic solutions of (7.23). The essential new ideas for the vector case is Lemma 7.1 and it is a nontrivial task (although only computational) to obtain the vector field on the center manifold.

In problems of transmission of light through a ring cavity (see the references), the following model has been proposed:

$$(7.28) \qquad \left(\epsilon_m \frac{d}{dt} + 1\right)\cdots\left(\epsilon_1 \frac{d}{dt} + 1\right)y(t) = g(y(t-1),\lambda),$$

where each $\epsilon_j > 0$ is a small parameter. In the supplementary remarks of Chapter 11, we have sufficient conditions for the existence of a slowly oscillating periodic solution of (7.28). These conditions are satisfied if the ϵ_j are sufficiently small and the map $y \to g(y, \lambda)$ undergoes a generic supercritical period-doubling bifurcation at $(y, \lambda) = (0, 0)$. Is it possible to determine the limit of this solution as $\epsilon_j \to 0$, $j = 1, 2 \ldots, m$, from Theorem 7.3?

If we scale the ϵ_j as $\epsilon_j = \epsilon\alpha_j^{-1}$, $j = 1, 2, \ldots, m$, and let $x_1 = y$, $x_j = \epsilon\alpha_j^{-1}\dot{x}_{j-1} + x_{j-1}$, $j = 2, 3, \ldots, m$, then we obtain equivalent equations that are a special case of (7.23). It is possible to use Theorem 7.3 to prove the following.

Theorem 7.4. *Consider the Equation* (7.28) *with* $\epsilon_j = \epsilon\alpha_j^{-1}$, $j = 1, 2, \ldots, m$. *If the scalar map* $y \to g(y, \lambda)$ *undergoes a generic period-doubling bifurcation at* $(y, \lambda) = (0, 0)$ *and* $f(x, \lambda) = \mathrm{col}(g(x_1, \lambda), g(x_1, \lambda))$, $x = \mathrm{col}(x_1, x_2)$, *then the map* $x \to f(x, \lambda)$ *undergoes a generic period-doubling bifurcation at* $(x, \lambda) = (0, 0)$ *and the corresponding system* (7.23) *for* (7.28) *satisfies* $(\mathrm{H}_1) - (\mathrm{H}_5)$ *with* $r_0 = \sum_{j=1}^{m} \alpha_j^{-1}$. *Therefore, the conclusions in Theorem* 7.3 *are valid.*

Let us now consider a further generalization of Equation (7.23) consisting of a matrix delay differential equation coupled with a matrix difference equation, a so-called *hybrid system*

$$
\begin{aligned}
\epsilon\dot{x}(t) + Ax(t) &= Af(y(t), \lambda) \\
y(t) &= g(x(t-1), y(t-1), \lambda),
\end{aligned}
\tag{7.29}
$$

where $\epsilon > 0, \lambda$ are small real parameters, $x \in \mathbb{R}^m$, $y \in \mathbb{R}^n$ are vectors, the $m \times m$ matrix A has an inverse and the functions $f(y, \lambda)$ and $g(x, y, \lambda)$ are smooth vector-valued functions.

For $\epsilon = 0$, we obtain the map on $\mathbb{R}^n$ defined by

$$
y \in \mathbb{R}^n \mapsto G^\lambda(y) \stackrel{\text{def}}{=} g(f(y, \lambda), y, \lambda) \in \mathbb{R}^n.
\tag{7.30}
$$

Suppose that Equation (7.30) undergoes a generic supercritical period doubling at $\lambda = 0$ with the period-two points being $d_{1\lambda}, d_{2\lambda}$. In $\mathbb{R}^m \times \mathbb{R}^n$, we have, for $\epsilon = 0$, the square wave $(x^\epsilon(t), y^\epsilon(t))$, $t \in \mathbb{R}$, which alternately takes on the values $(f(d_{1\lambda}, \lambda), G_\lambda(d_{1\lambda}))$ and $(f(d_{2\lambda}, \lambda), G_\lambda(d_{2\lambda}))$ on intervals of length one. Since the bifurcation is supercritical, this function will be stable if we impose a few additional conditions on the function f. Therefore, one would expect that there is a solution of Equation (7.29) for ϵ small that will be close to the square wave. Under appropriate hypotheses, this is true. The first step in attacking the problem is to use scaling to eliminate the parameter ϵ. However, for the results that have been obtained so far, the next step in the proof is completely different from the preceding one. It involves methods more functional analytical in nature and uses concepts of exponential dichotomies. See the supplementary remarks for references that contain a precise statement with the proof.

Other problems more general than Equation (7.29) arise in a very natural way. For example, suppose that we consider a scalar equation with two delays:

$$\epsilon \dot{x}(t) = -x(t) + f(x(t-1), x(t-\sigma), \lambda) \tag{7.31}$$

where $\sigma \geq 1$. There is essentially nothing known about the relationship between the solutions of (7.31) for $\epsilon > 0$ small and the corresponding solutions of the difference equation:

$$x(t) = f(x(t-1), x(t-\sigma), \lambda). \tag{7.32}$$

If σ is irrational, then Equation (7.32) is an infinite-dimensional problem and, of course, none of the ideas mentioned seem to shed much light on the problem. On the other hand, if σ is rational, then Equation (7.32) can be considered as a map on a finite-dimensional space and, at least, we may speak of generic period doubling. It certainly would be of interest to know something about the implications for $\epsilon = 0$. One possible approach would be the following. For simplicity, suppose that $\sigma = 2$. If we define $x(t-1) = y(t)$, then Equation (7.32) is equivalent to the hybrid system:

$$\begin{aligned} \epsilon \dot{x}(t) + x(t) &= f(x(t-1), y(t-1), \lambda) \\ y(t) &= x(t-1). \end{aligned} \tag{7.33}$$

If we could extend these theories to these systems, then we will have at least solved the problem of period doubling.

12.8 Averaging

In this section, we review some of the results on the application of the method of averaging for RFDE. We begin with a brief review of the results and methods that are used in ODE.

12.8.1 Averaging in ODE

For $\epsilon > 0$ a small parameter, we consider the ODE

$$\dot{x} = f(\frac{t}{\epsilon}, x) \tag{8.1}$$

where $f(\tau + 1, x) = f(\tau, x)$ for all τ, x, is continuous in (t, x) and is continuously differentiable in x. It is possible to extend the following remarks to functions f with a more general dependence on τ (for example, almost periodic in τ uniformly for x in bounded sets), and we discuss this case only to avoid technical difficulties. Along with Equation (8.1), we consider also the *averaged equation*

$$\dot{y} = f_0(y), \tag{8.2}$$

where

$$f_0(y) = \int_0^1 f(\tau, y)d\tau. \tag{8.3}$$

If $\epsilon > 0$ is sufficiently small, it is possible to make a transformation of variables in Equation (8.1), which is periodic in t of period 1 and close to the identity, to obtain a new ODE for which the vector field is close to the averaged vector field. More specifically, if we let $x = z + \epsilon u(\frac{t}{\epsilon}, z)$, where

$$u(s, x) = \int_0^s [f(\tau, x) - f_0(x)]\, d\tau,$$

then

$$\dot{z} = f_0(z) + g(\frac{t}{\epsilon}, z, \epsilon), \tag{8.4}$$

where

$$g(\tau, z, 0) = 0, \quad g(\tau, z, \epsilon) = g(\tau + 1, z, \epsilon). \tag{8.5}$$

The first classical result on averaging asserts that we can keep a solution of Equation (8.1) close to a special function associated with a solution of the averaged equation for as long as we want if we choose ϵ sufficiently small. More precisely, we have

Theorem 8.1. *Let $x(t)$ be a solution of Equation (8.1) with $x(0) = x_0$ and let $y(t)$ be a solution of Equation (8.2) with $y_0 = y_0(x_0)$ chosen to satisfy the equation $x_0 = y_0 + \epsilon u(0, y_0)$. If $y(t)$ is bounded for $t \geq 0$, then, for any $\eta > 0$, $L > 0$, there is an $\epsilon_0 = \epsilon_0(\eta, L) > 0$ such that for $0 < \epsilon < \epsilon_0$, we have*

$$|x(t) - x^*(t)| \leq \eta \quad \text{for } 0 \leq t \leq L, \tag{8.6}$$

where $x^*(t) = y(t) + \epsilon u(\frac{t}{\epsilon}, y(t))$.

It is also possible to obtain some qualitative results that are valid on the infinite time interval $[0, \infty)$, for example,

Theorem 8.2. *If Equation (8.2) has a hyperbolic equilibrium point x_0, then there is an ϵ-periodic solution $x^*(t, \epsilon)$ of Equation (8.1), $x(\tau, 0) = x_0$, which is hyperbolic and has the same stability properties as the equilibrium point x_0 of Equation (8.2).*

If Equation (8.2) *has a hyperbolic periodic orbit* γ, *the Equation* (8.1) *has a hyperbolic invariant manifold* $M_\epsilon \subset \mathbb{R} \times \mathbb{R}^n$ *such that* $M_0 = \mathbb{R} \times \gamma$, *the cross section* $M_{\epsilon t}$ *of* M_ϵ *at time* t *is* ϵ*-periodic in* t; *that is, there exists a hyperbolic invariant torus of Equation* (8.1).

We remark that if we let $t \mapsto \epsilon t$ in Equation (8.1), we obtain the more classical equation

$$\dot{x} = \epsilon f(t, x) \tag{8.7}$$

this is encountered so often in the theory of nonlinear oscillations. Of course, these results can be easily translated to the solutions of Equation (8.7) and the averaged equation

$$\dot{y} = \epsilon f_0(y). \tag{8.8}$$

12.8.2 Averaging in RFDE

It is possible to extend the results of the previous section to the RFDE

$$\dot{x}(t) = f(\frac{t}{\epsilon}, x_t), \tag{8.9}$$

where $f(\tau + 1, \phi) = f(\tau, \phi)$ for all $(\tau, \phi) \in \mathbb{R} \times C$, is continuous in (τ, ϕ) and is continuously differentiable in ϕ. The *averaged equation* is

$$\dot{y}(t) = f_0(y_t), \tag{8.10}$$

where

$$f_0(\phi) = \int_0^1 f(\tau, \phi)\, d\tau. \tag{8.11}$$

To be able to obtain results similar to those in the previous section, we will use the variation-of-constants formula in C, considering RFDE (8.9) as a perturbation of the zero vector field:

$$\dot{x}(t) = 0, \qquad t > 0. \tag{8.12}$$

If we let $T_0(t)$ be the semigroup on C generated by Equation (8.12),

$$T_0(t)\phi(\theta) = \begin{cases} \phi(t+\theta) & \text{for } t+\theta \le 0, \\ \phi(0) & \text{for } t+\theta > 0, \end{cases} \tag{8.13}$$

then the solution of RFDE (8.9) with initial value ϕ at $t = 0$ can be represented as

$$x_t = T_0(t)\phi + \int_0^t d[K_0(t,s)] f(\frac{s}{\epsilon}, x_s) \tag{8.14}$$

where $K_0(t,s)(\theta) = \int_0^s X_0(t+\theta-\alpha)\,d\alpha$, $-r \le \theta \le 0$ and

$$X_0(t) = \begin{cases} 0, & t < 0; \\ I, & t \ge 0. \end{cases}$$

Fix $T > 0$ and define the following transformation of variables $x_t = \mathcal{F} z_t$ with $\mathcal{F} : BC([0,T], C) \to BC([0,T], C)$ and

$$\mathcal{F}v(t) = v(t) - \epsilon A_0 \int_0^t d[K(t,\tau)]u(\frac{\tau}{\epsilon}, v(\tau)) + \epsilon u(\frac{t}{\epsilon}, v(t)) \tag{8.15}$$

where $u(s,\phi) = \int_0^s [f(\tau,\phi) - f_0(\phi)]\,d\tau$ and A_0 is the generator associated with Equation (8.12).

It is easy to verify that the transformation $x_t = \mathcal{F} z_t$ given by (8.15) is well defined, periodic of period 1, and close to the identity; that is, there is a constant C and an $\epsilon_0 > 0$ such that for $0 \le \epsilon \le \epsilon_0$, the difference

$$\sup_{0 \le t \le T} |z_t - x_t| < \epsilon C.$$

Next we derive the integral equation for z_t when x_t satisfies Equation (8.14). If we substitute (8.15) into (8.14) and rewrite the expression, we obtain

$$\begin{aligned} z_t = T_0(t-s)z_0 &+ \int_0^t d[K_0(t,\tau)]f_0(z_\tau) \\ &+ \int_0^t d[K_0(t,\tau)]n(\frac{\tau}{\epsilon}, z_\tau) \end{aligned} \tag{8.16}$$

where $x_0 = z_0 + \epsilon u(0, z_0)$ and

$$n(\frac{\tau}{\epsilon}, v) = -\epsilon D_\phi u(\frac{\tau}{\epsilon}, v)\frac{dv}{dt} + f(\frac{\tau}{\epsilon}, \mathcal{F}v) - f(\frac{\tau}{\epsilon}, v). \tag{8.17}$$

To prove that y_t is a solution of the averaged equation up to terms of order ϵ, it remains to analyze the nonlinearity $n(t,v)$. A simple estimate yields

Lemma 8.1. *For $v \in BC^1([0,T], C)$ there is a constant $C > 0$ such that*

$$\Big|\int_0^t d[K_0(t,\tau)]\, N(\frac{\tau}{\epsilon}, v(\tau), \epsilon)\, d\tau\Big| \le C\big(\epsilon|v|_1 + |\mathcal{F}v - v|\big). \tag{8.18}$$

So if $x_t = \mathcal{F} z_t$, then z_t is a solution of the averaged equation up to terms of order ϵ. As a first application, we compare the solution x_t of Equation (8.9) with $x_0 = \phi \in C$ with the approximate solution $x_t^* \stackrel{\text{def}}{=} \mathcal{F} y_t^*$ where y_t^* is a solution of the averaged equation with $y_0^* = \psi$ and $\phi = \psi + \epsilon u(0, \psi)$.

Theorem 8.3. *If for $\phi \in C$, the solution y_t^* of the averaged equation* (8.10) *with $y_0 = \psi$ and $\phi = \psi + \epsilon u(0, \psi)$ is uniformly bounded for $t \geq 0$, then for any η and L, there is an ϵ_0 such that for $0 < \epsilon < \epsilon_0$, the difference*

$$|x_t - x_t^*| \leq \eta \qquad \text{for} \quad 0 \leq t \leq L. \tag{8.19}$$

where $x_t^* = \mathcal{F} y_t^*$.

With the transformation theory mentioned earlier, we can use the methods from the theory of invariant manifolds to obtain

Theorem 8.4. *If y_0 is a hyperbolic equilibrium point of the averaged equation* (8.10)*, then there exist positive constants ϵ_0, η, such that for $0 < \epsilon \leq \epsilon_0$, there is an ϵ-periodic solution $x^*(t, \epsilon)$ of Equation* (8.9)*, $x^*(\cdot, 0) = y_0$, which is hyperbolic, has the same stability properties as y_0, and is unique in the set $\{ x \in \mathbb{R}^n : |x - y_0| < \eta \}$.*

If y_0 is hyperbolic and uniformly asymptotically stable, then the unique ϵ-periodic solution is hyperbolic, and uniformly asymptotically stable and there are positive constants ρ, C, γ such that if $x(t, \phi)$ (resp. $y(t, \phi)$) is the solution of (8.9) *(resp.* (8.10)*) through $(0, \phi)$ and $|\phi - y_0| < \rho$, then, for $t \geq 0$, we have*

$$\begin{aligned} |x(t, \phi) - x^*(t, \epsilon)| &\leq C e^{-\gamma t}, \\ |x(t, \phi) - y(t, \phi)| &< \eta. \end{aligned} \tag{8.20}$$

It is possible also to consider attractors.

Theorem 8.5. *If the averaged equation* (8.10) *has a local attractor $\mathcal{A}_0$, then there is an $\epsilon_0 > 0$ such that for $0 < \epsilon \leq \epsilon_0$, the Poincaré map for Equation* (8.9) *has a local attractor $\mathcal{A}_\epsilon$ and* dist $(\mathcal{A}_\epsilon, \mathcal{A}_0) \to 0$ *as $\epsilon \to 0$.*

As an example illustrating the results, we consider the equation

$$\dot{x}(t) = -x(t) + b \frac{x(t-r)}{1 + x(t-r)^n} \tag{8.21}$$

where n is a fixed even integer and $b > 0$ is a parameter.

The solution map is point dissipative and thus there exists a global attractor $\mathcal{A}_\epsilon$. It is known (at least numerically) that for $n \geq 8$, there exists a $b_0 > 0$ such that for $b \geq b_0$, there is some chaotic motion on $\mathcal{A}$. Let us consider the following class of rapidly oscillating disturbances of (8.21):

$$\dot{x}(t) = -x(t) + b \frac{\alpha \cos(t/\epsilon) + x(t-r)}{1 + (\cos(t/\epsilon) + x(t-r))^n}, \tag{8.22}$$

where $\alpha > 0$ is a constant that measures the energy of the perturbation. It is possible to prove the following result: *For $\epsilon > 0$ sufficiently small, the* *attractor* $\mathcal{A}_\epsilon$ *of the Poincaré map for Equation* (8.22) *is just a singleton provided that* $\alpha > \max\{2b, 3\}$. This result implies that high-frequency perturbations can eliminate complicated motion on the attractor. The proof of the result consists of averaging Equation (8.22), estimating the resulting nonlinear vector field, and using a Razumikhin-type theorem to obtain the existence of a globally attracting equilibrium point for the averaged equation. Theorem 8.5 completes the proof.

Averaging also has been discussed for the equation

$$\dot{x}(t) = \epsilon f(t, x), \tag{8.23}$$

where $\epsilon > 0$ is a small parameter. If f_0 is defined by Equation (8.11), then the averaged equation is the ODE

$$\dot{y}(t) = \epsilon f_0(\tilde{y}), \quad \tilde{y}(\theta) = y \quad \text{for } \theta \in [-r, 0]. \tag{8.24}$$

Results similar to Theorem 8.4 are available, but the proofs that have been given follow a different approach. If we consider RFDE (8.23) as a perturbation of

$$\dot{x}(t) = 0 \cdot x_t, \tag{8.25}$$

then the decomposition in C, $x_t = \tilde{I}z(t) + w_t$, $\tilde{I}(\theta) = I$, the identity, for $\theta \in [-r, 0]$, for the linear equation (8.25) implies that w_t approaches zero faster than any exponential. We can now use the invariant manifold theory to show that the flow for RFDE (8.23) in any given bounded set is equivalent to the flow defined by an ODE

$$\dot{z} = \epsilon g(t, z, \epsilon), \quad g(t, z, 0) = f(t, \tilde{z}).$$

The classical averaging procedure can be applied to this ODE.

If we let $t \mapsto t/\epsilon$ in RFDE (8.23) and let $x(t/\epsilon) = y(t)$, then we obtain the equation

$$\dot{y}(t) = f(\frac{t}{\epsilon}, y_{t,\epsilon}),$$

where $y_{t,\epsilon}(\theta) = y(t+\epsilon\theta)$, $\theta \in [-r, 0]$. This is an equation with a small delay, but it is rapidly oscillating in t. Therefore, it reasonable to expect that it should be possible to obtain these results for RFDE (8.23) by using the transformation theory that we used for (8.9), but this has not been done.

12.9 Infinite delay

Suppose that $0 \le r \le \infty$ is given. If $x : [\sigma - r, \sigma + A) \to \mathbb{R}^n$, $A > 0$, is a given function, then for each $t \in [s, \sigma + A)$, $\theta \in [-r, 0]$, we define, as usual, $x_t(\theta) = x(t + \theta)$. It is understood here that $[\sigma - r, \sigma + A) = (-\infty, \sigma + A)$ if $r = \infty$. In the theory of RFDE,

$$\dot{x}(t) = f(t, x_t), \tag{9.1}$$

the choice of the space for the initial data, the *phase space*, is never completely clear. For each particular application, a decision is made that is believed to reflect the important aspects of the problem under investigation. In the case of finite delay ($r < \infty$), the solution of (9.1) is required to be a continuous function for $t \geq \sigma$. Therefore, after one delay interval r, the state $x_t, t \geq \sigma + r$, belongs to the space of continuous functions. As a consequence, the choice of the phase space is not so important from the point of view of the qualitative theory. However, in specific applications, it is convenient to have other phase spaces. For example, in control theory, the space $L^2((-r, 0), \mathbb{R}^n) \times \mathbb{R}^n$ is frequently encountered. In this setting, the problem is formulated in a Hilbert space, which leads to the adaptability of many classical results to RFDE. This has proved to be particularly useful in linear control and identification problems. On the other hand, for nonlinear problems, we do not have, at this time, a theory in this space that can be used to develop a general qualitative theory. This is probably due to the fact that the requirement that the solutions are differentiable with respect to the initial data puts very severe requirements on the nonlinearities.

If the delay interval is infinite, then the state x_t at time t always contains the initial data. As a consequence, the introduction of a new phase space in a particular application requires a new and separate development of the theory. On the other hand, it is possible to give an abstract and axiomatic definition of a phase space for which many of the fundamental and desired properties hold. We present an axiomatic framework that will permit the development of the fundamental theory of existence, uniqueness, continuation, continuous dependence, differentiability with respect to initial data and parameters, etc. In addition, we need the abstract properties to imply something about the global behavior of orbits; for example, when are bounded orbits precompact, when is stability in $\mathbb{R}^n$ equivalent to stability in the function space, etc? In this way, we will gain a better understanding of the equations and at the same time avoid too much repetition.

We first remark that our axioms prevent the norm in the space from imposing any differentiability properties on the initial functions. In the applications, it is convenient at times to require the initial functions to belong to a Banach space of functions that have some derivatives with specified properties. However, if we consider all differential equations whose right-hand sides are continuous or continuously differentiable in such a space, then the equations will be of neutral type: that is, the derivatives of the independent variable will contain delays. The theory for such systems should be developed, but it will require more sophistication than the one described later for RFDE.

The phase space $\mathcal{B}$ for RFDE with infinite delay is a linear space, with a seminorm $|\cdot|_{\mathcal{B}}$ mapping $(-\infty, 0]$ into the finite-dimensional Banach space $E = \mathbb{R}^n$ or $\mathbb{C}^n$. The first two axioms on $\mathcal{B}$ are motivated by the fact that we want a solution of (9.1) to be continuous to the right of the initial time and we desire certain continuity properties of the solutions.

(A) There is a positive constant H and functions K, $M : \mathbb{R}^+ \to \mathbb{R}^+$, with K continuous and M locally bounded, such that for any $\sigma \in \mathbb{R}$, $a > 0$, if $x : (-\infty, \sigma + a) \to E$, $x_\sigma \in \mathcal{B}$, and x is continuous on $[\sigma, \sigma + a)$, then for every $t \in [\sigma, \sigma + a)$, the following conditions hold:

(i) $x_t \in \mathcal{B}$,

(ii) $|x(t)|_E \leq H|x_t|_{\mathcal{B}}$,

(iii) $|x_t|_{\mathcal{B}} \leq K(t-\sigma)\sup\{\,|x(s)|_E : \sigma \leq s \leq t\,\} + M(t-\sigma)|x_\sigma|_{\mathcal{B}}$.

(A_1) For the function x in (A), x_t is a $\mathcal{B}$-valued continuous function for $t \in [\sigma, \sigma + a)$.

We remark that the elements in a space $\mathcal{B}$ satisfying these axioms may satisfy $|\phi - \psi|_{\mathcal{B}} = 0$ and the functions may not be pointwise equal on $(-\infty, 0]$. However, from condition A(ii), if $|\phi - \psi|_{\mathcal{B}} = 0$, then $\phi(0) = \psi(0)$.

Let us give examples of spaces that satisfy (A) and (A_1). For any continuous positive function g on $(-\infty, 0]$, let

$$C_g = \{\,\phi \in C((-\infty, 0], \mathbb{R}^n) : |\phi|_g < \infty\,\}$$

where

$$|\phi|_g \stackrel{\text{def}}{=} \sup\{\,\frac{|\phi(\theta)|}{g(\theta)} : -\infty < \theta \leq 0\,\}.$$

Let

$$UC_g = \{\,\phi \in C_g : \frac{\phi}{g} \text{ is uniformly continuous on } (-\infty, 0]\,\}$$

$$LC_g = \{\,\phi \in C_g : \lim_{\theta \to -\infty} \frac{\phi(\theta)}{g(\theta)} \text{ exists in } \mathbb{R}^n\,\}.$$

$$LC_g^0 = \{\,\phi \in C_g : \lim_{\theta \to -\infty} \frac{\phi(\theta)}{g(\theta)} = 0\,\}.$$

For the special case where $g(\theta) = e^{-\gamma\theta}$, where $\gamma > 0$ is a constant, we define $C_\gamma \stackrel{\text{def}}{=} LC_g$. For $g = 1$, we obtain the following classical spaces:

$$BC = \{\,\phi \in C((-\infty, 0], \mathbb{R}^n) : \sup|\phi(\theta)| < \infty\,\}$$

$$BU = \{\,\phi \in BC : \phi \text{ is uniformly continuous on } (-\infty, 0]\,\}$$

$$LC = \{\,\phi \in BC : \lim_{\theta \to -\infty} \phi(\theta) \text{ exists in } \mathbb{R}^n\,\}$$

$$LC_0 = \{\,\phi \in BC : \lim_{\theta \to -\infty} \phi(\theta) = 0\,\}$$

It is possible to prove that the spaces UC_g, LC_g, LC_g^0 with the function g nonincreasing satisfy the axioms (A) and (A_1). In particular, this is true for the spaces BU, LC, LC_0, and C_γ. The space BC satisfies (A) but not (A_1).

If we now suppose, for example, that $f(t, \phi)$ in (9.1) is continuous in $\mathbb{R} \times \mathcal{B}$, continuously differentiable in ϕ, is locally bounded, and the space $\mathcal{B}$ satisfies axioms (A) and (A_1), then for any $(\sigma, f) \in \mathbb{R} \times \mathcal{B}$, it is possible to prove the local existence and uniqueness of the solution $x(t, \sigma, \phi)$ of (9.1), defined on an interval to the right of σ and $x_\sigma(\cdot, \sigma, \phi) = \phi$. Furthermore, the solution is continuously differentiable in ϕ. We define $T(t, \sigma)\phi = x_t(\cdot, \sigma, \phi)$ and refer to $T(t, s)$ as the *solution operator* of (9.1). For simplicity in the presentation, we assume that $T(t, \sigma)$ is defined for all $t \geq \sigma$. From the assumptions on f in (9.1), the mapping $T(t, \sigma)$ is a bounded map for each $t \geq s$.

To describe some further properties of the solution operator, we need some more axioms.

(B) The space $\mathcal{B}$ is complete.

We say that a sequence of functions $\phi^n \in \mathcal{B}$ *converges compactly* on $(-\infty, 0]$ to a function ϕ on $(-\infty, 0]$ if the sequence converges uniformly on compact subsets of $(-\infty, 0]$.

(C_1) If $\{ \phi^n \} \subset \mathcal{B}$ is a Cauchy sequence in $\mathcal{B}$ with respect to the seminorm and if ϕ^n converges compactly to ϕ on $(-\infty, 0]$, then ϕ is in $\mathcal{B}$ and $|\phi^n - \phi|_{\mathcal{B}} \to 0$ as $n \to \infty$.

For $\phi \in \mathcal{B}$, the symbol $\hat{\phi}$ denotes the equivalence class $\{ \psi : |\psi - \phi|_{\mathcal{B}} = 0 \}$ and $\widehat{\mathcal{B}}$ denotes the quotient space $\{ \hat{\phi} : \phi \in \mathcal{B} \}$, which becomes a normed linear space with the norm $|\hat{\phi}|_{\widehat{\mathcal{B}}} = |\phi|_{\mathcal{B}}$. Axiom (B) is equivalent to saying that $\widehat{\mathcal{B}}$ is a Banach space.

It is rather surprising that one of the basic properties of the map $T(t, s)$ for $t > s$ is determined by the trivial RFDE in $\mathcal{B}$:

$$\dot{x}(t) = 0. \tag{9.2}$$

We assume that $\mathcal{B}$ satisfies the axioms (A), (A_1), (B), and (C_1). Let $S(t)$, $S(0) = I$, be the solution operator of (9.2) and let $S_0(t)$ be the restriction of $S(t)$ to the closed subspace

$$\mathcal{B}_0 = \{ \phi \in \mathcal{B} : \phi(0) = 0 \}. \tag{9.3}$$

The operator $S_0(t) : \mathcal{B}_0 \to \mathcal{B}_0$ and satisfies the inequality

$$|S_0(t)|_{\mathcal{B}} \leq M(t), \tag{9.4}$$

where $M(t)$ is the function in axiom (A(iii)). Let $\widehat{S}(t)$ and $\widehat{S}_0(t)$, $t \geq 0$, be, respectively, the induced operators on $\widehat{\mathcal{B}}$ and $\widehat{\mathcal{B}}_0$. These are C_0-semigroups of operators.

Let us recall that the α-measure of noncompactness of a bounded linear operator A on a Banach space X is defined by $\alpha(A) = \inf\{ k : \alpha(AB) \leq k\alpha(B)$ for all bounded sets $B \subset X \}$. Also recall that $r_e(A)$ denotes the radius of the essential spectrum of A. Let

$$\beta = \lim_{t\to\infty} \frac{1}{t} \log \alpha\big(\widehat{S}(t)\big). \tag{9.5}$$

An important result is the following.

Lemma 9.1. $r_e(\widehat{S}(t)) = e^{\beta t} \le |S_0(t)|_{\mathcal{B}}$ *for all* $t \ge 0$.

As an example, we remark that it can be shown that

$$r_e(\widehat{S}(t)) \le \sup\{ \frac{g(\theta + t)}{g(\theta)} : -\infty < \theta \le t \} \qquad \text{if} \quad \mathcal{B} = C_g \tag{9.6}$$

$$r_e(\widehat{S}(t)) = e^{-\gamma t} \qquad \text{if} \quad \mathcal{B} = C_\gamma. \tag{9.7}$$

Lemma 9.2. *Suppose that* $\mathcal{B}$ *satisfies the axioms* (A), (A_1), (B), *and* (C_1) *and* $K(t)$ *is bounded for* $t \ge 0$. *Then the solution operator* $T(t,\sigma)$ *of Equation* (9.1) *can be written as*

$$T(t,\sigma)\phi = S(t-\sigma)\phi + U(t,\sigma)\phi, \quad t \ge \sigma, \tag{9.8}$$

where the operator $U(t,\sigma)$ *on* $\mathcal{B}$ *is completely continuous.*

Lemma 9.2 is a consequence of our axioms and the representation of $U(t,\sigma)$ as

$$[U(t,\sigma)\phi](\theta) = \begin{cases} 0, & \text{if } t+\theta < \sigma, \\ \int_\sigma^{t+\theta} f(s, T(s,\sigma)\phi)\, ds, & \text{if } t+\theta > \sigma. \end{cases}$$

It is interesting to consider the implications of Lemma 9.2 for the existence of compact global attractors for autonomous equations. The same remarks will hold for the Poincaré map of an equation that is periodic in time. Suppose that $f \in C^1(\mathcal{B}, \mathbb{R}^n)$ is a locally bounded map and consider the autonomous equation

$$\dot{x}(t) = f(x_t) \tag{9.9}$$

on the space $\mathcal{B}$ satisfying all of the previous axioms. Let $T(t)$ be the solution operator of Equation (9.9) with $T(0) = I$.

Theorem 9.1. *Suppose that* $\mathcal{B}$ *satisfies the axioms* (A), (A_1), (B), *and* (C_1) *and* $K(t)$ *is bounded for* $t \ge 0$ *and* $M(t) \to 0$ *as* $t \to \infty$. *If* $T(t)$ *is point dissipative and positive orbits of bounded sets are bounded, then there is a compact global attractor for Equation* (9.9).

The hypothesis on $M(t)$, (9.1), Lemma 9.1, and (9.4) imply that $r_e(S(t)) \to 0$ as $t \to \infty$. From Lemma 9.2, we infer that $T(t), t \ge 0$, is an α-contraction. One also shows that the ω-limit set of any bounded set is a compact invariant set and then the conclusion follows from Theorem 4.3.3.

The corresponding result for the Poincaré map of a periodic system will yield the existence of a compact global attractor and the existence of a fixed point. Therefore, there will be a periodic solution of the RFDE of the same period as the coefficients of the vector field.

The theory of linear equations can be developed on the Banach space $\widehat{\mathcal{B}}$. To be somewhat more specific, suppose that $\mathcal{B}$ satisfies the axioms (A), (A_1), (B), and (C_1) and consider the linear autonomous equation

$$\dot{x}(t) = Lx_t, \tag{9.10}$$

where $L : \mathcal{B} \to E$ is a bounded linear operator. Equation (9.10) generates a strongly continuous semigroup $T(t)$, $t \geq 0$, on $\mathcal{B}$. Let A be the infinitesimal generator. We can define the operators $\widehat{T}(t)$ and $\widehat{A}$ on $\widehat{\mathcal{B}}$ induced by the operators $T(t)$ and A and given by the formulas $\widehat{T}(t)\hat{\phi} = T(t)\phi$, $\widehat{A}\hat{\phi} = A\phi$ for all $\phi \in \hat{\phi}$. Then $\widehat{T}(t)$ is a strongly continuous semigroup on $\widehat{\mathcal{B}}$ and $\widehat{A}$ is the infinitesimal generator.

The type number of $\widehat{T}(t)$ is denoted by α_L and is given by

$$\alpha_L = \lim_{t\to\infty} \frac{1}{t} \log |\widehat{T}(t)|_{\mathcal{B}} = \inf_{t>0} \frac{1}{t} \log |\widehat{T}(t)|_{\mathcal{B}} \tag{9.11}$$

and the spectral radius $r_\sigma(\widehat{T}(t))$ of $\widehat{T}(t)$ is given by $e^{t\alpha_L}$. If we let $P\sigma(\widehat{A})$ denote the point spectrum of $\widehat{A}$, then it is possible to show that

$$P\sigma(\widehat{A}) = \{\, \lambda \in \mathbb{C} : \exists b \in \mathbb{C}, b \neq 0, e^{\lambda\cdot} b \in \mathcal{B}, \lambda b - Le^{\lambda\cdot} b = 0 \,\}. \tag{9.12}$$

Furthermore,

$$\alpha_L = \max\{\beta, \sup\{\, \mathrm{Re}\,\lambda : \lambda \in P\sigma(\widehat{A}) \,\}\},$$

where β is given in (9.5).

It is possible to continue in this way to obtain all of the decomposition theory of Chapter 7 and, therefore, we have at our hands all of the machinery for the local theory that we had for the case of finite delay. We do not pursue this any further and recommend the references for details and further references.

12.10 Supplementary remarks

The definition of functional differential equations on manifolds as given in Section 1 (as well as many of the examples) is due to Oliva [1,2,3]. A more complete presentation and proofs of many of the results are in Hale, Magalhães, and Oliva [1]. Theorem 1.2 was stated by Kurzweil [4] (the first complete proof was given by Mallet-Paret [2]). Theorem 1.3 was first proved by Kurzweil [1,2,4] where he also presented other interesting results for RFDE near ordinary differential equations (see also Kurzweil [3,5]). Theorem 1.4 is due to Oliva [4]. For the complete proof of the assertion in Example 1.7 about the limit of $T_F(t)$ as indicated in the text, see Hale and Raugel [2]. Another proof is in Hale, Magalhães, and Oliva [1]. For a different proof of Theorem 1.3 and generalizations that permit the consideration of structural stability and generic one-parameter bifurcations near equilibrium points, see Magalhães [5,6,7].

Mallet-Paret [4] proved that the compact attractor has finite Hausdorff dimension in a separable Hilbert space. Mañe [1] proved the more general results in Theorems 2.1 and 2.2. Theorems 2.3 and 2.4 are due to Mallet-Paret [4] and generalize the corresponding results of Cartwright [1,2] for ordinary differential equations.

The stability result (Theorem 3.1) on Morse-Smale systems and a proof were given in an unpublished work of Oliva [5] and is reproduced in Hale, Magalhães, and Oliva [1]. Theorem 3.3 is due to Hale and Rybakowski [1].

Some good references for the generic theory of ordinary differential equations are Abraham and Robbin [1], Markus [1], Nitecki [1], Peixoto [1] and Smale [1]. A very readable proof of the Kupka-Smale theorem is given by Peixoto [2]. The theorem was first proved by Kupka [1] and Smale [2], but Markus [2] had previously announced some partial results.

The generic theory of RFDE initiated from the important contribution of Oliva [1]. In this paper, he began the generalization of the Kupka-Smale theorem by proving that the sets $\mathcal{G}_0(K)$ and $\mathcal{G}_1(K)$ of RFDE(F) with $F \in \mathcal{X}^k$ such that the critical points are nondegenerate and hyperbolic, respectively, are open and dense. Oliva proved (officially announced in Oliva [2]) the result that the sets $\mathcal{G}_{3/2}(T, K)$ and $\mathcal{G}_2(T, K)$ of RFDE(F) with $F \in \mathcal{X}^k$ with nondegenerate and hyperbolic, respectively, periodic orbits in K of period $\leq T$ are open. For the completion of the proof of Theorem 4.1, it was, therefore, necessary to prove density, which is the most difficult part. Mallet-Paret [1] proved the density by the ingenious proof outlined in the text. The generic results on the differential difference equation $\dot{x}(t) = f(x(t), x(t-1))$ also are due to Mallet-Paret [3].

It is of interest to note that ideas from generic theory have been used by Chow and Mallet-Paret [2] to define an index for periodic orbits of RFDE as was done by Fuller [1] for ordinary differential equations. Fuller's index can be used to obtain a new class of periodic solutions of certain equations, for example, the equation

$$\dot{x}(t) = -[\alpha x(t-1) + \beta x(t-2)]f(x(t)).$$

The generic theory for NFDE has received very little attention. We mention the paper of de Oliveira [1] in which he proved that the sets $\mathcal{G}_0(K)$ and $\mathcal{G}_1(K)$ are generic.

Theorem 5.1 is due to Nussbaum [1]. Theorems 5.2 and 5.3 are due to Hale and Oliva [1]. Theorems 5.4 and 5.5 and Propositions 5.2 and 5.3

are due to Sternberg [1]. Theorem 5.6 is a consequence of an abstract lower semicontinuity result for gradient systems due to Hale and Raugel [1].

Theorems 6.1–6.3 and the remarks on periodic solutions in Section 6 are due to Mallet-Paret [5]. For linear equations, the observation that the number of zeros per unit interval (the discrete Liapunov functional V) does not increase with time goes back to Mishkis [1]. The same property holds also for scalar parabolic partial differential equations in one space variable (see Nickel [1], Matano [1]).

Cao [1] has generalized the definition of the discrete Liapunov functional in the text in such a way as to be able to characterize the small solutions (those that approach zero faster than any exponential) of linear nonautonomous equations. Cao [2] has used this functional also to show that there can be no small solutions of an analytic delay differential equation $\dot{x}(t) = F(x(t), x(t-1))$ provided that $\partial F(x, y)/\partial y$ is not zero if $(x, y) = (0, 0)$.

Dynamical systems for which there exist such discrete Liapunov functions (generalizing the number of zeros of a function) have many very interesting properties. For example, for the ordinary differential equation $\dot{x}(t) = f(x)$, $x \in \mathbb{R}^n$, for which the matrix $\partial f(x)/\partial x$ is of Jacobi type (the matrix is tridiagonal with the off-diagonal elements positive), there is a discrete Liapunov function that is given by the number of sign changes in the vector x. This property can be used to prove that hyperbolicity of equilibrium points implies that the stable and unstable manifolds intersect transversally (Fusco and Oliva [1]). Oliva, Kuhl, and Magalhães [1] have extended this result to diffeomorphism with oscillatory Jacobians.

We refer to the papers of Fiedler and Mallet-Paret [1], Fiedler [1] and Fusco and Oliva [1] and the references therein for other aspects of this exciting area of research.

Kaplan and Yorke [2] were the first to use the projection of an orbit of a delay differential equation onto a plane to obtain the existence of a periodic orbit and some of the asymptotic properties of special solutions. This was done for slowly oscillating solutions and a very special nonlinear equation.

Fiedler and Mallet-Paret [2] showed that the connecting orbits $C_N^{N^*}$ exist for $N < N^*$ and McCord and Mischaikow [1] gave the general result that $C_K^N \neq \emptyset$ for all $K < N$.

As remarked in the text, the flow on a Morse set $\mathcal{A}_N$ may be very complicated. The numerical experiments of Mackey and Glass [1], Farmer [1], Chow and Green [1] and Hale and Sternberg [1] clearly indicate this fact. The numerical computations of Hale and Sternberg [1] were designed to test the hypothesis that the chaotic motion was a consequence of the creation of a transversal intersection of the stable and unstable manifolds of a periodic orbit. Some theoretical results exhibiting classes of delay differential equations that possess a hyperbolic periodic orbit with its stable and unstable manifolds having nonempty transversal intersection may be found in Walther [10], an der Heiden and Walther [1] and Hale and Lin [1].

Equation (7.1) has served as a model for many applications, including physiological control systems (Glass and Mackey [1], an der Heiden and Mackey [1], Lasota [1], Mackey and Glass [1], Mackey and an der Heiden [1], Wazewska-Czyzewska and Lasota [1]), optically bistable devices and the transmission of light through a ring cavity (Berre et al. [1], Derstine et al. [1,2], Gibbs et al. [1], Hopf et al. [1], Ikeda [1], Ikeda, Daido, and Akimoto [1], Ikeda, Kondo, and Akimoto [1], Ikeda and Matsumoto [1], Malta and Ragazzo [1]) and population dynamics (Blythe et al. [1], Gurney et al. [1], Hoppensteadt [1]).

Propositions 7.1 and 7.2 are due to Ivanov and Sharkovsky [1]. Theorem 7.1 is due to Mallet-Paret and Nussbaum [1,2] (see also Mallet-Paret and Nussbaum [4]). They also give an explanation of the Gibbs' phenomenon mentioned in the text. For a given function f, it is a nontrivial task to verify that the hypotheses of Theorem 7.1 are satisfied. Mallet-Paret and Nussbaum [3] have given ranges of the parameters for which the hypotheses are satisfied for each of the following functions:

$$f_1(x) = \mu - x^2, \quad f_2(x) = x^3 - \mu x, \quad f_3(x) = -\mu[\sin(x+\alpha) - \sin\alpha],$$
$$f_4 = \mu x^\nu e^{-x}, \; x \geq 0, \quad f_5(x) = \frac{\mu x^\nu}{x^\lambda + 1}, \quad x \geq 0.$$

Theorem 7.2 for the supercritical case was conjectured by Chow and Mallet-Paret [3] and proved by Chow and Huang [1] by a method different from the one outlined in the text. The proof in the text is due to Chow, Hale, and Huang [1]. The subcritical case is due to Hale and Huang [2]. All of the results mentioned in Section 7 for the matrix case are due to Hale and Huang [3].

Equation (7.28) has served as a model of transmitted light through ring cavities with several chambers (Vallee, Dubois, Coté, and Delisle [1], Valee and Marriott [1]) as well as some problems in physiology (an der Heiden [1]). The system (7.29) has been used by Ikeda [1], Ikeda, Daido, and Akimoto [1] as a model of a ring cavity containing a nonlinear dielectric medium for which part of the transmitted light is fed back into the medium. For some precise results on the existence of periodic solutions of (7.29) with a supercritical period doubling for (7.30), see Chow and Huang [1].

One of the difficulties in the proofs outlined in Section 7 is the determination of the first few terms in the Taylor series of the vector field on a center manifold. In the papers referred to earlier, there is a general pattern that is followed to do these computations, but it is perhaps not easily recognized by a nonexpert. Recently, Faria and Magalhães [1,2] have developed the theory of normal forms for functional differential equations and have given systematic methods for the computation of the normal forms. These methods can be used for the determination of the approximate vector field on the center manifold. They have used these methods to discuss the Bogdanov-Takens singularity and the Hopf bifurcation.

Since the flow for an RFDE evolves in an infinite-dimensional space, it is perhaps to be expected that all of the complications that occur in ordinary differential equations must appear in an RFDE. Of course, this is true if we make the dimension of the RFDE very large. On the other hand, if the dimension of the RFDE is fixed, say at $n = 1$, it is not clear that this is the case. In fact, this is not the case if the RFDE is a differential delay equation with one delay. It is therefore interesting to investigate in more detail the types of flows that can be realized by RFDE of fixed dimension. Not too much is known, but there are some local results. To be more precise, let us suppose that the linear scalar RFDE

$$\dot{x}(t) = Lx_t \tag{10.1}$$

has m eigenvalues (counting multiplicity) on the imaginary axis and let us consider the perturbed linear system

$$\dot{x}(t) = Lx_t + f(x_t) \tag{10.2}$$

where f is a C^∞-function from C to $\mathbb{R}$ and is small. For each given f, we can determine a center manifold $CM(f)$ of (10.2). The flow on $CM(f)$ is determined by an ordinary differential equation

$$\dot{y} = By + Y(f, y), \tag{10.3}$$

where $y \in \mathbb{R}^m$, the eigenvalues of the $m \times m$ matrix B are purely imaginary and coincide with the eigenvalues of (10.1) that are on the imaginary axis and the m-vector function $Y(f, \cdot)$ vanishes when $f = 0$. The problem is to determine the range of the mapping $f \in C^\infty(C, \mathbb{R}) \mapsto Y(f, \cdot) \in \mathbb{R}^m$; that is, describe those vector fields that can be realized on the center manifold.

Since the RFDE is a scalar equation, it is possible to show that (10.3) is equivalent to an mth-order scalar equation:

$$z^{(m)} + a_1 z^{(m-1)} + \cdots + a_m z = G(f, z^{(m-1)}, \ldots, z). \tag{10.4}$$

The results of Hale [25,26] imply that for any given k-jet $j_k g$ of a function $g : \mathbb{R}^m \to \mathbb{R}$, there exists a function $f : C \to \mathbb{R}$ such that the k-jet of $G(f, z_m, \ldots, z_1)$ coincides with $j_k g$. Furthermore, there exist constants $r_1 < r_2 < \ldots < r_{m-1} \leq r$ and a function $F : \mathbb{R}^m \to \mathbb{R}$ such that $f(\phi) = F(\phi(0), \phi(-r_1), \ldots, \phi(-r_{m-1}))$, that is, a differential delay equation. It was asserted in Hale [25,26] that all vector fields could be realized, but the proof only yields the above information. Rybakowski [4,5], using the Nash-Moser implicit function theorem, has shown that every vector field can be realized in the following sense: for every $m \geq 17$ and every C^{m+15}-function g, $g : \mathbb{R}^m \to \mathbb{R}$, with $g(0) = 0$ there is a C^m-delay differential equation F with $m - 1$ delays, $F(0) = 0$, such that $G(F, z^{(m-1)}, \ldots, z) = g(z^{(m-1)}, \ldots, z)$.

For the case in which (10.1) and (10.2) are FDE in $\mathbb{R}^n$, Faria and Magalhães [3] extended the results of Hale [25,26] by showing that all k-jets on a center manifold can be realized if n is larger or equal to the largest number of Jordan blocks associated with each of the eigenvalues of the matrix B. The proof uses their theory of normal forms (Faria and Magalhães [1,2]). Under these assumptions, Rybakowski [5] has shown that all vector fields can be realized under the same differentiability assumptions mentioned in the previous paragraph.

For scalar equations Faria and Magalhães [4] use the theory of normal forms to determine the restrictions that are imposed on the vector fields when the number of delays in the nonlinearity is less than $m-1$. There are no restrictions for the generic Hopf bifurcation or the Bogdanov-Takens singularity with a double-zero eigenvalue. There are restrictions when there are two purely imaginary and one zero eigenvalue.

For the classical method of averaging in ODE, the reader may consult, for example, Bogoliubov and Mitropolsky [1] or Hale [21]. The averaging theory for RFDE with rapid oscillations in the time variable and the details of the example (8.16) are due to Hale and Verduyn Lunel [1,2]. The theory also is applicable to parabolic partial differential equations. The results for Equation (8.17) were first given by Hale [5] and extended earlier work of Halanay [3].

The first axiomatic approach for equations with infinite delay was given by Coleman and Mizel [1,2,3] (see also Coleman and Dill [1], Coleman and Mizel [4,5], Coleman and Owen [1], Lima [1], and Leitman and Mizel [1,2,3]) for a special class of fading memory spaces. The beginnings of the general abstract theory of phase spaces appeared for the first time in Hale [27], but there were only a few axioms, no proofs, and, therefore, several points of confusion and omission. The more complete theory was developed by Hino [1,2,3], Naito [1,2,3], Hale [20], Hale and Kato [1], Schumacher [1,2], Shin [1,2]. The recent book of Hino, Murakami, and Naito [1] contains almost all of the earlier works as well as a more extensive theory and applications to stability theory and the existence of periodic and almost periodic solutions. This work contains an extensive bibliography to the other literature and methods. The presentation in the text also is based on this book. Makay [2] has given further interesting remarks on the determination of stability by using Liapunov functionals. For a theory of dependence of solutions on the memory function, see Hines [1] and references therein.

There are many problems in FDE that we have not addressed in these notes that certainly are important and deserve to be studied in detail. Partial differential equations for which there are delays in time occur frequently in modeling. For the basic theory of existence and uniqueness of solutions; see, for example, Fitzgibbon [1], Thieme [1], Webb [1,2,3], and the references therein. For some of the models that occur in age-dependent populations, interesting new ideas are required in the development of the theory (see Thieme [1], Webb [4]).

Travis and Webb [1,2], Mitropolskii and Fodčuk [1], Mitropolskii and Korenevskii [1,2], and Dombrovskii [1] have used the generalization to such equations of the decomposition theory of Chapter 7 and made applications to stability theory and the existence of invariant manifolds. Memory [1,2] has given a complete theory of stable and unstable manifolds near an equilibrium point.

In ecological models, we encounter systems of the form

$$\begin{aligned}\frac{\partial u_i(x,t)}{\partial t} &= d_i \Delta u_i(x,t) \\ &\quad + b_i u_i(x,t)[1 - \sum_{j=1}^{n} c_{ij} \int_{-r}^{0} u_j(x,t+s)d\eta_{ij}(s)],\end{aligned} \tag{10.5}$$

in a smooth bounded domain Ω with boundary conditions. We consider only the nonnegative solutions of Equation (10.5). With an appropriate generalization of the notion of negative feedback and restrictions on the coefficients, this equation defines a monotone dynamical system. An extensive theory has been developed in this direction (for results and further references, see, Martin and Smith [1,2], Smith [3], and Smith and Thieme [1,2,3]). For similar situations where the coefficients are periodic or almost periodic, see Tang and Kuang [1,2].

A special case of Equation (10.5) is the scalar equation

$$u_t - d\Delta u(x,t) = u(x,t)[1 - u(x,t-\tau)] \tag{10.6}$$

with either Dirichlet or Neumann boundary conditions. In case $\Omega = (0,1)$ (one space dimension), Luckhaus [1] has shown that (10.6) is point dissipative in the L^2-norm for all choices of the positive parameters d, τ. For a smooth bounded general domain Ω and a fixed value of d, he also proved the same result if $\tau < \tau_0$, where τ_0 is sufficiently small. Friesecke [1] has shown that there are positive constants d_0, τ_0 such that (10.6) is point dissipative if $d > d_0$ and $\tau < \tau_0$. He proves also the following surprising fact if the diffusion coefficient is too small and the delay is too large: *If* dim $(\Omega) \geq 2$, *then there exist positive constants* d_1, δ_1 *such that* (10.6) *is not point dissipative in the region*

$$\{\,(d,\tau) : d < d_1, \tau > \frac{\delta_1}{\sqrt{\delta_1(d_1 - d)}}\}.$$

In this region of parameters for (d,τ) *and in the set of nonnegative solutions, there is an open nonempty subset of initial values for which the* L^1*-norm of each solution tends exponentially to* ∞ *as* $t \to \infty$.

Another variant of population models is a one-dimensional diffused version of Wright's equation:

$$u_t - du_{xx} = -(\frac{\pi}{2} + \mu)u(x,t-1)[1 + u(x,t)] \tag{10.7}$$

in the interval $(0,1)$ with Neumann boundary conditions. Yoshida [1] and Morita [1] have discussed the existence and stability of the spatially independent periodic orbit that arises through a Hopf bifurcation as a function of the parameters d, μ. For a fixed value of $\mu > 0$, Morita [1] has shown that this solution is unstable for $d < d_0$, with d_0 sufficiently small. Memory [3] has shown there is a positive constant $d_1 > d_0$ at which another Hopf bifurcation from zero occurs, resulting in an unstable, spatially varying, periodic solution. She also shows how to destabilize the original periodic solution (as d is decreased) before this bifurcation occurs by replacing the term $u(x, t-1)$ by $u(x, t-1) + hu^3(x, t-1)$ for appropriate h. This shows that the global attractor can exhibit interesting dynamic behavior.

Recently, stochastic RFDE have received some attention. The method of averaging has been extended to the case of stochastic evolutionary equations by Seidler and Vrkoč [1], and Maslowski, Seidler, and Vrkoč [1,2]. The results here also overlap with the averaging procedure of Section 12.8.

For questions in stochastic RFDE related to the topics discussed in these notes—existence, uniqueness, stability, Liapunov exponents, variation-of-constants formula, stable manifolds—see Ito and Nisio [1], Mizel and Trutzer [1], Mohammed [1,2], Mohammed and Scheutzow [1], Mohammed, Scheutzow, and Weizsäcker [1] and Scheutzow [1] and the extensive references in Scheutzow [2].

Freidlin [1], Freidlin and Wentzell [1], and Ventsel and Freidlin [1] (same people) have given an extensive theory of large deviations for Gaussian processes with values in Hilbert spaces. They have applied these results to the study of random perturbations of ordinary differential equations. In particular, for an ordinary differential equation with a globally stable equilibrium point 0 and any neighborhood V of 0, they use quasi-potentials to determine the most likely point of escape from V. Langevin, Oliva, and de Oliveira [1] have extended such results to random perturbations of RFDE. Similar results have been given for NFDE by de Oliveira [1]. Galves, Langevin, and Vares [1] have considered similar problems for maps when the attractor is one-dimensional with three fixed points (similar problems for differential equations had been considered by Freidlin and Wentzell [1]).

In recent years, there have been many papers devoted to the oscillatory properties of the solutions of scalar delay differential equations. For linear autonomous equations, it is a general rule that a necessary and sufficient condition for solutions to be oscillatory is that no roots of the characteristic equation be real. In the nonautonomous case, only sufficient conditions have been given. There are also several results on autonomous nonlinear equations that are related to stability and instability properties of solutions. The reader may consult the book of Gyori and Ladas [1] and the proceedings of a recent conference (Graef and Hale [1]) for details and references.

Appendix
Stability of characteristic equations

The purpose of this appendix is to give methods for determining when the roots of a characteristic equation are in the left half-plane. The most general results are due to Pontryagin [1] for the zeros of characteristic equations of the form $P(z, e^z) = 0$ where $P(x, y)$ is a polynomial in x, y. Pontryagin gave necessary and sufficient conditions for all solutions of $P(z, e^z)$ to lie in the left half-plane. To obtain the results, he extended the methods used in proving the Routh-Hurwitz criterion for the zeros of a polynomial to be in the left half-plane. We state the results of Pontryagin without proof and give applications to a few specific equations.

Suppose $P(z, w)$ is a polynomial in z, w,

$$P(z, w) = \sum_{m=0}^{r} \sum_{n=0}^{s} a_{mn} z^m w^n. \tag{A.1}$$

We call $a_{rs} z^r w^s$ the *principal term* of the polynomial if $a_{rs} \neq 0$ and if for each other term $a_{mn} z^m w^n$ with $a_{mn} \neq 0$, we have either $r > m$, $s > n$, or $r = m$, $s > n$, or $r > m$, $s = n$. Clearly, not every polynomial has a principal term.

If $w = e^z$, then $P(z, e^z) = 0$ corresponds to the characteristic equation for the scalar differential difference equation

$$\sum_{m=0}^{r} \sum_{n=0}^{s} a_{mn} \frac{d^m}{dt^m} x(t+n) = 0. \tag{A.2}$$

The equation $P(z, e^z)$ could also correspond to a matrix system of differential difference equations. One important thing to notice is that the only characteristic equations that can be discussed by the methods of this appendix are those for which the delays have ratios that are rational. One can then change the time variable to obtain integer delays.

In Equation (A.2), let

$$x(t) = y_1(t),\ dx(t)/dt = y_2(t), \ldots,\ d^{r-1}x(t)/dt^{r-1} = y_r(t).$$

Then Equation (A.2) can be written as the system

$$\dot{y}_{j-1}(t) = y_j(t), \qquad j = 2, 3, \ldots, r$$

$$\sum_{n=0}^{s} a_{rn}\dot{y}(t+n) = \sum_{m=0}^{r-1}\sum_{n=0}^{s} a_{mn}y(t+n). \tag{A.3}$$

To say that $P(z, w)$ has a principal term is equivalent to saying that System (A.3) is a neutral differential difference equation according to the definition in Chapter 9. For neutral equations, we have previously remarked that all zeros of $P(z, e^z) = 0$ must have real parts bounded above. The fact that these equations are the only ones for which this is true is a consequence of the following result.

Theorem A.1. *If the polynomial $P(z, w)$ has no principal term, then the equation $P(z, e^z) = 0$ has an infinity of zeros with arbitrarily large real parts.*

The basic results for applications are the next theorems.

Theorem A.2. *Let $\Delta(z) = P(z, e^z)$ and suppose $P(z, w)$ is a polynomial with principal term $a_{rs}z^r w^s$. All of the zeros of $\Delta(z)$ have negative real parts if and only if*

(i) *The complete vector $\Delta(iy)$ rotates in the positive direction with a positive velocity for y ranging in $(-\infty, \infty)$.*

(ii) *For $y \in [-2k\pi, 2k\pi]$, $k \geq 0$ an integer, there is an $\epsilon_k \to 0$ as $k \to \infty$ such that $\Delta(iy)$ subtends an angle $4k\pi s + \pi r + \epsilon_k$.*

Theorem A.3. *Let $\Delta(z) = P(z, e^z)$ where $P(z, w)$ is a polynomial with principal term. Suppose $\Delta(iy)$, $y \in \mathbb{R}$ is separated into its real and imaginary parts, $\Delta(iy) = F(y) + iG(y)$. If all zeros of $\Delta(z)$ have negative real parts, then the zeros of $F(y)$ and $G(y)$ are real, simple, alternate, and*

$$G'(y)F(y) - G(y)F'(y) > 0 \tag{A.4}$$

for $y \in \mathbb{R}$. Conversely, all zeros of $\Delta(z)$ will be in the left half-plane provided that either of the following conditions is satisfied:

(i) *All the zeros of $F(y)$ and $G(y)$ are real, simple, and alternate and Inequality (A.4) is satisfied for at least one y.*

(ii) *All the zeros of $F(y)$ are real and for each zero, Relation (A.4) is satisfied.*

(iii) *All the zeros of $G(y)$ are real and for each zero, Relation (A.4) is satisfied.*

One other result is needed for the applications. Suppose $f(z, u, v)$ is a polynomial in z, u, v with real coefficients that has the form

(A.5) $$f(z,u,v) = \sum_{m=0}^{r}\sum_{n=0}^{s} z^m \phi_m^{(n)}(u,v)$$

where $\phi_m^{(n)}(u,v)$ is a homogeneous polynomial of degree n in u, v. The *principal term* in the polynomial $f(z,u,v)$ is the term $z^r\phi_r^{(s)}(u,v)$ for which either $r > m$, $s > n$ or $r = m$, $s > n$ or $r > m$, $s = n$ for all other terms in (A.5).

Let $z^r\phi_r^{(s)}$ denote the principal term of $f(z,u,v)$ in (A.5), let $\phi_*^{(s)}(u,v)$ denote the coefficient of z^r in $f(z,u,v)$,

$$\phi_*^{(s)}(u,v) = \sum_{n=0}^{s} \phi_r^{(n)}(u,v),$$

and let

$$\Phi_*^{(s)}(z) = \phi_*^{(s)}(\cos z, \sin z).$$

Theorem A.4. *Let $f(z,u,v)$ be a polynomial with principal term $z^r\phi_r^{(s)}(u,v)$. If ϵ is such that $\Phi_*^{(s)}(\epsilon+iy) \neq 0$, $y \in \mathbb{R}$, then, for sufficiently large integers k, the function $F(z) = f(z, \cos z, \sin z)$ will have exactly $4ks + r$ zeros in the strip $-2k\pi + \epsilon \leq \operatorname{Re} z \leq 2k\pi + \epsilon$. Consequently, the function $F(z)$ will have only real roots if and only if, for sufficiently large integers k, it has exactly $4ks + r$ roots in the strip $-2k\pi + \epsilon \leq \operatorname{Re} z \leq 2k\pi + \epsilon$.*

The following result is due to Hayes [1] with the proof based on Bellman and Cooke [1].

Theorem A.5. *All roots of the equation $(z+a)e^z + b = 0$, where a and b are real, have negative real parts if and only if*

(A.6) $$\begin{aligned} &a > -1 \\ &a + b > 0 \\ &b < \zeta \sin\zeta - a\cos\zeta \end{aligned}$$

where ζ is the root of $\zeta = -a\tan\zeta$, $0 < \zeta < \pi$, if $a \neq 0$ and $\zeta = \pi/2$ if $a = 0$.

Proof. If $\Delta(z) = (z+a)e^z + b$; $\Delta(iy) = F(y) + iG(y)$, $y \in \mathbb{R}$, then

(A.7) $$\begin{aligned} F(y) &= a\cos y - y\sin y + b \\ G(y) &= a\sin y + y\cos y. \end{aligned}$$

Necessity. From Theorem A.3, the zeros of $G(y)$ must be real and simple. If $g(y,u,v) = uy + av$, $G(y) = g(y, \cos y, \sin y)$, then the function $\Phi_*^{(s)}(z)$ in Theorem A.4 is $\cos z$ and we may take $\epsilon = 0$. For k sufficiently large, Theorem A.4 implies the function $G(y)$ has exactly $4k+1$ zeros for $-2k\pi \leq y \leq 2k\pi$.

The equation $G(y) = 0$ is equivalent to the equation

$$y = -a \tan y. \tag{A.8}$$

We must have $a \neq -1$, for otherwise, the equation $G(y) = 0$ has a triple root at $y = 0$, which contradicts Inequality (A.4). If $a < -1$, then there is only one root in $[-\pi, \pi]$ and, exactly one root in any interval $[n\pi, (n+1)\pi]$, $n \neq 0, -1$. Therefore, there are exactly $4k - 1$ roots on the interval $[-2k\pi, 2k\pi]$ for any integer k. This contradicts Theorem A.4 since we should have $4k+1$. Therefore, $a > -1$.

If $a > -1$, then there is exactly one root y_n in each interval $(n\pi, \pi(n+1))$ and no other roots except the root $y = 0$ for $n = 0$, $n = -1$. Let $y_0 = 0$. Let us now check Inequality (A.4) at the zeros of G. To do this, first observe that $G'(y)$ is given by

$$G'(y) = a \cos y + \cos y - y \sin y.$$

If $y = 0$, then $G'(0)F(0) = (a+1)(a+b) > 0$ implies $a + b > 0$. It is easy to verify that, for any $y = y_n \neq 0$, $(\sin y)G'(y) = -y + \frac{1}{2}\sin 2y$. Therefore, $(\sin y)G'(y) > 0$ for all $y \in (0, \infty)$: Since $y_1 \in (0, \pi)$, we have $\sin y_1 > 0$. Therefore, $G'(y_1) < 0$ and Relation (A.4) is valid at $y = y_1$ if and only if $F(y_1) < 0$. This is precisely the relation $b < y_1 \sin y_1 - a \cos y_1$. This completes the proof of the necessity.

Sufficiency. Assume $a > -1$. As before, Theorem A.4 implies all the roots of $G(y) = 0$ are real. Observe that $y_n = y_{-n}$ and, thus, we need only check Condition (iii) of Theorem A.3 for $n \geq 0$. For $y = 0$, we have already observed that $G'(0)F(0) = (a+1)(a+b) > 0$ if $a > -1$, $a+b > 0$. We also have observed that $(\sin y)G'(y) < 0$ for all $y \in (0, \infty)$. Observe first that the last of Inequalities (A.6) is equivalent to $b < \sqrt{a^2 + \zeta^2}$. At any $y = y_n$ with n odd, we have

$$F(y) = \sqrt{a^2 + y^2} + b \qquad \text{if } a > 0$$
$$F(y) = -\sqrt{a^2 + y^2} + b \qquad \text{if } a < 0.$$

If $a > 0$ and (A.6) is satisfied, then $F(y) > 0$. But, if $a > 0$, then $\sin y_{2k+1} < 0$, $G'(y_{2k+1}) > 0$ and Condition (A.4) is satisfied. If $a < 0$ and Conditions (A.6) are satisfied, then $F(y) < 0$. But, if $a < 0$, then $\sin y_{2k+1} > 0$, $G'(y_{2k+1}) < 0$ and Condition (A.4) is satisfied. The roots y_{2k} of G are treated in a similar manner to complete the proof of the theorem. □

For the equation $(z + a)e^{zr} + b = 0$, Boese [1] has given a more explicit stability chart in terms of $r < r_0(a, b)$.

Theorem A.6. *All roots of the equation $(z^2 + az)e^z + 1 = 0$ have negative real parts if and only if $a > (\sin\zeta)/\zeta$ where ζ is the unique root of the equation $\zeta^2 = \cos\zeta$, $0 < \zeta < \pi/2$.*

Proof. If $\Delta(z) = (z^2 + az)e^z + 1$, $\Delta(iy) = F(y) + iG(y)$, $y \in \mathbb{R}$, then

$$\begin{aligned} F(y) &= -y^2 \cos y - ay \sin y + 1, \\ G(y) &= -y^2 \sin y + ay \cos y. \end{aligned} \tag{A.9}$$

If $g(y, u, v) = -y^2 v + yu$, then $G(y) = g(y, \cos y, \sin y)$ and the function $\Phi_*^{(s)}(z)$ in Theorem A.4 is $(-\sin y)$. Therefore, we may take $\epsilon = \pi/2$ in Theorem A.4. From Theorem A.4, all zeros of $G(y)$ are real if and only if there are $4k+2$ real zeros of $G(y)$ in the interval $[-2k\pi + \pi/2, 2k\pi + \pi/2]$ for k a sufficiently large integer. Observe that $G(0) = 0$ and $y \neq 0$, $G(y) = 0$ is equivalent to the equation

$$y = a \cot y. \tag{A.10}$$

Therefore, $G(y) = 0$ has $4k + 2$ roots on the interval $[-2k\pi + (\pi/2), 2k\pi + (\pi/2)]$ if and only if $a > 0$. Theorem A.4 implies the zeros of $G(y)$ are real if and only if $a > 0$.
Necessity. If all zeros of $\Delta(z)$ are in the left half-plane, then we must have Inequality (A.4) satisfied at the zeros of $G(y)$; that is, $G'(y)F(y) > 0$ for all y such that $G(y) = 0$. Since

$$G'(y) = -(2 + a)y \sin y - (y^2 - a) \cos y$$

we have $G'(0)F(0) = a > 0$. Using Equation (A.10), we obtain

$$\begin{aligned} G'(y) &= \frac{-y \sin y}{a}\,(y^2 + a^2 + a) \\ F(y) &= 1 - \frac{y \sin y}{a}\,(y^2 + a^2). \end{aligned}$$

Therefore, the sign of $G'(y)F(y)$ is determined by the sign of the expression

$$h(y) = \frac{\sin^2 y}{a^2}\,(y^2 + a^2) - \frac{\sin y}{ay}.$$

From Equation (A.10), we observe that

$$h(y) - 1 - \frac{\sin y}{ay}$$

at any zero $y \neq 0$ of $G(y)$. Thus, we must have $a > (\sin y)/y$ for all solutions y of Equation (A.10). It is clear that this requires restricting a so that $a > (\sin y_1(z))/y_1(a)$ where $y_1(a)$ is the unique root of $G(y) = 0$, $0 < y < \pi/2$. Let ζ be the unique root of $\zeta^2 = \cos \zeta$, $0 < \zeta < \pi/2$. One can now check that $a > \zeta$ is equivalent to the last statement. This proves necessity.
Sufficiency. One easily reverses the steps and the proof of the theorem is complete. □

Bibliography

Abraham, R. and J. Robbin [1] *Transversal Mappings and Flows.* Benjamin, 1967.

Alt, W. [1] Some periodicity criteria for functional differential equations. *Manuscripta math. 23* (1978), 295–318; [2] Periodic solutions of some autonomous differential equations with variable time delay. *Lect. Notes Math. 730* (1979), 16–31.

an der Heiden, U. [1] Periodic solutions of a nonlinear second order differential equation with delay. *J. Math. Anal. Appl. 70* (1979), 599–609.

an der Heiden, U. and M. C. Mackey [1] The dynamics of production and destruction: analytic insight into complex behavior. *J. Math. Biol. 16* (1982), 75–101.

an der Heiden, U. and H.-O. Walther [1] Existence of chaos in control systems with delayed feedback. *J. Differential Eqns. 47* (1983), 273–295.

Artola, M. [1] Sur les perturbations des équations d'évolution. Application à des problémes de retard. *Annales Ec. Norm. Sup. 2* (1969), 137–253.

Artstein, Z. [1] On continuous dependence of fixed points of condensing maps. *Dynamical Systems—An International Symposium*, 73–76, Academic Press, 1976.

Asner, B. A. [1] New constructions for pointwise degenerate systems. IFAC, 6th Triennial World Conf. 1 (1975), 9.6.1–9.6.5.

Anser, B. A. and A. Halanay [1] Pointwise degenerate second-order delay differential systems. *Anal. Univ. Bucharesti, Mat. Mec. 22* (1973), 45–60; [2] Algebraic theory of pointwise degenerate delay differential systems. *J. Differential Eqns. 14* (1973), 293–306; [3] Delay-feedback using derivatives for minimal time linear control processes. *J. Math. Anal. Appl. 48* (1974), 257–262; [4] Non-controllability of time-invariant systems using one-dimensional linear delay feedback. *Rev. Roumaine Sci. Tech. Ser. Electrotechnique et Énergétique 18* (1973), 283–293.

Avellar, C. E. and J. K. Hale [1] On the zeros of exponential polynomials. *J. Math. Anal. Appl. 13* (1980), 434–452.

Babin, A. B. and M. I. Vishik [1] *Attractors in Evolutionary Equations* (in Russian) Nauka, Moscow, 1989.

Bailey, H. R. and E. B. Reeve [1] Mathematical models describing the distribution of I^{131}-albumin in man. *J. Lab Clin. Med. 60* (1962), 923–943.

Bailey, H. R. and M. Z. Williams [1] Some results on the differential difference equation $\dot{x}(t) = \sum_{i=0}^{N} A_i x(t - T_i)$. *J. Math. Anal. Appl. 15* (1966), 569–587.

Baiocchi, C. [1] Teoremi di esistenza e regolarita per certe classi di equazioni differenziali astratte. *Ann. Math. Pura Appl.* (*4*) 72 (1966), 365–418; [2] Sulle equazioni differenziali astratte lineari del primo e del secondo ordine negli spazi di Hilbert. *Ann. Mat. Pura. App.* (*4*) 76 (1967), 233–304.

Banks, H. T. [1] *Modeling and Control in the Biomedical Sciences.* Lecture Notes in Biomathematics, Vol. 6, Springer-Verlag, 1975;

Banks, H. T. and G. Kent [1] Control of functional differential equations to target sets in functions space. *SIAM J. Control 10* (1972), 567–593.

Banks, H. T. and A. Manitius [1] Projection series for retarded functional differential equations with applications to optimal control problems. *J. Differential Eqns. 18* (1975), 296–332.

Barbu, V. and S. Grossman [1] Asymptotic behavior of linear integral differential systems. *Trans. Amer. Math. Soc. 173* (1972), 277–289.

Barnea, D. I. [1] A method and new results for stability and instability of autonomous functional differential equations. *SIAM J. Appl. Math. 17* (1969), 681–697.

Bartosiewicz, Z. [1] Density of images of semigroup operators for linear neutral functional differential equations. *J. Differential Eqns. 38* (1980), 161–175.

Bellman, R. and K. Cooke [1] *Differential Difference Equations.* Academic Press, 1963; [2] Stability theory and adjoint operators for linear differential difference equations. *Trans. Amer. Math. Soc. 92* (1959), 470–500; [3] Asymptotic behavior of solutions of differential difference equations. *Mem. Amer. Math. Soc. 35* (1959); [4] On the limit solutions of differential difference equations as the retardation approaches zero. *Proc. Nat. Acad. Sci. 45* (1959), 1026–1028.

Bellman, R. and J. M. Danskin [1] A survey of the mathematical theory of time lag, retarded control, and hereditary processes. The Rand Corporation, R–256, 1954.

Berre, M. L., Ressayre, E., Tallet, A. and H. M. Gibbs, High dimension chaotic attractors of a nonlinear ring cavity. *Phys. Rev. Lett. 56* (1986), 274–277.

Bhatia, N. and O. Hajek [1] *Local Semi-Dynamical Systems.* Lecture Notes in Math., vol. 90, Springer-Verlag, 1969.

Billotti, J. E. and J. P. LaSalle [1] Periodic dissipative processes. *Bull. Amer. Math. Soc. 6* (1971), 1082–1089.

Biroli, M. [1] Solutions presque périodiques d'une équation et d'une inéquation parabolique avec terme de retard nonlineáire. I, II, III *Atti Accad. Naz. Lincei Rend. Cl. Sci. Fis. Mat. Natur.* (*8*) *48* (1970), 576–580; *ibid* (*8*) *49* (1970), 23–26 (1971); *ibid* (*8*) *49* (1970), 175–179 (1971).

Blythe, S. P., Nisbet, R. M. and W. S. C. Gurney [1] Instability and complex dynamic behavior in population models with long time delays. *Theor. Pop. Biol. 2* (1982), 147–176.

Boas, R. [1] *Entire Functions.* Academic Press, New York, 1954.

Boese, F. G. [1] Some stability charts and stability conditions for a class of difference-differential equations. *Z. Angew. Math. Mech. 67* (1987), 56–59.

Boffi, V. and R. Scozzafava [1] Sull'equazione funzionale lineare $f'(x) = -A(x)f(x-1)$, *Rend. Math. e Appl. (5) 25* (1966), 402–410; [2] A first–order linear differential difference equation with N delays. *J. Math. Anal. Appl. 17* (1967), 577–589.

Bogoliubov, N. N. and Y. A. Mitropolsky [1] *Asymptotic Methods in the Theory of Nonlinear Oscillations*, Gordon and Breach, New York, 1961.

Brayton, R. [1] Nonlinear oscillations in a distributed network. *Quart. Appl. Math.* *24* (1976), 289–301; [2] Small signal stability criterion for electrical networks containing lossless transmission lines. *IBM J. Res. Dev.* *12* (1968), 431–440.

Brayton, R. and R. A. Willoughby [1] On the numerical integration of a symmetric system of differential difference equations of neutral type. *J. Math. Anal. Appl.* *18* (1967), 182–189.

Browder, F. [1] On a generalization of the Schauder fixed-point theorem. *Duke Math. J.* *26* (1959), 291–303; [2] A further generalization of the Schauder fixed-point theorem. *Duke Math. J.* *32* (1965), 575–578.

Brumley, W. E. [1] On the asymptotic behavior of solutions of differential difference equations of neutral type. *J. Differential Eqns.* *7* (1970), 175–188.

Burd, V. S. and Ju. S. Kolesov [1] On the dichotomy of solutions of functional differential equations with almost periodic coefficients. *Sov. Math. Dokl.* *11* (1970), 1650–1653.

Burton, T. A. [1] Uniform asymptotic stability in functional differential equations. *Proc. Am. Math. Soc.* *38* (1978), 195–200; [2] *Volterra Integral and Differential Equations.* Academic Press, New York, 1983; [3] *Stability and Periodic Solutions of Ordinary and Functional Differential Equations*, Academic Press, New York, 1983.

Cao, Y. [1] The discrete Lyapunov function for scalar differential delay equations. *J. Differential Eqns.* *87* (1990), 365–390; [2] Non-existence of small solutions for scalar differential delay equations. *Preprint* 1992.

Carr, J. [1] *Applications of Centre Manifold Theory.* Springer-Verlag, New York, 1981.

Cartwright, M. L. [1] Almost-periodic flows and solutions of differential equations. *Proc. London Math. Soc. (3) 17* (1967), 355–380. Corringenda *(3) 17* (1967), 768; [2] Almost-periodic differential equations and almost-periodic flows, *J. Differential Eqns.* *5* (1969), 167–181.

Cesari, L. [1] Nonlinear oscillations in the frame of alternative problems, vol. I, 29–50. *Dynamical Systems—An International Symposium.* Academic Press, 1976; [2] Alternative methods in nonlinear analysis, *Intern. Conf. Diff. Eqns.* Los Angeles, September 1974, 95–148. Academic Press, 1975.

Chafee, N. [1] The bifurcation of one or more closed orbits from an equilibrium point of an autonomous differential equation. *J. Differential Eqns.* *4* (1968), 661–679; [2] A bifurcation problem for functional differential equations of finitely retarded type. *J. Math. Anal. Appl.* *35* (1971), 312–348.

Charrier, P. [1] Linear delay differential systems—Controllability to a function space. 31–33 of *Proc. Symp. Differential-Delay and Functional Equations.* Univ. Warwick, July 1972.

Chary, K. S. R. [1] A note on functional differential equations of neutral type. *Proc. Nat. Acad. Sci. India Sect. A* *43* (1973), 271–278.

Chow, S. [1] Remarks on one-dimensional delay differential equations. *J. Math. Anal. Appl.* *41* (1973), 426–429; [2] Existence of periodic solutions of autonomous functional differential equations. *J. Differential Eqns.* *15* (1974), 350–378.

Chow, S.-N. and D. Green, Jr. [1] Some results on singular delay-differential equations. In *Chaos, Fractals and Dynamics* (P. Fischer and W. R. Smith, Eds.) Marcel Dekker, 1986, 161–182.

Chow, S. and J. K. Hale [1] Strongly limit compact maps, *Funk. Ekv. 17* (1974), 31–38; [2] Periodic solutions of autonomous equations. *J. Math. Anal. Appl. 66* (1978), 495–506.

Chow, S.-N., Hale, J. K. and W. Huang [1] From sine waves to square waves in delay equations. *Proc. Roy. Soc. Edinburgh 120A* (1992), 223–229.

Chow, S.-N. and W. Huang [1] Singular perturbation problems for a system of differential difference equations. Submitted to *J. Differential Eqns.*

Chow, S.-N. and J. Mallet-Paret [1] Integral averaging and bifurcation. *J. Differential Eqns. 26* (1977), 112–159; [2] The Fuller index and global Hopf bifurcation. *J. Differential Eqns. 29* (1978), 66–85; [3] Singularly perturbed delay differential equations. In *Coupled Nonlinear Oscillators* (Eds. J. Chandra and A. Scott), North Holland, 1983.

Chow, S.-N. and H.-O. Walther [1] Characteristic multipliers and stability of periodic solutions of $\dot{x}(t) = g(x(t-1))$. *Trans. Am. Math. Soc. 307* (1988), 127–142.

Clément, Ph., Diekmann, O., Gyllenberg, M., Heijmans, H.J.A.M. and H.R. Thieme [1] Perturbation theory for dual semigroups. I. The sun-reflexive case. *Math. Ann. 277* (1988), 709–725.

Coffman, C. V. and J. J. Schäffer [1] Linear differential equations with delays. Existence, uniqueness, growth, and compactness under natural Caratheodory conditions. *J. Differential Eqns. 16* (1974), 26–44; [2] Linear differential equations with delays: Admissibility and conditional exponential stability. *J. Differential Eqns. 9* (1971), 521–535.

Coleman, B. D. and H. Dill [1] On the stability of certain motions of incompressible materials with memory. *Arch. Rat. Mech. Anal. 30* (1968), 197–224.

Coleman, B. D. and V. J. Mizel [1] Norms and semigroups in the theory of fading memory. *Arch. Rat. Mech. Anal. 2* (1966), 87–123; [2] On the general theory of fading memory. *Arch. Rat. Mech. Anal. 29* (1968), 18–31; [3] On the stability of solutions of functional differential equations. *Arch. Rat. Mech. Anal. 30* (1968), 173–196; [4] A general theory of dissipation in materials with memory. *Arch. Rat. Mech. Anal. 27* (1968), 255–274; [5] Existence of entropy as a consequence of asymptotic stability. *Arch. Rat. Mech. Anal. 25* (1967), 243–270.

Coleman, B. D. and D. R. Owen [1] On the initial-value problem for a class of functional differential equations. *Arch. Rat. Mech. Anal. 55* (1974), 275–299.

Comincioli, V. [1] Ulteriori osservazioni sulle soluzioni del problema periodico per equazioni paraboliche lineari con termini di perturbazioni. *Ist. Lombardo Acad. Sci. Lett. Rend. A 104* (1970), 726–735.

Cooke, K. L. [1] Functional differential equations: Some models and perturbation problems. *Differential Equations and Dynamical Systems (Proc. Int. Symp. Mayaguez, P. R.* (1965), 167–183. Academic Press, 1967; [2] Linear functional differential equations of asymptotically autonomous type. *J. Differential Eqns. 7* (1970), 154–174; [3] The condition of regular degeneration for singularly perturbed linear differential difference equations. *J. Differential Eqns. 1* (1965), 39–94.

Cooke, K. and J. Ferreira [1] Stability conditions for linear retarded functional differential equations. *J. Math. Anal. Appl. 96* (1983), 480–504.

Cooke, K. and D. Krumme [1] Differential difference equations and nonlinear initial-boundary-value problems for linear hyperbolic partial differential equations. *J. Math. Anal. Appl. 24* (1968), 372–387.

Cooke, K. L. and K. R. Meyer [1] The condition of regular degeneracy for singular perturbed systems of linear differential difference equations. *J. Math. Anal. Appl. 14* (1966), 83–106.

Cooke, K. L. and S. M. Verduyn Lunel [1] Distributional and small solutions for linear time-dependent delay equations, Vrije Universiteit Amsterdam Rapport WS-393 (1992), to appear *Differential Integral Eqn.*

Cooke, K. L. and J. A. Yorke [1] Equations modelling population growth, economic growth, and gonorrhea epidemiology, 35–55 in *Ordinary Differential Equations*, L. Weiss, Ed., Academic Press, 1972.

Coppel, W. A. [1] *Stability and Asymptotic Behavior in Differential Equations.* Heath Mathematical Monographs, Boston, 1965.

Corduneanu, C. [1] *Integral Equations and Stability of Feedback Systems.* Academic Press, 1973; [2] Sur la stabilité des systèmes perturbés à argument retardé. *An. Sti. Univ. "Al. I. Cuza" Iaşi Sect. I. a Mat. (N.S.) 11* (1965), 99–105.

Cruz, M. A. and J. K. Hale [1] Existence, uniqueness and continuous dependence for hereditary systems. *Annali Mat. Pura Appl. (4) 85* (1970), 63–82; [2] Stability of functional differential equations of neutral type. *J. Differential Eqns. 7* (1970), 334–355; [3] Asymptotic behavior of neutral functional differential equations. *Arch. Rat. Mech. Anal. 34* (1969), 331–353; [4] Exponential estimates and the saddle-point property for neutral functional differential equations. *J. Math. Anal. Appl. 34* (1971), 267–288.

Cunningham, W. J. [1] A nonlinear differential difference equation of growth. *Proc. Nat. Acad. Sci. U.S.A. 40* (1954), 709–713.

Dafermos, C. [1] On the existence and asymptotic stability of solutions to the equations of linear thermoelasticity. *Arch. Rat. Mech. Anal. 29* (1968), 241–271; [2] An abstract Volterra equation with applications to linear viscoelasticity. *J. Differential Eqns. 7* (1970), 554–569; [3] Semiflows associated with compact and uniform processes. *Math. Sys. Theory 8* (1974), 142–149.

Darbo, G. [1] Punti uniti in transformazioni a condominio non compatto. *Rend. Sem. Math. Univ. Padova 24* (1955), 84–92.

Datko, R. [1] An algorithm for computing Liapunov functionals for some differential difference equations. In *Ordinary Differential Equations* (L. Weiss, Ed.), (1972), 387–398, Academic Press; [2] Linear autonomous neutral differential equations in a Banach space. *J. Differential Eqns.* (1977), 258-274; [3] Remarks concerning the asymptotic stability and stabilization of linear delay differential equations. *J. Math. Anal. Appl. 111* (1985), 571–584.

Delfour, M. C. and A. Manitius [1] The structural operator F and its role in the theory of retarded systems I. *J. Math. Anal. Appl. 73* (1980), 466–490; [2] The structural operator F and its role in the theory of retarded systems II. *J. Math. Anal. Appl. 74* (1980), 359–381.

de Nevers, K. and K. Schmitt [1] An application of the shooting method to boundary-value problems for second-order delay equations. *J. Math. Anal. Appl. 36* (1971), 588–597.

Derstine, M. W., Gibbs, H. M., Hopf, F. A. and D. L. Kaplan [1] Bifurcation gap in a hybrid optically bistable system. *Phys. Rev. A, 26* (1982), 3720–3722; [2] Alternate paths to chaos in optical bistability. *Ibid 27* (1983), 3200–3208.

Diekmann, O. [1] Volterra integral equations and semigroups of operators. MC Report TW 197 Centre for Mathematics and Computer Science, Amsterdam; [2] A duality principle for delay equations. In *Equadiff 5* (M. Gregas, Ed.) Teubner Texte zur Math. 47 (1982), 84–86; [3] Perturbed dual semigroups and delay equations. In *Dynamics of Infinite Dimensional Systems* (S.-N. Chow and J. K. Hale, Eds.) Springer-Verlag, Series F: 37 (1987), 67–74.

Diekmann, O. and S. A. van Gils [1] The center manifold for delay equations in the light of suns and stars. *Preprint*, 1990.

Diekmann, O., van Gils, S. A., Verduyn Lunel, S.M. and H.O. Walther [1] *Delay Equations: Complex, Functional, and Nonlinear Analysis.* Springer-Verlag, New York, to appear.

Diestel, J. and J. J. Uhl [1] *Vector Measures.* Amer. Math. Soc., Math. Surveys 15, 1973.

Doetsch, G. [1] *Handbuch der Laplace-Transformation Band I*, Birkhäuser, Basel, 1950.

Dombrovskii, V. A. [1] The stability of periodic solutions of systems with distributed parameters and lag. [Russian] *Ukrain. Mat. Z. 24* (1972), 161–170.

Driver, R. D. [1] A functional differential system of neutral type arising in a two-body problem of classical electrodynamics. 474–484. *Nonlinear Differential Equations and Nonlinear Mechanics.* Academic Press, 1963; [2] Existence and continuous dependence of solutions of a neutral functional differential equation. *Arch. Rat. Mech. Anal. 19* (1965), 149–166; [3] *Ordinary and Delay Differential Equations.* Springer-Verlag, New York, 1977.

Dugundji, J. [1] An extension of Tietze's theorem. *Pac. J. Math. 1* (1951), 353–367.

Dunkel, G. [1] Single-species model for population growth depending on past history. *Seminar on Differential Equations and Dynamical Systems*, 92–99. Lecture Notes in Math. vol. 60. Springer-Verlag, 1968.

El'sgol'tz, L. E. [1] *Qualitative Methods in Mathematical Analysis. Trans. Math. Mono., vol. 12*, Amer. Math. Soc. 1964; [2] *Introduction to the Theory of Differential Equations with Deviating Arguments.* Holden-Day, 1966.

El'sgol'tz, L. E. and S. P. Norkin [1] *Introduction to the Theory and Applicaton of Differential Equations with Deviating Arguments.* Translated from Russian by J. L. Casti, Academic Press, 1963.

Ergen, W. K. [1] Kinetics of the circulating fuel nuclear reaction. *J. Appl. Phys. 25* (1954), 702–711.

Faria, T. and L. T. Magalhães [1] Normal forms for retarded functional differential equations and applications to Bogdanov singularity. *Preprint* 1991; [2] Normal forms for retarded functional differential equations with parameters and applications to Hopf bifurcation. *Preprint* 1991; [3] Realization of ordinary differential equations by retarded functional differential equations in neighborhoods of equilibrium points. *Preprint* 1992; [4] Restrictions on the possible flows of scalar retarded functional differential equations in neighbourhoods of singularities. *Preprint* 1992.

Farkas, M. and G. Stépán [1] On perturbation of the kernel in infinite delay systems. *ZAMM 72* (1992), 153–156.

Farmer, J. D. [1] Chaotic attractors of an infinite dimensional dynamical system. *Phys. D 4* (1982), 366–393.

Fennell, R. and P. Waltman [1] A boundary-value problem for a system on nonlinear functional differential equations. *J. Math. Anal. Appl. 26* (1969), 447–453.

Fiedler, B. [1] Discrete Lyapunov functionals and ω-limit sets. *M^2AN 23* (1989), 415–431.

Fiedler, B. and J. Mallet-Paret [1] A Poincaré-Bendixson theorem for scalar reaction diffusion equations. *Arch. Rat. Mech. Anal. 107* (1989), 325–345; [2] Connections between Morse sets for delay differential equations. *J. Reine Angew. Math. 397* (1989), 23–41.

Fink, A. M. [1], *Almost Periodic Differential Equations.* Lect. Notes Math. vol. 377 (1974), Springer-Verlag.

Fitzgibbon, W. E. [1] Semilinear functional differential equations in Banach space. *J. Differential Eqns. 29* (1978), 1–14.

Fodčuk, V. I. [1] Integral varieties for nonlinear differential equations with retarded arguments. [Russian] *Ukrain. Mat. Z. 21* (1969), 627–639; [2] Integral manifolds for nonlinear differential equations with retarded arguments. [Russian] *Differentialniye Uravenija 6* (1970), 798–808.

Franke, J. M. and H. W. Stech [1], Extensions of an algorithm for the analysis of nongeneric Hopf bifurcations, with applications to delay-differential equations. In *Lect. Notes Math. 1475* (1991), 161–175.

Freidlin, M. J. [1] Semilinear PDE and limit theorems for large deviations. *Preprint* 1991.

Freidlin, M. J. and A. D. Wentzell [1] *Random Perturbations of Dynamical Systems.* Springer-Verlag, 1985.

Friesecke, G. [1] Exponentially growing solutions for a delay diffusion equation with negative feedback. *J. Differential Eqns. 98* (1992), 1–18.

Fuller, F. B. [1] An index of fixed-point type for periodic orbits. *Am. J. Math. 89* (1967), 133–148.

Fusco, G. and W. M. Oliva [1] Jacobi matrices and transversality. *Proc. Royal Soc. Edinb. 109A* (1988), 231–241.

Galves, A., Langevin, R. and M. E. Vares [1], Tunneling for randomly perturbed Morse-Smale systems. *IMPA* (1989). *Preprint.*

Gantmacher, F. R. [1] *Theory of Matrices,* vol. II. Chelsea, 1960.

Gerstein, V. M. [1] On the theory of dissipative differential equations in a Banach space. [Russian] *Funk. Anal. i Prilŏzen, 4* (1970), 99–100.

Gerstein, V. M. and M. A. Krasnoselskii [1] Structure of the set of solutions of dissipative equations [Russian] *Dokl. Akad. Nauk SSSR 183* (1968), 267–269.

Gibbs, H. M., Hopf, F. A., Kaplan, D. L. and R. L. Shoemaker [1] Observation of chaos in optical bistability. *Phys. Rev. Letters. 46* (1981), 474–477.

Ginzburg, R. E. [1] An application of the method of Liapunov functions to the investigation of oscillations in nonlinear systems with lag. [Russian] *Differentialniye Uravnenija 7* (1971), 1903–1905; [2] Oscillations of linear systems with autonomous self-regulating lag [Russian] *Differentialniye Uravnenija 6* (1970), 1257–1264.

Glass, L. and M. C. Mackey [1] Pathological conditions resulting from instabilities in physiological control systems. *Ann. New York Acad. Sci. 316* (1979), 214–235.

Gohberg, I. C. and E. I. Sigal [1] An operator generalization of the logarithmic residue theorem and the theorem of Rouché. *Mat. Sb. 84* (1971) 609–629 (Russian) (*Math. USSR Sb. 13* (1971), 603–625).

Gopalsamy, K. [1] *Stability and Oscillations in Delay Differential Equations of Population Dynamics.* Kluwer Academic Publishers, Dordrecht, 1992.

Graef, J. R. and J. K. Hale [1] *Oscillation and Dynamics in Delay Equations.* Contemporary Math. 129, Amer. Math. Soc., Providence, 1992.

Grafton, R. [1] A periodicity theorem for autonomous functional differential equations. *J. Differential Eqns. 6* (1969), 87–109; [2] Periodic solutions of certain Leinard equations with delay. *J. Differential Eqns. 11* (1972), 519–527; [3] Periodic solutions of Leinard equations with delay: Some theoretical and experimental results, 321–334, in *Delay and Functional Differential Equations and Their Applications.* Ed. K. Schmitt, Academic Press, 1972.

Granas, A. [1] The theory of compact vector fields and some of its applications to topology of function spaces (I). *Rozprawy Mat. 30* (1962), 93 pp.

Greiner, G. and M. Schwarz [1] Weak spectral mapping theorems for functional differential equations. *J. Differential Eqns. 94* (1991), 205–216.

Grimm, L. J. and K. Schmitt [1] Boundary-value problems for differential equations with deviating arguments. *Aequationes Math. 4* (1970), 176–190; [2] Boundary-value problems for delay differential equations. *Bull. Amer. Math. Soc. 74* (1968), 997–1000.

Grimmer, R. and G. Seifert [1] Stability properties of Volterra integrodifferential equations. *J. Differential Eqns. 19* (1975), 142–166.

Grippenberg, G., Londen, S.-O. and O. Staffans [1] *Volterra Integral and Functional Equations*, Encyclopedia of Mathematics and Its Applications. Cambridge University Press, 1990.

Gromova, P. S. and A. M. Zverkin [1] On trigonometric series whose sums are continuous unbounded functions on the real axis—Solutions of equations with retarded arguments. [Russian] *Differentialniye Uravnenija 4* (1968), 1774–1784.

Grossberg, S. [1] A prediction theory for some nonlinear functional differential equations. I. Learning of lists. *J. Math. Anal. Appl. 21* (1968), 643–694. II. Learning of patterns. *Ibid. 22* (1968), 422–490; [2] Learning and energy-entropy dependence in some nonlinear functional differential equations. *Bull. Amer. Math. Soc. 75* (1969), 1238–1242.

Grossman, S. F. [1] Stability in n-dimensional differential delay equations. *J. Math. Anal. Appl. 40* (1972), 541–546.

Gumowski, I. [1] Sur le calcul des solutions périodiques de l'equation de Cherwell-Wright. *C.R. Acad. Sci. Paris. Sér. A–B 268* (1969), A157–A159.

Gurney, W. S. C., Blythe, S. P. and R. M. Nisbet [1] Nicholson's blowflies revisited. *Nature 287* (1980), 17–21.

Gyori, I. and G. Ladas, *Oscillation Theory of Delay Differential and Delay Difference Equations with Applications.* Oxford Univ. Press, 1991.

Habets, P. [1] Singular perturbations of functional differential equations. *Proc. VIIth Int. Conf. on Nonlinear Oscillations*, Berlin, Sept. 1975.

Hadeler, K. P. and J. Tomiuk [1] Periodic solutions of difference differential equations. *Arch. Rat. Mech. Anal. 65* (1977), 87–95.

Hahn, W. [1] On difference differential equations with periodic coefficients. *J. Math. Anal. Appl. 3* (1961), 70–101.

Haddock, J. R., Krisztin, T. Terjécki, J. and J. H. Wu [1] An invariance principle of Lyapunov-Razumikhin type for neutral functional differential equations. *J. Differential Eqns.* To appear.

Halanay, A. [1] *Differential Equations, Stability, Oscillations, Time Lags.* Academic Press, 1966. *Teoria Calitativa a Ecuatilior Diferentiale* [Rumanian], Editura Acad. Rep. Populaire Romine, 1963; [2] Almost periodic solutions for a class of nonlinear systems with time lag. *Rev. Roumaine Math. Pures*

Appl. 14 (1969), 1269–1276; [3] On the method of averaging for differential equations with retarded argument. *J. Math. Anal. Appl. 14* (1966), 70–76.

Halanay, A. and J. Yorke [1] Some new results and problems in the theory of differential delay equations. *SIAM Review 13* (1971), 55–80.

Halanay, A. and D. Wexler [1] *Qualitative Theory of Sampled-Data System.* [Rumaniam] *Editura Acad. RSR*, Bucharest, 1968. Russian edition in 1971 published by Mir, Moscow.

Hale, J. K. [1] Forward and backward continuation for neutral functional differential equations. *J. Differential Eqns. 9* (1971), 168–181; [2] Sufficient conditions for stability and instability of autonomous functional differential equations. *J. Differential Eqns. 1* (1965), 452–482; [3] Asymptotic behavior of the solutions of differential difference equations. *Proc. Int. Symp. Nonlin. Vibrations*, IUTAM, Kiev, 1963, vol. 2. 409–426; [4] Linear functional differential equations with constant coefficients. *Contributions to Differential Equations 2* (1963), 291–319; [5] Averaging methods for differential equations with retarded arguments. *J. Differential Eqns. 2* (1966), 57–73; [6] Linearly asymptotically autonomous functional differential equations. *Rend. Circ. Palermo (2) 15* (1966), 331–351; [7] *Applications of Alternative Problems.* CDS Lecture Notes 71–1, Brown Univ., Div. of Appl. Math., 1971; [8] Behavior near constant solutions of functional differential equations. *J. Differential Eqns. 15* (1974), 278–294; [9] Critical cases for neutral functional differential equations. *J. Differential Eqns. 10* (1971), 59–82; [10] Solutions near simple periodic orbits of functional differential equations. *J. Differential Eqns. 9* (1970), 126–183; [11] *Functional Differential Equations.* Appl. Math. Sci., vol. 3, Springer-Verlag, 1971; [12] Continuous dependence of fixed points of condensing maps. *J. Math. Anal. Appl. 46* (1974), 388–394; [13] Parametric stability in difference equations. *Bol. Un. Mat. It. (4) 11* Supp. (1975), 209–214; [14] Smoothing properties of neutral equations. *An. Acad. Brasil Ci. 45* (1973), 49–50; [15] A class of neutral equations with the fixed–point property. *Proc. Nat. Acad. Sci. U.S.A. 67* (1970), 136–137; [16] α-contractions and differential equations. *Equations Differentielles et Fonctionelles Nonlineaires*, 15–42. Hermann, Paris, 1973; [17] Oscillations in neutral functional differential equations. In *Nonlinear Mechanics.* C.I.M.E., June 1972; [18] Stability of linear systems with delays. In *Stability Problems.* C.I.M.E., June 1974; [19] Functional differential equations of neutral type. *Dynamical Systems—An International Symposium,* 179–194. Academic Press, 1976; [20] Functional differential equations with infinite delays. *J. Math. Anal. Appl. 48* (1974), 276–283; [21] *Ordinary Differential Equations*, Wiley, 1969; [22] *Functional Differential Equations*, Springer-Verlag, 1977; [23] *Asymptotic Behavior of Dissipative Systems*, Amer. Math. Soc., 1988; [24] Nonlinear oscillations in equations with delays. In *Nonlinear Oscillations in Biology* (Ed. F. C. Hoppensteadt), Lecture in Applied Math. 17(979), 157–185. Am. Math. Soc.; [25] Flows on centre manifolds for scalar functional differential equations. *Proc. Royal Soc. Edinb. 101A* (1985), 193–201; [26] Local flows for functional differential equations. *Contemporary Math. 56* (1986), 185–192; [27] Dynamical systems and stability. *J. Math. Anal. Appl. 26* (1969), 39–69; [28] Introduction to dynamic bifurcation. In *Lect. Notes Math. 1057* (1984), 106–151. Springer-Verlag.

Hale, J. K. and W. Huang [1] Global geometry of the stable regions for two delay differential equations. *J. Math. Anal. Appl.*. to appear; [2] Period doubling in singularly perturbed delay equations. *J. Differential Eqns.* to appear; [3] Square and pulse waves in matrix delay differential equations. *Dynamic Systems and Applications 1* (1992), 51–70.

Hale, J. K., Infante, E. F. and F.-S. P. Tsen [1] Stability in linear delay equations. *J. Math. Anal. Appl. 105* (1985), 533–555.

Hale, J. K. and A. F. Ivanov [1] On a high order differential delay equation. *J. Math. Anal. Appl.*, to appear.

Hale, J. K. and A. Ize [1] On the uniform asymptotic stability of functional differential equations of neutral type. *Proc. Amer. Math. Soc. 28* (1971), 100–106.

Hale, J. K. and J. Kato [1] Phase space for retarded equations with infinite delay. *Funk. Ekvacioj 21* (1978), 11–41.

Hale, J. K. and L. A. C. Ladeira [1] Differentiability with respect to delays. *J. Differential Eqns. 92* (1991), 14–26; [2] Differentiability with respect to delays for a retarded reaction-diffusion equation. *Nonlinear Anal.*, to appear.

Hale, J. K., LaSalle, J. P. and M. Slemrod [1] Theory of a general class of dissipative processes. *J. Math. Anal. Appl. 39* (1972), 177–191.

Hale, J. K. and X.-B. Lin [1] Examples of transverse homoclinic orbits in delay equations. *Nonlinear Anal. 10* (1986), 693–709; [2] Symbolic dynamics and nonlinear semiflows. *Annali de Mat. pura appl.* (IV) CXLIV (1986), 229–260.

Hale, J. K. and O. Lopes [1] Fixed-point theorems and dissipative processes. *J. Differential Eqns. 13* (1973), 391–402.

Hale, J. K., Magalhães, L. and W. Oliva [1] *An Introduction to Infinite Dimensional Dynamical Systems—Geometric Theory*, Springer-Verlag, 1984.

Hale, J. K. and P. Martinez-Amores [1] Stability in neutral equations. *J. Nonl. Anal. Theory Meth. Appl. 1* (1977), 161–172.

Hale, J. K. and J. Mawhin [1] Coincidence degree and periodic solutions of neutral equations. *J. Differential Eqns. 15* (1974), 295–307.

Hale, J. K. and K. R. Meyer [1] A class of functional equations of neutral type. Mem. Amer. Math. Soc., No. 76, 1967.

Hale, J. K. and W. Oliva [1] One-to-oneness for linear retarded functional differential equations. *J. Differential Eqns. 20* (1976), 28–36.

Hale, J. K. and C. Perelló [1] The neighborhood of a singular point for functional differential equations. *Contributions to Differential Equations 3* (1964), 351–375.

Hale, J. K. and G. Raugel [1] Lower semicontinuity of attractors of gradient systems and applications. *Ann. Mat. pura appl.* (IV) CLIV (1989), 281–326; [2] Convergence in gradient-like systems and applications. *J. Appl. Math. Phys. (ZAMP) 43* (1992).

Hale, J. K. and K. Rybakowski [1] On a gradient-like integro-differential equation, *Proc. Roy. Soc. Edinburgh 92A* (1982), 77–85.

Hale, J. K. and J. Scheurle [1] Smoothness of bounded solutions of nonlinear evolution equations. *J. Differential Eqns. 56* (1985), 142–163.

Hale, J. K. and N. Sternberg [1] Onset of chaos in differential delay equations. *J. Comp. Phys. 77* (1988), 271–287.

Hale, J. K. and S. M. Verduyn Lunel [1] The effect of rapid oscillations in the dynamics of delay equations. In *Mathematical Population Dynamics* (Eds. O. Arino, D. E. Axelrod, and M. Kimmel) (1991), 211–216, Marcel-Dekker; [2] Averaging in infinite dimensions. *J. Integral Equations Appl. 2* (1990), 463–494.

Harband, J. [1] On the asymptotic stability of solutions to an equation of car following. Tech. Rpt. Math. 13, Univ. Negev, August 1972.

Hastings, S. P. [1] Backward existence and uniqueness for retarded functional differential equations. *J. Differential Eqns. 5* (1969), 441–451.

Hausrath, A. [1] Stability in the critical case of purely imaginary roots for neutral functional differential equations. *J. Differential Eqns. 13* (1973), 329–357.

Hayes, N. D. [1] Roots of the transcendental equation associated with a certain differential difference equations. *J. London Math. Soc. 25* (1950), 226–232.

Henry, D. [1] Linear autonomous neutral functional differential equations. *J. Differential Eqns. 15* (1974), 106–128; [2] Small solutions of linear autonomous functional differential equations. *J. Differential Eqns. 8* (1970), 494–501; [3] The adjoint of a linear functional differential equation and boundary-value problems. *J. Differential Eqns. 9* (1971), 55–66; [4] Linear autonomous neutral functional differential equations. *J. Differential Eqns. 15* (1974), 106–128; [5] *Geometric Theory of Semilinear Parabolic Equations*, Lect. Notes Math., Vol. 840, 1981, Springer-Verlag; [6] *Variedades invariantes perto dum ponto fixo.* Lecture Notes, Univ. São Paulo, Brasil, 1983; [7] Topics in Analysis. *Pub. Mat. UAB 31* (1987), 29–84.

Hetzer, G. [1] Some applications of the coincidence degree for k-set contractions to functional differential equations of neutral type. *Comment. Math. Univ. Carolinae 16* (1975), 121–138.

Hille, E. and R. Phillips [1] *Functional Analysis and Semigroups.* Amer. Math. Soc. Colloq. Publ., vol. 31, 1957.

Hines, W. [1] *Ph.D. thesis*, Georgia Institute of Technology, 1992.

Hino, Y. [1] Continuous dependence for some functional differential equations. *Tohoku Math. J. 23* (1971), 565–571; [2] Asymptotic behavior of solutions of some functional differential equations. *Tohoku Math. J. 22* (1970), 98–108; [3] On stability of some functional differential equations. *Funkcial. Ekvac. 14* (1971), 47–60.

Hino, Y., Murakami, S. and T. Naito [1] *Functional Differential Equations with Infinite Delay*, Lect. Notes Math. 1473, Springer-Verlag, 1991.

Hirsch, M., Pugh, C. and M. Shub [1] *Invariant Manifolds*, Lect. Notes Math 583, Springer-Verlag, 1977.

Hopf, E. [1] Abzweigung einer Periodischen Lösung eines Differential Systems, *Berichen Math. Phys. Kl. Säch. Akad. Wiss. Leipzig 94* (1942), 1–22.

Hopf, F. A., Kaplan, D. L., Gibbs, H. M. and R. L. Shoemaker [1] Bifurcation to chaos in optical stability. *Phys. Rev. A, 25* (1982), 2172–2182.

Hoppensteadt, F. C. [1] Mathematical theories of population: Demographics, genetics and epidemics. *Regional Conf. in Appl. Math. 120* (1975), SIAM, Philadelphia.

Hoppenstadt, F. and P. Waltman [1] A problem in the theory of epidemics, I and II, *Math. Biosciences 9* (1970), 71–91. MR 44, 7083; *Ibid. 12* (1971), 133–145.

Horn, W. A. [1] Some fixed-point theorems for compact mappings and flows on a Banach space. *Trans. Amer. Math. Soc. 149* (1970), 391–404.

Huang, W. [1] Generalization of Liapunov's theorem in a linear delay system. *J. Math. Anal. Appl. 142* (1989), 83–94.

Huang, Y. S. and J. Mallet-Paret [1] Asymptotics of the spectrum for linear periodic differential delay equations. *Preprint* 1992; [2] The infinite dimensional version of the Floquet theorem. *Preprint* 1992.

Hughes, D. K. [1] Variational and optimal control problems with delayed argument. *J. Opt. Theory Appl. 2* (1968), 1–14.

Ikeda, K. [1] Multiple-valued stationary state and its instability of the transmitted light in a ring cavity. *Opt. Comm. 30* (1979), 257–261.

Ikeda, K., Daido, H. and O. Akimoto [1] Optical turbulence: chaotic behavior of transmitted light from a ring cavity. *Phys. Rev. Lett. 45* (1980), 709–712.

Ikeda, K., Kondo, K. and O. Akimoto [1] Successive higher harmonic bifurcations in systems with delayed feedback. *Phys. Rev. Lett. 49* (1982), 1467–1470.

Ikeda, K. and K. Matsumoto [1] High dimensional chaotic behavior in systems with time delayed feedback. *Physica 29D* (1987), 223–235.

Imaz, C. and Z. Vorel [1] Generalized ordinary differential equations in Banach space and applications to functional equations. *Bol. Soc. Mat. Mexicana 10* (1966), 47–59.

Infante, E. F. and W. B. Castelan [1] A Liapunov functional for a matrix difference differential equation. *J. Differential Eqns. 29* (1978), 439–451.

Infante, E. F. and M. Slemrod [1] Asymptotic stability criteria for linear systems of differential difference equations of neutral type and their discrete analogues. *J. Math. Anal. Appl. 38* (1972), 399–415.

Israelson, D. and A. Johnson [1] Application of a theory for circummutations to geotropic movements. *Physiologia Plantorium. 21* (1968), 282–291; [2] Phase-shift in geotropical oscillations—A theoretical and experimental study. *Physiologia Plantorium. 22* (1969), 1226–1237.

Itô, K. and M. Nisio [1] On stationary solutions of a stochastic differential equation. *J. Math. Kyoto Univ. 4* (1964), 1–75.

Ivanov, A. F. and A. N. Sharkovsky [1] Oscillations in singularly perturbed delay equations. *Dynamics Reported (New Series) 1* (1991), 165–224.

Izé, A. [1] Asymptotic stability implies uniform asymptotic stability in periodic retarded equations. Personal communication; [2] Linear functional differential equations of neutral type asymptotically autonomous. *Ann. Mat. Pura Appl. (4) 96* (1973), 21–39.

Izé, A. and N. de Molfetta [1] Asymptotically autonomous neutral functional differential equations with time dependent lag. *Dynamical Systems—An International Symposium,* 127–132, Academic Press, 1976.

Johnson, A. and H. G. Karlsson [1] A feedback model for biological rhythms. I. Mathematical description and basic properties of the model. *J. Theor. Biol. 36* (1972), 153–174.

Jones, G. [1] Hereditary structure in differential equations. *Math. Systems Theory, 1* (1967), 236–278; [2] Stability and asymptotic fixed-point theory. *Proc. Nat. Acad. Sci. U.S.A. 53* (1965), 1262–1264; [3] The existence of critical points in generalized dynamical systems, 7–19. *Seminar on Differential Equations and Dynamical Systems.* Lecture Notes in Math., vol. 60, Springer-Verlag, 1968; [4] The existence of periodic solutions of $f'(x) = -\alpha f(x-1)[1+f(x)]$. *J. Math. Anal. Appl. 5* (1962), 435–450; [5] On the nonlinear differential difference equation $f'(x) = -\alpha f(x-1)[1+f(x)]$. *J. Math. Anal. Appl. 4* (1962), 440–469; [6] Periodic motions in Banach space and applications to functional differential equations. *Contributions to Differential Equations 3* (1964), 75–106.

Jones, G. and Yorke, A. [1] The existence and nonexistence of critical points in bounded flows. *J. Differential Eqns. 6* (1969), 238–246.

Kaashoek, M.A. and S. M. Verduyn Lunel [1] Characteristic matrices and spectral properties of evolutionary systems. *Trans. Amer. Math. Soc. 334* (1992), 479–517; [2] An integrability condition on the resolvent for hyperbolicity of the semigroup. *J. Differential Eqns.*, to appear.

Kamenskii, G. A. [1] On the inverse operator of the shift operator over trajectories of equations with retardation. [Russian] *Trudy Sem. Teorii Diff. Urav. Otkl. Argumentom. 9* (1975), 87–92.

Kamenskii, G. A., Norkin, S. B., and L. E. El'sgol'tz [1] Certain directions in the development of the theory of differential equations with retarded arguments. [Russian] *Trudy Sem. Teor. Diff. Urav. Otkl. Argumentom. 6* (1968), 3–36.

Kaplan, J. and J. Yorke [1] Ordinary differential equations which yield periodic solutions of differential delay equations. *J. Math. Anal. Appl. 48* (1974), 317–325; [2] On the stability of a periodic solution of a delay differential equation. *SIAM J. Math. Anal. 6* (1975), 268–282.

Kappel, F. [1] Some remarks to the problem of degeneracy for functional differential equations, 463–472 of Janssens, Mawhin, Rouche, Ed., *Equations Differentielles et Fonctionnelles Non Linéairès.* Hermann, 1973; [2] Degeneracy of functional differential equations, 434–448. *Proc. Int. Conf. on Diff. Eqns.* H. A. Antosiewicz, Ed. Academic Press, 1975; [3] On degeneracy of functional differential equations. *J. Differential Eqns. 22* (1976), 250–267; [4] Laplace-transform methods and linear autonomous functional differential equations. *Ber. math.-stat. Sektion* Forschungszentrum Graz. No. 64 (1976).

Kappel, F. and H. K. Wimmer [1] An elementary divisor theory for autonomous linear functional differential equations. *J. Differential Eqns. 21* (1976), 134–147.

Kappel, F. and K. P. Zhang [1] A neutral functional differential equation with nonatomic difference operator. *J. Math. Anal. Appl. 113* (1986), 311–346.

Kato, J. [1] On Liapunov-Razumikhin type theorems for functional differential equations. *Funcial. Ekvac. 16* (1973), 225–239; [2] On the existence of O-curves II, *Tohoku Math. J. 19* (1967), 126–140; [3] Asymptotic behaviors in functional differential equations. *Tohoku Math. J. (2) 18* (1966), 174–215.

Klein, E. [1] An application of nonlinear retarded differential equations to the circummutation of plants. M.Sc. thesis, Brown University, June 1972.

Kobyakov, I. I. [1] Conditions for negativity of the Green's function of a two-point boundary value problem with deviating arguments. [Russian] *Differentialniye Uravenija 8* (1972), 443–452.

Kolmanovskii, V.B. and V.R. Nosov [1] *Stability of Functional Differential Equations.* Academic Press, San Diego, 1986.

Kolmanovskii, V.B. and A. Myshkis [1] *Applied Theory of Functional Differential Equations.* Kluwer Academic Publishers, Dordrecht, 1992.

Konovalov, Ju. P. [1] The almost-periodic solutions of quasiharmonic systems with time lag. [Russian] *Izv. Vysś, Ućebn. Zaved. Matematika (89)* (1969), 62–69.

Kovač, Ju. I. and L. I. Savčenko [1] A boundary-value problem for a nonlinear system of differential equations with retarded arguments. [Russian] *Ukrain. Mat. Z. 22* (1970), 12–21.

Koval, B. O. and E. F. Čarkov [1] Necessary and sufficient conditions of the absolute asymptotic stability of linear systems of differential equations with constant lag. [Russian] *Dopovidi Akad. Nauk Ukrain. RSR Ser. A* (1972), 506–509, 573.

Krasnoselskii, M. A. [1] *The Operator of Translation Along the Trajectories of Differential Equations.* Amer. Math. Soc., Providence, RI, 1968; [2] *Positive Solutions of Operator Equations*, Noordhoff, Groningen, 1964.

Krasovskii, N. [1] *Stability of Motion,* Moscow, 1959. Translation, Stanford University Press, 1963; [2] On the stabilization of unstable motions by additional forces when the feedback loop is incomplete. [Russian] *Prikl. Mat. Mek. 27* (1963), 641–663. TPMM 971–1004.

Kuang, Y. and H. L. Smith [1] Slowly oscillating periodic solutions of autonomous state-dependent delay equations. *Nonlinear Anal.*, to appear; [2] Periodic solutions of differential delay differential equations with threshold-type delays. In *Oscillations and Dynamics in Delay Equations*, Contemporary Math. *129* (1992), 153–176.

Kupka, I. [1] Contribution à la Théorie des champs génériques. *Contributions to Differential Equations 2* (1963), 457–484. *Ibid. 3* (1964), 411–420.

Kuratowski, C. [1] Sur les espaces complets. *Fund. Math. 15* (1930), 301–309.

Kurzweil, J. [1] Invariant manifolds, I. *Comm. Math. Univ. Carolinae 11* (1970), 309–336; [2] Invariant manifolds for flows. *Differential Equations and Dynamical Systems.* Academic Press, 1967, 431–468; [3] Invariant manifolds in the theory of functional differential equations. *Int. Conf. Nonlinear Osc.* E. Berlin, September 1974; [4] Global solutions of functional differential equations. In Lecture Notes in Math. vol. 144, Springer-Verlag, 1970; [5] On a system of operator equations. *J. Differential Eqns. 11* (1972), 364–375.

Kwapisz, M. [1] On the existence of periodic solutions of functional differential equations. [Russian] *Trudy Sem. Teor. Diff. Urav. Otk. Arg. 7* (1969), 43–54; [2] On quasilinear differential difference equations with quasilinear conditions. *Math. Nach. 43* (1970), 215–222.

Ladyzenskaya, O. A. [1] A dynamical system generated by the Navier-Stokes equation. *Zap. Nauchn. Sem. Leningrad Otdel. Mat. Inst. Steklov.* (*LOMI*) *27* (1972), 91–115; [2] On the determination of minimal global attractors for the Navier-Stokes and other partial differential equations. *Russian Mathematical Surveys 42:6* (1987), 27–73.

Laksmikantham, V. and S. Leela [1] *Differential and Integral Inequalities*, vol. 2. Academic Press, 1969.

Langevin, R., Oliva, W. M. and J. C. F. de Oliveira [1] Retarded functional differential equations with white noise perturbations. *Univ. São Paulo* (1989). *Preprint.*

LaSalle, J. P. [1] A study of synchronous asymptotic stability. *Annals of Mathematics, 65* (1957), 571–581; [2] An invariance principle in the theory of stability. *International Symposium on Differential Equations and Dynamical Systems.* Academic Press, 1967, 277.

Lasota, A. [1] Ergodic problems in biology. *Astérique 50* (1977), 239–250.

Leitman, M. J. and V. J. Mizel [1] On linear hereditary laws. *Arch. Rat. Mech. Anal. 38* (1970), 45–68; [2] On fading memory spaces and hereditary integral equations. *Arch. Rat. Mech. Anal. 55* (1974), 18–51; [3] Asymptotic stability and the periodic solutions of $x(t) + \int_{\infty}^{t} a(t-s)g(s,x(s))\,ds = f(t)$. *J. Math. Anal. Appl. 66* (1978), 606–625.

Leung, A. [1] Periodic solutions for a prey-predator differential delay equation. *Preprint.*

Levin, J. J. and J. Nohel [1] On a nonlinear delay equation. *J. Math. Anal. Appl. 8* (1964), 31–44.

Levinger, B. W. [1] A folk theorem in functional differential equations. *J. Differential Eqns. 4* (1968), 612–619.

Levinson, N. [1] Transformation theory of nonlinear differential equations of the second order. *Annals of Math. 45* (1944), 724–737; [2] A second-order differential equation with singular solutions. *Ann. Math. 50* (1949), 126–153.

Lillo, J. C. [1] Backward continuation of retarded functional differential equations. *J. Differential Eqns. 17* (1975), 349–360; [2] Oscillatory solutions of the equation $y'(x) = m(x)y(x - n(x))$. *J. Differential Eqns. 6* (1969), 1–36; [3] Periodic differential difference equations. *J. Math. Anal. Appl. 15* (1966), 434–441; [4] The Green's function for periodic differential difference equations. *J. Differential Eqns. 4* (1968), 373–384; The Green's function for nth-order periodic differential difference equations. *Math. Sys. Theory 5* (1971), 13–19; [5] Periodic perturbations of the nth-order differential difference equations. *Amer. J. Math. 94* (1972), 651–675.

Lima, P. [1] Hopf bifurcation in equations with infinite delays. Ph.D. thesis, Brown University, Providence, R. I., 1977.

Littlewood, J. E. [1] On nonlinear differential equations of the second order: IV. The general equation $\ddot{y} + kf(y)\dot{y} + g(y) = bkp(\phi)$, $\phi = t + a$. *Acta Mathematica, 98* (1957).

Lizano, M. [1] The sunflower equation. *Preprint.*

London, W. P. and J. A. Yorke [1] Recurrent epidemics of measles, chickenpox, and mumps I: Seasonal variation in contact rates. *Amer. J. Epid. 98* (1973), 453–468. II: Systematic differences in rates and stochastic effects. *Amer. J. Epid. 98* (1973), 469–482.

Lopes, O. [1] Stability and forced oscillations in nonlinear distributed networks. *Preprint*; [2] Existencia e estabilidade de oscilaçoes forcadas de equações diferenciais funcionais. Dep. Mat. Inst. Cienc. Mat. São Carlos, Brasil, 1975. Tese para o Concurso de Livre Docência; [3] Periodic solutions of perturbed neutral differential equations. *J. Differential Eqns. 15* (1974), 70–76; [4] Positive eigenvectors and periodicity theorem for autonomous retarded equations. Trabalhos de Matematica, no. 85, Universidade de Brasilia, July 1974; [5] Asymptotic fixed-point theorems and forced oscillations in neutral equations. Ph.D. Thesis, Brown University, Providence, RI, June 1973; [6] Forced oscillations in nonlinear neutral differential equations. *SIAM J. Appl. Math. 29* (1975), 196–207.

Luckhaus, S. [1] Global boundedness for a delay differential equation. *Trans. AMS 294* (1986), 767–774.

MacCamy, R. C. [1] Exponential stability for a class of functional differential equations. *Arch. Rat. Mech. Anal. 40* (1971), 120–138; [2] Nonlinear Volterra equations on a Hilbert space. *J. Differential Eqns. 16* (1974), 373-393; [3] Stability theorems for a class of functional differential equations. *SIAM J. Appl. Math. 30* (1976), 557–576.

Mackey, M. C. and U. an der Heiden [1] Dynamical diseases and bifurcations: Understanding functional disorders in physiological systems. *Funkt. Biol. Med. 156* (1982), 156–164.

Mackey, M. C. and L. Glass [1] Oscillation and chaos in physiological control systems. *Science 197* (1977), 287–289.

Magalhães, L. T. [1] Exponential estimates for singularly perturbed linear functional differential equations. *J. Math. Anal. Appl. 103* (1984), 443–460; [2] Invariant manifolds for singularly perturbed linear functional differential equations. *J. Differential Eqns. 54* (1984), 310–345; [3] Convergence and boundary layers in singularly perturbed linear functional differential equations. *J. Differential Eqns. 54* (1984), 295–310; [4] The asymptotics of solutions of singularly perturbed functional differential equations: Distributed and concentrated delays are different. *J. Math. Anal. Appl. 105* (1985), 250–257; [5] Invariant manifolds for functional differential equations close to ordinary differential equations. *Funk. Ekv. 28* (1985), 57–82; [6] Persistence and smoothness of hyperbolic invariant manifolds for functional differential equations. *SIAM J. Math. Anal. 18* (1987), 670–693; [7] Shadow ordinary differential equations for retarded functional differential equations with small delays. *Preprint*; [8] The spectra of linearization of dissipative infinite dimensional dynamical systems. *Preprint.*

Mahaffy, J. M. [1] Geometry of the stability region for a differential equation with two delays. *Preprint* 1990.

Makay, G. [1] On the asymptotic stability in terms of two measures for functional differential equations. *Nonlinear Anal. 16* (1991), 721–727; [2] On the asymptotic stability of the solutions of functional differential equations with infinite delay. *J. Differential Eqns.*, to appear.

Mallet-Paret, J. [1] Generic periodic solutions of functional differential equations. *J. Differential Eqns. 25* (1977), 163–183; [2] Generic and qualitative properties of retarded functional differential equations. Meeting Func. Diff. Eqns. Braz. Math. Soc. São Carlos, July 1975; [3] Generic properties of retarded functional differential equations. *Bull. Amer. Math. Soc. 81* (1975), 750–752; [4] Negatively invariant sets of compact maps and an extension of a theorem of Cartwright. *J. Differential Eqns. 22* (1976), 351–348; [5] Morse decompositions for delay-differential equations. *J. Differential Eqns. 72* (1988), 270-315.

Mallet-Paret, J. and R. Nussbaum [1] Global continuation and asymptotic behavior for periodic solutions of a differential-delay equation. *Ann. Mat. pura appl. (IV) CXLV* (1986), 33–128; [2] Global continuation and complicated trajectories for periodic solutions of a differential delay equation. In *Proc. Symp. Pure Math. 45* (1986), 155–167. Am. Math. Soc.; [3] A differential-delay equation arising in optics and physiology. *SIAM J. Math. Anal. 20* (1989), 249–292; [4] A bifurcation gap for a singularly perturbed delay equation. In *Chaotic Dynamics and Fractals* (Eds. M. F. Barnsley and S. G. Demko (1986), 263–286. Academic Press; [5] Boundary layer phenomena for differential-delay equations with state-dependent time lags, I. Arch. Rat. Mech. Anal. 120 (1992), 99–146; [6] Ibid, II. *Preprint* 1991.

Malta, C. P. and C. G. Ragazzo [1] Bifurcation structure of scalar differential delayed equations. *Int. J. Bif. Chaos 1* (1991), 657–665.

Mañe, R. [1] On the dimension of the compact invariant sets of certain nonlinear maps. In *Lecture Notes in Math. 898* (1981), 230–242, Springer-Verlag.

Manitius, A. [1] Completeness and F-completeness of eigenfunctions associated with retarded functional differential equations, *J. Differential Eqns. 35* (1980), 1–29.

Mansurov, K. [1] Stability of linear systems with delay. [Russian] *Studies in Differential Equations and Their Applications,* 190–199. Izdat. "Nauka," Alma–Alta, 1965.

Markus, M. [1] *Lectures in Differentiable Dynamics*. Regional Conference Series in Math., No. 3 (1971); [2] Generic properites of differential equations. *Int. Symp. on Nonlinear Differential Equations and Nonlinear Mechanics.* Academic Press, 1963, p. 22.

Marriott, C. Vallée, R. and C. Delisle [1] Analysis of a first order delay differential equation containing two delays. *Phys. Rev. A, 40* (6) (1989), 3420–3428.

Marsden, J. E. and M. F. MacCracken [1] *The Hopf Bifurcation and Its Applications.* Appl. Math. Sci., Vol. 19, Springer-Verlag.

Martin, R. H. and H. L. Smith [1] Abstract functional differential equations and reaction diffusion systems. *Trans. Amer. Math. Soc.* to appear; [2] Reaction diffusion systems with time delays: Monotonicity, invariance, comparison and convergence. *J. Reine Angew. Math.* XXX (1990).

Martynyuk, D. I. [1] *Lectures on the Theory of Stability of Solutions of Systems with Retardations.* [Russian] Inst. Mat. Akad. Nauk. UK. SSR, Kiev, 1971.

Maslowski, B., Seidler, J. and I. Vrkoč [1] An averaging principle for stochastic evolution equations II. *Math. Bohemica 116* (1991), 191–224; [2] Integral continuity and stability for stochastic hyperbolic equations. *Česk. Akad. Věd, Mat. Ústav. Preprint* 1991.

Matano, H. [1] Nonincrease of the lap number of a solution for a one dimensional semilinear parabolic equation. *J. Fac. Sci. Univ. Tokyo Sect. IA Math. 29* (1982), 401–441.

Mawhin, J. [1] Periodic solutions of nonlinear functional differential equations. *J. Differential Eqns. 10* (1971), 240–261; [2] Nonlinear perturbations of Fredholm mappings in normed spaces and applications to differential equations. Lecture notes, Univ. de Brasilia, May 1974; [3] Topology and nonlinear boundary-value problems, vol. I, 51–82. *Dynamical Systems—An International Symposium.* Academic Press, 1976.

McCord, C. and K. Mischaikow [1] On the global dynamics of attractors for scalar delay equations. CDSNS Ga. Tech Preprint 1992.

Medzitov, M. [1] Two-point boundary-value problem for a nonlinear differential equation with retarded argument. [Russian] *Trudy Sem. Teor. Diff. Urav. Otk. Arg. 7* (1969), 178–182.

Melvin, W. R. [1] A class of neutral functional differential equations. *J. Differential Eqns. 12* (1972), 524–534; [2] Topologies for neutral functional differential equations. *J. Differential Eqns. 13* (1973), 24–32; [3] Some extensions of the Karsnoselskii fixed-point theorems. *J. Differential Eqns. 11* (1972), 335–348; [4] Stability properties of functional differential equations. *J. Math. Anal. Appl. 48* (1974), 749–763.

Memory, M. [1] Stable and unstable manifolds for partial functional differential equations. *Nonlinear Anal.*, to appear; [2] Invariant manifolds for partial functional differential equations. In *Mathematical Population Dynamics* (Eds. O. Arino, D. E. Axelrod, and M. Kimmel) (1991), 223–232, Marcel-Dekker; [3] Bifurcation and asymptotic behavior of solutions of a delay differential equation with diffusion. *SIAM J. Math. Anal. 20* (1989), 533–546.

Mikolajska, Z. [1] Une remarque sur les solutions bornées d'une équation différo-differéntielle nonlinéaire. *Ann. Polon. Math. 15* (1964), 23–32.

Miller, R. K. [1] *Nonlinear Volterra Integral Equations.* Benjamin, 1971; [2] Linear Volterra integrodifferential equations as semigroups. *Funkcialaj Ekvaciaj, 17* (1974), 39–55.

Miller, R. K. and G. Sell [1] Topological dynamics and its relation to integral equations and nonautonomous systems. In *Dynamical Systems* (Eds. Cesari, Hale, Lasalle), Vol. 1, Academic Press, 1976, 223–249.

Minorsky, N. [1] Self-excited oscillations in dynamical systems possessing retarded actions. *J. Appl. Mech. 9* (1942), 65–71; [2] *Nonlinear Oscillations,* D. Van Nostrand Company, Inc., Princeton, 1962.

Minsk, A. F. [1] The second method of Liapunov for equations of neutral type. [Russian] *Trudy Sem. Teor. Diff. Urav. Otkl. Arg. 6* (1968), 78–109; [2] Absolute stability of nonlinear systems of automatic regulators of neutral type. [Russian] *Trudy Sem. Teor. Diff. Urav. Otkl. Argum.* *7* (1969), 92–106.

Mishkis, A. D.[1] *Lineare Differentialgleichungen mit nacheilenden Argumentom,* Deutscher Verlag. Wiss. Berlin, 1955. Translation of the 1951 Russian edition *Linear Differential Equations with Retarded Arguments.* [Russian] Izdat. "Nauka" Moscow, 1972. MR 50, 5135; [2] General theory of differential equations with a retarded argument. Amer. Math. Soc. Transl. no. 55 (1951). Uspehi Mat. Nauk (N.S.) 4 (33) (1949), 99–141.

Mishkis, A. D. and L. E. El'sgol'tz [1] The state and problems of the theory of differential equations with perturbed arguments. [Russian] *Uspehi Mat. Nauk 22* (134) (1967), 21–57.

Mitropolskii, Ju. A. and V. I. Fodčuk [1] Asymptotic methods of nonlinear mechanics applied to nonlinear differential equations with retarded arguments. [Russian] *Ukrain. Mat. Z. 18* (1966), 65–84.

Mitropolskii, Yu. and D. G. Korenevskii [1] The investigation of nonlinear oscillations in systems with distributed parameters and time lag. [Russian] *Mathematical Physics, No. 4* (1968), 93–145. Naukova Dumka, Kiev, 1968; [2] An application of asymptotic methods to systems with distributed parameters and lag. [Russian] *Prikladna. Meh. 5* (1969), vyp. 4, 1969.

Mizel, V. J. and V. Trutzer [1] Stochastic hereditary equations: Existence and asymptotic stability. *J. Integral Equations 7* (1984), 1–72.

Mohammed, S. E. A. [1] *Stochastic Functional Differential Equations.* Pitman, Boston, 1984; [2] The Lyapunov spectrum and stable manifolds for stochastic linear delay equations. *Stochastics 29* (1990), 89–131.

Mohammed, S. E. A. and M. K. R. Scheutzow [1] Lyapunov exponents and stationary solutions for affine stochastic delay equations. *Stochastics 29* (1990), 259–283.

Mohammed, S. E. A., Scheutzow, M. K. R. and H. v. Weizsäcker [1] Hyperbolic state space decomposition for a linear stochastic delay equation. *SIAM J. Control and Opt. 24* (1986), 543–551.

Morita, Y. [1] Destabilization of periodic solutions arising in delay diffusion systems in several space dimensions. *Japan J. Appl. Math. 1* (1984), 39–65.

Moreno, C.J. [1], The zeros of exponential polynomials. *I. Comp. Math. 26* (1973), 69–78.

Mosjagin, V. V. [1] A boundary-value problem for a differential equation with lagging argument in a Banach space. [Russian] *Leningrad Gos. Ped. Inst. Učen. Zap. 387* (1968), 198–206.

Myjak, J. W. [1] A boundary-value problem for nonlinear differential equations with a retarded argument. *Ann. Polon. Math. 27* (1973), 133–142.

Naito, T. [1] Integral manifolds for linear functional differential equations on some Banach space. *Funkial. Ekvac. 13* (1970), 199–213; [2] On autonomous linear functional differential equations with infinite retardations. *J. Differential Eqns. 21* (1976), 297–315; [3] Adjoint equations of autonomous linear functional differential equations with infinite retardations. *Tohoku Math. J. 28* (1976), 135–143.

Neustadt, L. W. [1] On the solutions of certain integral-like operator equations. Existence, uniqueness, and dependence theorems. *Arch. Rat. Mech. Anal. 38* (1970), 131–160.

Neves, A.F., H. Ribeiro and O. Lopez [1] On the spectrum of evolution operators generated by hyperbolic systems. *J. Funct. Anal. 67* (1986), 320–344.

Nickel, K. [1] Gestaltaussagen über Lösungen papabolischer Differentialgleichungen. *J. Reine Angew. Math. 211* (1962), 78–94.

Nitecki, Z. [1] *Differentiable Dynamics.* MIT Press, 1972.

Noonburg, V. W. [1] Bounded solutions of a nonlinear system of differential delay equations. *J. Math. Anal. Appl. 33* (1971), 66–76.

Norkin, S. B. [1] *Differential Equations of the Second Order with Retarded Arguments.* [Russian] Moscow; 1965. English translation by K. Schmitt and L. J. Grimm. Trans. Math. Mono. vol. 31. Amer. Math. Soc. 1972.

Nosov, V. R. [1] Linear boundary-value problems with small lags. *Diff. Urav. 3* (1967), 1025–1028; [2] Periodic solutions of systems of linear equations of general form with retarded arguments. [Russian] *Differentialniye Uravnenija 7* (1971), 639–650; [3] Periodic solutions of quasilinear functional differential equations. [Russian] *Izv. Vysš. Učebn. Zved. Matematica* 1973 (132), 55–62; [4] Periodic solutions of quasilinear functional differential equations. [Russian] *Izv. Vysš. Učebn. Zaved. Mat.* 1975 (132), 55–62.

Nussbaum, R. [1] Periodic solutions of analytic functional differential equations are analytic. *Mich. Math. J. 20* (1973), 249–255; [2] Some asymptotic fixed-point theorems. *Trans. Amer. Math. Soc. 171* (1972), 349–375; [3] Periodic solutions of some nonlinear autonomous functional differential equations. *Ann. Math. Pura Appl. 10* (1974), 263–306; [4] Periodic solutions of some nonlinear autonomous functional differential equations, II. *J. Differential Eqns. 14* (1973), 368–394; [5] A global bifurcation theorem with applications to functional differential equations. *J. Functional Anal. 19* (1975), 319–339; [6] The range of periods of periodic solutions of $x'(t) = -\alpha f(x(t-1))$. *J. Math. Anal. Appl. 58* (1977), 280-292; [7] Global bifurcation of periodic solutions of some autonomous functional differential equations. *J. Math. Anal. Appl. 55* (1976), 699-725; [8] Existence and uniqueness theorems for some functional differential equations of neutral type. *J. Differential Eqns. 11* (1972), 607–623; [9] A Hopf global bifurcation theorem for retarded functional differential equations. *Trans. Amer. Math. Soc. 238* (1978), 139–163; [10] Periodic solutions of nonlinear autonomous functional differential equations. *Lect. Notes Math. 730* (1979), 283–325; [11] Differential delay equations with two time delays. *Mem. Am. Math. Soc. 16*, 1978.

Oliva, W. M. [1] Functional differential equations on compact manifolds and an approximation theorem. *J. Differential Eqns. 5* (1969), 483–496; [2] Functional differential equations—Generic theory. *Proc. Int. Symp. Diff. Eqn. Dyn. Syst.*, Brown University, August 1974, *Dynamical Systems—An International Symposium*, vol. 1, 195–209. Academic Press, 1976; [3] Some open questions in the geometric theory of retarded functional differential equations. *Proc. 10th Braz. Colloq. Math.*, Poços de Caldas, July 1975; [4] The behavior at infinity and the set of global solutions of retarded functional differential equations. Meeting on Func. Diff. Eqns. Braz. Math. Soc., São Carlos, July 1975; [5] Stability of Morse-Smale maps. *Preprint*, 1982.

Oliva, W. M., Kuhl, N. M. and L. T. Magelhães [1] Diffeomorphism of $\mathbb{R}^n$ with oscillatory Jacobian. Personal communication.

de Oliveira, J. C. F. [1] Quasipotencial relativo a un poço. Univ. São Paulo (1989). *Preprint.*

Onuchic, N. [1] On a criterion of instability for differential equations with time delay, 339–342. *Periodic Orbits, Stability and Resonance*, Ed. by G. E. O. Giacaglia, Reidel Publ., Dordrecht, 1970; [2] On the asymptotic behavior of the solutions of functional differential equations. In *Differential Equations and Dynamical Systems,* 223–233, Academic Press, 1967; [3] Asymptotic behavior of a perturbed linear differential equation with time lag on a product space. *Ann. Mat. Pura Appl. (4) 86* (1970), 115–124.

Pavel, N. [1] On the boundedness of the motions of a periodic process. *Atti. Accad. Naz. Lincei Rend. Cl. Sci. Fis. Mat. Natur. (8) 54* (1973), 25–33.

Pazy, A. [1] *Semigroups of Linear Operators and Applications to Partial Differential Equations.* Springer-Verlag, 1983.

Pecelli, G. [1] Dichotomies for linear functional differential equations. *J. Differential Eqns. 9* (1971), 555–579.

Peixoto, M. [1] Qualitative theory of differential equations and structural stability. *Differential Equations and Dynamical Systems.* Academic Press, 1967, 469–480; [2] On an approximation theorem of Kupka-Smale. *J. Differential Eqns. 3* (1967), 214–227.

Perelló, C. [1] Cualidades del retrato fase que se preservanal introducir un retraso pequeño en el tiempo. *Acta Mexicana Ci. Tecn. 3* (1969), 12–30; [2] Periodic solutions of differential equations with time lag containing a small parameter. *J. Differential Eqns. 4* (1968), 160–175; [3] A note on periodic solution of nonlinear differential equations with time lag. In *Differential Equations and Dynamical Systems*; 185–187, Academic Press, 1967.

Perez, J. F., Malta, C. P. and F. A. B. Coutinho [1] Qualitative analysis of oscillations in isolated populations of flies. *J. Theor. Biol. 71* (1978), 505–514.

Pesin, Ya. B. [1] On the behavior of a strongly nonlinear differential equation with retarded argument. *Diff. Urav. 10* (1974), 1025–1036.

Pitt, H. R. [1] A theorem on absolutely convergent Fourier series. *J. Math. Phys. 16* (1937), 191–195.

Pliss, V. A. [1] *Nonlocal Problems of the Theory of Nonlinear Oscillations,* Academic Press, 1966. (Translation of 1964 Russian edition.)

Pontryagin, L. S. [1] On the zeros of some elementary transcendental functions. [Russian] *Izv. Akad. Nauk SSSR, Ser. Mat. 6* (1942), 115–134. English translation in *Amer. Math. Soc. Transl. (2) 1* (1955), 95–110.

Popov. V. M. [1] Pointwise degeneracy of linear, time invariant, delay differential equations, *J. Differential Eqns. 11* (1972), 541–561; [2] Delay-feedback, time-optimal, linear time-invariant control systems. *Ordinary Differential Equations*, 1971 NRL–MRC Conference, Academic Press, 1972, 545–552.

Prokopev, V. P. and S. N. Shimanov [1] On stability in the critical case of two zero roots for systems with lags. [Russian] *Differentialniye Uravnenija 2* (1966), 453–462.

Ragazzo, C. G. and C. P. Malta [1] Mode selection on a differential equation with two delays arising in optics. *Preprint* 1990.

Razumikhin, B. S. [1] On the stability of systems with a delay. [Russian] *Prikl. Mat. Meh. 20* (1956), 500–512; [2] Application of Liapunov's method to problems in the stability of systems with a delay. [Russian] *Automat. i Telemeh. 21* (1960), 740–749.

Razvan, V. [1] *Stabilitatea Absoluta a Sistemelor Automatĕ cu Intirziere.* Ed. Acad. Rep. Soc. Romania, Bucharest, 1975; [2] Absolute stability of a class of control processes described by functional differential equations of neutral type. *Equations Differentielles et Fonctionelles Nonlineaires.* Hermann, 1973.

Reissig, R., Sansone, G., and R. Conti [1] *Nichtlineare Differential Gleichungen Höherer Ordnung.* Cremonese, 1969.

Repin, Yu. M. [1] Quadratic Liapunov functionals for systems with delay. [Russian] *Prikl. Mat. Meh. 29* (1965), 564–566; TPMM 669–67; [2] On conditions for the stability of systems of differential equations for arbitrary delays. *Uchen. Zap. Ural. 23* (1960), 31–34.

Rubanik, V. P. [1] *Oscillations of Quasilinear Systems with Retardation.* [Russian] Nauk, Moscow, 1969.

Ruiz-Clayessen, J. [1] Effect of delays on functional differential equations. *J. Differential Eqns. 20* (1976), 404–440.

Rybakowski, K. P. [1] On the homotopy index for infinite-dimensional semiflows, *Trans. Amer. Math. Soc. 269* (1982), 351–382; [2] An introduction to homotopy index theory in noncompact spaces, Appendix in Hale, Magalhães, and Oliva [1]; [3] *The Homotopy Index Theory on Metric Spaces with Applications to Partial Differential Equations*, Universitext, Springer-Verlag, 1987; [4] Realization of arbitrary vector fields on center manifolds for parabolic Dirichlet BVPs. *J. Differential Eqns.* to appear; [5] Realization of arbitrary vector fields on center manifolds for functional differential equations. *J. Differential Eqns.*, to appear.

Sabbagh, L. D. [1] Variational problems with lags. *J. Optimization Theory Appl. 3* (1969), 34–51.

Sacker, R. J. and G. R. Sell [1] Existence of dichotomies and invariant splittings for linear differential equations. *J. Differential Eqns. 15* (1974), 429–458.

Sadovskii, B. N. [1] Limit compact and condensing operators. [Russian] *Uspehi Math. Nauk 27* (163) (1972), 81–146. *Russian Math. Surveys*, 85–146.

Saupe, D. [1] Global bifurcation of periodic solutions of some autonomous differential delay equations. In *Forschungsschwerpunkt Dynamische Systeme* Rep. No. 71, Univ. Bremen, Bremen, July, 1982; [2] Accelerated PL-Continuation methods and periodic solutions of parameterized differential delay equations, Ph.D. dissertation, Univ. Bremen, Bremen, 1982 (German).

Schäffer, J. J. [1] Linear differential equations with delays: Admissibility and conditional exponential stability, II. *J. Differential Eqns. 10* (1971), 471–484.

Scheutzow, M. K. R. [1] Qualitative behavior of stachastic delay equations with a bounded memory. *Stochastics 12* (1986), 41–80; [2] *Stationary and periodic stochastic differential systems.* Habilitationsschrift, Univ. Kaiserslautern, 1988.

Schumacher, K. [1] Existence and continuous dependence for differential equations with unbounded delay. *Arch Rat. Mech. Anal. 67* (1978), 315–335; [2] Dynamical systems with memory on history spaces with monotonic seminorms. *J. Differential Eqns. 34* (1979), 440–463.

Seidler, J. and I. Vrkoč [1] An averaging principle for stochastic evolution equations. *Časopis pěst. mat. 111* (1990), 240–263.

Seifert, G. [1] Global asymptotic stability for functional differential equations. Personal communication; [2] Positive invariant closed sets for systems of delay differential equations. *J. Differential Eqns. 22* (1976), 292–304; [3] Positively invariant closed sets for systems of delay differential equations. *J. Differential Eqns. 22* (1976) 292–304.

Sell, G. [1] *Lectures on Topological Dynamics and Differential Equations*, Van Nostrand-Reinhold, Princeton, N. J., 1971; [2] *Linear Differential Systems*, Lecture Notes, Univ. Minnesota, 1975.

Sentebova, E. Yu. [1] Method of quasilinearization for ordinary differential equations with retarded arguments. [Russian] *Differentialniye Uravnenija 8* (1972), 2260–2263.

Shimanov, S. N. [1] On the instability of the motion of systems with retardations. [Russian] *Prikl. Mat. Meh. 24* (1960), 55–63; TPMM 70–81; [2] On stability in the critical case of a zero root with time lag. [Russian] *Prikl. Mat. Meh. 23* (1959), 836–844; [3] On the theory of linear differential equations with retardations. [Russian] *Differentialniye Uravnenija 1* (1965), 102–116; [4] On the theory of linear differential equations with periodic coefficients and time lag. [Russian] *Prikl. Math. Meh. 27* (1963), 450–458; TPMM 674–687; [5] On the vibration theory of quasilinear systems with time lag. [Russian] *Prikl. Mat. Meh. 23* (1959), 836–844.

Shin, J. S. [1] An existence theorem of functional differential equations with infinite delay in a Banach space. *Funk. Ekvacioj 30* (1987), 225–236; [2] Existence of solutions and Kamke's theorem for functional differential equations in Banach spaces. *J. Differential Eqns. 81* (1989), 294–312.

Sigueira Marconato, S.A. and C.E. Avellar [1] Difference equations with delays depending on time. *Applicable Analysis*, to appear.

Silkowskii, R. A. [1] Star-shaped regions of stability in hereditary systems. Ph.D. thesis, Brown University, Providence, RI, June 1976.

Slater, M. and H. S. Wilf [1] A class of linear differential difference equations. *Pac. J. Math. 10* (1960), 1419–1427.

Slemrod, M. [1] A hereditary partial differential equation with application in the theory of simple fluids. *Preprint*; [2] Nonexistence of oscillations in a nonlinear distributed network. *J. Math. Anal. Appl. 36* (1971), 22–40.

Smale, S. [1] Differentiable dynamical systems. *Bull. Amer. Math. Soc. 73* (1967), 747–817; [2] Stable manifolds for differential equations and diffeomorphisms. *Ann. Scuola Nomale Sup. Pisa 18* (1963), 97–116.

Smith, H. L. [1] On periodic solutions of delay integral equations modeling epidemics and population growth. Ph.D. thesis, Univ. of Iowa, May 1976; [2] Structures polulation models, threshold-type delay equations and functional differential equations. In *Delay and Differential Equations*, World Scientific, 1992, 57–64; [3] Monotone semiflows generated by functional differential equations. *J. Differential Eqns. 66* (1987), 420–442.

Smith, H. L. and H. R. Thieme [1] Quasi convergence and stability for strongly order preserving semiflows. *SIAM J. Math. Anal. 21* (1990), 673–692; [2] Convergence for strongly order preserving semiflows. *SIAM J. Math. Anal. 22* (1991); [3] Strongly order preserving semiflows generated by functional differential equations. *J. Differential Eqns. 93* (1992), 332–363.

Solodovnikov, V. V. [1] *Techniceskaya Kibernatika.* [Russian] vol. 2, Chap. XI, Moscow, Masinostroenie, 1967.

Somolinos, A. [1] Stability of Lurie-type functional equations. *J. Differential Eqns. 26* (1977), 191–199.

Staffans, O.J. [1] A neutral FDE with stable D-operator is retarded. *J. Differential Eqns. 49* (1983), 208–217; [2] Semigroups generated by a neutral functional differential equation. *SIAM J. Math. Anal. 17* (1986), 46–57.

Starik, L. K. [1] Coupled quasilinear oscillating systems with a source of energy involving retarded connections. [Russian] *Trudy Sem. Teorii Diff. Urav. Otk. Arg. 3* (1965), 119–132.

Stech, H. W. [1] The Hopf bifurcation: stability result and application. *J. Math. Anal. Appl. 71* (1979), 525–546; [2] Nongeneric Hopf bifurcations in functional differential equations. *SIAM J. Appl. Math. 16* (1985), 1134–1151.

Stephan, B. H. [1] On the existence of periodic solutions of $z'(t) = -az(t - r + \mu k(t, z(t))) + F(t)$. *J. Differential Eqns. 6* (1969), 408–419; [2] Periodic solutions of differential equations with almost constant time lag. *J. Differential Eqns. 8* (1970), 554–563.

Sternberg, N. [1] One-to-oneness of the solution map in retarded functional differential equations, *J. Differential Eqns. 85* (1990), 201–213; [2] A Hartman-Grobman theorem for a class of retarded functional differential equations *J. Math. Anal. Appl.* submitted; [3] A Hartman-Grobman theorem for maps. *Partial and Functional Differential Equations*, Longman, 1992.

Stokes, A. P. [1] On the stability of integral manifolds of functional differential equations. *J. Differential Eqns. 9* (1971), 405–49; [2] A Floquet theory for functional differential equations. *Proc. Nat. Acad. of Sci. U.S.A. 48* (1962), 1330–1334; [3] On the approximation of nonlinear oscillations. *J. Differential Eqns. 12* (1972), 535–558; [4] On the stability of a limit cycle of an autonomous functional differential equation. *Contributions to Differential Equations 3* (1964), 121–139; [5] Some implications of orbital stability in Banach spaces. *SIAM J. Appl. Math. 17* (1969), 1317–1325; [6] Local coordinates around a limit cycle of functional differential equations. *J. Differential Eqns. 24* (1977), 153–172.

Strygin, V. V. [1] The bifurcation of quasistationary periodic solutions of differential difference equations. [Russian] *Differentialniye Uravnenija 10* (1974), 1332–1334.

Táboas, P. [1] Periodic solutions of a planar delay equation. *Proc. Royal Soc. Edinburgh* or *New Directions in Differential Equations and Dynamical Systems*, 573–589, Royal Soc. Edinburgh, 1991.

Tang, B. and Y. Kuang [1] Asymptotic behavior for a class of delayed nonautonomous Lotka Volterra type equations. *Preprint* 1991; [2] Existence and uniqueness of periodic solutions of periodic functional differential systems. *Preprint* 1991.

Taylor, A. F. [1] *Introduction to Functional Analysis.* John Wiley and Sons, 4th ed., 1964.

Temam, R. [1] *Infinite Dimensional Dynamical Systems*, Springer-Verlag, 1988.

Thieme, H. R. [1] Semiflows generated by Lipschitz perturbations of non-densely defined operators. *Preprint* 1990.

Travis, C. C. and G. F. Webb [1] Partial differential equations with deviating arguments in the time variable. *Preprint*; [2] Existence and stability for partial differential equations. *Trans. Amer. Math. Soc. 200* (1974), 395–418.

Tychonov, A. [1] Sur les équations fonctionelles de Volterra et leurs applications á certains problèmes de la physique mathematique. *Bull. de l'Univ. d'Etat de Moscou, Ser. Internat. Sect. A, 1* (1938), 1–25.

Vallee, R., Dubois, P. Coté, M. and C. Delisle [1] *Phys. Rev A36* (1987), 1327.

Vallee, R. and C. Marriott [1] Analysis of an Nth-order nonlinear differential delay equation. *Phys. Rev. A39* (1989), 197–205.

Ventsel, A. D. and M. I. Freidlin [1] On small random perturbations of dynamical systems. *Russian Math. Surveys 25* (1970), 1–55.

Verduyn Lunel, S. M. [1] A sharp version of Henry's theorem on small solutions. *J. Differential Eqns. 62* (1986), 266–274; [2] Exponential type calculus for linear delay equations. Centre for Mathematics and Computer Science, Tract No. 57, Amsterdam, 1989; [3] Series expansions and small solutions for Volterra equations of convolution type. *J. Differential Eqns. 85* (1990), 17–53; [4] The closure of the generalized eigenspace of a class of infinitesimal generators. *Proc. Roy. Soc. Edinburgh Sect. A 117* (1991), 171–192; [5] Small solutions and completeness for linear functional differential equations. In *Oscillations and Dynamics in Delay Equations*, (Ed. J.R. Graef and J.K. Hale), Contemporary Mathematics 129, Amer. Math. Soc. 1992; [6] About completeness for a class of unbounded operators, Report Ga Tech, 1993; [7] Series expansions for functional differential equations. In preparation 1993.

Volterra, V. [1] Sur la théorie mathématique des phénomènes héréditaires. *J. Math. Pures Appl. 7* (1928), 249–298; [2] *Théorie Mathématique de la Lutte pour la Vie*, Gauthier-Villars, Paris, 1931; [3] Sulle equazioni integrodifferenziali della teorie dell'elasticita, *Atti Reale Accad. Lincei 18* (1909), 295.

Walther, H. O. [1] Existence of a nonconstant periodic solution of a nonlinear nonautonomous functional differential equation representing the growth of a single species population. *J. Math. Biol. 1* (1975), 227–240; [2] On a transcendental equation in the stability analysis of a population growth model. *J. Math. Biol. 3*; [3] Stability of attractivity regions for autonomous functional differential equations. *Manuscripta Math. 15* (1975), 349–363; [4], *Über Ejektivität und periodische Lösungen bei autonomen Funktionaldifferentialgleichungen mit verteilter Verzögerung*, Habilitationsschrift, Univ. München, 1977; [5] On instability, ω-limit sets and periodic solutions of nonlinear autonomous differential delay equations. *Lect. Notes Math. 730* (1979), 489–503; [6] Density of slowly oscillating solutions of $\dot{x}(t) = -f(x(t-1))$. *J. Math. Anal. Appl. 79* (1981), 127–140; [7] Bifurcation from periodic solutions in functional differential equations. *Math. Z. 182* (1983), 269–325; [8] An invariant manifold of slowly oscillating solutions for $\dot{x}(t) = -\mu x(t) + f(x(t-1))$. *J. Reine Angew. Math. 414* (1991), 67–112; [9] Unstable manifolds of periodic orbits of a differential delay equation. In Oscillations and Dynamics in Delay Equations, (Eds. J.R. Graef and J.K. Hale), Contemporary Mathematics 129, Amer. Math. Soc. 1992; [10] Homoclinic solutions and chaos in $\dot{x}(t) = f(x(t-1))$. *Nonlinear Anal. 5* (1981), 775–788.

Waltman, P. [1] *Deterministic Threshold Models in the Theory of Epidemics.* Lecture Notes in Biomathematics, Vol. 1, Springer-Verlag, 1974.

Waltman, P. and J. W. Wong [1] Two-point boundary-value problems for nonlinear functional differential equations. *Trans. Amer. Math. Soc. 164* (1972), 39–54.

Wangersky, P. J. and W. J. Cunningham [1] Time lag in prey-predation population models. *Ecology 38* (1957), 136–139; [2] Time lag in population models. *Cold Spring Harbor Symposia on Quantitative Biology 22* (1957).

Wazewska-Czyzewska, M. and A. Lasota [1] Mathematical models of the red cell system (in Polish). *Matematyka Stosowana 6* (1976), 25–40.

Webb, G. F. [1] Autonomous nonlinear functional differential equations and nonlinear semigroups. *J. Math. Anal. Appl. 46* (1974), 1–12; [2] Asymptotic stability for abstract functional differential equations. *Proc. Amer. Math. Soc. 54* (1976), 225–230; [3] Functional differential equations and nonlinear semigroups in L^p-spaces. *J. Differential Eqns. 20* (1976), 71–89; [4] *Theory of Nonlinear Age-Dependent Population Dynamics.* Marcel Dekker, New York 1985.

Weiss. L. [1] On the controllability of delay differential systems. *SIAM J. Control 5* (1967), 575–587.

Wexler, D. [1] Solutions periodiques des systems lineaires a argument retarde. *J. Differential Eqns. 3* (1967), 336–347.

Widder, D.V. [1] *The Laplace Transform.* Princeton University Press, Princeton, 1946.

Winston, E. and J. A. Yorke [1] Linear delay differential equations whose solutions become identically zero. *Rev. Roum. Math. Pures Appl. 14* (1969), 885–887.

Wright, E. M. [1] A functional equation in the heuristic theory of primes. *The Mathematical Gazette, 45* (1961), 15–16; [2] A nonlinear differential difference equation *J. Reine Angew. Math. 194* (1955), 66–87; [3] Linear differential difference equations. *Proc. Camb. Phil. Soc. 44* (1948), 179–185.

Xie, X. [1] The multiplier equation and its application to S-solutions of a differential delay equation. *J. Differential Eqns. 95* (1992), 259–280.

Yorke, J. A. [1] Noncontinuable solutions of differential delay equations. *Proc. Amer. Math. Soc. 21* (1969), 648–652; [2] Asymptotic stability for one-dimensional differential delay equations. *J. Differential Eqns. 7* (1970), 189–202.

Yoshida, K. [1] *Functional Analysis,* 6th ed. Springer-Verlag, 1980.

Yoshida, K. [1] The Hopf bifurcation and its stability for semilinear diffusion equations with time delay arising in ecology. *Hiroshima Math. J. 12* (1982), 321–348.

Yoshizawa, T. [1] *Stability Theory by Liapunov's Second Method.* Math. Soc. Japan, 1966; [2] *Stability Theory and the Existence of Periodic Solutions and Almost-Periodic Solutions.* Applied Math. Sciences, vol. 14, 1975. Springer-Verlag.

Ziegler, H. J. [1] Sur une note de M. C. Perelló sur des solutions périodiques d'equations fonctionnnelles différentielles. *C. R. Acad. Sci. Paris Ser. A–B 267* (1968), A783–A785.

Zivotovskii, L. A. [1] Absolute stability of the solutions of differential equations with retarded arguments. [Russian] *Trudy Sem. Teor. Diff. Urav. Otkl. Arg. 7* (1969), 82–91.

Zmood, R. B. and N. H. McClamroch [1] On the pointwise completeness of differential difference equations. *J. Differential Eqns. 12* (1972), 474–486.

Zverkin, A. W. [1] Dependence of the stability of solutions of linear differential equations with lagging argument upon the choice of the intial moment. [Russian] *Vestnik Mos. Univ. Ser. Mat. Meh. Astr. Fiz. Him. 5* (1959), 15–20; [2] The pointwise completeness of systems with lag. [Russian] *Differentialniye Uravnenija 9* (1973), 430–436, 586–587; [3] Expansion of solutions of differential difference equations in series. [Russian] *Trudy Sem. Teor. Diff. Urav. Otkl. Argumentom 4* (1967), 3–50; [4] The connection between boundedness and stability of solutions of linear systems within infinite number of degrees of freedom. [Russian] *Differentialniye Uravnenija 4* (1968), 366–367; [5] The completeness of a Floquet-type system of solutions for equations with retardation. [Russian] *Differentialniye Uravnenija 4* (1968), 474–478; [6] Appendix to Russian Translation of Bellman and Cooke, *Differential Difference Equations,* Mir, Moscow, 1967.

Zverkin, A. M., Kamenskii, G. A. Norkin, S. B. and L. E. El'sgol'tz [1] Differential equations with retarded arguments I. [Russian] *Uspehi Mat. Nauk 17* (1962), 77–164; II, *Trudy Sem. Teor. Diff. Urav. Otkl. Argumentom, 2* (1963), 3–49.

Index

List of symbols

Applied Mathematical Sciences

(continued from page ii)

52. *Chipot:* Variational Inequalities and Flow in Porous Media.
53. *Majda:* Compressible Fluid Flow and System of Conservation Laws in Several Space Variables.
54. *Wasow:* Linear Turning Point Theory.
55. *Yosida:* Operational Calculus: A Theory of Hyperfunctions.
56. *Chang/Howes:* Nonlinear Singular Perturbation Phenomena: Theory and Applications.
57. *Reinhardt:* Analysis of Approximation Methods for Differential and Integral Equations.
58. *Dwoyer/Hussaini/Voigt (eds):* Theoretical Approaches to Turbulence.
59. *Sanders/Verhulst:* Averaging Methods in Nonlinear Dynamical Systems.
60. *Ghil/Childress:* Topics in Geophysical Dynamics: Atmospheric Dynamics, Dynamo Theory and Climate Dynamics.
61. *Sattinger/Weaver:* Lie Groups and Algebras with Applications to Physics, Geometry, and Mechanics.
62. *LaSalle:* The Stability and Control of Discrete Processes.
63. *Grasman:* Asymptotic Methods of Relaxation Oscillations and Applications.
64. *Hsu:* Cell-to-Cell Mapping: A Method of Global Analysis for Nonlinear Systems.
65. *Rand/Armbruster:* Perturbation Methods, Bifurcation Theory and Computer Algebra.
66. *Hlavácek/Haslinger/Necasl/Lovísek:* Solution of Variational Inequalities in Mechanics.
67. *Cercignani:* The Boltzmann Equation and Its Applications.
68. *Temam:* Infinite Dimensional Dynamical Systems in Mechanics and Physics.
69. *Golubitsky/Stewart/Schaeffer:* Singularities and Groups in Bifurcation Theory, Vol. II.
70. *Constantin/Foias/Nicolaenko/Temam:* Integral Manifolds and Inertial Manifolds for Dissipative Partial Differential Equations.
71. *Catlin:* Estimation, Control, and the Discrete Kalman Filter.
72. *Lochak/Meunier:* Multiphase Averaging for Classical Systems.
73. *Wiggins:* Global Bifurcations and Chaos.
74. *Mawhin/Willem:* Critical Point Theory and Hamiltonian Systems.
75. *Abraham/Marsden/Ratiu:* Manifolds, Tensor Analysis, and Applications, 2nd ed.
76. *Lagerstrom:* Matched Asymptotic Expansions: Ideas and Techniques.
77. *Aldous:* Probability Approximations via the Poisson Clumping Heuristic.
78. *Dacorogna:* Direct Methods in the Calculus of Variations.
79. *Hernández-Lerma:* Adaptive Markov Processes.
80. *Lawden:* Elliptic Functions and Applications.
81. *Bluman/Kumei:* Symmetries and Differential Equations.
82. *Kress:* Linear Integral Equations.
83. *Bebernes/Eberly:* Mathematical Problems from Combustion Theory.
84. *Joseph:* Fluid Dynamics of Viscoelastic Fluids.
85. *Yang:* Wave Packets and Their Bifurcations in Geophysical Fluid Dynamics.
86. *Dendrinos/Sonis:* Chaos and Socio-Spatial Dynamics.
87. *Weder:* Spectral and Scattering Theory for Wave Propagation in Perturbed Stratified Media.
88. *Bogaevski/Povzner:* Algebraic Methods in Nonlinear Perturbation Theory.
89. *O'Malley:* Singular Perturbation Methods for Ordinary Differential Equations.
90. *Meyer/Hall:* Introduction to Hamiltonian Dynamical Systems and the N-body Problem.
91. *Straughan:* The Energy Method, Stability, and Nonlinear Convection.
92. *Naber:* The Geometry of Minkowski Spacetime.
93. *Colton/Kress:* Inverse Acoustic and Electromagnetic Scattering Theory.
94. *Hoppensteadt:* Analysis and Simulation of Chaotic Systems.
95. *Hackbusch:* Iterative Solution of Large Sparse Systems of Equations.
96. *Marchioro/Pulvirenti:* Mathematical Theory of Incompressible Nonviscous Fluids.
97. *Lasota/Mackey:* Chaos, Fractals and Noise: Stochastic Aspects of Dynamics, 2nd ed.
98. *de Boor/Höllig/Riemenschneider:* Box Splines.
99. *Hale/Lunel:* Introduction to Functional Differential Equations.
100. *Arnol'd/Constantin/Feigenbaum/Golubitsky/Joseph/Kadanoff/Kreiss/McKean/Marsden/Teman:* Trends and Perspectives in Applied Mathematics.
101. *Nusse/Yorke:* Dynamics: Numerical Explorations.
102. *Chossat/Iooss:* The Couette-Taylor Problem.

CPSIA information can be obtained at www.ICGtesting.com
Printed in the USA
LVOW07*1115210713

343877LV00012B/537/A